# TRADE, INV ON
# AND THE
# T



The last two decades have seen great economic change in Asia and this has impacted upon the vexed question of access to affordable healthcare and medicines in many Asian states. In this book Locknie Hsu examines the issue of access to medicines in Asia from a fresh perspective which embraces trade and investment law, innovation, intellectual property law, competition policy and public health issues. Hsu explores the key evolving legal issues in these areas, including ASEAN integration, free trade agreement negotiations (such as those for the TPP), bilateral investment agreements, investor-State disputes and significant court decisions. The book goes on to present proposals for steps to be taken in addressing access to medicines in Asia and will be useful to academic researchers, regulators, lawmakers and global organizations involved in the issues surrounding access to affordable healthcare and medicines.

LOCKNIE HSU is Professor of Law at Singapore Management University, where she has been the recipient of five teaching awards. Previously she has worked with the Singapore Ministry of Trade and Industry as a legal consultant on trade issues including those arising under WTO and free trade agreement issues. She received her legal training from the National University of Singapore and Harvard University and is a member of the Singapore Bar. She is currently Chair of the Trade and Investment Group and a member of the Working Group on Harmonisation of Commercial Laws of the ASEAN Law Association, and co-rapporteur of the Committee on Sustainable Development and the Green Economy in International Trade Law, International Law Association. She is also a member of the editorial board of the Journal of World Investment and Trade.

CAMBRIDGE INTERNATIONAL TRADE AND ECONOMIC LAW

*Series editors*



As the processes of regionalization and globalization have intensified, there have been accompanying increases in the regulations of international trade and economic law at the levels of international, regional and national laws.

The subject matter of this series is international economic law. Its core is the regulation of international trade, investment and cognate areas such as intellectual property and competition policy. The series publishes books on related regulatory areas, in particular human rights, labour, environment and culture, as well as sustainable development. These areas are vertically linked at the international, regional and national level, and the series extends to the implementation of these rules at these different levels. The series also includes works on governance, dealing with the structure and operation of related international organizations in the field of international economic law, and the way they interact with other subjects of international and national law.

*Books in the series*

*Trade, Investment, Innovation and their Impact on Access to Medicines: An Asian Perspective*
Locknie Hsu

*The Law, Economics and Politics of International Standardisation*
Panagiotis Delimatsis

*The WTO and International Investment Law: Converging Systems*
Jürgen Kurtz

*Export Restrictions on Critical Minerals and Metals: Testing the Adequacy of WTO Disciplines*
Ilaria Espa

*Optimal Regulation and the Law of International Trade: The Interface between Societal Values and WTO Law*
Boris Rigod

*The Social Foundations of World Trade: Norms, Community and Constitution*
Sungjoon Cho

*Public Participation and Legitimacy in the WTO*
Yves Bonzon

*The Challenge of Safeguards in the WTO*
Fernando Piérola

*General Interests of Host States in International Investment Law*
Edited by Giorgio Sacerdoti, Pia Acconci, Mara Valenti and Anna De Luca

*The Law of Development Cooperation: A Comparative Analysis of the World Bank, the EU and Germany*
Philipp Dann

*WTO Disciplines on Subsidies and Countervailing Measures: Balancing Policy Space and Legal Constraints*
Dominic Coppens

*Domestic Judicial Review of Trade Remedies: Experiences of the Most Active WTO Members*
Müslüm Yilmaz

*International Organizations in WTO Dispute Settlement: How Much Institutional Sensitivity?*
Marina Foltea

*Public Services and International Trade Liberalization: Human Rights and Gender Implications*
Barnali Choudhury

*The Law and Politics of WTO Waivers: Stability and Flexibility in Public International Law*
Isabel Feichtner

*African Regional Trade Agreements as Legal Regimes*
James Thuo Gathii

*Liberalizing International Trade after Doha*
David Gantz

*Processes and Production Methods (PPMs) in WTO Law: Interfacing Trade and Social Goals*
Christiane R. Conrad

*Non-Discrimination in International Trade in Services: 'Likeness' in WTO/GATS*
Nicolas Diebold

*The Law, Economics and Politics of Retaliation in WTO Dispute Settlement*
Edited by Chad P. Bown and Joost Pauwelyn

*The Multilateralization of International Investment Law*
Stephan W. Schill

*Trade Policy Flexibility and Enforcement in the WTO: A Law and Economics Analysis*
Simon A.B. Schropp

# TRADE, INVESTMENT, INNOVATION AND THEIR IMPACT ON ACCESS TO MEDICINES

## An Asian Perspective

LOCKNIE HSU

*School of Law, Singapore Management University*

CAMBRIDGE
UNIVERSITY PRESS

**CAMBRIDGE**
**UNIVERSITY PRESS**

University Printing House, Cambridge CB2 8BS, United Kingdom

One Liberty Plaza, 20th Floor, New York, NY 10006, USA

477 Williamstown Road, Port Melbourne, VIC 3207, Australia

314-321, 3rd Floor, Plot 3, Splendor Forum, Jasola District Centre, New Delhi - 110025, India

79 Anson Road, #06-04/06, Singapore 079906

Cambridge University Press is part of the University of Cambridge.

It furthers the University's mission by disseminating knowledge in the pursuit of education, learning and research at the highest international levels of excellence.

www.cambridge.org
Information on this title: www.cambridge.org/9781107421035



First published 2016
First paperback edition 2018

*A catalogue record for this publication is available from the British Library*

ISBN 978-1-107-07273-2 Hardback
ISBN 978-1-107-42103-5 Paperback

*To Basil*
*and my dear, late parents*

# CONTENTS

# FIGURES

# TABLES

# FOREWORD

The world has witnessed several monumental shifts in international intellectual property (IP) law over the course of the last two decades. Before the World Trade Organisation Agreement on Trade-Related Aspects of Intellectual Property Rights (TRIPS), states had almost unbridled discretion in defining the parameters of their domestic IP rights protection. The level of protection accorded to patents affecting pharmaceutical industries in developing countries was often seen as underwhelming. Indeed, pharmaceutical inventions were non-patentable in more than forty countries. With most major pharmaceutical corporations located in developed countries, it was in the economic interest of developing countries to allow the generic drug market to flourish, thus keeping the cost of pharmaceuticals low and helping to maintain adequate access to affordable medicine. On the other hand, developed countries had the opposite concern of seeking to afford their pharmaceutical corporations market exclusivity for breakthrough inventions. If profits were eroded by the emergence of generic drugs in developing countries, this might also chill the appetite to invest in research and development.

The successful conclusion of TRIPS in 1995 was an important milestone for the protection of IP rights on a global scale, setting minimum standards for the protection of IP rights and striking a balance of sorts between the pharmaceutical industry's interest in recouping the hefty research and development costs associated with drug inventions and the health priorities of developing countries seeking affordable medicine. Given the high stakes involved on all sides, it is perhaps inevitable that TRIPS incorporated various exceptions and ambiguities. As a result post-TRIPS international IP norms may have fallen short of the levels originally envisaged by the developed countries. The evolution of international IP law since TRIPS has been characterized by the efforts of developed countries to achieve the standards of protection they had anticipated or perhaps hoped for from TRIPS but on a piecemeal, bilateral basis rather than within a single multilateral forum. In a sense, TRIPS has become just the beginning of the quest to harmonize international IP law and to establish new and higher patent protection benchmarks for pharmaceuticals.

Over the last two decades, there have been important developments with efforts to dilute the exceptions and flexibilities TRIPS had accorded to developing countries by striking fresh bargains conferring trade benefits for enhanced IP protection under bilateral Free Trade Agreements (FTAs). Known as "TRIPS-Plus" provisions, these IP protection commitments go beyond the obligations WTO members had agreed to multilaterally under TRIPS. Much has been written in this regard, but few have significantly shed light on the implications of TRIPS-Plus provisions across Asia, and in this regard, I consider Locknie's book a welcome addition to the existing body of literature on the subject.

Locknie's book is also noteworthy as it comes at a time where we are beginning to see the balance of IP trade slowly shift from the United States to Asia. In the last decade especially, innovation has come increasingly to be recognized as an important plank supporting economic development, and as such the adoption of TRIPS-Plus provisions can sometimes be seen as a means not only to secure trade benefits but also to encourage innovation. Conversely, an aggressive push for strict IP protection by developed countries may need to be moderated as they become increasingly reliant on Asian innovations.

The book first sets the scene with an overview of the diverse issues arising from the intersection between trade and investment law (an area in which Locknie has written extensively), IP law and public health. Locknie then highlights some of the potential pitfalls that TRIPS may pose to Asian countries due to its latent ambiguities, in particular the controversial issue of whether genetic material is patentable, and it offers Asian policy-makers useful recommendations pertaining to this debate by drawing upon lessons emerging from significant recent litigation in the US, EU and Australia. The book also provides a comprehensive and detailed survey of the existing legal frameworks that Asian nations have adopted towards various matters including the patentability of living matter, patentability of new uses of existing medicine, patent extension rules, patent linkage and data exclusivity protection laws and regulations.

Locknie's insightful analysis of the divergent scope and content of the legal frameworks across Asian nations in relation to these issues will certainly be of great use to governments considering the re-evaluation of their patent regimes as well as to pharmaceutical companies intending to invest in Asia.

Policy-makers will also find useful a variety of exception clauses Locknie suggests might be included in FTAs. For instance, her detailed analysis of how the failure to include a *Bolar* exception in an FTA (which specifically exempts certain acts from constituting patent infringement) could delay the entry to the market of generic pharmaceuticals in an Asian party is astute and succinct. Finally, her study of the diverse and myriad set of TRIPS-Plus provisions in existing FTAs that Asian nations have entered into with their counterparties is useful in yielding ideas and options for countries aspiring to enter into similar FTAs while seeking to maintain an appropriate balance between national trade interests and the public interest in ensuring access to affordable medicine.

I am certain her scholarly discussion and careful review of the range of options available on the negotiating table will go a long way to level the playing field for all parties involved. I welcome this book and congratulate Locknie on having completed a remarkable and very useful piece of scholarship.

*Sundaresh Menon*
*Chief Justice*
*Supreme Court of the Republic of Singapore*

# PREFACE

Additional web-based resources, such as links to further relevant Asian/ASEAN materials, will be available at www.cambridge.org/9781107072732.

The primary *raison d'etre* of this book is to contribute an Asian voice to the debate on trade, investment, innovation and their impact on access to medicines. After reading many admirable works in the field, it was clear to me that a work that would present salient issues cutting across various areas of law and policy, particularly to the Asian policy-maker, lawyer or judge, was not yet in existence. I therefore embarked upon what turned out to be a most educational journey of research on those issues, on Asian approaches (or lack thereof) on intellectual property rights and treaty strategies, and on a number of global, topical developments, that can deeply affect access to medicines. An Asian view was also thought to be timely given that aging (with its concomitant ailments and need for affordable medicines) is a serious issue in many Asian countries, including China, Japan and Singapore. A third reason for presenting an Asian view is that Asian participation in bilateral and regional economic agreements – be they in the form of free trade agreements (FTAs) or investment agreements, has been growing rapidly. Some of these contain legal commitments which could affect how a state may (or may not) regulate the patenting, pricing, distribution of and reimbursement for medicines. Such commitments may also exceed the requirements of the WTO's Agreement on Trade-Related Aspects of Intellectual Property Rights (TRIPS), which can pose challenges for developing countries in Asia which aim to keep medicines affordable. Such agreements may also open the door to the possibility of investors taking up investor-state arbitration against signatory states which act or regulate in relation to medicines. This links the access to medicines debate directly to another contentious, global debate on the value and shortcomings of such investment arbitrations. Finally, while intellectual property, trade and investment policy issues play a significant role in the present debate, several other types of factors affect access to medicines in Asia, including corruption, the state of national drug procurement processes and the availability of general infrastructure such as roads, transport systems and medical facilities.

The objective of this book is to present within the Asian and global contexts some significant factors which impact on the access question, the existing legal "construct" within which states operate, economic treaty obligations and the exceptions which can be critical, and examples of investment dispute problems which may "travel" into the pharmaceutical domain. Already, certain judicial decisions of Canada on patentability, as they relate to medicines, have been raised as the subject matter of one foreign investor's claim. Patentability of certain subject matter, such as the products of biotechnology involving human genetic material, is not universally accepted. For some countries in Asia, such as Cambodia, Lao PDR and Myanmar, patent law itself is a relatively new area altogether. There remain important areas of uncertainty in Asian laws – such as patentability of such newer types of subject matter – which require clear thought, debate and policy by national governments. Ideally, these areas should be examined thoroughly before governments embark on economic agreements with economically advanced partners. Regional groups such as ASEAN, in particular, have an opportunity to shape a regional perspective on this and other aspects of the access to medicines debate.

Another relatively less debated area in the access discussion – at least in many parts of Asia – is the application of competition law to pharmaceutical patent settlement agreements which may impede competition by generic drug producers. While there have been some significant decisions of late in the United States and the European Union in this context, there have been hardly any in Asia, except for a recent decision of the Korean Supreme Court on a pharmaceutical settlement agreement that was found to be anti-competitive. Again, this is an area in which there is room for national regulators in Asia to study the issues and make clear, strong policies.

This book also aims to provide some innovative suggestions and linkages to further advance access to medicines, particularly in Asia. These suggestions range from treaty negotiation approaches to national policy changes and funding considerations of R&D of medicines. They are not intended to be "magic bullets" but, rather, to be ideas that can be used to contribute to the pool of numerous, excellent initiatives and ideas which already exist. To paraphrase a famous Chinese saying (on medicines!), 良药苦口利于病, 忠言逆耳利于行(*liang yao ku kou li yu bing, zong yan ni er li yu xing*): just as a bitter medicine may be good for curing an ailment, earnest words that inconvenience may be good for producing sound actions. It is hoped that this book provides a number of "earnest words" and helpful recommendations.

*Locknie Hsu*
*Singapore*

# ACKNOWLEDGEMENTS

Given the scope of work undertaken, I am obviously indebted to a large number of persons who took the time to assist and to discuss various aspects of the subject matter with me. I am humbled by the modesty of my own knowledge and have thoroughly enjoyed the learning journey. I wish to thank first of all the Honourable Chief Justice of the Republic of Singapore, Sundaresh Menon, who very kindly took the time to review some of my ideas to write the Foreword (in very encouraging terms indeed) despite his extremely busy schedule. I would also like to thank the Singapore Management University (SMU) for the research support for a related, foundational article on public health; Dean Yeo Tiong Min, SMU School of Law, for his support; the SMU law librarians, Charlotte Gill and Rita Lam, for their prompt and ever-cheerful assistance; and my research assistant, Rachel Loke, for the invaluable research work she did. I thank the following for their generosity in time, provision of information and for connecting me with others for my research: Wolf Meier-Ewert, Adelina Mendoza and Dolores Halloran (WTO); Nuno Pires de Carvalho, Marco Aleman and Aida Dolotbaeva (WIPO); Kimani Goddard (formerly of WIPO); Dr Peter Beyer and Dr Manisha Shridhar (WHO); Anders Jessen (Head of Unit for Public Procurement and Intellectual Property, D-G Trade, European Commission); the Honorable Justice George Wei (Singapore); Chu Swee Yeok (EDBI, Singapore), Kelvin Wong and Helen Chou (Economic Development Board, Singapore); Toh Han Li and Lee Jwee Nguan (Competition Commission of Singapore); Associate Professor Tracey Evans Chan (Ministry of Health, Singapore/National University of Singapore); Serene Wee (Singapore Academy of Law); Jon Pender and Jung-Wook Kim (GlaxoSmithKline); Peter Thomsen (Novartis); the Novartis Institute for Tropical Diseases, Singapore; Judit Rius and Yuanqiong Hu (*Médecins Sans Frontières*); Benjamin Gaw (Drew & Napier); Nadine Mansour (McKinsey & Company); Jo-Ann See (Amica Law LLC); Jimmy Hsu (Eagle Asia Partners); Professor Benn McGrady

(Georgetown Law School); Professor Hi-Taek Shin and Professor Dukgeun Ahn (Seoul National University); Dr Calvin Ho (Yong Loo Lin School of Medicine, National University of Singapore); Professor Terry Kaan (University of Hong Kong); Professor Sakda Thanitcul and Professor Orabhund Panuspattha (Chulalongkorn University, Thailand); Professor Nguyen Thanh Tu (International Law Department, Ministry of Justice, Vietnam); Professor Dr Tran Viet Dung (Dean, International Law Faculty, Ho Chi Minh City University of Law, Vietnam) ; Associate Professor Nithat Sirichotiratana (Mahidol University, Thailand); Archie Alexander; and Kannikar Kijtiwatchakul (FTA Watch, Thailand). I also acknowledge with deep appreciation the encouragement and friendship that have been extended to me by the following in my journey of learning in international economic law: Emeritus Professor Ernst-Ulrich Petersmann (European University Institute), Professor Peter Van den Bossche (Appellate Body Chair and Member, WTO) and Professor Petros Mavroidis (Columbia Law School, European University Institute and University of Neuchatel). I am also very grateful to Kim Hughes, Helen Francis and Fleur Jones of Cambridge University Press for their valuable guidance in the initial and editing processes. I apologize if I have inadvertently omitted anyone, and any errors are my own.

Last but most certainly not least, I am grateful to God for this journey, and I wish to thank my husband, Basil, for his unstinting support and encouragement, and my siblings.

# 1

# Economic integration, Asia and medicines

## 1.1 Introduction – globalization and Asia

### *1.1.1 Asia in the twenty-first century*

The last two decades have seen great economic change in many Asian states. Many have emerged from the post-World War II (and postcolonial) era to establish relatively successful economic and legal infrastructures. They have, to different extents, taken advantage of the globalization of supply chains and trade liberalization trends to become global providers (and consumers) of a variety of raw materials, processing input and finished goods. Many Asian states attract and host large numbers of transnational corporations that trade in goods and services due to the availability of labour (skilled and unskilled), land, physical and technological infrastructure, or a combination of these. Asian states are, with few exceptions, World Trade Organization (WTO) members and therefore participants in multilateral efforts at dismantling trade barriers. In the context of this book, the focus will be mainly on states making up the Association of Southeast Asian Nations (ASEAN) and their major trade partners, namely China, India, Korea and Japan.

With China joining the WTO in 2001, the number of people represented at the organization immediately grew by 1.5 billion, covering 97% of the world population.[1] The domestic economic reforms brought about by the entry of this economic juggernaut into the multilateral scheme of trade rules have helped to fuel trade and investment not only in China, but also in those countries that trade with it. India has developed into an economic powerhouse in parallel, with a series of significant economic reforms in the 1990s.[2]

[1] WTO, *Doha Declarations*, opening page, available at: www.wto.org/english/res_e/booksp_e/ddec_e.pdf (accessed 25 May 2015).

[2] See Shalendra D. Sharma, *China and India in the Age of Globalization* (Cambridge University Press, 2009).

Within ASEAN, more recently, political and economic reforms occurring in Myanmar have led to a great deal of trade and investment interest in the country. More broadly, ASEAN is in the process of an ambitious regional integration exercise – the formation of an ASEAN Economic Community (AEC). This will see the formation of a single market and production base in a region with a 600-million strong population. The process began in earnest as early as in 1992, when ASEAN members resolved to establish an ASEAN Free Trade Area (AFTA). Since then, ASEAN has actively sought further integration on three fronts – economic, political-security and socio-cultural – with a target of establishing the ASEAN Economic Community (AEC) by the end of 2015.

At the WTO, a number of Asian states have become regular users of the WTO dispute settlement system – both as claimants and as respondents – including China and India. As one of two economic "powerhouses" with large economies in Asia, India has been active both in treaty and policy debate work in the area of trade. In particular, India has been emerging as a significant pharmaceutical producer, hosting a number of generic drug manufacturers and stoutly defending flexibilities under the WTO Trade-Related Aspects of Intellectual Property Rights (TRIPS) agreement, which allows members a certain amount of leeway in regulating pharmaceutical products.

The vexed question of access to affordable healthcare and medicines affects many countries, with those in Asia being no exception. A confluence of factors heightens the urgency of dealing with this question in Asia.

First, most countries in Asia are members of the WTO's multilateral rules system. Apart from the trade liberalization work at the WTO, there has been a rapid emergence of free trade agreements (FTAs) in many Asian countries. FTA activity in Asia has been spurred in part by the interest of the US and EU, and by intra-Asia arrangements.[3] ASEAN's

[3] The interest of the US in Asia was demonstrated during the administration of former US president George W. Bush. In 2002, about a year after taking office, Bush visited Asia twice in four months. The visit covered China, South Korea and Japan; see http://fpc.state.gov/documents/organization/9185.pdf. President Obama – continuing with the declared "pivot" to Asia – visited Southeast Asia recently during his second term in office, in November 2012, when he visited Myanmar amidst its implementation of various reform plans, as well as Cambodia and Thailand. The Bush Administration used the Trade and Investment Framework Agreement (TIFA) approach to trade and investment negotiations with Asia, as a preparatory step towards creating treaty arrangements with more concrete liberalization commitments. The US has signed TIFAs with ASEAN as a whole, and with Brunei Darussalam, Cambodia, Indonesia, Malaysia, Myanmar, the Philippines, Thailand and Vietnam. The US signed an FTA with Singapore in 2001. The EU, too, has separately

individual members have been involved in a number of bilateral FTAs, and ASEAN itself has also entered into FTAs with several external trade partners. Some of these FTAs carry significant implications for trade in pharmaceutical products. At the moment, ASEAN is negotiating with six of these partners to form a Regional Comprehensive Economic Partnership (RCEP). Beyond ASEAN, Northeast Asian countries – China, Korea and Japan – are presently in negotiations for an FTA. In South Asia, India has also signed a number of bilateral FTAs and is currently negotiating one with the EU. Several Asian states, including Brunei Darussalam, Malaysia, Singapore and Vietnam, have been engaged in negotiations for the Trans-Pacific Partnership Agreement (TPP).[4] According to the WTO, Asian members have been the most active participants in FTA negotiations.[5]

Secondly, rapidly aging populations and rising health costs in many parts of Asia – not necessarily all with high economic levels – are placing greater demands on healthcare systems. These factors contribute to increasing attention being paid to pressing healthcare needs and costs.[6] The availability of affordable healthcare and pharmaceuticals for the aged and those requiring long-term medication will increasingly be a priority issue that Asian states have to handle, in order to meet domestic expectations and needs.

With ongoing economic integration, what are the implications for trade in pharmaceutical goods in Asia – is such trade affected adversely or positively? This is an important question as pharmaceutical companies look to Asia for growth opportunities and, in doing so, seek to influence the paradigm of regulatory rights over their products.[7]

embarked on trade and investment negotiations with ASEAN, with the result of completed negotiations for a bilateral FTA so far with Singapore; see generally, EU, available at: http://ec.europa.eu/trade/policy/countries-and-regions/countries/singapore/ (accessed 19 March 2015).

4 The other eight negotiating parties are Australia, Canada, Chile, Japan, Mexico, New Zealand, Peru and the United States. In early October 2015, the parties announced that negotiations had been successfully concluded and that the TPP had been signed; next steps will include ratification and implementation by the parties.

5 WTO, *World Trade Report 2011*, p. 57; see also *WTO* Press Release, *Asia is at the Leading Edge of New Trade Pacts, Says World Trade Report 2011* (27 July 2011), available at: www.wto.org/english/news_e/pres11_e/pr635_e.htm (accessed 4 September 2014).

6 See J. Menon and A. C. Melendez, Aging in Asia, Trends, Impacts and Responses, *ASEAN Economic Bulletin*, Vol. 26, No. 3 (2009), 293–305. See also www.imf.org/external/pubs/ft/wp/2006/wp06272.pdf; www.adb.org/sites/default/files/economics-wp281.pdf.

7 For an overview of globalization and the pharmaceutical industry, see Claudio Casadio Tarabusi and Graham Vickery, *Globalization in the Pharmaceutical Industry*, Parts I

## 1.2 Aging Asia: healthcare needs

In a list of ten "mega-trends" in global healthcare identified by the *Harvard Business Review*, aging populations have been named as one of the phenomena which are set to overwhelm healthcare systems.[8]

Asia's populations, as part of this global phenomenon, face healthcare challenges in light of aging, increased life expectancies and expectations of healthcare (see Table 1.1). These have a significant impact on adequacy of healthcare infrastructures and on healthcare costs. In China alone, the number of persons expected to be 60 years or older by 2050 is 437 million.[9] In Southeast Asia's largest country, Indonesia, the figure is 70 million.

As of 2000, an estimated 53% of the world's oldest persons live in Asia, with this percentage expected to increase further by 2050. Table 1.2 elaborates on this increase.

In terms of absolute numbers, by 2050, the number of persons 60 years or older in China will be 437 million, and in India 324 million. Even in tiny Singapore – with one of the most rapidly aging populations in Asia – it would be about 1.7 million.[10] China reportedly has 185 million elderly (those over 60), with an estimated 34% of the population being elderly by 2050.[11] The percentage growth of persons aged 65 and above between 2000 and 2050 is estimated to be a staggering 314%.[12] With *more than half* of the world's aging population already living in Asia, healthcare

and II, *Int. J. Health Serv.*, Vol. 28, No. 1 (1998), 67–105; and *Int. J. Health Serv.*, Vol. 28, No. 2 (1998), 281–303.

8 Karen Dillon and Steve Prokesch, Megatrends in Global Healthcare, *Harvard Business Review*, available at: http://hbr.org/web/extras/insight-center/health-care/globaltrends/1-slide (accessed 16 June 2014).

9 United Nations, *World Population Aging 2000–2050*, Chapter 2, *Magnitude and Speed of Population Aging* p. 11. See also, generally, *Asian Governments Must Find Strategies to Deal with Aging Populations, South China Morning Post* editorial, 13 October 2015, available at: www.un.org/esa/population/publications/worldageing19502050/pdf/119japan.pdf (accessed 13 October 2015).

10 United Nations, *World Population Aging 2000–2050*, Annex III, *Profiles of Aging*, available at: www.un.org/esa/population/publications/worldageing19502050/countriesorareas.htm (accessed 13 October 2015). For Japan, those aged 60 and above in 2050 are expected to make up 42.3% of the population; see www.un.org/esa/population/publications/worldageing19502050/pdf/119japan.pdf (accessed 13 October 2015).

11 See China Health and Retirement Longitudinal Study, available at: http://charls.ccer.edu.cn/zh-CN, and Charles Riley, *New Portrait of China's 185 Million Seniors*, CNN, 31 May 2013 (accessed 28 June 2014); see http://money.cnn.com/2013/05/31/news/economy/china-elderly/.

12 See *The Future of Population in Asia*, East West Center, Table 1 at p. 83, available at: www.eastwestcenter.org/fileadmin/stored/misc/FuturePop08Ageing.pdf (accessed 24 March 2014).

Table 1.1. *Asian life expectancy*

| Country | Life expectancy[a] | Population (2010)[b] |
|---|---|---|
| Brunei Darussalam | 78 | 401,000 |
| Cambodia | 63 | 14 m |
| Indonesia | 69 | 240 m |
| Lao PDR | 67 | 6.3 m |
| Malaysia | 74 | 28 m |
| Myanmar | 65 | 51 m |
| Philippines | 69 | 93 m |
| Singapore | 81 | 5 m |
| Thailand | 74 | 66 m |
| Vietnam | 75 | 89 m |
| China | 73 | 1.3bn |
| India | 65 | 1.2bn |
| Japan | 83 | 127 m |
| Korea | 81 | 48 m |

*Source:* Compiled from United Nations data.

[a] At birth, data for 2011, collated from https://data.un.org/Data.aspx?d=SOWC&f=inID%3A97.

[b] Figures from UN, available at: http://esa.un.org/wpp/Excel-Data/population.htm (accessed 3 April 2015).

Table 1.2. *Estimated and projected population aged 60 years or over in Asia for 1970, 2000 and 2050, by percentage of world population*

| Region | 1970 | 2000 | 2050 |
|---|---|---|---|
| Asia | 45 | 53 | 63 |

*Source:* Excerpted by this author from Table 2 in Barry Mirkin and Mary Beth Weinberger, *The Demography of Aging* available at: www.un.org/esa/population/publications/bulletin42_43/weinbergermirkin.pdf (accessed 24 March 2014), who drew from the United Nations, *The Sex and Age Distribution of the World Populations; the 1998 Revision, Volume II: Sex and Age.* For a breakdown of figures within Asian countries, see Table 2A in Jayant Menon, *Aging in Asia: Trends, Impacts and Responses*, Asia Development Bank, Working Paper Series on Regional Economic Integration No. 25, February 2009.

costs there – including pharmaceutical costs – can be expected to increase significantly in the coming years. As medical experts have observed:

> The elderly are the main consumers of most types of medicines because they are affected by illness much more often than are young and middle-aged adults, and by a much wider range of illnesses, most notably those that reflect degeneration of central nervous, cardiovascular and musculoskeletal systems.[13]

The link between the cost of medicines and aging is therefore of great relevance in Asia, with its vast greying populations.

## 1.3 Diseases in Asia and the Pacific region

### *1.3.1 Communicable diseases*

Communicable diseases in Asia include HIV/AIDS, tuberculosis and malaria.[14] About 95% of HIV-positive people live in developing countries,[15] while 90% of HIV-positive people in the Asia-Pacific region live in China, India, Indonesia, Thailand, Vietnam, Malaysia and Myanmar.[16] New HIV infections in Indonesia alone increased 48% between 2005 and 2013.[17] India accounts for 51% of AIDS-related deaths in the Asia-Pacific region.[18] While AIDS-related mortality has declined in Cambodia, Thailand, India and Myanmar between 2005 and 2013, it has increased in the same period in Indonesia, Pakistan, Malaysia and Nepal.[19] In Indonesia, the increase is particularly egregious – 427%.

[13] See John P. Griffin, John Posner and Geoffrey R. Barker, eds., *The Textbook of Pharmaceutical Medicine*, 7th ed. (Wiley-Blackwell, 2013), at p. 151.

[14] See WHO website information, available at: www.who.int/bulletin/volumes/88/3/09-065540/en/ (accessed 30 April 2014).

[15] Para. 116, United Nations, Report of the Secretary-General, *Roadmap to towards the Implementation of the Millennium Development Goals*, A/56/3266 September 2001, available at: www.unmillenniumproject.org/goals/core_mdgs.htm and *AIDS Global Fact Sheet*, 20th International AIDS Conference, 20–25 July 2014, Melbourne, Australia, available at: www.aids2014.org/Default.aspx?pageId=712 (accessed 21 July 2014).

[16] *Fact Sheet, HIV and AIDS in Asia and the Pacific*, 20th International AIDS Conference, 20–25 July 2014, Melbourne, Australia, available at: www.aids2014.org/Default.aspx?pageId=712 (accessed 21 July 2014).

[17] UNAIDS, *Fact Sheet 2014*. See also John McBeth, Health Needs to Become Priority, *The Straits Times*, 23 August 2014, p. A41.

[18] *Ibid.* [19] UNAIDS, *The Gap Report* (2014), pp. 60–62.

As for treatment, only Cambodia and Thailand have 50% of all those with HIV on antiretroviral treatment.[20] In Indonesia, only about 8% of those affected have access to treatment.[21] Notably, in 2013 the leading cause of death for HIV-positive people was tuberculosis.[22]

Certain infectious diseases affecting particularly poor countries – including those in Asia – such as tuberculosis and malaria, have not, until relatively recently, been the target of pharmaceutical companies due to a lack of financial incentives. This has led to this group of diseases becoming known as "neglected tropical diseases".[23] This lack of attention to such diseases has led to calls for more urgent research and development on drugs for such diseases.

Hepatitis C, a blood-borne viral disease that can lead to acute and chronic infection, is prevalent in Central and East Asia.[24] To compound the problem, there is often co-infection of this disease with HIV (e.g. through the sharing of needles in injected drug use) in parts of Asia.[25]

### 1.3.2 *Chronic diseases*

Non-communicable, chronic diseases such as cardiovascular diseases, chronic respiratory diseases, diabetes and cancer are the main causes of death in the Southeast Asian region, causing as much as 55% of deaths, with 34% of such deaths being those of persons below the age of 60.[26]

[20] Note 17 above. In Vietnam, changes are being made to the availability of ARV medicines; see Circular No. 32/2013/TT-BYT, available at: www.asemconnectvietnam.gov.vn/Law.aspx?ZID1=10&ID1=2&MaVB_id=2346 (accessed 11 October 2015) and generally, *HIV Aid to be Cut in 2017*, 22 August 2015, http://vietnamnews.vn/society/274824/hiv-aid-to-be-cut-in-2017.html (accessed 11 October 2015).

[21] Note 19 above, p. 60. [22] Note 17 above.

[23] See *The Textbook of Pharmaceutical Medicine*, note 13 above, at p. 21.

[24] *WHO Fact Sheet*, available at: www.who.int/mediacentre/factsheets/fs164/en/ (accessed 23 August 2014). See also Chapter 8, in relation to Egypt and the use of differential and negotiated pricing.

[25] See International Harm Reduction Association and Eurasian Harm Reduction Network, available at: www.ihra.net/files/2013/05/11/HRI_ERHN_HCV.pdf (accessed 23 August 2014), citing N. Walsh, P. Higgs and N. Crofts, Recognition of Hepatitis C Virus Coinfection in HIV-positive Injecting Drug Users in Asia, *J. Acquir. Immune Defic. Syndr.*, Vol. 45 (2007), 363–365. For a discussion on the costs and issues of availability of a new Hepatitis C drug in Asia, see Chapter 7.

[26] See World Health Organization information, available at: www.searo.who.int/entity/noncommunicable_diseases/en/ and *Action Plan for the Prevention and Control of Noncommunicable Diseases in South-East Asia, 2013–2020*, p. 2. In 2013, health experts

In China, the most prevalent non-communicable diseases include cardiovascular disease, chronic obstructive pulmonary disease, diabetes and lung cancer.[27] India, too, faces the challenge of dealing with chronic cardiopulmonary and respiratory diseases and diabetes.[28]

A further dimension is the high incidence of certain cancers in Asia – lung and bronchus cancers are the most common.[29] Further, according to Cancer Research UK, the highest rates of stomach cancer in the world occur in Eastern Asia (China, Japan, Republic of Korea, Democratic Republic of Korea and Mongolia).[30]

## 1.4 Poverty in Asia

According to the 2011 report of the International Fund for Agricultural Development (IFAD), a specialized agency of the United Nations, despite progress in reducing poverty, 1.4 billion people remain very poor, that is, living on less than US$1.25 a day.[31]

In the East Asian and Pacific region, it has been reported that while less than 10% of the region live on US$1.25 a day, close to half a billion people still live on US$2 a day.[32] In Southeast Asia, it has been estimated that in 2010, 76 million live on *less than* US$1.25 a day,[33] while 208 million live on less than US$2 a day.[34] In China alone, for 2010, the number of people

formed the ASEAN Non-Communicable Diseases Network to deal with these diseases; see *Straits Times* report, 8 April 2013.

[27] China's Health Challenge: Control of Chronic Diseases, *The Lancet*, Vol. 378, No. 9790 (2011), p. 456.

[28] Vikram Patel, Somnath Chatterji, Dan Chisholm, Shah Ebrahim, Gururaj Gopalakrishna, Colin Mathers, Viswanathan Mohan, Dorairaj Prabhakaran, Ravilla D. Ravindran and K. Srinath Reddy, Chronic Diseases and Injuries in India, *The Lancet*, Vol. 377 (2011), pp. 413–428.

[29] American Cancer Society, *Global Cancer Facts & Figures*, 2nd ed. (2008), at p. 5.

[30] See Cancer Research UK, available at: www.cancerresearchuk.org/cancer-info/cancerstats/types/stomach/incidence/uk-stomach-cancer-incidence-statistics#geog (accessed 1 May 2014).

[31] IFAD, *Rural Poverty Report 2011*, p. 3. See also the ASEAN Framework Plan of Action on Rural Development and Poverty Eradication for 2011–2015 (FAP), p. 5.

[32] World Bank, available at: www.worldbank.org/en/region/eap/overview (accessed 27 June 2014).

[33] Asian Development Bank, *Poverty in Asia and the Pacific: An Update*, August 2011, at p. 6; figure excludes Brunei Darussalam, Myanmar and Singapore, and is computed from figures from the source. See also generally, UNESCAP, available at: www.unescap.org/stat/data/syb2011/I-People/Income-poverty-and-inequality.asp (accessed 27 June 2014).

[34] *Ibid.*, p. 10; computed from figures from the source.

living on less than US$2 a day is estimated to be 243 million, while the number in India is 812 million.

In addition, many of the poor in Asia live in rural areas. In Lao PDR, for example, classified as a least developed country, more than three quarters of the population live in rural areas, with the poverty rate in upland areas being as high as 43%.[35] Poverty, coupled with poor health infrastructure and other problems, can contribute to issues in access to healthcare and to medicines.

Out-of-pocket healthcare spending by patients can also form an obstacle to access to treatment. According to the Organisation for Economic Co-operation and Development (OECD), out-of-pocket payments represent *more than 50%* of the total expenditure on health in countries such as the Philippines, Singapore and Vietnam, while they reach 71% in Myanmar[36] and 86.4% in India.[37]

## 1.5 The growth of National Health Expenditure

According to the United Nations:

> *Per capita* health expenditure, both public and private, tends to increase with population aging[38] (Italics emphasis added)
>
> (Original footnote omitted)

Given the aging profile in Asia, described earlier, it is not surprising that health expenses are on the rise. In addition, developing countries are becoming increasingly important markets in pharmaceuticals; it has been estimated that in 2009, Brazil, China, India, Mexico, Russia, South Korea and Turkey contributed 51% of global pharmaceutical growth.[39]

The annual growth rate of pharmaceutical expenditures in major Asian countries is shown in Table 1.3.

[35] IFAD rural poverty portal, available at: www.ruralpovertyportal.org/country/home/tags/laos (accessed 27 June 2014).

[36] OECD, *Health at a Glance, Asia/Pacific 2014*, p. 82.

[37] India Department of Pharmaceuticals, *Annual Report 2011–2012*, http://pharmaceuticals.gov.in/annualreport2012.pdf, at p. 23 (accessed 11 August 2014). For out-of-pocket expenditures on private healthcare worldwide, see World Bank, available at: http://data.worldbank.org/indicator/SH.XPD.OOPC.ZS (accessed 28 August 2014).

[38] UN, *World Population Aging Report 2013*, p. 48.

[39] Brian Tempest, *The Structural Changes in the Global Pharmaceutical Marketplace and Their Possible Implications for Intellectual Property*, Policy Brief No. 10, July 2011, UNCTAD-ICTSD Project on IPRs and Sustainable Development, at p. 2.

Table 1.3. *Growth in pharmaceutical expenses in Asia (2000–2010)*

| Country | Income group | Average annual growth rate in real pharmaceutical spending per capita in % |
|---|---|---|
| Brunei Darussalam | High | 0.82 |
| Cambodia | Low | 0.0 |
| Indonesia | Low–mid | 6.61 |
| Lao PDR | Low–mid | 8.93 |
| Malaysia | Up–mid | 7.16 |
| Myanmar[a] | Low | 15.14 |
| Philippines | Low–mid | 8.76 |
| Singapore | High | 11.66 |
| Thailand | Up–mid | 9.58 |
| Vietnam | Low–mid | 10.16 |
| China | Up–mid | 9.64 |
| India | Low–mid | 2.99 |
| Japan | High | 2.32 |
| Korea | High | 7.48 |

*Source:* Compiled by author from OECD, *Health at a Glance: Asia/Pacific 2014.*[b]

[a] For a discussion of Myanmar's healthcare system in recent times, see Benjamin Shobert, *Healthcare in Myanmar*, 19 August 2013, Forbes, www.forbes.com/sites/benjaminshobert/2013/08/19/healthcare-in-myanmar/.

[b] OECD, *Health at a Glance: Asia/Pacific 2014 – Measuring Progress towards Universal Health Coverage*. For information on pharmaceutical expenditure as a percentage of countries' GDP, see *World Medicines Situation 2011, Medicine Prices.*

The data and profiles in Table 1.3 suggest that demand for pharmaceutical products in Asia is expected to continue growing for many countries in the coming years. In China, public expenditure on pharmaceuticals was, as at 2007, a significant 42.7% of the country's total health expenditure.[40] In 2009, the Chinese government announced a blueprint for reforming the healthcare system in order to make medicines and healthcare safe, available and affordable, under universal healthcare

[40] WHO, *World Medicines Situation 2011, Annex – Medicine Expenditures*, pp. 12–13. See also, Qiang Sun, Michael A. Santoro, Qingyue Meng, Caitlin Liu and Karen Eggleston, Pharmaceutical Policy in China, *Health Affairs*, Vol. 27, No. 4 (July/August 2008), 1042–1050, at 1043, and generally.

coverage, with a target date of 2020.[41] After the issue of the blueprint, a number of follow-up measures were taken, including the establishment of the National Essential Medicines System (NEMS), under which a revised National Essential Medicines List (EML) comprising 307 medicines was issued; listed medicines would be subject to price control and be covered by insurance reimbursement. A "zero mark-up" policy would apply so that essential medicines will be available at public primary healthcare facilities at cost.[42]

Table 1.4 summarizes public healthcare costs in a number of Asian countries.

Despite the data in Table 1.4 – and the obvious interest of policy-makers in containing pharmaceutical and healthcare cost, the legal literature on the pharmaceutical regulation regimes in Asia (and ASEAN) emanating from this part of the world has been, surprisingly, relatively sparse.[43]

### *1.5.1 Singapore: examples of recent measures dealing with healthcare expenses for a rapidly aging population*

Singapore has one of the most rapidly aging populations in Asia. As early as in 1993, the Singapore government had issued a White Paper on

[41] See Zhao Litao and Huang Yanjie, China's Blueprint for Healthcare Reform, *East Asian Policy*, Vol. 2, No. 1 (2010), 51–59, Sarah L. Barber, Baobin Huang, Budiono Santoso, Richard Laing, Valerie Paris and Chunfu Wu, The Reform of the Essential Medicines System in China: a Comprehensive Approach to Universal Coverage, *Journal of Global Health*, Vol. 3, No. 1 (June 2013), 1–9, and Qian Jiwei, China's National Essential Medicine System, EAI Background Brief No. 666, 13 October 2011.

[42] See Barber et al., *The Reform of the Essential Medicines System in China*, Yang Li, Cui Ying, Guo Sufang, Philippa Brant, Li Bind and David Hipgrave, Evaluation, in Three Provinces, of the Introduction and Impact of China's National Essential Medicines Scheme, *Bull. World Health Organ*, Vol. 91 (2013), 184–194, World Bank, Financing, Pricing and Utilization of Pharmaceuticals in China: The Road to Reform, *China Health Policy Notes* (2010) at especially at pp. 15–17, and Wei Cheng, Yunyun Fang, Dehui Fan, Jing Sun, Xuefeng Shi and Jiao Li, The Effect of Implementing "Medicines Zero Mark-up Policy" in Beijing Community Health Facilities, *Southern Med. Rev.*, Vol. 5, No. 1 (1 July 2012), 53–56.

[43] An example of a significant work explaining the drug regulatory systems of ASEAN is that of Sauwakon Ratanawijitrasin, *Drug Regulation and Incentives for Innovation: The Case for ASEAN*, available at: www.who.int/intellectualproperty/studies/drug_regulation/en/ (accessed 5 February 2014). See generally, World Health Organization, *The World Medicines Situation 2011 – Global Health Trends: Global Burden of Disease and Pharmaceutical Needs*, available at: http://apps.who.int/medicinedocs/en/m/abstract/Js20036en/ (accessed 30 June 2014). For various WHO publications on medicines in Asia, see generally SEARO – SE Asia Regional Office – Asia medicines reports, available at: http://apps.who.int/medicinedocs/en/cl/CL8.1.1.6/clmd,50.html (accessed 30 June 2014).

Table 1.4. *Healthcare costs (public) – selected Asian countries*

| Country/territory | Healthcare costs as % of GDP (2012 unless otherwise indicated) | Government expenditure on health as a % of total % of total government expenditure (2012) |
|---|---|---|
| China | 5.4 | 12.5 |
| Hong Kong SAR | 5.1[a] | Not available |
| India | 4.1 | 9.1 |
| Japan | 10.1 | 19.4 |
| Korea | 7.5 | 13.6 |
| Mongolia | 6.3 | 9.0 |
| Pakistan | 2.7 | 4.7 |
| Sri Lanka | 3.2 | 6.4 |
| Taiwan | 6.9[b] | Not available |
| | **ASEAN** | |
| Brunei Darussalam | 2.3 | 6.60 |
| Cambodia | 5.4 | 6.7 |
| Indonesia | 3.0 | 6.9 |
| Lao PDR | 2.9 | 6.1 |
| Malaysia | 4.0 | 5.8 |
| Myanmar | 1.8 | 1.5 |
| Philippines | 4.6 | 10.3 |
| Singapore | 4.7 | 11.4 |
| Thailand | 3.9 | 14.2 |
| Vietnam | 6.6 | 9.5 |

*Source:* Compiled by author from WHO data at http://apps.who.int/gho/data/node.main.75 and World Bank data at http://data.worldbank.org/indicator/SH.XPD.TOTL.ZS unless otherwise indicated (accessed 21 February 2014 unless otherwise indicated).

[a] Figure for April 2010 to March 2011, source: www.fhb.gov.hk/statistics/en/statistics/health_expenditure.htm (accessed 21 February 2015).

[b] From Taiwan's Ministry of Health and Welfare webpage, *Universal Health Coverage in Taiwan*, p. 1, www.nhi.gov.tw/English/webdata/webdata.aspx?menu=11&menu_id=290&webdata_id=2974&WD_ID=290 (accessed 1 April 2014); however, see Ministry of Health and Welfare, *Taiwan Public Health Report 2013*, pp. 19–20, which states the figure as 6.6%, www.mohw.gov.tw/EN/Ministry/DM2_P.aspx?f_list_no=475&fod_list_no=5198&doc_no=46673 (accessed 21 February 2015).

Affordable Healthcare[44] in order to address rising healthcare costs. This White Paper has formed the basis of discussion and policy in the ensuing years, resulting in various initiatives being established. According to the White Paper, as of 1993, the National Health Expenditure (NHE) stood at 3.1% of Singapore's GDP. Compared with 2002, the 2011 NHE had increased by about 22%.[45]

Public financing of the healthcare system in Singapore is underpinned by a strong policy of individual responsibility and community support.[46] Healthcare is therefore financed by a system comprising provident fund schemes, co-payment through patients' out-of-pocket funds and pharmaceutical subsidies (such as those for medicines on the Ministry of Health's Standard Drug List, or SDL)[47]; and for the needy, various other forms of assistance are available. The SDL of subsidized medicines is reviewed annually.[48] The SDL, as a list of medicines eligible for subsidies, differs from the WHO's essential drug list, and was not made public until relatively recently.[49] To access expensive drugs which are not in the SDL

[44] Available at the Ministry of Health website, available at: www.moh.gov.sg/content/moh_web/home/Publications/Reports/1993/affordable_healthcare.html (accessed 24 March 2014). Singapore's healthcare financing system includes MediSave, a compulsory national medical savings plan; MediShield, a low-cost catastrophic medical insurance scheme; MediFund, a medical endowment fund for the needy; and ElderShield, a medical insurance scheme for severe disability.

[45] See the Parliamentary Speech by the Minister for Health, 13 May 2013. For an overview of Singapore's healthcare financing system, including the roles of Medisave, Medishield and Medifund, see the speech of the minister at the World Health Assembly, Geneva, Switzerland, 21 May 2012; Keynote Speech at the *Healthcare in Asia Conference*, and for an overview of the medication subsidies system, see the minister's speech in Parliament, 7 March 2012, see esp. para. 32–38 (see generally, www.moh.gov.sg).

[46] See Central Provident Fund (CPF) website, available at: http://mycpf.cpf.gov.sg/CPF/my-cpf/Healthcare/PvdHC2.htm (accessed 30 June 2014). The CPF operates the government's healthcare financing schemes, MediSave, MediShield and MediFund. MediShield Life, which aims to provide universal healthcare for Singaporeans, is currently being implemented. See also government statements such as those in Hetty Musfirah Abdul Khamid, Healthcare Co-payment Promotes Self-Reliance: Grace Fu, *ChannelNewsAsia* (14 April 2013), available at: www.channelnewsasia.com/news/singapore/govt-prepared-to-take/638184.html (accessed 30 June 2014).

[47] The list of medicines in the SDL is available at: www.moh.gov.sg/content/moh_web/home/costs_and_financing/schemes_subsidies/drug_subsidies.html (accessed 27 June 2014).

[48] See www.healthxchange.com.sg/News/Pages/Subsidised-drug-list-reviewed-every-year.aspx (accessed 27 June 2014).

[49] In his Parliamentary Speech of 4 March 2011, the Minister for Health explained further regarding the SDL: "the SDL list, or what is called the Essential Drug List in Australia and some other countries. I want to clarify that the two are different. In many countries, with national health insurance system, the Essential Drug List is the list of drugs available. In

but are clinically necessary and appropriate, the Medication Assistance Fund (MAF) provides financial assistance for Singaporeans in public hospitals.[50] In 2011, the fund was doubled to S$20 million.[51] As of December 2012, more than S$6 million of MAF assistance had been provided.[52]

The following explanation summarizes the Singapore approach to medicines and their subsidies:

> The government provides drug subsidies to improve the affordability of commonly used medications. The list of subsidized medications, known as the "standard drug list" (SDL), contains drugs that have been assessed to be cost-effective and essential. Most SDL drugs are capped at a price of $1.40 per week for subsidized patients in the public sector, while the rest are subsidized at 50% of the retail price.

other words, if a drug is on the list, only then will the insurers pay. If a drug is not on the list, you cannot even get the drug. Whereas our SDL is for the purpose of deciding on the degree of subsidy in public hospitals. So if a drug is not on the list, it does not mean the drug is not available. It is available; you can buy it from Guardian or even from our hospital pharmacies except that they have got to pay for it, without subsidy, that is all. So that is the reason why we did not think it necessary to make it a public document but there is nothing secretive about the list. If you ask any government doctor, he will know what the list is. [The opposition member of Parliament] mentioned that some doctors are ignorant about it. If it is private doctors, I suppose they ought to be ignorant because they do not work in public hospitals."

50 See generally, www.moh.gov.sg/content/moh_web/home/costs_and_financing/schemes_subsidies/drug_subsidies.html (accessed 27 June 2014). In a Parliamentary Speech on 14 February 2011, the Minister for Health explained the system as follows:

> Our Standard Drug List (SDL) was modelled after the World Health Organization's (WHO) Essential Drug List, with modifications to suit local disease profiles and practice. It consists of drugs assessed to be cost-effective and essential to the provision of medical care to all Singaporeans. The assessment is done objectively and professionally by the Drug Advisory Committee (DAC) which comprises a few senior doctors and chaired by the Director of Medical Services.
>
> The DAC, with inputs from other clinicians, reviews the SDL on a yearly basis to take into account changes in clinical practice and advances in medical science. It considers three main factors when determining whether a drug should enter into the SDL:
>
> (i) whether the drug is essential for the treatment of medical conditions that are important causes of morbidity and mortality in Singapore;
> (ii) whether the drug offers a major improvement in terms of efficacy and effectiveness, as compared to existing standard drugs; and
> (iii) whether there is sufficient evidence of long-term safety and cost-benefits of the drug.

51 Parliamentary Speech, Minister for Health, 4 March 2011.

52 Statement in Parliament, Minister for Health, 13 May 2013.

> We recently raised the subsidy for some SDL drugs used in the treatment of chronic conditions to 75%, for patients with per capita household income of $1,500 or less, halving the cost for them. The previous subsidy was 50%. Now we have raised it to 75%. In addition, the Medication Assistance Fund (MAF) assists Singaporeans who need selective but expensive drugs that are not on the SDL with up to 75% subsidy, also up from 50% previously. The MAF has also been expanded to include other non-standard drugs that are not in the list, if they are assessed to be clinically necessary and appropriate for the patients.
>
> Together, the SDL and MAF help to keep drugs affordable for subsidized patients, often significantly below their cost. We also constantly review the SDL to include more drugs to help patients. In fact, we will be adding about 20 new drugs into the SDL and MAF this year. . . .
>
> A key driver for the increase in healthcare cost is improvements in medical care – technology, new and improved drugs, better treatments, breakthroughs in surgical techniques – that improve quality of life and extend life. This is good for patients and their families. However, as a society, we cannot afford to support and subsidize all new treatments "at all cost". New does not necessarily mean better. We need to consider what appropriate and cost-effective treatment is.[53]

There are no direct price controls on medicines in Singapore. The government uses various indirect means to influence prices such as public procurement through a system of competitive tenders for medicines to be supplied to public hospitals. This also achieves savings through bulk purchasing.

In addition, generic medicines are usually more affordable than brand-name medicines. Table 1.5 shows, for examples, the comparative prices of two generic and brand-name medicines (as of May 2015)[54].

In 2014, the Singapore government committed to increasing subsidies for certain medicines and the number of medicines covered.[55] It announced that it would increase its direct healthcare spending from S$4 billion in 2011 to S$7.5 billion in 2014, with the figure projected to reach S$12 billion by 2020.[56] In addition, in 2014, the government

[53] Reply to questions in Parliament, Minister for Health, 7 March 2012.

[54] Data from Pharmaceutical Society of Singapore website, www.pss.org.sg/know-your-medicines/medicines-price-lists/drug-prices-common-chronic-diseases#.VUwoUvAuzcy (accessed 8 May 2015).

[55] See speech by the Singapore Minister for Health in Parliament, 12 March 2014. Pharmaceuticals for public health facilities are purchased through a General Procurement Office; see the Minister's speech on 7 March 2012 (see www.moh.gov.sg).

[56] See healthcare announcements made in the Singapore Parliament relating to the 2014 Budget (www.moh.gov.sg) and statement by the Minister for Finance, 5 March 2014, reported at: www.channelnewsasia.com/news/specialreports/budget2014/news/

Table 1.5. *Examples of prices of brand-name and generic medicines*

| Medicine | Generic (S$ per tablet/ capsule) | Brand-name (S$ per tablet/capsule) |
|---|---|---|
| FRUSEMIDE 40 mg tablet | 0.06–0.10 | (LASIX) 0.48–0.52 |
| PRAZOSIN HCL 1 mg tablet | 0.14–0.25 | (MINIPRESS) 0.61–0.65 |

introduced for elderly Singaporeans, that is, those who were at least 16 years old in 1965 – the year of Singapore's independence, a series of healthcare measures under the *Pioneer Generation Package*, including enhanced subsidies for such elderly citizens for drugs on the SDL at subsidized specialist outpatient clinics and polyclinics.[57]

## 1.6 Medicines, R&D and Asia – an overview

Access to and affordability of medicines is a growing concern in Asia, particularly as more states in the region introduce universal health coverage and experience the accompanying financial burden. A number of events have also helped to more clearly focus on the access issue in this part of the world, such as the influenza outbreak in 2009, the use (or threat of use) of compulsory or government-use licences in a small number of Asian countries, the actions of global drug companies in Thailand in relation to newer anti-retroviral (ARV) drugs, and India's grant of its first compulsory licence, which was upheld by the Indian Supreme Court in 2013.[58]

healthcare-spending-to/1021468.htm, with transcript of the Minister's speech in Parliament (accessed 30 June 2014).

[57] Singapore Government Budget 2014 website: www.singaporebudget.gov.sg/budget_2014/pd.aspx#s2 (accessed 30 June 2014); also see www.moh.gov.sg/content/moh_web/home/pressRoom/pressRoomItemRelease/2014/roll-out-of-enhanced-medication-subsidies-on-1-january-2015.html (accessed 25 September 2015).

[58] During the H5N1 outbreak, Indonesia declined to provide viral samples to the WHO's Global Influenza Surveillance Network (GISN), unless the country's access to the relevant vaccine was addressed; see Richard Husada and Raymond R. Tjandrawinata, The Healthcare System and the Pharmaceutical Industry in Indonesia, in *The New Political Economy of Pharmaceuticals – Production, Innovation and TRIPS in the Global South*, ed. Hans Löfgren and Owain David Williams (Palgrave Macmillan, 2013), at pp. 145–147, and Adam Kamradt-Scott and Sungwon Yoon, *The Global Health Governance of Pandemic Influenza – A Snapshot of Asia's Contribution, in Asia's Role in Governing Global Health*, ed. Kelley Lee, Tikki Pang and Yeling Tan (Routledge, 2013), pp. 100–101.

Several Asian countries are embracing innovation and R&D as key economic drivers in a globalized world of fast-changing technology and new opportunities. The more advanced economies, such as China, India, Japan, Korea, Singapore and Thailand, are leveraging on their stock of scientific expertise, infrastructure and ability to fund innovation initiatives. China and India are among the largest suppliers of active pharmaceutical ingredients (APIs) for the global pharmaceutical production.[59] China's market share is 18% while that of India is 10%.[60] India has a well-developed generic medicines manufacturing industry. At the same time, India is a recipient of outsourced R&D work from other countries.[61] Among other areas, China has also announced plans to develop its manufacturing and creative capabilities by 2025 in biomedicines and high-performance medical apparatus, under its "Made in China 2025" strategy.[62]

Pricing of medicines in developing countries – including those in Asia – is often higher than elsewhere. According to the UN:

On the use of compulsory licences in Asia, see Chapter 5. On events in Thailand, see Sakda Thanitcul and Matthew Lim Braslow, *Thai J. Pharm. Sci.*, Vol. 37, No. 2 (April–June 2013), 61–83, and Sinfah Tunsarawuth, Thailand Avoids Compulsory Licence on Cancer Drug; Three More Drugs Undecided, *Intellectual Property Watch*, 31 January 2008, www.ip-watch.org/2008/01/31/thailand-avoids-compulsory-licence-on-cancer-drug-3-more-drugs-undecided/ (accessed 4 April 2015).

Also, for an account of the problems relating to access in South Korea to Glivec (also known as Gleevec), a drug of Novartis, a Swiss pharmaceutical company, see Bong-min Yang and Hye-young Kwon, *TRIPS and New Challenges for the Pharmaceutical Sector in South Korea*, in *The New Political Economy of Pharmaceuticals – Production, Innovation and TRIPS in the Global South*, pp. 209–210. See also Glivec Debate Poses Vital Issues, *Joongang Daily*, 18 May 2002, Stephanie Strom and Matt Fleischer-Black, Drug Maker's Vow to Donate Cancer Medicine Falls Short, *The New York Times*, 5 June 2003, and *South Korea Court OKs Pricing of Novartis' Anti-Leukemia Drug*, 22 January 2010, www.firstwordpharma.com/node/490057#axzz3WJjEc3Ad. The Glivec drug was also the subject of India's first compulsory licence, upheld by the Supreme Court in *Novartis AG v Union of India & Ors* (see Chapter 5 for a discussion of this case).

59 See for example www.mdtvalliance.org/ (accessed 22 August 2014).

60 As at 2013; see www.businessworld.in/en/storypage/-/bw/the-rise-of-api-barons/r745982.37494/page/0 (accessed 22 August 2014).

61 See generally, M. Mazumdar, *Performance of Pharmaceutical Companies in India: A Critical Analysis of Industrial Structure, Firm Specific Resources, and Emerging Strategies, Chapter 2: Overview of the Pharmaceutical Industry in India* (Springer 2013). See also Government of India Pharmaceuticals Report, 2011–2012, Department of Pharmaceuticals, Ministry of Chemicals and Fertilizers, http://pharmaceuticals.gov.in/annualreport2012.pdf (accessed 4 September 2014).

62 Zhao Yi Nan, *China Unveils Ambitious Plans to Upgrade Manufacturing Power*, 26 March 2014, http://english.gov.cn/premier/news/2015/03/26/content_281475077781817.htm (accessed 25 May 2015).

> Prices for essential medicines in low and lower middle-income countries were, on average, *3.3 times higher* than international reference prices in public sector facilities and *5.7 times higher* in private sector facilities. *Essential medicines are available* in only *57 per cent* of public sector facilities and *65 per cent* of private facilities (in selected developing countries).[63]
>
> [Italic emphasis added]

Governments in such countries therefore need to balance innovation, trade and investment policies with policies to ensure healthcare affordability.

### *1.6.1 ASEAN: pharmaceutical production*

The pharmaceutical sector has been identified as one of the priority sectors for integration activity by ASEAN.[64] A number of steps have been taken to facilitate trade in this sector. At the technical level, for example, ASEAN has already developed a mutual recognition agreement (MRA) on good manufacturing practice (GMP) inspection for medicinal products, as well as an ASEAN Common Technical Dossier (ACTD).[65] Within ASEAN, countries with capacity for pharmaceutical production include Cambodia, Indonesia, Lao PDR, Malaysia, the Philippines, Singapore, Thailand and Vietnam, with some of these (such as Singapore and Thailand) having capacity to manufacture advanced pharmaceuticals.[66] As an LDC member of the WTO, Cambodia has been

[63] UN Millennium Development Goals Fact Sheet, Goal 8; www.un.org/millenniumgoals/pdf/Goal_8_fs.pdf (accessed 21 July 2014). On HIV medicines, see WHO, *Increasing Access to HIV Treatment in Middle-Income Countries – Key Data on Prices, Regulatory Status, Tariffs and the Intellectual Property Situation* (2014); www.who.int/phi/publications/WHO_Increasing_access_to_HIV_treatment.pdf (accessed 21 July 2014).

[64] See *ASEAN Framework Agreement for the Integration of Priority Sectors*, Vientiane, 29th November 2004, *ASEAN Sectoral Integration Protocol for Healthcare*, and Part XVIII, Roadmap for Integration of Healthcare Sector (www.asean.org).

[65] See information on the ASEAN Pharmaceutical Sector, at www.asean.org/news/item/asean-sectoral-mras (accessed 19 March 2015).

[66] ASEAN-NDI, *Strategic Business Plan* (2013), Annex 5. There have been reports that pharmaceutical firms have been moving from India to Southeast Asia, due to the attractiveness of ASEAN: www.asiabriefing.com/news/2013/07/survey-pharmaceutical-companies-leaving-india-to-enter-asean/ and http://articles.economictimes.indiatimes.com/2013-07-04/news/40371911_1_indian-pharma-pharma-industry-mr-d-s-rawat. See also WHO, country profiles and monitoring of pharmaceutical situations, www.who.int/medicines/areas/coordination/coordination_assessment/en/index1.html (accessed 4 September 2014).

granted until 1 January 2016 to introduce and implement a patent system for pharmaceutical products in accordance with the TRIPS Agreement.[67]

Innovation centres have sprung up in the region to take advantage of pro-innovation policies, of local scientific talent and expertise, and of the medical markets available in the region and to possibly develop medicines of particular relevance to Asian patients.[68] Singapore and Thailand have been specifically promoting innovation in their health sectors, to build reputations as medical tourism hubs as well as centres of innovation for biomedicine.[69] Singapore, for example, has dedicated substantial funds to promoting the island as a biomedical hub.[70]

Thailand is a producer of both prescription and generic medicines, and has a relatively large pharmaceutical industry with 714 pharmaceutical companies operating in the Kingdom.[71] It exports pharmaceutical

[67] Doha Ministerial Declaration on TRIPS and Public Health, para. 7. As a result, LDCs such as Cambodia are not obliged, with respect to pharmaceutical products, to implement or apply Sections 5 and 7 of Part II of the TRIPS Agreement or to enforce rights provided for under these Sections until 1 January 2016. On other LDC transitional provisions, see generally, www.wto.org. On Cambodia's pharmaceutical industry, see Samnang Chea and Hach Sok, Cambodia's Membership in the WTO and the Implications for Public Health, *Yale J. H. Pol., L. & Eth.*, Vol. 4, No. 2 (2004), 363–372.

[68] In China, for example, Swiss medicines manufacturer AstraZeneca established the AstraZeneca Innovation Center China in 2012; see *Asian Innovation: A Unique Opportunity for the Pharmaceutical Industry*, Fall 2012, Drug Discovery World. In Singapore, a key government-backed agency which promotes biomedical innovation is the Agency for Science and Technology (A*STAR, previously the National Science and Technology Board or NSTB); the agency works with research institutes and biomedical experts, with a view to developing medicines that meet the needs of Asian patients; see report of the Economic Development Board (EDB), www.edb.gov.sg/content/edb/en/news-and-events/news/singapore-business-news/Opinion/developing-drugs-for-asia.html (accessed 3 July 2014). One such research institute established in Singapore is the Novartis Institute for Tropical Diseases, which was set up in 2002.

[69] See generally OECD, *Innovation in Southeast Asia*, 2013, www.oecd-ilibrary.org/docserver/download/9212031e.pdf?expires=1409196035&id=id&accname=ocid194282&checksum=FDA074C4CF016F744E5ED09F987E093A (accessed 28 August 2014). For further remarks on Asia as a base of innovation, see, for example, the speech by Mr Lim Chuan Poh, Chairman, Agency for Science and Technology (A*STAR), at the Technology Transfer Summit Asia 2013: www.a-star.edu.sg/Media/News/Speeches/ID/1969/Opening-Address-By-Mr-Lim-Chuan-Poh-Chairman-ASTAR-At-The-Technology-Transfer-Summit-Asia-2013.aspx (accessed 3 July 2014). On the role of A*STAR, see next note.

[70] See EDB website, www.edb.gov.sg/content/edb/en/industries/industries/pharma-biotech.html and Heather L. Van Epps, Singapore's Multimillion Dollar Gamble, *J. Exp. M.*, Vol. 203, No. 5 (2006), 1139–1142, at: www.ncbi.nlm.nih.gov/pmc/articles/PMC2121196/ (accessed 3 July 2014). See also Chapter 7.

[71] *Thailand Pharmaceutical Market Update*, 2013, www.pharmaphorum.com/articles/thailand-pharmaceutical-market-update-2013 (accessed 4 September 2014).

products to the ASEAN region and beyond. Between 2006 and 2011, its pharmaceutical exports reportedly rose by 66% while its imports of such products rose by 65%, with Vietnam, Myanmar, Belgium, Cambodia and Malaysia taking up 57% of these exports.[72] The government Pharmaceutical Organization (GPO) of Thailand, which operates under the Ministry of Public Health, is a large producer of pharmaceutical products, both for export and domestic consumption.[73] It had, in the past, expressed interest in serving the particular markets in the newer ASEAN member countries – Cambodia, Lao PDR, Myanmar and Vietnam (CLMV).[74] Such policies of focus, the increasing economic integration in ASEAN and imminent establishment of the AEC, could together further facilitate trade in pharmaceuticals between Thailand and fellow ASEAN member countries. Malaysia has identified the pharmaceutical industry as a new economic growth area, and efforts have been made to attract investments in this field. A number of Malaysian companies produce generic medicines, such as antibiotics, painkillers and injectables, and these exist alongside foreign-owned companies that manufacture pharmaceuticals in Malaysia, as well as multinational companies that are licensed importers (such as Novartis and Eli Lilly).[75] Malaysia is also developing its halal pharmaceutical industry rigorously.[76] Singapore has identified a number of strategic areas as key economic drivers. It has devoted significant public resources on building the infrastructure to propel it to become a hub for biomedical R&D and healthcare services

[72] Investment Opportunities Ripen in Thai Pharmaceuticals Industry, *Thailand Investment Review*, Vol. 21, No. 8 (August 2011), Board of Investment (BOI), at www.boi.go.th/tir/issue/201108_21_8/cover.htm (accessed 10 March 2015). See also Jakkrit Kuanpoth, TRIPS-Plus Policies and the Pharmaceutical Industry in Thailand, in Ricardo Meléndez-Ortiz and Pedro Roffe, *Intellectual Property and Sustainable Development: Development Agendas in A Changing World* (Edward Elgar Publishing Ltd., 2009), at p. 150.

[73] See generally, Thailand's GPO website at: www.intergpomed.com/ (accessed 10 March 2015).

[74] See interview of Dr Witit Artavatkun, then Managing Director, GPO, reported in *Pharmaboardroom*, 4 May 2012, at http://pharmaboardroom.com/interviews/interview-with-witit-artavatkun-managing-director-the-government-pharmaceutical-organiza tion-gpo-tha (accessed 10 March 2015). The work of the GPO has however been affected in part due to the firing of at least two Managing Directors, including this former MD and Suwach Siasiriwattana; see www.nationmultimedia.com/national/GPO-agrees-to-fire-managing-director-for-negligenc-30206401.html and www.bangkokpost.com/news/general/455776/nopporn-tipped-as-gpo-managing-director (accessed 10 March 2015).

[75] Malaysia Investment Development Authority, *Guide on Pharmaceutical Industry in Malaysia*, 2013, p. 4, www.mida.gov.my/env3/uploads/Publications_pdf/Guide_on_Pharmaceutical/Pharmacuetical_2013.pdf (accessed 22 August 2014).

[76] See Chapter 8.

in Asia.[77] Major innovator pharmaceutical companies with global operations are located within Singapore, both for the purposes of manufacture of products and to take advantage of the specialized and general infrastructure, geographical location and policies of the city state.

While some Asian countries may also wish to consider promoting local production of generic pharmaceutical products to increase access to medicines for its people it can be a challenging task to do so.[78]

### *1.6.2 Traditional medicines and traditional knowledge in Asia*

Traditional medicine, comprising the use of herbal and medicinal plants and traditional knowledge and practices indigenous to various cultures, plays an important role in Asian countries. Asian countries are a rich source of such medicines and knowledge. From Chinese herbal medicines and acupuncture to Indonesian natural remedies such as *jamu*, there is a great variety of therapeutic products and methods that have been in use for centuries, are of commercial value and form the target of commercial exploitation. Indeed, traditional medicines and use of genetic resources is the subject of international attention and within ASEAN.[79]

[77] See Economic Review Committee Report (2003), recommendations of the Healthcare Services Working Group, at www.mti.gov.sg/ResearchRoom/Pages/Reports-of-the-Economic-Review-Committee.aspx, and generally, www.mti.gov.sg/AboutMTI/Pages/Economic%20Review%20Committee.aspx and Singapore Economic Development Board, www.edb.gov.sg/content/edb/en/industries/industries/pharma-biotech.html (accessed 19 March 2015).

[78] See, for example, *Report of the Special Rapporteur on the right of everyone to the enjoyment of the highest attainable standard of physical and mental health, Anand Grover, on access to medicines*, paras. 15–17, A/HRC/23/42, available at: www.ohchr.org/Documents/HRBodies/HRCouncil/RegularSession/Session23/A-HRC-23-42_en.pdf (accessed 23 February 2015). Such promotion has been suggested with respect to China given, *inter alia*, the dominance of public health providers in the market there; see World Bank, *A Generic Drug Policy as Cornerstone to Essential Medicines in China*, pp. 32–33 http://siteresources.worldbank.org/HEALTHNUTRITIONANDPOPULATION/Resources/281627-1285186535266/AGenericDrugPolicy.pdf (accessed 23 February 2015).

[79] The Nagoya Protocol on Access to Genetic Resources and the Fair and Equitable Sharing of Benefits Arising from Their Utilization, which addresses the use of genetic resources and traditional medicines, enters into force on 12 October 2014. See also generally, WHO, *Traditional Medicines Strategy, 2014–2023*, www.who.int/medicines/publications/traditional/trm_strategy14_23/en/and ASEAN, www.asean.org/news/asean-secretariat-news/item/harmonising-regulation-of-traditional-medicines-and-health-supplements-ahead-of-asean-integration and Health Sciences Authority, Singapore, www.hsa.gov.sg/content/hsa/en/Health_Products_Regulation/Complementary_Health_Products/Overview/ASEAN_Harmonization_of_Traditional_Medicines_and_Health_Supplements.html (accessed 4 September 2014).

It has been said that for 80% of the population in Asia and Africa, these often form the main source of primary healthcare products.[80] In many systems, such as those in ASEAN, efforts are being made to integrate traditional medicines into the main healthcare regulatory regime, to ensure quality, safety and efficacy of these medicines.[81] At the global level, the WHO aims to promote access to safe and high-quality traditional medicines and their integration into national healthcare systems.[82]

## 1.7 Economic integration activities in Asia and health matters: an overview

The WTO's TRIPS Agreement heralded a new era of global intellectual property rules. While the Paris Convention and Berne Convention had existed long before the WTO was established, it was the TRIPS Agreement which took international IP regulations to new heights, with the increased detail and scope of coverage, and availability of the WTO dispute settlement to which aggrieved WTO members could now turn. With the increased legal commitments and potential for WTO scrutiny, WTO member countries have placed a great deal of emphasis on the question of what "flexibilities" TRIPS offers. A large part of the multilateral debate – which Chapter 2 will elaborate on –revolved around this issue, to decipher what policy space remains, post-TRIPS. Among other topics, access to medicines in emergencies or situations of extreme urgency in member countries has been part of this debate at the WTO, resulting in the 2001 Doha Declaration on TRIPS and Public Health, and changes to the TRIPS regime on compulsory licensing.[83]

[80] Opening speech by the then ASEAN Secretary-General, Surin Pitsuwan, at the 2nd Conference on Traditional Medicines in ASEAN, Hanoi, Vietnam, 31 October 2010; www.asean.org/resources/2012-02-10-08-47-56/speeches-statements-of-the-former-secretaries-general-of-asean/item/opening-speech-by-dr-surin-pitsuwan-secretary-general-of-asean-at-the-2nd-conference-on-traditional-medicine-in-asean-countries (accessed 27 June 2014).

[81] See, for example, the Kuala Lumpur Declaration on Traditional Medicine in ASEAN, 28 November 2012, ASEAN e-Health Bulletin, No. 2, July–December 2012, at: www.asean.org/resources/publications/asean-publications/item/asean-e-health-bulletin-towards-harmonisation-of-traditional-medicine-practices, and the Bangkok Declaration on Traditional Medicine in ASEAN, 1 September 2009 (accessed 27 June 2014).

[82] See *WHO Traditional Medicines Strategy 2014–2023;* www.who.int/medicines/publications/traditional/trm_strategy14_23/en/ (accessed 27 June 2014). This updates and builds on the *WHO Traditional Medicines Strategy 2002–2005.*

[83] Members have agreed to a waiver of Article 31(f) of the TRIPS Agreement in paragraph 6 of the Declaration (the "Paragraph 6 Waiver"), pending TRIPS amendment; see the WTO

At the bilateral and regional levels, TRIPS provisions have spawned progeny in the form of copy-cat provisions in FTAs, and more importantly, provisions that go far beyond what TRIPS itself requires (hence, the term, "TRIPS-plus" commitments). For example, some FTAs have resulted in limiting the flexibility as to what may be excluded from patentability, while others have imposed stricter requirements for the protection of confidential data (such as clinical tests data) than what TRIPS requires. Such provisions, called "TRIPS-plus" provisions, in effect provide rights holders with a degree of protection exceeding that under TRIPS. Moreover, because such FTAs typically permit investor-state arbitration, a rights holder gains a direct avenue of complaint against the state for any claims of violation of these provisions. Such provisions have important implications for the questions of access to medicines and more broadly, public health regulatory space.

### *1.7.1 Trade, investment and health in regional FTA negotiations*

The US–Singapore FTA (USSFTA) and Korea–US FTA are two examples of FTAs containing TRIPS-plus provisions which have an impact on trade in medicines in the signatory countries. Examples of such provisions and their implementation are discussed in Chapters 4 and 5.

Elsewhere, TRIPS-plus provisions have formed the subject of heated debate and in some cases, been impediments to progress in FTA negotiations. Examples include the negotiations for an EU–India FTA, a Thailand–EU FTA[84] and a US–Thailand FTA.

In the negotiations for a Trans-Pacific Partnership Agreement (TPP), five of the twelve participant countries are Asian, with four of these being ASEAN members as well. The TPP negotiations have included discussions on IPRs and pharmaceutical products, and while an outcome is not known at the time of writing, the TPP, if concluded, is likely to present important provisions on these matters.

website: www.wto.org/english/tratop_e/trips_e/implem_para6_e.htm (accessed 16 June 2014). The waiver is aimed at increasing access – via a combination of compulsory licensing and permission to export (in a departure from Article 31(f)) – for members which have no capacity to manufacture necessary medicines themselves.

[84] See for example, www.nationmultimedia.com/business/Thai-EU-FTA-talks-centre-on-R&D-medicine-sector-30215504.html.

## 1.8 TRIPS-plus requirements and Asia – an introduction

Access to medicines and barriers to such access are a complicated matter. While a great deal of negative publicity has arisen over patents, trade treaties which extend patent and related rights, and pharmaceutical company practices in their respective roles in contributing to such barriers, several other factors are at play which influence or impede access to medicines. As it has been observed:

> While the international patent system is is in need of improvement, it is overly simplistic to blame drug patents and the global pharmaceutical industry for the access problem. The reality surrounding the challenges of access to medicines is morecomplex. In the debate over barriers to access, the culprits include corruption, poverty, poor infrastructure, taxes and tariffs, and pharmaceutical counterfeiting.[85]

Other relevant factors that have been identified are as follows:

> Many factors contribute to the problem of limited access to essential medicines. Unavailability can be caused by logistical supply and storage problems, substandard drug quality, inappropriate selection of drugs, wasteful prescription and inappropriate use, inadequate production, and prohibitive prices. Despite the enormous burden of disease, drug discovery and development targeted at infectious and parasitic diseases in poor countries has virtually ground to a standstill because drug companies in developed and developing nations simply cannot recoup the cost of R&D for products to treat diseases that exist in developing countries.[86]

One might elaborate further on these lists by adding, for example, certain practices of pharmaceutical companies, in terms of marketing tactics and patent registration and licensing strategies.

As mentioned above, poverty in Asia can be a significant barrier to better access to healthcare; poor national resources and infrastructure can compound the problem. In addition, corruption – which can lead to misuse or loss of healthcare goods such as medicines, or abuse of government procurement of such goods for public health

[85] Kristina M. Lybecker, The Economics of Access to Medicines: Meeting the Challenges of Pharmaceutical Patents, Innovation and Access for Global Health, *Harv. Int. L. J. Online*, Vol. 53 (2011), pp. 25–43, at p. 26.

[86] Ellen F. M. 't Hoen, *TRIPS, Pharmaceutical Patents and Access to Essential Medicines: Seattle, Doha and Beyond*, at p. 42, www.who.int/intellectualproperty/topics/ip/tHoen.pdf (accessed 12 August 2014).

institutions – is also a factor in many Asian countries.[87] Finally, populations – particularly in rural regions with poor healthcare facilities or poor infrastructure which impedes access to such facilities, such as in parts of Cambodia,[88] Lao PDR[89] and Myanmar[90] – may face additional obstacles towards essential healthcare services and medicines.[91]

Nonetheless, patents and related rights, especially where they are extended via trade treaties beyond what international agreements such as the TRIPS Agreement require, can be significant barriers to access.

While patents protect the originator's rights to exclusive use and exploitation of the invention, the parallel track of marketing approval laws and regulations protects the state's authority to regulate marketing and sales of the patented product. Considerations in this track differ from those of patenting in that inventiveness is not the focus. Rather, marketing approval authorities are concerned with the safety and efficacy of a pharmaceutical product, based on research and filings information available. Here, TRIPS-plus treaty provisions have also emerged as a potential barrier which can delay or block competition in the pharmaceutical sector. Chapter 2 will explain the debate further

[87] See, for example, *Asian Intelligence*, No. 871, 20 March 2013 report by the Political and Economic Risk Consultancy (PERC), at: www.asiarisk.com/library.html (accessed 26 June 2014). As for corruption in procurement of medicines, see the WHO's initiative at: www.who.int/mediacentre/news/notes/2006/np31/en/ (accessed 26 June 2014). In some countries, such as China, anti-corruption enforcement actions have recently been taken against pharmaceutical company executives. See, for example, Megha Rajagopalan and Kazunori Takada, *Chinese Police Charge British Former Head of GSK in China with Bribery*, 14 May 2014, available at: www.reuters.com/article/2014/05/14/us-china-glaxos mithkline-idUSBREA4D03720140514 and Jamil Anderlini, Patty Waldmeir, Andrew Jack, George Parker and Caroline Biham, Police Accuse GlaxoSmithKline China Head of "Ordering" Bribes, *The Financial Times*, 14 May 2014 (accessed 26 June 2014). With regard to Thailand, see U. Tumkosit, *Two case studies of corruption in medicine and medical supplies procurement in the Ministry of Public Health: II – a framework of relationships between civil society and good governance*, www.eldis.org/go/home&id=11488&type=Document#.U6wJ67FLVcp and The "Strong Thailand" Project at the Ministry of Public Health, *Thai Health* (2010).

[88] See *Rural Poverty*, www.ruralpovertyportal.org/country/home/tags/cambodia and WHO and Cambodia Ministry of Health, *Health Service Delivery Profile (2012)*.

[89] See Asian Development Bank, *Lao People's Democratic Republic Rural Access Roads Project*, pp. 17 and 28, www.ilo.org/public/english/employment/recon/eiip/. . ./irap_laos3.pdf (accessed 19 March 2015).

[90] See World Bank, www.worldbank.org/en/country/myanmar/overview (accessed 19 March 2015).

[91] Factors such as distance and lack of transport to health facilities are relevant; see e.g. Karin Wiedenmeyer, *Access to Medicines – Medicines Supply: Lessons from Tanzania and Mozambique* (2004), p. 6.

while Chapter 3 will cite examples of such extended rights, and how they might affect the Asian systems.

### 1.9 New legal risks – potential of investor-state claims

With the rapid growth of FTAs (most of which carry investor-state dispute arbitration mechanisms) and bilateral investment treaties (BITs), the risk of legal complaints based on violations (or alleged violations) by investment host states is on the rise. Furthermore, many such treaties include intellectual property rights and related rights – among a host of other broadly worded rights – within their definitions of protected investments which could form the subject matter of investor-state dispute settlement (ISDS). Where an FTA contains such terms, as well as TRIPS-plus patent provisions for pharmaceutical products, signatory host states will find themselves having to deal with reduced policy- and lawmaking space. Chapter 4 addresses trends in such treaties, especially those signed by the Asian countries, as well as draws lessons from ISDS disputes and complaints elsewhere. Chapter 5 further addresses the importance of exceptions in such treaties to retain valuable policy space in the area of public health, particularly those permitted under the much-debated "flexibilities" of the TRIPS Agreement. Examples are also given of attempts in some recent Asian treaties to include such exception or flexibility-related provisions.

### 1.10 Syncopation between the law, science and social policy

The law often lags scientific advancements. One author has described the norms of science and of patent law to have shown "mutual avoidance" in the past.[92] Several reasons exist. First, there is often no clear direction as to how to deal with evolving technologies, and age-old concepts and policies sometimes struggle to keep pace with scientific advancements. An example is in the area of regulating proprietary rights over and use of human genes, where scientific possibilities have gone far ahead of the law. Many national laws have not fully and clearly expressed national policies on evolving issues in scientific inventions and discoveries, particularly in new technologies, with possible wide and long-term implications, such as gene technologies, life forms and nanotechnology, while others are in the

[92] Rebecca S. Eisenberg, Proprietary Norms and the Norms of Science in Biotechnology Research, *Yale L.J.*, Vol. 97 (1987–1988), 177, at pp. 180–181.

process of doing so. Some of these gaps and evolving legal positions, such as in the area of patentability, are discussed in Chapter 3. Apart from the speed of technological development, the rate of increase of healthcare costs also often outpaces the policy-making process. A challenge for the makers of such policy is the need to ensure a coherent and supportive legal framework that best reflects the overall public interest, whether in the economic or social spheres, while taking into account scientific strides and other relevant global developments in a timely manner.

Secondly, while the gaps and lag in national laws persist, national legal institutions – such as judiciaries – are being called upon in specific disputes to make pronouncements with policy implications beyond their function.[93] It may not be the competence of such judiciaries which may be at issue; rather, the question is the appropriateness of leaving such decisions to them rather than having them decided by elected representatives through proper national debate.

The result is, therefore, at times, a syncopation between the law, science and regulatory policy. This syncopation may lead to treaty negotiators of more advanced economies making use of legal and informational arbitrage, to the detriment of treaty partners with less capacity to evaluate (or resist) implications of commitments that are built on regulatory and policy gaps.

## 1.11 The role of international organizations

Increasingly, international organizations involved in the debate on access to medicines, working within their respective mandates, collaborate to address relevant issues. Intellectual property rights issues – particularly as they relate to public health – occupy all three organizations regularly now.

Within the UN, a number of bodies deal with the interface of IP and public health, including the United Nations Conference on Trade and Development (UNCTAD), United Nations Programme on HIV/AIDS (UNAIDS) and United Nations Development Programme (UNDP).

The WHO, as the leading international organization focusing on public health issues, has been responsible for a large number of initiatives and studies in relation to access to medicines. Through a number of World Health Assembly (WHA) Resolutions, international

[93] See, for example, the Canadian Supreme Court's candid remarks in *Commissioner of Patents v President and Fellows of Harvard College*, discussed in Chapter 3.

policy-making on making medicines more affordable and safe has been greatly advanced.[94]

A significant milestone was the development of a common list of reference for pharmaceuticals which would be accepted as indispensable. The WHO published its first model list of "essential" drugs in 1977, with revisions made every 2 years thereafter.[95] In 1978, the 31st World Health Assembly (in resolution WHA31.32) requested the Director General, *inter alia*, "to continue to identify the drugs and vaccines which, in the light of scientific knowledge, are *indispensable for primary healthcare and control of diseases prevalent in the population . . .*"[96]

> In 2002, an Expert Committee report further defined essential medicines as follows:
>
> **Definition:** Essential medicines are those that satisfy the priority healthcare needs of the population.[97]

Following this definition was a key implication:

> **Purpose:** Essential medicines are intended to be available within the context of functioning health systems *at all times, in adequate amounts*, in the appropriate dosage forms, with *assured quality*, and *at a price the individual and the community can afford*.[98]
>
> [Italic emphasis added]

WHO has, since 1977, issued an Essential Medicines List (EML) as a guide for national policy-makers.[99] The EML, which is updated every 2 years, is used by many countries as the basis for their own national formularies or lists, for purposes including provision of subsidies or for reimbursements. The WHO also regularly designs a global medicines

94 See, for example, 61st WHA 2008, Global strategy and plan of action on public health, innovation and intellectual property, http://apps.who.int/gb/ebwha/pdf_files/A61/A61_R21-en.pdf and 67th WHA, May 2014. The WHA has also addressed the TRIPS-health interface: see also WHA Resolutions on TRIPS, www.who.int/medicines/areas/policy/trips_assembly_doc/en/. See also WHA Resolution A67/B/CONF./6 on Access to Medicines WHO Secretariat Report, *Revised procedure for updating WHO's Model List of Essential Drugs*, (2001) EB 109/8.

95 WHO website, www.who.int/medicines/publications/essentialmedicines/en/; and WHO report, *Regional Workshop on Recent Developments in Essential Medicines* (2008), Colombo, Sri Lanka, 30 October to 2 November 2007, at p. 9.

96 The WHA adopted resolution A67/B/CONF./6 on Access to Medicines WHO Secretariat Report, *Revised procedure for updating WHO's Model List of Essential Drugs*, (2001) EB 109/8, available at: www.who.int/selection_medicines/list/en/ (accessed 16 May 2014).

97 *Ibid.* 98 *Ibid.*

99 See WHO, at www.who.int/medicines/publications/essentialmedicines/en/ (accessed 8 May 2015).

strategy to continually address changing needs and developments.[100] The EML was recently updated to include a number of new medicines, including sofosbuvir, which is used in treating hepatitis C, and others used for treating cancer and tuberculosis.[101]

In 2013, the WHO, World Intellectual Property Organization (WIPO) and WTO collaborated to publish a landmark trilateral study on the interface of intellectual property rights, public health and trade, reflecting further the growing interfaces of these areas, divergent views and policies on them, and resulting tensions.[102] Other international organizations also aim to address these issues within their spheres of work and influence.[103]

## 1.12 The global and regional context

### *1.12.1 Global public health and access to medicines discussions*

Healthcare goals have also been discussed and addressed at the global level in non-trade settings. The following discussion provides an overview of multilateral venues and objectives which have progressively been addressing access to healthcare and medicines as well as the related debates.

As early as in 1978, the International Conference on Primary Healthcare had declared that health is "a fundamental human right", emphasizing that primary healthcare should be affordable.[104]

[100] See WHO, *Continuity and Change – Implementing the third WHO Medicines Strategy 2008–2013*, www.who.int/medicines/areas/policy/en/. The WHO has also issued the Global Strategy and Plan of Action on Public Health, Innovation and Intellectual Property (2011), and the Guideline on Pharmaceutical Pricing Policies. Regional medicines strategy plans have also been issued; see, for example, the Western Pacific region access to medicine strategy and *Financing Drugs in SE Asia*, 1998 Southeast Asian Region (SEARO) report.

[101] See WHO news release, 8 May 2015, www.who.int/mediacentre/news/releases/2015/new-essential-medicines-list/en/ (accessed 11 May 2015). On sofosbuvir, see also the discussion in Chapter 7.

[102] *Promoting Access to Medical Technologies and Innovation: Intersections between Public Health, Intellectual Property and Trade* (2013).

[103] See, for example, World Bank website on pharmaceutical policy, http://web.worldbank.org/WBSITE/EXTERNAL/TOPICS/EXTHEALTHNUTRITIONANDPOPULATION/EXTHSD/0,contentMDK:22426764~menuPK:376799~pagePK:148956~piPK:216618~theSitePK:376793,00.html, and the Asian Development Bank website, www.adb.org/features/public-private-pharmacy-services and ADB Guidebook on Private-Public Partnerships in Pharmacy (2013): www.adb.org/publications/guidebook-public-private-partnership-pharmacy (accessed 5 September 2014).

[104] See paras. I and VI respectively of the *Declaration of Alma-Ata, International Conference on Primary Health Care*, Alma-Ata, USSR, 6–12 September 1978; WHO website, www.who.int/publications/almaata_declaration_en.pdf (accessed 8 August 2014).

The Ottawa Charter for Health Promotion of 1986 was adopted at the first International Conference for Health Promotion.[105] A key part of this Charter is its statement of responsibility of stakeholders (such as government, non-government and voluntary organizations, professional and social groups and health personnel) "to mediate between differing interests in society for the pursuit of health".

Other declarations promoting access to health and medicines include the Jakarta Declaration of 1997[106] and the 2011 UN General Assembly Political Declaration on the Prevention and Control of Non-Communicable Diseases. Access to medicines for non-communicable diseases (NCDs) has also been the subject of international discussions.[107]

The UN's Millennium Development Goals (MDG) includes a number of global targets to improve human lives at various levels. Goal 8E aims at developing a "global partnership for development" and includes the following target:

> In cooperation with pharmaceutical companies, provide access to affordable essential drugs in developing countries.[108]

Goal 6, in particular, aims to "combat AIDS/HIV, malaria and other diseases".[109]

[105] WHO website, www.who.int/healthpromotion/conferences/previous/ottawa/en/ (accessed 8 August 2014).

[106] See www.who.int/healthpromotion/milestones_jakarta.pdf.

[107] See UN General Assembly, *Political Declaration of the High-level Meeting of the General Assembly on the Prevention and Control of Non-communicable Diseases*, A/66/L.1. www.un.org/ga/search/view_doc.asp?symbol=A/66/L.1; see Hans V. Hogerzeil, Jonathan Liberman, Veronika J. Wirtz, Sandeep P. Kishore, Sakthi Selvaraj, Rachel Kiddell-Monroe, Faith N. Mwangi-Powell, Tido vonSchoen-Angerer, on behalf of The Lancet NCD Action Group, Promotion of Access to Essential Medicines for Non-communicable Diseases: Practical Implications of the UN Political Declaration, *The Lancet*, Vol. 381, No. 9867 (2013), pp. 680–689.

[108] See UN Millennium Project website, www.unmillenniumproject.org/goals/gti.htm#goal8 (accessed 21 July 2014).

[109] The specific targets under this Goal are:
"Target 6A. Have halted by 2015 and begun to reverse the spread of HIV/AIDS
Target 6B. Achieve, by 2010, universal access to treatment for HIV/AIDS for all those who need it"; see *ibid.*, and www.who.int/topics/millennium_development_goals/diseases/en/ (accessed 21 July 2014). In September 2015, UN member states further adopted the 2030 Agenda for Sustainable Development and the Sustainable Development Goals, which includes the goal of ensuring healthy lives and well-being for all at all ages; see www.unosd.org/index.php?page=view&type=13&nr=46&menu=177 and https://sustainabledevelopment.un.org/topics (both accessed 11 October 2015). Specifically, one of the associated targets under Goal 3 is stated as follows: "3.8 Achieve universal health coverage, including financial risk protection, access to

The WHO supports these goals through its essential medicines and pharmaceutical policies such as devising and updating its Medicines Strategy.[110] In dealing with the interface of trade and public health, the WHO summarized the link between trade liberalization and affordable medicines as follows at the WTO Hong Kong Ministerial Conference in 2005:

> The liberalization of trade promoted by WTO can affect health in multiple ways. Sometimes the impact is direct and the effect is obvious, as when a disease crosses a border together with traded goods. Other times the effects are more indirect. For example, reducing trade tariffs may impact prices for medical equipment and health related products; changing international rules concerning patent protection affects access to essential medicines, diagnostic devices and technology transfer; and liberalizing trade in health services can affect national health systems.[111]

The UN Sustainable Development Goals address R&D for medicines in the context of developing country needs and supports their right to make use of the flexibilities in the TRIPS Agreement.[112]

Another dimension in the debate relates to the discussion on access to medicines in the context of rights to healthcare. In June 2011, the United Nations Human Rights Council adopted the resolution on *The Right of Everyone to the Enjoyment of the Highest Attainable Standard of Physical*

quality essential healthcare services and access to safe, effective, quality and affordable essential medicines and vaccines for all".

110 Since conception of the essential medicines list in 1977, the WHO has revised the list annually. The period of its second Medicines Strategy was 2004–2007. See WHO, *Continuity and Change, Implementing the Third Medicines Strategy 2008–2013*, (2009).

111 Statement of the World Health Organization on International Trade and Health at the WTO Ministerial Conference, Sixth Session, 13–18 December 2005, Hong Kong; text available at: www.who.int/trade/WHO_Statement_Hong_Kong.pdf?ua=1 (accessed 21 July 2014). See also Parts D and F, *WTO Agreements and Public Health: A Joint Study by the WHO and the WTO Secretariat* (Geneva, Switzerland, 2002).

112 Para. 3b states:

> 3.b. support research and development of vaccines and medicines for the communicable and non-communicable diseases that primarily affect developing countries, provide access to affordable essential medicines and vaccines, in accordance with the Doha Declaration which affirms the right of developing countries to use to the full the provisions in the TRIPS agreement regarding flexibilities to protect public health and, in particular, provide access to medicines for all.

UN, http://sustainabledevelopment.un.org/focussdgs.html. For the background on this, see *Outcome Document*, in *Rio+20 Report*, A/CONF.216/16, 20–22 June 2012, esp. paras. 141–143 (accessed 5 August 2014).

*and Mental Health.*[113] In June 2013, the Council further elaborated on this resolution by focusing on the issue of access to medicines, adopting the resolution on *Access to Medicines in the Context of the Right of Everyone to the Enjoyment of the Highest Attainable Standard of Physical and Mental Health.*[114]

### *1.12.2 Regional healthcare priorities and actions*

#### 1.12.2.1 ASEAN integration and harmonization in trade in pharmaceuticals

Twelve priority sectors for integration have been identified by ASEAN for expedited action and initiatives.[115] The healthcare sector is one such area, with its own distinct "roadmap" for action, covering a number of trade aspects ranging from customs and logistics, to standards and conformity for pharmaceuticals.[116] Under the ASEAN Trade in Goods Agreement (ATIGA), all ASEAN members other than CLMV (those who joined ASEAN most recently), had to ensure that tariffs in the priority integration sectors as set out in "Schedule A" of their individual tariff schedules were eliminated by 2010; until 2015, CLMV would have to do the same.[117] While the majority of pharmaceutical products are now subject to zero tariff commitments in ASEAN, there are some notable exceptions including some applicable to important medicines for the treatment of HIV/AIDS.[118]

ASEAN members have been working on harmonization of various measures affecting pharmaceutical trade and regulation. Examples include the initiatives of the ASEAN Consultative Committee on Standards and Quality (ACCSQ) which was established in 1992, and its Pharmaceutical Product Working Group (PPWG).[119] These include the implementation of the ASEAN Common Technical Document (ACTD) and ASEAN Common Technical Requirements (ACTR), a Mutual

113 A/HRC/17/L.16. 114 A/HRC/23/L.10/Rev.1.

115 See ASEAN Framework Agreement for the Integration of Priority Sectors, Vientiane, Lao PDR, 29 November 2004, ASEAN Framework (Amendment) Agreement for the Integration of Priority Sectors, Cebu, Philippines, 8 December 2006.

116 ASEAN website: www.asean.org.

117 Article 19(2), ATIGA. The members' individual ATIGA tariff liberalization schedules (including Schedule A items) are available at: www.asean.org/news/item/annex-2-tariff-schedules (accessed 7 May 2015).

118 See Chapter 3.

119 See PPWG website, www.asean.org/communities/asean-economic-community/item/accsq-pharmaceutical-product-working-group (accessed 17 March 2015).

Recognition Arrangement (MRA) for Good Manufacturing Practices (GMP) for Inspection of Manufacturers of Medicinal Products, and more recently, an MRA addressing bioequivalence reports of pharmaceuticals.[120] The further removal and reduction of tariff barriers and implementation of trade facilitation activities pursuant to the goals of the AEC could contribute to an improved flow of medicines within ASEAN.

#### 1.12.2.2 ASEAN regional health actions related to trade in medicines

ASEAN integration efforts have been proceeding on the economic, political security and sociocultural fronts. Although ASEAN has been in existence since 1967, it had for several years functioned as a relatively loose grouping without a formal constitutional document. In an effort to better define the role and powers of ASEAN through formal legal means, the members signed the landmark ASEAN Charter in 2007, with the Charter entering into force in December 2008.[121]

Within Asia, regional initiatives to address access to medicines and healthcare exist for ASEAN member states. In particular, ASEAN members have been moving towards the provision of universal healthcare in the region.[122] In 2012, the ASEAN leaders issued a landmark document, the ASEAN Human Rights Declaration. Article 29 of the Declaration states as follows:

> "29. (1) Every person has the *right* to the enjoyment of the highest attainable standard of physical, mental and reproductive health, *to basic and affordable healthcare services, and to have access to medical facilities.*"
>
> [Italic emphasis added][123]

[120] See ASEAN website, www.asean.org/news/asean-secretariat-news/item/asean-develops-mutual-recognition-arrangement-for-bioequivalence-study-reports (accessed 17 March 2015).

[121] See ASEAN website, www.asean.org/asean/asean-charter/asean-charter (accessed 8 August 2014).

[122] See the highly informative work in Hoang Van Minh, Nicola Suyin Pocock, Nathorn Chaiyakunapruk, Chhea Chhorvann, Ha Anh Duc, Piya Hanvoravongchai, Jeremy Lim, Don Eliseo Lucero-Prisno, III, Nawi Ng, Natalie Phaholyothin, Alay Phonvisay, Kyaw Min Soe and Vanphanom Sychareun, *Progress Toward Universal Healthcare Coverage in ASEAN*, Glob. Health Action, Vol. 7 (2014), available at: www.ncbi.nlm.nih.gov/pmc/articles/PMC4256544/pdf/GHA-7-25856.pdf (accessed 7 May 2015). See also, Indonesia Aims for Universal Health Care by 2019, *The Asian Scientist*, 19 March 2013, available at: www.asianscientist.com/2013/03/health/indonesia-aims-universal-health-care-2019/ (accessed 7 May 2015).

[123] ASEAN Human Rights Declaration, adopted on 19 November 2012, in Phnom Penh, Cambodia.

Within ASEAN, therefore, there has been a formal, explicit elevation of affordable healthcare and medical facilities as a recognized right of each ASEAN citizen. The importance of such access has been taken further in subsequent instruments that build ASEAN as an integrated community. Public health concerns and collaboration fall within the purview of the ASEAN Socio-Cultural Community (ASCC). The ASCC has set out a number of objectives in its blueprint for integration. The ASEAN Strategic Framework on Health Development (2010–2015) was developed under this blueprint. One of the goals under the Framework is access to health care and promotion of healthy lifestyles, which includes as a strategic objective, affordable health care, medical services and medicine.[124] The ASCC sets out a number of specific health-related goals and initiatives (in the form of 55 "health action lines").[125]

[124] The Framework was adopted by ASEAN Health Ministers at the 10th ASEAN Health Ministers Meeting in 2010; see www.asean.org/communities/asean-socio-cultural-community/item/joint-statement-of-the-10th-asean-health-ministers (accessed 28 April 2014). Para. B4 of the Framework relates to health care and promotion of healthy lifestyles. The text of the ASCC, which contains the 55 health "action lines", is available at: See also generally www.globinmed.com/index.php?option=com_content&view=article&id=103894:operationalisation-of-asean-strategic-framework-on-health-development-2010-2015&catid=265&Itemid=316 (accessed 28 April 2014).

[125] The ASCC "action lines" include the following:

v. Enhance awareness on the impact of regional/global trade policies and economic integration on health and develop possible strategies to mitigate their negative impacts through regional workshops and seminars, advocacy, sharing of studies and technical documents;

vi. Develop strategies for ASEAN to strengthen capacity and competitiveness in health related products and services, including in the pharmaceutical sector; . . .

xi. Promote the sharing of best practices in improved access to health products including medicines for people in ASEAN . . .

xxii. Promote rational use of drug, especially on prescription of antibiotic;

xxiii. Exchange of information and experience on drug price control to access essential drugs in all ASEAN Member States; and

xxiv. Promote the exchange of experiences among ASEAN Member States on public health policy formulation and management

In more specific terms, the ASCC *Plan of Action* provides for the following as one of its targeted concerns in ASEAN:

> Ensuring access to safe, quality and affordable medicines by building ASEAN capacity and competitiveness in pharmaceutical as well as traditional medicines and complementary and alternative medicines.
>
> (Full text available at: www.asean.org/news/item/the-asean-socio-cultural-community-ascc-plan-of-action (accessed 3 April 2015).)

The ASEAN Expert Group on Pharmaceutical Development is tasked with implementation of the objectives and actions in B.4, and is implementing its current Work Plan for 2010–2015. ASEAN is also moving towards universal healthcare for its citizens.[126]

Harmonization on pharmaceutical regulation has made limited but important progress.[127] At the Health Ministers' level, the issues of accessibility and affordability of medicines are regularly discussed, with observable commitments having been made on a gradual basis. In 2014, in particular, the ASEAN Health Ministers' joint statement included the following important commitments:

> 2. We pledge to strengthen primary healthcare in order to achieve universal health coverage in AMS and increase access to primary healthcare for ASEAN people. We ensure the availability and timely provision of essential medicines and cost effective health technologies at all levels of health facilities to achieve Goal 3 of Sustainable Development Goals: ensure healthy lives and promote well-being for all at all ages. We commit to build up sufficient capacity of well-trained, motivated health workers, especially at community level to provide services to meet health needs of people in AMS.
>
> ...
>
> 7. We commit to promote access to good, safe, quality, efficacious and affordable essential medicines within the national healthcare systems as well as rational use of medicines in AMS. We also agree to facilitate research and cross country information sharing on strategies to increase access to medicines including pricing policy and the use of Trade-Related Aspects of the Intellectual Property Rights (TRIPS) flexibilities particularly for high-cost essential drugs. We encourage exchange of experience in integrating safe, effective and quality Traditional Medicine, Complementary and Alternative Medicine (TM/CAM) into the national healthcare system, where applicable, and across other sectors.
>
> ...
>
> 12. We commit to advocate Health in All Policies (HiAP) to ASEAN Political-Security Community (APSC) and ASEAN Economic Community (AEC), so that they may collectively tackle social injustice and health inequity that cause ill health. We are fully aware that health development is a shared responsibility, therefore close collaboration with

[126] Joint Statement, 11th ASEAN Health Ministers Meeting, 12 July 2012, Phuket, Thailand, para. 3, at www.asean.org/images/2012/Economic/AHMM/11th%20AHMM%20Joint%20Statement%20Final.pdf (accessed 19 March 2015).

[127] See Chapters 2 and 3, and also generally, *Harmonization and Standardization of the ASEAN Medical Industry*, Mason School of Business, The College of William & Mary, 2010, available at: http://prema.or.th/uploads/premanews_detail/75_asean_harmonizationstandardizationbymasonschool.pdf (accessed 4 April 2015).

non-health sector and participation of the people, communities and institutions are prerequisite to achieve healthy ASEAN.[128]

While some of the above commitments appear weaker (e.g. with regard to TRIPS flexibilities in para. 5) than others, they represent important steps forward in building a common ASEAN stance on access to medicines.

In the area of intellectual property policy, while ASEAN has developed an *IP Action Plan*[129] there is no explicit linkage between IP development and treatment of pharmaceutical patents and compulsory licensing by members. Directions for initiatives on affordable healthcare are also not expressly linked to the IP policy. There has been no publicly articulated common FTA negotiating stance on provisions which may affect pharmaceutical patents, TRIPS-plus provisions in FTAs or pricing of medicines. ASEAN leaders did however recently expressly resolve to ensure access to affordable healthcare and affordable medicines.[130]

In the area of ARVs, ASEAN has established a Framework for Increasing Access to Antiretrovirals and Diagnostic Reagents.[131] Its objectives are as follows:

- To increase access to ARV drugs and diagnostic reagents through enhancing regional cooperation;
- To forge a common position in ASEAN to increase access to ARV drugs and diagnostic reagents, through various mechanisms of drug supplies management, that is, pool procurement, special negotiating and local manufacturing capacity; and
- To use as guidelines for further ASEAN cooperation in this area.

Its strategies include the following:

- Establish national and regional committee of experts for regional cooperation

128 Joint Statement, 12th ASEAN Health Ministers Meeting 18 September 2014, Hanoi, Vietnam; see also Chapter 8.

129 *ASEAN Intellectual Property Rights Action Plan 2011–2015*. The *Action Plan* does aim to develop a "strong negotiating position" via a "minimum" negotiating framework for IPRs, though (para. 25).

130 *Bali Concord III*; see also discussion in Chapter 8. The ASEAN-Network for Drugs, Diagnostics, Vaccines and Traditional Medicine Innovation (ASEAN-NDI) was launched in 2009. This network launched an ambitious plan in June 2013, the ASEAN-NDI Strategic Business Plan, to allow the network to be a prime facilitator of R&D and innovation collaboration within ASEAN. See www.asean-ndi.org/index.php?option=com_filecabinet&view=files&id=4&Itemid=17 (accessed 23 February 2015).

131 See www.asean.org/archive/Final_Framework_Access_to_ARV.pdf (accessed 23 February 2015).

- Strengthen Communication and Information Management System
- *Promote:*
  - *the adoption of TRIPS safeguards into national legislation (based on review of intellectual property rights laws)*
  - *the actual use of safeguards that already exist in the law*
- Develop a regional action plan to include different options for drug procurement
- Fast-track registration of ARV drugs
- Advocate selected, specific issues related to increasing access to ARVs
- Build a monitoring and evaluation system
- Build capacity building on:
  - IPR and access to medicines,
  - logistics management of medicines and diagnostics, and
  - ethical aspects of the use, prescription and trade in medicines.
- Establish a research system.[132]

[Italic emphasis added]

Some harmonization initiatives relating to pharmaceutical products are already in place, including the ASEAN Common Technical Dossier and the Sectoral Mutual Recognition Arrangement for Good Manufacturing Practice (GMP) Inspection of Manufacturers of Medicinal Products.[133] In addition, an ASEAN Agreement on Medical Device Directive (AMDD) to harmonize regulation of medical devices was recently signed.[134]

ASEAN also has health-related collaborative activities with external trade partners, such as those under the auspices of ASEAN Plus Three.[135]

[132] At p. 3, *ibid.* Although the above framework targets access to antiretrovirals, it is not a binding a legal document and appears to be a loose and weak framework, there are two pertinent observations. First, it could offer an important and useful building block for considering what may be done with regard to access to other essential medicines in the region. For example, ASEAN could explore the above initiatives for other high-priority diseases and examine how they may be used or adapted to accelerate cooperation and access to medicines for such other diseases. Second, the above strategies do not explain how ASEAN member states might need to coordinate their approaches – particularly with regard to the emphasized strategies above – in the face of FTA and other trade pressures from external trade partners, such as those in the context of the TPP negotiations.

[133] See ASEAN website, www.asean.org (accessed 3 April 2015).

[134] Signed 21 November 2014; see www.asean.org/communities/asean-economic-community/category/agreements-declarations-13 (accessed 3 April 2015).

[135] See, for example, the Joint Statement of *ASEAN Plus Three* Health Ministers:

> The ASEAN Plus Three Health Ministers acknowledged the importance of coordinating trade and health policies in pursuing economic growth with equity. The Plus Three Ministers welcomed the planned work by ASEAN

#### 1.12.2.3 Domestic health policies on access in ASEAN

At the national level, increasing healthcare costs – including the cost of medicines – occupy ASEAN governments.[136] Healthcare is a national priority, particularly as a number of Asian Constitutions, such as those indicated below, contain either a right to health, healthcare or some variant of this.[137] In addition, several have in place detailed national policy plans on access to healthcare and medicines.[138] The following extracts from ASEAN member states' constitutional and legislative documents provide important illustrations of the importance placed on healthcare in national laws.

Article 28H of the Indonesian Constitution provides as follows[139]:

> Health Ministers to mount a regional initiative that would bring together health, trade and other relevant policy-makers and stakeholders in the region to discuss and develop consultative and inclusive multi-sectoral approaches for integration of healthcare into development agendas.

The *ASEAN Plus Three Cooperative Work Plan 2013–2017* includes the following area of cooperation:

> 4.11.8 Increase collaboration on prioritised activities to improve accessibility to safe, quality, affordable and effective medicines and to promote the rational use of medicines including containing anti-microbial drug resistance.

See also the Joint Statement on Universal Healthcare www.asean.org/images/archive/documents/5th%20ASEAN%20Plus%20Three%20Joint%20Statement_Final.pdf (accessed 23 February 2015).

136 On the issue of costs of medicines in China, see, for example, Physician, Heal Thyself, *The Economist*, 1 February 2014, www.economist.com/news/china/21595431-medicines-are-over-prescribed-and-overpriced-physician-heal-thyself. For examples of ASEAN member states' national health policies see: http://healthspace.asia/profiles/blogs/asean-nations-health-policies-and-strategies-documents-update-in- (accessed 23 February 2015).

137 See generally, WHO, *The Right to Health in the Constitutions of Member States of the World Health Organization South-East Asia Region*, (2011).

138 See, for example, China's 12th Five-Year Plan, 2011–2015 (www.wpro.who.int/health_services/china_nationalhealthplan.pdf), India's 12th Five-year Plan 2012–2016 (http://planningcommission.gov.in/plans/planrel/12thplan/pdf/12fyp_vol3.pdf), 10th Malaysia Plan – Country Health Plan, 2011–2015 (www.moh.gov.my/images/gallery/Report/Country_health.pdf), the *Philippines Medicines Policy 2011–2016* (www.ncpam.doh.gov.ph/), Singapore's Master Health Plan 2020 and 2014 updates (see www.moh.gov.sg), and the 11th Thai National Health Development Plan, 2012–2016 (www.nationalplanningcycles.org/sites/default/files/country_docs/Thailand/11ththailandnational_health_development_plan.pdf) (accessed 5 October 2014).

139 Constitution of the Republic of Indonesia, including amendments until the 4th amendment (2002).

(1) Every person shall have the right to live in physical and spiritual prosperity, to have a home and to enjoy a good and healthy environment, and *shall have the right to obtain medical care.*
(2) Every person shall have the right *to receive facilitation and special treatment* to have the same opportunity and benefit in order to achieve equality and fairness. . . .

[Italic emphasis added]

For the Philippines, the following provisions in the Constitution of 1987 are pertinent.[140]

Article II Section 15 provides:

> Section 15. The State shall protect and promote the *right to health* of the people and instill health consciousness among them.

Article XIII Section 11 provides:

> Section 11. The State shall adopt an integrated and comprehensive approach to health development which shall endeavor *to make essential goods, health and other social services available to all the people at affordable cost.* There shall be priority for the needs of the underprivileged, sick, elderly, disabled, women, and children. The State shall endeavor to provide free medical care to paupers.
>
> [Italic emphasis added]

In 1988, the Philippines also enacted the Generic Drugs Act 1988 to promote the use of generic medicines. Section 2 of the Act provides as follows:

> Section 2. Statement of Policy – It is hereby declared the policy of the State:
>
> *To promote, encourage and require the use of generic terminology in the importation, manufacture, distribution, marketing, advertising and promotion, prescription and dispensing of drugs;*
>
> *To ensure the adequate supply of drugs with generic names at the lowest possible cost and endeavor to make them available for free to indigent patients;*
>
> To encourage the extensive use of drugs with generic names through a rational system of procurement and distribution;
>
> To emphasize the scientific basis for the use of drugs, in order that health professionals may become more aware and cognizant of their therapeutic effectiveness; and
>
> To promote drug safety by minimizing duplication in medications and/or use of drugs with potentially adverse drug interactions.
>
> [Italic emphasis added]

[140] 1987 Constitution of the Republic of the Philippines.

Section 6 requires all medical practitioners to write prescriptions using generic names, with discretion to include the brand name "if so desired".[141]

In 2008, the Universally Cheaper and Quality Medicines Act[142] was enacted to promote access to medicines. In 2010, a separate National Center for Pharmaceutical Access and Management (NCPAM), was established within the Department of Health to better implement this and other measures promoting access to medicines.[143]

Article 38 of Vietnam's Constitution provides as follows:

> 1. Everyone has the *right to health protection and care*, and to equality in the use of medical services, and has the obligation to comply with regulations on the prevention of disease and medical examination or treatment.
> 2. Acts threatening the life or health of other persons and the community are prohibited.
>
> [Italic emphasis added]

Article 367 of the Myanmar's Constitution provides as follows:[144]

> Every citizen shall, in accord with the health policy laid down by the Union, have the right to healthcare.

Further, to achieve the healthcare-related objectives of the Constitution, including that in Article 367, the Myanmar National Health Plan (2011–2016) includes the following aim:

> To make quality basic/essential medicines, vaccines and traditional medicine available adequately.[145]

[141] There have however been reports of poor compliance with the requirement, low awareness of the law and a public perception that generic medicines are of inferior quality to brand-name medicines, and the legislation in its present form has been criticized.See, for example, statements by Congressman, the Hon. Philip A. Pichay, in attempting to introduce amendments in 2012. See also *Lack of Awareness Deems Generics Law Underutilized*, 26 February 2014, *Medical Observer* (citing a PIDS-DoH study titled *The Impact of the Cheaper Medicines Act on Households in Metro Manila*), and Paolo Romero, *Congress Urged to Review Generics Act, The Philippine Star*, 3 November 2012, quoting Rep. Rodolfo Albano.

[142] Republic Act No. 9502.

[143] See generally the NCPAM website, www.ncpam.doh.gov.ph/. The site also contains a "Drug Price Watch" table of comparative pricing information of certain medicines in public and private facilities which is updated from time to time.

[144] Constitution of the Republic of the Union of Myanmar (2008).

[145] Myanmar, *Health in Myanmar* (2013), at p. 24.

Emerging from decades of isolation, Myanmar has in the last few years embarked on major reforms, including a fourfold increase in healthcare spending as well as aiming to achieve universal healthcare by 2030.[146]

In the case of Thailand, s. 51 of the previous Constitution of 2007 provided as follows[147]:

> **Section 51.** A person shall enjoy an equal right to receive public health services which are appropriate and up to the quality, and the indigent shall have the right to receive free medical treatment from public health centres of the State. A person has the right to receive public health services from the State, which shall be provided thoroughly and efficiently. A person has the right to be appropriately protected by the State against harmful contagious diseases, and to have such diseases eradicated, without charge and in a timely manner.

The above provide an overview of the place of healthcare in the constitutional and legislative sphere of these countries, which in turn forms an important backdrop against which the issues in the following chapters may be better appreciated.

## 1.13 Economic reform and IPRs in Asia

Patent laws have a relatively recent history in Asia, especially in some Southeast Asian countries.[148] Within ASEAN, there is a great diversity in economies, ranging from LDCs (Cambodia, Lao PDR and Myanmar) to high-income countries such as Singapore. As all ASEAN members are WTO members, necessarily, all need to subscribe to the minimum standards of the WTO. Within this framework, however, national

[146] See Tim Evans, *Myanmar, Setting A Goal for Universal Healthcare*, World Bank, 4 April 2014. The main legislation regulating medicines in Myanmar is the National Drug Law, The State Law and Order Restoration Council Law No. 7/92, 20 October 1992. Regarding the regulation of pharmaceuticals in Myanmar, see generally Thida Aye and James Finch, *Pharmaceuticals in Myanmar – Law and Procedure* (2000) 4 S J I C L 115–147.

[147] Constitution, B.E. 2550 (2007). Note that after the military *coup d'état* on 22 May 2014, the 2007 Constitution was repealed by the National Council for Peace and Order (NCPO), and a new interim Constitution, B.E. 2557 (2014) has been enacted, with a view to a new Draft Constitution being prepared (see unofficial translation at: www.isranews.org/isranews-article/item/31533-translation.html). The interim Constitution does not contain a public health provision such as s. 51.

[148] On patent legislation in certain Asian countries, see generally: WHO, *Review of Patent Legislation – India, Indonesia, Sri Lanka, Thailand*: http://apps.who.int/medicinedocs/en/m/abstract/Js20125en/.

emphases and attitudes on IPR protection may vary.[149] Singapore, on one hand, aims to be an IP hub with strong IP laws and enforcement. Other countries, on the other hand, may have only recently enacted IP laws, and are learning to implement and enforce these.

Cambodia, Lao PDR and Vietnam only joined the WTO in 2004, 2013 and 2007, respectively, and each has embarked on a number of economic liberalization and legislative reforms. For countries with LDC status, such as Cambodia and Lao PDR, the WTO members' Declaration on the TRIPS Agreement and Public Health extends the deadline for implementing TRIPS protection for pharmaceutical products to 1 January 2016. It also permits LDCs to obtain further extensions under Article 66.1 of TRIPS.[150] Cambodia's patent law is found in its *Law on Patents, Utility Models and Industrial Designs*, promulgated on 22 January 2003.[151] It committed at its accession to fully apply the TRIPS Agreement by 1 January 2007.[152]

Lao PDR, with its LDC status in the WTO, pledged to fully comply with all TRIPS requirements by 31 December 2016, and to avail itself of special and differential treatment provisions in the TRIPS Agreement and relevant Ministerial Declarations.[153] Patents are regulated under its *Intellectual Property Law of 2011*,[154] with exceptions to exclusive patent rights being provided for in Articles 53 and 54 of the latter law.[155]

[149] For an overview of Asian national IP laws, see generally *Intellectual Property in Asia: Law, Economics, History and Politics*, ed. Paul Goldstein and Joseph Straus, (Springer 2009) and Jakkrit Kuanpoth, *Patent Rights in Pharmaceuticals in Developing Countries* (Edward Elgar Publishing Ltd, 2010). See also generally, Verena Schuren, What a Difference a State Makes: Pharmaceutical Innovation after the TRIPs Agreement, *Bus. & Pol.*, Vol. 15, No. 2 (2013), 217–243, and ed. Kenneth C. Shadlen, Samira Guennif, Alenka Guzmán and N. Lalitha, *Intellectual Property, Pharmaceuticals and Public Health: Access to Medicines in Developing Countries* (Edward Elgar, 2011). On the role of IPRs in Singapore and in ASEAN, see, for example, Singapore's view, in the speech by the Singapore Minister for Law, 26 August 2014, www.ipos.gov.sg/MediaEvents/Readnews/tabid/873/articleid/285/category/Speeches/parentId/80/year/2014/Default.aspx (accessed 23 February 2015).

[150] Para. 7, *Declaration on the TRIPS Agreement and Public Health*, WT/MIN(01)/DEC/2, adopted 14 November 2001.

[151] NS/RKM/0103/005, text available at WIPO, www.wipo.int/edocs/lexdocs/laws/en/kh/kh002en.pdf (accessed 7 May 2015).

[152] Paras. 204–206, Cambodia, WTO Accession Working Party Report, WT/ACC/KHM/21.

[153] Paras. 176, 226–227, Lao PDR, WTO Accession Working Party Report, WT/ACC/LAO/45.

[154] Law No. 01/NA, 20 December 2011; text available at WIPO, www.wipo.int/wipolex/en/profile.jsp?code=LA (accessed 7 May 2015).

[155] Paras. 203–204, Working Party Report, note 153 above.

Myanmar was a party to the GATT 1947 and became a WTO member when the organization was formed. Emerging from decades of economic isolation from around 2007 onwards, Myanmar has been reviewing its laws. In the review it has enacted several new laws (such as the new Foreign Investment Law 2013) and is in the process of preparing or enacting others (such as an intellectual property law and an arbitration law).[156]

For countries that inherited the common law system, such as Brunei Darussalam, Singapore and Malaysia, the patent laws came about by adoption of statutes resembling those of England.[157]

China joined the WTO in December 2001 and enacted its present version of the patent law in 2008.[158]

India has a well-established patent legislation system, with its primary legislation being the Patents Act. Prior to 1999, India's law did not provide protection for pharmaceutical patents. The US and EU initiated separate WTO dispute settlement proceedings against India.[159] India's law was found to be non-compliant with the transitional provisions of TRIPS on the provision of exclusive marketing rights where a patent system was not yet available.[160] In 1999, India amended its law to provide for the requisite marketing approval system under Article 70 TRIPS.[161] In 2005, India enacted the Patents Act 2005 to provide for pharmaceutical patent protection.

### *1.13.1 ASEAN integration and IP initiatives*

The advent of the AEC spells increased trade and investment flows, and this has spurred collaborative activities among IP authorities of ASEAN.

[156] For recent reports on Myanmar's economic and legal reforms, see Myanmar, WTO Trade Policy Review Report, 2014, reports by the WTO Secretariat and by the Myanmar government, at: www.wto.org/english/tratop_e/tpr_e/tp393_e.htm (accessed 23 July 2014).

[157] The respective laws are the Brunei Patents Order 2011, Singapore Patents Act and Malaysia Patents Act.

[158] Patent Law of the People's Republic of China, Decision of the Standing Committee of the National People's Congress on Amending the Patent Law of the People's Republic of China (Adopted at the 6th Meeting of the Standing Committee of the Eleventh National People's Congress on December 27 2008); text available at: http://english.sipo.gov.cn/laws/lawsregulations/201101/t20110119_566244.html (accessed 23 July 2014).

[159] Section 5 of the Patents Act, 1970 excluded substances for use as medicines from being patented; see www.wipo.int/wipolex/en/text.jsp?file_id=128092 (accessed 23 July 2014).

[160] *India – Patent Protection for Pharmaceutical and Agricultural Chemical Products*, WT/DS/50/AB/R and WT/DS/79/R.

[161] For the 1999 amendment Act, see WIPO website: www.wipo.int/wipolex/en/details.jsp?id=7622 (accessed 23 July 2014).

While there is as yet no harmonization of IP laws and regulations, there have been concerted efforts to increase cooperation in areas such as patent examination.[162] Such cooperative efforts include the ASEAN Agreement on IP Cooperation and the ASEAN IPR Action Plan 2011–2015.[163] However, the IP and public health/access to medicines nexus does not seem to be explored in this arena. This could be because issues relating to public health and medicines fall not under the AEC, but under the ASCC.[164]

## 1.14 Conclusion

The rapidly aging populations within Asia and growing concern of governments over healthcare budgets are setting the scene for authorities to pay close attention to budgetary and affordability issues in this sector. The high degree of variance in economic development among Asian countries and divergences in regulatory rules and policies provide a challenge to those seeking to promote greater regional integration, such as in ASEAN.

International treaty instruments, such as the TRIPS Agreement and FTAs which contain obligations affecting states' regulatory work in relation to access of medicines, have given rise to a global debate on the role of such treaties, the institutions promoting them, and of the various players in the pharmaceutical chain, including drug innovators, generic producers, national drug regulatory authorities and intellectual property authorities. The next two chapters will provide an explanation of various aspects of this debate and the source and types of tensions faced by regulators, and the existing legal construct, particularly as it exists in Asia.

Several suggestions and proposals have been put forward from a variety of bodies. In addition, various new mechanisms targeting access to medicines by poor countries have been devised within the last decade, notably, such as the Medicines Patent Pool, which facilitates the licensing and sub-licensing of key medicines for generic production in poor

[162] For an excellent overview of ASEAN's IP activities, see Ng Siew Kuan, *ASEAN IP Harmonization: Striking the Delicate Balance*, *PILR*, Vol. 25, No. 1 (2013). For other cooperation initiatives, see also the ASEAN IP Portal at www.aseanip.org/News-Events/News/ASEAN-Intellectual-Property-Enhanced-with-New-Initiatives (accessed 23 February 2015).

[163] See ASEAN IP Portal, *ibid.*

[164] See ASEAN Socio-Cultural Community (ASCC) Blueprint, especially paras. B4 and B5.

countries. An example of a significant proposal is that made by the WHO Consultative Expert Working Group on Research and Development (CEWG) Report,[165] which advocates the creation of a new binding multilateral Convention on R&D. The proposal is based on the de-linking of R&D costs and pharmaceutical prices.[166] An inventory of such proposals and mechanisms is explored in Chapter 7, together with further suggestions at the treaty level and the policy-making level.

[165] See especially pp. 14, 53, 122–123, and introduction to the report. www.who.int/phi/cewg/en/.

[166] See pp. 9–10, 15, 37, 49, 53, 55–61, 63, 123 and 130, *ibid.* See also Final report – *Research and Development to Meet Health Needs of Developing Countries: Strengthening Global Financing and Coordination.*

# 2

# Access to medicines, innovation and the debate at the multilateral level

This chapter explains salient global and industry shifts which provide the background for the access to medicines debate, which comprises many "moving parts", and key elements of the debate. The focal areas include a revisiting of the notion of the patent bargain, the source of regulatory tensions and the need for balanced policy- and rule-making. First, a number of key developments affecting medicines are discussed below to provide the context for the present debate at the global level and for Asia.

## 2.1 Globalization, trade and public health dimensions: The context

"Globalization" has been explained as "the increased interconnectedness and interdependence of peoples and countries" and "is generally understood to include two interrelated elements: the opening of borders to increasingly fast flows of goods, services, finance, people and ideas across international borders; and the changes in institutional and policy regimes at the international and national levels that facilitate or promote such flows. It is recognized that globalization has both positive and negative impacts on development".[1]

Globalization, accompanied by technological advancements enabling a greater ease of transfer of goods, services and information, has affected the location of goods production, research and development (R&D) and skilled persons.[2] Critics of the globalization process have pointed to a number of gaps that have arisen, giving rise to an imbalance between economic and non-economic priorities. One argument, for example, is that at the global institutional level, institutions work within their own

[1] WHO, www.who.int/trade/glossary/story043/en/ (accessed 19 March 2015).

[2] See, generally, Samuel S. Kim, ed., *East Asia and Globalization* (Rowman & Littlefield Publishers, Inc., 2000).

mandates and, as a result, do not make policies with a view to global coherence. The reality, however, is that "actions taken in one field now increasingly affect outcomes elsewhere".[3]

Globalization has also spawned a new generation of cross-border legal agreements dealing with various aspects of international trade, both at the international (or multilateral) and regional levels. Multilateral trade agreements – both a driver and a product of globalization – have been seen by some to have been imbalanced, producing serious gaps and inequity.[4]

Globalized markets and trading have provided certain opportunities to various Asian countries. Domestic industrial and investment policies were, for some, already fuelling economic prosperity, and the large-scale, worldwide liberalization of trade associated with the advent of the WTO system in the mid-1990s opened doors further. With the greater facility and ease of moving goods, services, technology and capital across borders, countries have been able to redesign policies to ride the tide and be strategic manufacturers, suppliers and geographical hubs. Some have pushed these frontiers further by entering into additional liberalization arrangements such as FTAs and economic partnership agreements with specific targeted partners. Such agreements, however, often come at a price; for example, they may result in effects on public health issues, such

[3] Para. 352, Chapter 2, *A Fair Globalization: Creating Opportunities for All*, World Commission on the Social Dimensions of Globalization, International Labour Organization (2004).

[4] One such argument merits setting out here:

> The evolving system of multilateral agreements and rules has revealed a bias in agenda setting towards measures to expand markets. In contrast, only limited attention has been paid to measures to achieve a more balanced strategy for global growth and full employment. This is the essential underpinning for policies to achieve a more inclusive pattern of globalization. Together with this, it will be important to pursue complementary initiatives such as the development of a multilateral framework on the cross-border movement of people; measures to regulate global markets, including curbing anti-competitive practices in global production systems; avoidance of tax havens; correcting serious failures in the global financial market; and the development of new sources of funding for aid and global public goods.
>
> Most of the agreements that have been reached have been imbalanced. For example, in the multilateral trading system significant trade barriers remain in key sectors that are vital for expanding the exports of developing countries. In addition, the developing countries have, to their detriment, had to cede policy autonomy in important areas of development policy where they still need to develop their capacity.
>
> (*Ibid.*, paras. 354–5.)

as in the area of access to medicines. Part of the current debate has therefore been over the role of such agreements as being high-cost "Trojan horses", in that together with perceived economic and geopolitical benefits, come less apparent (at least initially, perhaps) social challenges such as reduced access to medicines.[5]

At the same time, the pharmaceutical industry has also had to evolve with a globalized world, as demand patterns change, competitive suppliers emerge, walls to movement of persons and technology become increasingly porous, and counterfeiting and regulatory changes have to be managed. Innovative output in the industry has also been diminishing, and with the expiry of patents associated with highly profitable or "blockbuster" medicines, the urgency of finding alternative revenue streams has been growing.

### *2.1.1 The changing face of the global pharmaceutical industry and the relevance to Asia*

One may point to certain significant developments in the last two decades which have had an impact on R&D and prices of medicines, especially in Asia. Some of these relate to the global pharmaceutical industry in general, while others pertain to Asian states, in particular those which have embarked on aggressive promotion of innovative activity in the pharmaceutical and bioscience fields.

First, the conventional approach of large pharmaceutical companies in the so-called "blockbuster era" centred on targeting discovery of medicines for large populations, which would result in blockbuster drugs reaping annual sales of US$1b or more.[6] The delivery of such blockbuster medicines brought large profits as the innovators behind them had exclusive rights to block others' manufacture and distribution, through patent rights. As such patents reached their expiry dates, competition by generic manufacturers would begin, leading to a possible fall in prices of the innovator medicines.[7] If no new blockbuster

[5] See, e.g., Ruth Lopert and Deborah Gleeson, The High Price of US Trade Agreements and Access to Medicines, *J. L. Med. & Ethics*, Vol. 41, No. 1 (Spring 2013), pp. 199–223.

[6] See, e.g., Stuart Carroll, Goodbye Blockbuster Medicines; Hello New Pharmaceutical Business Models, *Pharm. J.*, 5 June 2009.

[7] For information on the top five reported "blockbuster" drugs in 2014, with expected annual sales ranging between US$1.2b and US$19.6b, see: www.bloomberg.com/bw/articles/2013-11-14/2014-outlook-blockbuster-drugs-for-big-pharma (accessed 23 February 2015).

product was forthcoming to replace it in terms of profitability, these companies faced serious issues.[8] Large pharmaceutical companies have therefore been examining ways to "retool" and adapt to the new reality. Indeed, the shift has been described by one European regulator as the "obsolescence of a business model focusing on blockbusters" and as the emergence of "a new direction in research and development towards in particular the treatment of rare diseases, as well as biotechnology, the outsourcing of research, and the boom in cooperation between laboratories".[9]

Secondly, profits in developed country markets are declining as compared with those in emerging markets, leading pharmaceutical companies to pay more attention to emerging markets, such as those in Asia.[10]

Thirdly, research opportunities in newer areas such as biomedicines, biosimilars and personalized medicines are requiring exploration of new strategies, alliances and business models.[11] The role of new R&D and manufacturing facilities and partnerships in Asia also add to this new landscape. An example of a country in Asia aiming to take advantage of

[8] The rapid expiry of patents on blockbuster drugs without replacement by new drugs is often referred to as "patent cliffs". See, e.g., *Cliffhanger*, 3 December 2011, The Economist and generally, Stuart Carroll, Goodbye Blockbuster Medicines; Hello New Pharmaceutical Business Models, *Pharm. J.*, 5 June 2009. See also Iain M. Cockburn, The Changing Structure of the Pharmaceutical Industry, *Health Affairs*, Vol. 23, No. 1 (January 2004), 10–22. See also, EU Pharmaceutical Sector Inquiry, Final Report, available at: http://ec.europa.eu/competition/sectors/pharmaceuticals/inquiry/ (a number of follow-up reports are also available at this site) (accessed 16 March 2015), at p. 3.

[9] Website of the French Competition Authority, *Autorite de la Concurrence*, France, www.autoritedelaconcurrence.fr/user/standard.php?id_rub=483&id_article=2366 (accessed 23 December 2014). One author has also referred to two other "paradigm shifts" in medical research and commercialization: the first, a "targeting" of proteins for new medicines coupled with licensing of new partners, and the second, research on such gene or protein "targets" outpacing actual resultant medicines: Filip de Corte, *Licensing in the Medical Sector*, in *Gene Patents and Public Health*, in *Establissements Emile Bruyant*, ed. Geertrui Van Overwalle (2007), at pp. 92–96.

[10] The author is grateful for the information shared by Mr Jon Penner, GlaxoSmithKline, in a discussion on 27 October 2014 (notes are on file with the author).

[11] See, generally, *From Science to Operations – Questions, Choices and Strategies for Success in Biopharma*, ed. Ralf Otto, Alberto Santagostino and Ulf Schrader (McKinsey & Company, 2014). The author is grateful to Ms Nadine Mansour for kindly providing access to this useful publication. On personalized medicines (such as those based on genomic markers in individuals) see, e.g., the emphasis of US and EU regulators on this area, at US Food and Drug Administration (FDA), www.fda.gov/scienceresearch/special topics/personalizedmedicine/default.htm and EU, http://ec.europa.eu/research/health/policy-issues-personalised-medicine_en.html (accessed 12 March 2015).

these opportunities is Singapore, which has already identified such areas as future growth engines, setting up a new programme known as "Polaris", with funding of S$20 m.[12]

Developments such as the above have contributed to a shift in the business model approach in at least some pharmaceutical companies. Some industry players have been shifting emphases in business models, and diversifying and reorganizing their activities. Brand-name companies often also produce generics, and *vice versa*, while some companies may prefer to take over existing generic producers.

In the case of GlaxoSmithKline, it has chosen to adapt by shifting focus from short-term profits to increasing sales volume and sales value to grow market share over a longer term. It has also had to review its IP policies to recognize that the treatment of such rights may vary with different diseases. It has also had to adjust its approach with different national systems and needs, as well as a devise a specific strategy to address the access to medicines needs of least developed countries (LDCs) and developing countries.[13] It will be a challenge for companies with entrenched corporate cultures to follow suit, and it remains to be seen if other (large innovator) companies will adapt in similar or other ways.

Along with the changes in the industry, increased interest in the access debate has contributed to the creation of new information tools which raise public scrutiny over industry members. For instance, an *Access to Medicines Index* has been developed relatively recently to assess and compare the performance of major pharmaceutical companies across a number of factors such as pricing, patents policy and R&D activity.[14]

Such developments are of interest to Asia, both because of the global reach of decision-making affecting access to medicines and the resulting

[12] See Lynn Kan, Singapore Places Bets on Personalised Drugs, *The Straits Times*, 10 November 2012, and Radha Chitale, *POLARIS Initiative Hones in on Personalized Medicine*, January 2013.

[13] The author is grateful for the discussion with Mr Jon Penner, GlaxoSmithKline, see note 10 above. On LDCs and developing countries, the company has, e.g., established a unit to work on initiatives for these countries; see www.developingcountriesunit.gsk.com/Home (accessed 28 October 2014). Within Southeast Asia, these initiatives have been introduced in Cambodia, Lao PDR and Myanmar. GlaxoSmithKline is mentioned in the present context as a case study of adaptation, as the company has scored well – as number 1 out of a list of 20 companies evaluated – in the Access to Medicines Index of 2012; see www.accesstomedicineindex.org/ranking (accessed 28 October 2014). This Index assigns scores to pharmaceutical companies based on factors such as management, R&D, pricing, public policy, patents, capability and donations.

[14] *Ibid.*

impact on the cost of medicines in Asia, and as "players" in the R&D and production chain for such products.

### *2.1.2 The changing face of Asia, innovation and pharmaceutical industries*

States aim to make use of limited budgets to provide healthcare at affordable prices to their people. This is done through a variety of means, including negotiation of medicine prices with private provides (e.g., the innovators above), use of competitive tendering for medicines, price control mechanisms (direct and indirect) and the use of subsidies for healthcare goods and services.

In Asia, many governments provide some form of funding assistance for medicines such as subsidies. This ameliorates the burden on the consumer who otherwise may not be able to afford such medicines. As governments need to allocate resources to various needs, a healthcare budget requiring such support often finds itself in competition with the budget needs elsewhere, such as for the development of infrastructure or education. States therefore have to be highly mindful of the costs of the healthcare services and goods – such as medicines – that it aims to make more affordable.

At the same time, governments are mindful that they should not stymie innovative and entrepreneurial instincts of the innovators and potential innovators, and must provide enough incentives for their innovations to flourish. The argument is made, for example, that reduction of prices of pharmaceuticals would lead to significant disincentives for and reductions in R&D for new pharmaceuticals.[15]

In recent years, several Asian states have also shown strong commitments towards making innovation and biomedical research a key plank

[15] See, e.g., David R. Francis, National Bureau of Economic Research, *The Effect of Price Controls on Pharmaceutical Research*, www.nber.org/digest/may05/w11114.html (accessed 12 June 2014), citing Thomas Abbott and John Vernon, *The Cost of U. S. Pharmaceutical Price Reductions: A Financial Simulation Model of R&D Decisions* (NBER Working Paper No. 11114). India, e.g., has a system of national price controls for pharmaceuticals. For a study of price controls in OECD countries, see International Trade Administration, US Department of Commerce, *Pharmaceutical Price Controls in OECD Countries – Implications for US Consumers, Pricing, Research and Development, and Innovation* (2004). For a discussion of pricing options, see WHO, *Pharmaceuticals and Health Sector Reform in the Americas – An Economic Perspective* (1998), in Section 4.4.

and driver of their economic policies. Examples include China, India, Korea, Malaysia, Singapore and Thailand.[16] Several countries in Asia play host to both brand-name and generic manufacturers, and need to balance their interests in promoting economic growth, innovation and affordability of medicines. More broadly, policy-making, in terms of addressing investor interests and promoting innovation and affordability of health products and services, calls for a whole-of-government approach to achieve balance and coherence.

Within ASEAN, all members are signatories to the Paris Convention (except Myanmar) and the Patent Cooperation Treaty (except Cambodia and Myanmar).[17] All ten members are members of the WTO (and therefore signatories to the Agreement on Trade-Related Aspects of Intellectual Property Rights (TRIPS). China, Japan, Korea, India and Sri Lanka are parties to all three sets of obligations. China has also notified the Director-General of WIPO that the Paris Convention and PCT apply also to Hong Kong, and that the Paris Convention applies also to Macau. Chinese Taipei is party only to the WTO.[18]

In 1990, during the Uruguay Round (UR) negotiations for the TRIPS agreement, the five founding member states of ASEAN (Indonesia,

[16] On China, see information on *The National Guideline on Medium- and Long-term Program for Science and Technology Development (2006–2020)* issued by the State Council. See also generally, David Wertime, It's Official: China is Becoming a New Innovation Powerhouse, *Foreign Policy*, 6 February 2014. On Korea, see EU Mini-Country Report, Korea, 2011, ftp://ftp.ige.unicamp.br/pub/CT010/aula%208/sistema%20de%20inovacao%20coreano%20e%20institutos%20de%20pesquisa.pdf and Deok Soon Yim, *Korea's National Innovation System and the Science and Technology Policy*, ftp://ftp.ige.unicamp.br/pub/CT010/aula%208/sistema%20de%20inovacao%20coreano%20e%20institutos%20de%20pesquisa.pdf. On India, see *Science, Technology & Innovation Policy 2013*, www.dst.gov.in/sti-policy-eng.pdf (accessed 12 August 2014). On Malaysia, see *Malaysia's S&T Policy for the 21st Century*, Ministry of Science, Technology and Innovation website, www.mosti.gov.my/index.php?option=com_content&view=article&id=2032&Itemid=611&lang=en (accessed 12 August 2014). On Singapore and Thailand, see respectively, National Framework for Research, Innovation and Enterprise, National Research Foundation, www.nrf.gov.sg/innovation-enterprise/national-framework-for-research-innovation-and-enterprise and National Science Technology and Innovation Policy Office, www.sti.or.th/en/index.php/policya plans/sti-master-plan-2 and UNCTAD, http://unctad.org/meetings/en/Presentation/MYEM2_2014_Durongkaveroj_Pichet.pdf (accessed 12 August 2014).

[17] See WIPO list of signatories of the Patent Cooperation Treaty, Paris Convention and WTO treaties as on 5 June 2014, www.wipo.int/export/sites/www/pct/en/texts/pdf/pct_paris_wto.pdf (accessed 12 August 2014).

[18] This follows the WTO terminology; see www.wto.org/english/thewto_e/countries_e/chinese_taipei_e.htm (accessed 12 August 2014).

Malaysia, the Philippines, Singapore and Thailand) had expressed cautious views as follows[19]:

> **TRIPS**
>
> In the area of TRIPS, ASEAN notes that a text has been prepared and submitted by the Chairman on his own responsibility for Ministers' consideration in Brussels. The text is peppered with square brackets. But this cannot be avoided for the issues involved are sensitive ones, especially for developing countries who will be required to make most of the adjustments.
>
> Setting aside the issue of where to implement the agreement, which will be one of the major political decisions of Ministers, ASEAN agrees that there is a need to curb the unilateral interpretation of what constitutes effective intellectual property rights protection. ASEAN sees multilateral resolution of disputes as the only alternative. *But we are strongly opposed to any provision that disregards public interest, that undermines the integrity of our domestic legal systems, and that impose unreasonable and extremely burdensome demands on enforcement measures.*
>
> We would also like to further make known to everyone here our position on two other issues. One, there shall not be any dilution or exemption from the m.f.n. principle in the obligations to be agreed upon in TRIPs. And two, the transitional period and arrangements, and the extent of obligations for developing countries, shall take into full account their trade, finance, and development needs, and their administrative capacity to implement this agreement.
>
> [Italic emphasis added]

In ASEAN, IPR and innovation discussions often proceed separately from discussions on matters such as public health needs and implications. In the field of IP, for example, ASEAN members are working towards implementing an ASEAN IPR Action Plan 2011–2015.[20] Its five "strategic goals" are IP-centric with no explicit mention of related social aspects such as public health implications of IPRs.[21] Goal 3 does,

[19] Statement by the Representative of the Philippines, MTN.TNC/W/42, 29 November 1990.

[20] See www.asean.org/archive/documents/ASEAN%20IPR%20Action%20Plan%202011-2015.pdf (accessed 18 September 2014).

[21] The five goals are:

Strategic Goal 1:
A balanced IP system that takes into account the varying levels of development of Member States and differences in institutional capacity of national IP Offices to enable them to deliver timely, quality, and accessible IP services to promote the region as being conductive to the needs of users and generators of IP.

however, speak of utilization of IP as "a tool for innovation and development" and advancing the region's interests through "support for the transfer of technology to promote access to knowledge". Goal 4 sets out the aspiration for greater connectivity to address "stakeholders", and Initiative 25 under this Goal – to be led by the ASEAN Secretariat – is to develop a "strong negotiating position" in the form of a "minimum negotiating framework on IP" for ASEAN member states.[22] Separately, the *Krabi Initiative* of 2010, which addresses innovation in ASEAN, includes among its "tracks", "ASEAN Innovation for Global Markets" and "Science and Innovation for Life".[23] As part of the ASEAN 2015 Community vision, its rationales include: seeking a balance between competitiveness and human development and reinventing the ASEAN scientific community for a meaningful delivery of science, technology and innovation agenda in ASEAN.[24] While these themes and sub-themes are being further articulated they present opportunities and a challenge for thoughtful regional policy-making on the treatment of innovative scientific and technological activity in ASEAN, and how such policy could and should relate, for example, to ASEAN's other initiatives relating to affordable healthcare.

Strategic Goal 2:
Developed national or regional legal and policy infrastructures that address evolving demands of the IP landscape and AMSs participate in global IP systems at the appropriate time.

Strategic Goal 3:
The interests of the region are advanced through systematic promotion for IP creation, awareness, and utilization to ensure that IP becomes a tool for innovation and development; support for the transfer of technology to promote access to knowledge; and with considerations for the preservation and protection of indigenous products and services and the works of their creative people in the region.

Strategic Goal 4:
Active regional participation in the international IP community and with closer relationships with dialogue partners and institutions to develop the capacity of Member States and to address the needs of stakeholders in the region.

Strategic Goal 5:
Intensified cooperation among AMSs and increased level of collaboration among them to enhance human and institutional capacity of IP Offices in the region.

22 See p. 17, note 20 above.

23 See schematic overview of the Initiative at http://eeas.europa.eu/delegations/thailand/documents/thailande_eu_coop/science-and-technology/thailand_sti_policy_thailand-eu_workshop_11june2013_part1_2_dr_somchai_en.pdf (accessed 12 August 2014).

24 *Ibid.*

In pursuing innovation initiatives, monetization of innovative creations and acceleration of patent applications are logical sub-themes.[25] Singapore, for instance, unveiled its IP Hub Master Plan in 2013.[26] A relevant policy question therefore is how ASEAN innovation and IPR policies should relate to public health interests. For states with strong innovation policies, it is critical to balance the public health interest and the strong economic instinct to maximize the commercialization of innovation. The regulatory space needs to be managed in a manner that ensures that public health is not viewed as a secondary matter to economic imperatives; this is a balancing act which policy-makers increasingly have to grapple with in this new age of innovation-driven national progress in Asia. As one author has succinctly pointed out:

> It has been observed that a commitment to both innovation and access in biomedicine and biomedical research can be onerous.[27]

### *2.1.3 FTAs in Asia*

In terms of FTA activity, Asia has in the last two decades been in the forefront, engaging intra-Asia as well as with external partners.[28] This took place against the backdrop of the Doha Round of trade negotiations in the WTO, which, by 2000, were stalling. With these talks unable to move towards conclusive agreements, the FTA movement took on even greater momentum as states sought new and more open markets for their goods, services and investments. The financial crisis of 2007–2009 made the imperative to seek more trade even more urgent. Singapore, one of the most open trading nations in the world, led the vanguard, signing its

[25] See, e.g., Singapore's policy on monetization of IP assets, and the IP financing: www.mlaw.gov.sg/content/minlaw/en/news/speeches/Speech-by-Senior-Minister-of-State-for-Education-and-Law-Ms-Indranee-Rajah-at-the-Official-Opening-of-IP-101-on-Tuesday-8th-April-2014-at-Manulife-Centre.html (accessed 12 August 2014).

[26] IPOS website, www.ipos.gov.sg/Portals/0/Press%20Release/IP%20HUB%20MASTER%20PLAN%20REPORT%202%20APR%202013.pdf; see also Ministry of Law website, www.mlaw.gov.sg/news/press-releases/IP-hub-master-plan-launched.html (accessed 12 August 2014).

[27] Josephine Johnston and Angela A. Wasunna, *Patents, Biomedical Research and Treatments – Examining Concerns, Canvassing Solutions, Hastings Center Special Report*, January–February 2007, at p. s3.

[28] See WTO, *World Trade Report 2011*, at 57; see also WTO Press Release, *Asia is at the Leading Edge of New Trade Pacts of World Trade Report 2011* (27 July 2011) at http://wto.org/english/news (accessed 19 March 2015).

first FTA on 14 November 2000 with New Zealand.[29] As of March 2015, Singapore has signed 11 bilateral FTAs, 3 FTAs with external groups and 6 as a member of ASEAN.[30]

China, a fairly late entrant to the FTA arena, had by early 2015 signed 11 FTAs, with several others "under negotiation" or "under consideration".[31]

Against this backdrop, the US was pursuing its own trade agenda which included seeking ever higher intellectual property standards from its trade partners where possible. Since the TRIPS Agreement, the US has FTAs with at least 20 countries, with a professed aim of, *inter alia*, raising intellectual property protection standards through such agreements.[32]. Such proposals were reportedly made during the TPP negotiationswhich have now been completed.

## 2.2 Government reviews on the pharmaceutical sector and patents

Concerns over rising drug costs and the state of competition between the originator and generic drugs have prompted debate and reviews of pharmaceutical patent regulations in different jurisdictions. In 2011, for example, the EU published a highly significant report on its pharmaceutical sector arising in part from the need to control public budgets, examining issues such as patent filing practices, patent litigation and settlements, strategies relating to the life cycle of second-generation pharmaceutical products.[33] One of its recommendations was an increased role for competition law scrutiny.[34]

In 2012, Australia embarked on a review of pharmaceutical patents, and the Final Report was issued in May 2013.[35] As Australia is a

[29] See legal text at www.fta.gov.sg/fta_anzscep.asp?hl=9 (accessed 19 March 2015).

[30] See www.fta.gov.sg/sg_fta.asp (accessed 19 March 2015).

[31] See China Ministry of Commerce website, http://fta.mofcom.gov.cn/english/fta_qianshu.shtml (accessed 19 March 2015).

[32] See USTR website, https://ustr.gov/trade-agreements/free-trade-agreements (accessed 19 March 2015).

[33] EU *Pharmaceutical Sector Inquiry*, Final Report, available at: http://ec.europa.eu/competition/sectors/pharmaceuticals/inquiry/ (a number of follow-up reports are also available at this site) (accessed 16 March 2015).

[34] Para. 4.1, *Ibid.*

[35] *Pharmaceutical Patents Review, Final Report*, available at: www.ipaustralia.gov.au/pdfs/2013-05-27_PPR_Final_Report.pdf (accessed 16 March 2015). The aim of the review is stated as follows in the Preface of the Report: The Review is to examine whether Australia's patent system is effective in securing timely access to competitively priced

negotiating party to the TPP, certain concerns were raised in the Report about these negotiations and their possible impact on medicines.[36]

## 2.3 International patent commitments: from Paris to Uruguay and beyond

The negotiation of the TRIPS Agreement arose out of a failure of efforts to amend existing IP treaties such as the Paris Convention for the Protection of Industrial Property.[37] The TRIPS Agreement in its present form provides minimum requirements for national patent laws and undisclosed information, leaving members with a degree of discretion in areas of exclusion and in which express flexibility has been provided. As a number of developed country demands for certain protections were not met during the Uruguay Round (UR) negotiations for the TRIPS Agreement, the post-UR era has seen the growth of bilateral trade agreements which aim to incorporate provisions reflecting these unmet demands, as well as new demands.[38] As these provisions exceed the requirements under TRIPS, they are often referred to as "TRIPs-plus" provisions. Such provisions add to the access to medicines discussion as

pharmaceuticals and in supporting innovation and employment in the industry. An important part of the review is to examine the Australian provisions for extending the terms of eligible pharmaceutical patents.

36 See pp. vi, 48, 51, 55–57of the Report, *Ibid.* The Australian government's statement on the report is available at IP Australia website, *Ibid.* It states, in part:

> The government notes that the report is one of a number of reviews of the pharmaceutical system conducted during the term of the previous government. The government has no plans to respond to the report at this stage but may take information in the report into account when considering future policy. The views expressed and recommendations made in the report are those of the review panel and do not necessarily reflect government policy.

37 For a historical account of the TRIPS negotiations and motivating forces, see Chapter 1, Charles Clift, Why IPR Issues Were Brought to GATT: A Historical Perspective on the Origins of TRIPS, in *Research Handbook on the Protection of Intellectual Property under WTO Rules - Intellectual Property in the WTO*, Vol. 1, ed. Carlos M. Correa, (Edward Elgar Publishing Ltd, 2010). See also ed. Terence P. Stewart, GATT, The Uruguay Round - A Negotiating History, 1986–1994, Vol. 4, *The End Game* (Part 1), Kluwer Law International.

38 An example is "pipeline protection" for pharmaceuticals; see Clift, *ibid.*, at p. 19. See Also Sherry S. Marcellin, *The Political Economy of Pharmaceutical Patents: US Sectional Interests and the African Group at the WTO* (Ashgate Publishing Ltd, 2010), pp. 103–104.

they often provide broader and expanded protection to pharmaceutical patent holders, expanding their rights beyond what TRIPS patent protection requires. This gives such rights holders greater leverage in pricing their products, as well as extended periods in which protection applies. The pricing privilege and longer protection periods in turn affect competition and access. Examples of such provisions are discussed further below and in Chapter 4.

## 2.4 The promise of TRIPS: flexibilities for members

Members of the World Trade Organization (WTO) in 1995 commit to certain minimum standards of intellectual property protection and enforcement, primarily through the TRIPS.[39] These include the protection of patents. Along these commitments are also a number of flexibilities and exceptions.[40] Besides the WTO, these flexibilities are

[39] Text available at: www.wto.org/english/tratop_e/trips_e/t_agm0_e.htm. See also generally, WTO Gateway to IP and Access to Medicines: www.wto.org/english/tratop_e/trips_e/counciljun01_e.htm.

[40] The following are examples of provisions in the TRIPS Agreement which provide members with a degree of flexibility in IP law-making:

> **Article 7 Objectives**
>
> The protection and enforcement of intellectual property rights should contribute to the promotion of technological innovation and to the transfer and dissemination of technology, to the mutual advantage of producers and users of technological knowledge and in a manner conducive to social and economic welfare, and to a balance of rights and obligations.
>
> **Article 8 Principles**
>
> 1. Members may, in formulating or amending their laws and regulations, adopt measures necessary to protect public health and nutrition, and to promote the public interest in sectors of vital importance to their socio-economic and technological development, provided that such measures are consistent with the provisions of this Agreement.
> 2. Appropriate measures, provided that they are consistent with the provisions of this Agreement, may be needed to prevent the abuse of intellectual property rights by right holders or the resort to practices which unreasonably restrain trade or adversely affect the international transfer of technology.
>
> **Article 30 TRIPS Exceptions to Rights Conferred**
>
> Members may provide limited exceptions to the exclusive rights conferred by a patent, provided that such exceptions do not unreasonably conflict with a normal exploitation of the patent and do not unreasonably prejudice the legitimate interests of the patent owner, taking account of the legitimate interests of third parties.

being discussed and debated across a number of other multilateral fora, including the WIPO's Standing Committee on the Law of Patents,[41] UN Human Rights Council,[42] UNAIDS,[43] UNDP[44] and WHO.[45] The WHO, for example, in its Global Strategy and Plan of Action on Public Health, Innovation and Intellectual Property, explicitly refers to the TRIPS flexibilities.[46]

The *Bolar* exception, which permits a competitor to make limited quantities of a patented medicine solely for the purposes of obtaining marketing approval, was confirmed as being a permissible application of Article 30 in *Canada – Pharmaceutical Patents*.[47] The panel report in this case examined the application of Article 30 and its relationship with Article 27 in detail, while its observations on Articles 7 and 8 (which featured in the arguments of Canada, the EU and third parties including India and Thailand) were relatively limited.[48] Thailand's third party submission, for instance, argued that Articles 7 and 8 formed part of the context in which Article 30 should be read.[49] Another third party, Israel, argued that these two provisions "should serve as primary tools for interpretation of the Agreement".[50]

Article 30, and to a much more limited extent, Articles 7 and 8, were examined by a WTO panel in *Canada – Patent Protection of Pharmaceutical Products*, WTO panel report, WT/DS/114, 17 March 2000.

41 See, e.g., SCP documents at www.wipo.int/patents/en/topics/exceptions_limitations.htm and www.wipo.int/edocs/mdocs/scp/en/scp_20/scp_20_ref_summary.pdf, (accessed 23 February 2015).

42 See, e.g., discussions of the Council, 18–20 February 2015, at www.ohchr.org/Documents/Issues/SForum/SForum2015/PresentationTitlesAbstracts.pdf (accessed 19 March 2015).

43 See, e.g., www.unaids.org/sites/default/files/media_asset/JC2260_DOHA%2B10TRIPS_en_0.pdf (accessed 23 February 2015).

44 See, e.g., www.undp.org/content/dam/undp/library/hivaids/Using%20TRIPS%20Flexibility%20to%20improve%20access%20to%20HIV%20treatment.pdf (accessed 23 February 2015).

45 See, e.g., www.who.int/intellectualproperty/studies/TRIPS_flexibilities/en/ (accessed 23 February 2015).

46 See, e.g., paras. 8 and 12, and paras. 34–36 and 38 of its Annex; at http://apps.who.int/gb/ebwha/pdf_files/A61/A61_R21-en.pdf (accessed 7 May 2015).

47 *Canada – Patent Protection of Pharmaceutical Products*, WTO panel report, WT/DS/114, 17 March 2000.

48 See, e.g., panel report, *Ibid.*, para. 7.92.

49 Panel report, note 47 above, at para. 5.31.

50 Panel report, note 47 above, at para. 5.23.

Article 31 provides flexibility to use compulsory licensing, a form of compulsory use of a patented invention – such as a patented medicine – under the WTO rules.[51] However, with the cost of patented, brand-name medicines, including those for treating HIV/AIDS, being priced beyond the reach of several poor countries, in 2001, WTO members issued the Ministerial Declaration on TRIPS and Public Health.[52]

In 2003, several debates, WTO members agreed to extend the compulsory licensing process under Article 31 of TRIPS, through a decision to waive Article 31(f).[53] This extension of the Article 31 compulsory licensing process has come to be known as the "paragraph 6" procedure, and includes among other features a waiver of Article 31(f), which would have required production under a compulsory licence to predominantly domestic use.[54] This waiver is presently awaiting inclusion in TRIPS as a permanent amendment.[55] The new mechanism was adopted in the form of a waiver decision by WTO members, and currently an amendment to the TRIPS Agreement awaits ratification by a minimum number of members for it to become a permanent feature of the Agreement.[56] It has,

[51] Compulsory licensing in Asia will be discussed in Chapter 5. See generally, G. Van Overwalle, *Fair Use: A Workable Concept in EU Patent Law?*, in *Compulsory Licensing: Practical Experiences and Ways Forward*, ed. Hilty Reto S. and Kung-Chung Liu (Springer 2015). See also Susy Frankel and Jessica C. Lai, *Recognised and Appropriate Grounds for Compulsory Licences: Reclaiming Patent Law's Social Contract*, in the same title.

[52] See www.wto.org/english/thewto_e/minist_e/min01_e/mindecl_trips_e.htm.

[53] Decision of the General Council, WTO, *Implementation of Paragraph 6 of the Doha Declaration on the TRIPS Agreement and Public Health*, WT/L/540 and Corr.1, 30 August 2003. Article 31(f) requires that the compulsory licensing be utilized to largely supply one's domestic market. The decision waives this requirement, enabling WTO members to produce pharmaceuticals for export to eligible member countries which lack their own production capabilities.

[54] So called because the waiver was a decision to implement paragraph 6 of the Doha Declaration on the TRIPS Agreement and Public Health, WT/MIN(01)/DEC/2; paragraph 6 of that Declaration states:

6. We recognize that WTO members with insufficient or no manufacturing capacities in the pharmaceutical sector could face difficulties in making effective use of compulsory licensing under the TRIPS Agreement. We instruct the Council for TRIPS to find an expeditious solution to this problem and to report to the General Council before the end of 2002.

[55] See www.wto.org/english/news_e/pres05_e/pr426_e.htm.

[56] See General Council, WTO, *Decision on Amendment of the TRIPS Agreement*, WT/L/641, 6 December 2005.

however, so far only been invoked once, when Canada applied it to supply medicines to Rwanda.[57]

As there are excellent accounts elsewhere of the developments leading up to this waiver and the proposed amendment to the Agreement, these will not be recounted in detail here.[58]

The TRIPS Agreement requires members to protect undisclosed data of a party in respect of new chemical entities (NCEs), such as new pharmaceuticals, but permits the use of such data in circumstances where certain requirements are met. Article 39.3 provides:

> Members, when requiring, as a condition of approving the marketing of pharmaceutical or of agricultural chemical products which utilize new chemical entities, the submission of undisclosed test or other data, the origination of which involves a considerable effort, shall protect such data against unfair commercial use. In addition, Members shall protect such data against disclosure, *except where necessary to protect the public, or unless steps are taken to ensure that the data are protected against unfair commercial use.*
>
> [Italic emphasis added]

Interpretation of this provision is not settled, with questions having been raised with respect to, for example, the scope and meaning of "unfair commercial use".[59]

[57] See WTO, Annual Review of the Decision on the Implementation of Paragraph 6 of the Doha Declaration on the TRIPS Agreement and Public Health, 12 November 2008, IP/C/49.

[58] See, e.g., Daniel Gervais, *The TRIPS Agreement – Drafting History and Analysis*, 4th ed. (Sweet & Maxwell, 2012), pp. 54–75, and Cynthia M. Ho, *Complicated Compulsory Licenses: The Waiver/Article 31bis "Solution", in Access to Medicine in the Global Economy* (Oxford University Press, 2011). The major WTO reference documents in this connection are available at: www.wto.org/english/tratop_e/trips_e/pharmpatent_e.htm (accessed 19 March 2015).

[59] See, e.g., the views of the EU in discussions in the TRIPS Council, IP/C/W/280, at p. 4 and the views of the African Group, Barbados, Bolivia, Brazil, Cuba, Dominican Republic, Ecuador, Honduras, India, Indonesia, Jamaica, Pakistan, Paraguay, Philippines, Peru, Sir Lanka, Thailand and Venezuela in IP/C/W296, at pp. 8–9. Note, too, that during the negotiations on the draft text for TRIPS, there was disagreement as to whether trade secrets, later discussed as "undisclosed information" in TRIPS, should be included in the Agreement as intellectual property; see, e.g., Communication from Brazil, MTN.GNG/NG11/W/57, 11 December 1989, paras. 48–49; Chairman's Report to the Group of Negotiations on Goods, MTN.GNG/NG11/W/76, p. 1. See also the interesting discussion in Nuno Pires de Carvalho, *The TRIPS Regime of Antitrust and Undisclosed Information* (Kluwer Law International, 2008), pp. 222–223. For further drafts of the TRIPS Agreement, see Daniel Gervais, *Ibid.*, pp. 532–546.

## 2.5 Regulatory tensions

### *2.5.1 Identifying the interested parties and their interests: regulators, innovators, competitors and consumers*

The division of opinions and, more importantly, of positions stems from a difference in motivations and vested interests on the parts of various stakeholders. These motivations encompass profit, regulatory and humanitarian reasons. These make the division deep, and new solutions are urgently needed to resolve or, at least, ameliorate the differences. One way of viewing the divide is in terms of globalization and its relationship with private rights and public goods or goals.[60]

At the philosophical level, it is argued that medicines – at least "essential" medicines – are seen as public goods. So too, to some, is the R&D knowledge which arises in the quest for new medicines to treat diseases. An alternative frame of reference is that of a human right but even that itself gives rise to divergent views as to what/whose right one is talking about or should prevail. On one hand, it is argued that the right to affordable (essential?) medicines is a human right while on the other, inventors/patentees, too, assert that it is a human right to have the benefits of their creations protected.[61]

At the pragmatic level, it is argued that TRIPS-plus rights are eroding the TRIPS flexibilities given to WTO members, so that they are not in a position to provide effective, affordable medicines that are critical to their citizens.[62] This is seen to be the result of the increasing scope and

[60] See, e.g., the insightful discussion in Keith E. Maskus and Jerome H. Reichman, The Globalization of Private Knowledge Goods and the Privatization of Global Public Goods, *JIEL*, Vol. 7, No. 2 (2004), 279–320.

[61] Cynthia Ho, Unveiling Competing Patent Perspectives, *Hous. L. R.*, Vol. 46 (2009–2010), 1047 at pp. 1058–1076, discussing Thailand's laws and compulsory licensing actions as an example for the analysis. At p. 1114, after having examined the WTO rules on patents and compulsory licensing of medicines, Thailand's actions, and arguments of those holding what she refers to as the "uber-right" view of patents (see note 91, Prof. Ho frames the question pointedly: "If global rules can be easily jettisoned based on one side's argument that the rule is not proper policy, what then is the point of the rule of law?").

[62] At the World Health Assembly, May 2014, for example, Brazil's representative stated:

> IP should not be an obstacle to national policies on access to medicines. Countries should therefore be able to use them in a way that is conducive to national circumstances.

See also report of IP Watch, *World Health Assembly Approves Plan to Strengthen Access to Essential Medicines*, 24 May 2014, at: www.ip-watch.org/2014/05/24/world-health-assembly-approves-plan-to-strengthen-access-to-essential-medicines/ (accessed 26 May

reach of provisions in certain FTAs, where a *demandeur* state requires its FTA partner to provide for protection beyond that required under TRIPS. These are explained in the next chapter.

For reasons discussed below, the following appears to be an apt observation:

> Health policy is a key battleground in which "competing visions of the ethical and political basis of society, and of the nature of the economy, are fought out."[63]

Such "competing visions" are particularly clear in bilateral/regional trade and investment treaty negotiations. As will be elaborated upon in Chapters 3 and 4, the frequent requests by *demandeur* states for various forms of increased protection – often TRIPS-plus in nature – for pharmaceutical interests in such negotiations has attracted a great deal of debate and criticism. Such states argue that the heightened protections provide incentive for innovation in the field. Opponents and critics point to the WTO's TRIPS Agreement and its various "flexibilities", arguing that such demands can have the effect of eroding them, with the result that access to medicines is adversely affected in those countries which agree to the demands.[64] An example of a current "battleground" is in the arena of the Trans-Pacific Partnership (TPP) Agreement negotiations, in which

2014). See generally, H. Grosse-Kahn, The International Law Relationship between TRIPS and TRIPS-plus FTAs: Towards Safeguarding TRIPS Flexibilities? *J. Int. Prop. Law*, Vol. 18, No. 2 (2011), 327.

[63] Ministry of Social Affairs and Health, Finland, Kimmo Leppo, Eeva Ollila, Sebastián Peña, Matthias Wismar and Sarah Cook, ed., *Health in All Policies – Seizing Opportunities, Implementing Policies* (2013), citing M Mackintosh and M Koivusalo, eds., *Commercialization of Health Care: Global and Local Dynamics and Policy Responses* (Basingstoke: Palgrave Macmillan, 2005) (UNRISD Social Policy in a Development Context Series), p. xiv.

[64] On this latter view, see e.g., the recommendations (especially in paras. 70(c) and 71(d) in *Report of the Special Rapporteur on the right of everyone to the enjoyment of the highest attainable standard of physical and mental health, Anand Grover, on access to medicines*, para. 6, A/HRC/23/42. Paras. 70(c) and 71(d) state, respectively:

> 70. (c) *Use flexibilities* under the TRIPS Agreement to promote regional collaboration to pool resources and facilitate competitiveness of local production . . .
>
> 71. (d) *Resist trade policies that undermine* the ability of States to reimburse the price of essential medicines to local pharmaceutical companies.
>
> [Italic emphasis added]

Vietnam, a developing country and a relatively recent member of the WTO, is a participant.[65]

### 2.5.2 *End user interests*

Who are the "players" in this divided landscape? One might begin with the producers of medicines, since they are the ones who generate the subject matter of interest. However, for this discussion, the end users will be mentioned first, since they are the ones who create the demand for, and need, these medicines.

End users of medicines include not only the ultimate consumers of medicines, but are, for this purpose, inclusive of those that purchase medicines for them, such as government bodies that make use of formularies to acquire medicines for dispensing in public health institutions. This inclusion is made since, in terms of interest, these bodies share a common goal with consumers of medicines – obtaining effective medicines at affordable prices.

### 2.5.3 *Innovator interests*

The prices of patented medicines are usually determined by private parties who generally pursue a profit motive. There immediately

[65] At the time of writing, the text of the TPP was not yet publicly available. On the cautionary suggestions for Vietnam in relation to access to medicines and the TPP, see Addendum to the Report of the Special Rapporteur (*Ibid.*), A/HRC/20/15/Add. 2, 4 June 2012, in paras. 35–41, and recommendations for Vietnam in relation to the TPP negotiations, at paras. 62(f)-(g); Addendum available at: www.ohchr.org/Documents/HRBodies/HRCouncil/RegularSession/Session20/A-HRC-20-15-Add2_en.pdf (accessed 23 February 2015). These particular recommendations are for Vietnam to:

> (f) Increase transparency and ensure the participation of civil society and affected communities in the Trans-Pacific Partnership Agreement (TPPA) negotiation process in order to promote the full realization of the right to health and to strengthen the government's technical capacity in the area of access to medicines;
>
> (g) Ensure that, if Viet Nam accedes to the TPPA, it retains the ability to use TRIPS flexibilities and *does not accept TRIPS-plus measures*, and that intellectual property protections in the agreement do not expand protections beyond that which currently exist under Vietnamese law;
>
> (h) Explore carefully the possible development impacts of the TPPA, and in particular, its effect on the development of local generic pharmaceutical production in Viet Nam.
>
> [Italic emphasis added]

arises a tension between permitting private, market-driven prices and meeting public, social needs in the form of affordable medicines. A related tension arises from the belief that patents, in encouraging innovation, are sacrosanct and should be protected at all costs; when this belief is confronted by social need there is a dilemma.

### *2.5.4 State interests: measures of innovation, healthcare costs containment and regulatory space*

How does one measure innovation, and how does one measure the effect of patents on it? Measures of innovation tend to be "proxy" measures, such as measures of R&D activity and patent grant activity.[66]

Proponents of patent activity as a driver of innovation argue that patents undergird the innovation incentive. An example of such an argument can be found in the letter of the Innovation Alliance to U.S. Senate Judiciary Committee leadership dated 11 February 2014, as shown by the following excerpt:

> The U.S. patent system is a driving engine of our economy and a vital contributor to the nation's economic creativity and ingenuity. These rights have kept American inventors competitive internationally since the founding of our country, and they create countless jobs and significant prosperity every year. Weakening the protections of the patent system would surely result in the loss of America's creative, technological, and economic edge.[67]

Governments have to assess the role of patents in promoting innovation and integrate this assessment into healthcare planning and regulation.

[66] WIPO Secretariat, Standing Committee on the Law of Patents, *Study on the Role of Patent Systems in Promoting Innovative Medicines, and in Fostering the Technology Transfer Necessary to Make Generic and Patented Medicines Available in Developing Countries and Least Developed Countries*, paras. 4–10, 10 October 2014, SCP/21/8. See also Aaron S. Kess, Using Patent Data to Assess the Value of Pharmaceutical Innovation, *J. L. M. & Ethics* (2009), 176, Amir Attaran, How Do Patents and Economic Policies Affect Access to Essential Medicines in Developing Countries? *Health Affairs*, Vol. 23, No. 3 (2004), pp. 155–166, 2004, and Verena Schüren, What a Difference a State Makes: Pharmaceutical Innovation After the TRIPs Agreement *Bus. & Pol.* (2013), pp. 217–243.

[67] See http://innovationalliance.net/news-and-resources/innovation-alliance-letter-senate-judiciary-leadership-encouraging-caution-toward (accessed 23 February 2015). See also Richard Gilbert, Looking for Mr Schumpeter: Where Are We in the Competition-Innovation Debate? *Innov. Pol. & Econ.* (2006), 159, cited in Thomas Cheng, A Developmental Approach to the Patent-Antitrust Interface, *Nw. J. Int'l L. & Bus.*, Vol. 33, No. 1 (2012), 1–79, in note 35.

### 2.5.5 *The role of civil society and non-profit organizations*

Civil society groups and non-profit organizations have played an important role in the debate on access to medicines, including within Asia. Such groups and organizations provide an additional "voice" in the global debate, and some, such as *Medicins Sans Frontières* (Doctors Without Borders), are actors directly involved in supplying medicines to various countries as well as in advocacy for greater access.[68] Within Asia, national civil society groups have also been prominent in some countries in influencing the access to medicines debate and decision-making.[69] Some organizations go beyond advocacy and advising authorities, to mount litigation to oppose patent applications.[70]

## 2.6 The place of patents in the debate

The present legal and policy construct has revolved a number of key "prisms", including the innovation-incentive–IP (patent) approach and a reward or contract analysis.[71] These approaches relate the innovative activity of researchers to the idea that there is a "patent bargain" between

[68] See, e.g., MSF at www.msfindia.in/access-campaign-india (accessed 5 April 2015).

[69] See, e.g., Kyujin Shim, Impact of Social Media on Power Relations of Korean Health Activism, *Media & Comm.*, Vol. 2, No. 2 (2014), 72–83, explaining the role of the Korea Leukemia Patient Group in the discussions with Novartis and the Korean government on prices of Glivec, used in treating leukaemia. Civil society groups have also been contributors to the access debate and negotiations in India and Thailand; see for example, Nathan Ford, David Wilson, Onanong Bunjumnong and Tido von Schoen Angerer, The Role of Civil Society in Protecting Public Health over Commercial Interests in Thailand, *The Lancet* Vol. 363, No. 9408 (2004), pp. 560–563, Suwit Wibulpolprasert, Vichai Chokevivat, Cecilia Oh and Inthira Yamabhai, *Government Use Licenses in Thailand: The Power of Evidence, Civil Movement and Political Leadership*, (explaining the role of Thai civil society groups in requesting access to didanosine (DDI), a drug patented by Bristol-Myers Squibb) (2011) *Global Health*, Vol. 7, p. 32. See also generally, Nathan Ford, Patents, Access to Medicines and the Role of Non-Governmental Organizations, *J. Gen. Med.*, Vol. 1, No. 2 (January 2004), pp. 137–145, and civil society's effort at a "declaration", i.e. *The Bangkok Declaration on Free Trade Agreements and Access to Medicines* (2011), at http://blogs.mediapart.fr/blog/plondeix/120511/bangkok-declaration-free-trade-agreements-access-medicines (accessed 5 April 2015).

[70] For example, Initiative for Medicines, Access & Knowledge (I-MAK) filed an opposition to Gilead's patent application in India for sofosbuvir; see I-MAK website, www.i-mak.org/i-mak-blog-updates/ (accessed 20 August 2014). See also Chapters 5 and 6, discussing the case of Thailand's compulsory licensing actions, which saw highly active civil society involvement.

[71] See, e.g., Dominic Guellec, Patents as an Incentive to Innovate, in *The Economics of the European Patent System: IP Policy for Innovation and Competition*, ed. Dominique Guellec and Bruno van Pottelsberghe de la Potterie (Oxford University Press, 2007),

such researchers and the state that grants them patent rights for their inventions. More will be said later about this concept and its evolution in the age of proliferation of trade and investment treaties. Other perspectives of the debate include bargaining theory,[72] a "perspectives" approach[73] and a law and research norms analysis.[74] At the outset, it should be noted that the patent system is but one facet of the present access debate.

### *2.6.1 The innovation-incentive approach*

The incentive theory takes the view that the valuable rights under a patent monopoly provide the incentive for manufacturers of medicines to invest in the necessary R&D to generate new and useful medicines for society. Such a view naturally has its limitations, since the patent incentive can lead patent holders to make use of their monopoly to control both prices and supply of medicines.[75] Market-driven and profit-maximization motivations alone do not necessarily guarantee that public needs for medicines are adequately met. The attraction of the patent monopoly – that is, the incentive – with its power to exclude competitors for a period of time may also lead to over-inclusive patent claims, and to innovators focusing only on profitable medicines, that is, medicines that will reap most profits from the wealthiest markets. This has been one reason for the existence of "neglected" diseases.[76] Loss of innovative productivity in

and Vincenzo Denicolò, Luigi and Alberto Franzoni, The Contract Theory of Patents, *Int'l Rev. Law and Econ.*, Vol. 23 (2004), 365–380.

72 Daniel Benoliel and Bruno Salama, Towards an Intellectual Property Bargaining Theory – The post-WTO Era, *U. Pa. J. Int'l L.*, Vol. 32, No. 1 (2010), pp. 265–368.

73 Cynthia M. Ho, *A New Approach to the Compulsory License Conundrum, in Patent Law in Global Perspective*, ed. Okediji and Bagley, note 2 above.

74 See, e.g., Rebecca S. Eisenberg, Proprietary Rights and the Norms of Science in Biotechnology Research, *Yale L. J.*, Vol. 97(1987–1988) , 177, and Arti K. Rai, Regulating Scientific Research: Intellectual Property Rights and the Norms of Science, *Nw. U. L. Rev.*, Vol. 94 (1999–2000), 77.

75 Some have even argued that the patents system may slow innovation; see, e.g., Edwin Cameron, Patents and Public Health – Principles, Politics and Paradox, in ed. David Vaver, *Intellectual Property Rights – Critical Concepts in Law*, Vol. IV (Routledge, 2006), 437–469, especially at pp. 442–445 and 458. See also Paul Edward Geller, *An International Patent Utopia?*, in the same title, Vol. V, 233–246, especially at pp. 233–237.

76 Diseases may be neglected as they exist in markets where the ability to pay for medicines is poor. The WHO, e.g., lists 17 "neglected tropical diseases": see www.who.int/neglected_diseases/diseases/en/ (accessed 6 September 2014). For information on neglected diseases affecting Southeast Asia in particular, see www.who.int/bulletin/volumes/88/3/09-072322/en/ (accessed 6 September 2014).

recent years, together with patent "cliffs" occurring with the expiry of a number of best-selling (or blockbuster) medicines (mentioned earlier) and the rising costs of R&D,[77] has caused pharmaceutical innovators to rethink how best to continue to grow their profits and what new strategies can be crafted. Such pressures exacerbate the limitations of the incentive model. Clearly, the patent incentive alone cannot and does not serve adequately societal needs.

With respect to the narrower debate on the patent bargain, Kali Murray, for example, proposes a participatory patent framework to produce a "participatory patent bargain".[78] The element of participation is certainly important and this framework shifts the emphasis towards greater stake holder input into the process of formulating the new patent bargain. It is suggested however that this is but one facet of the present framing of the patent bargain.

The present legal construct of the patent bargain is, however, no longer one built out of purely a national law assuming a natural innovation–disclosure–patent protection–further innovation and public benefit cycle, or what might be called the classical patent bargain. Instead, we have today two versions of an internationalized patent bargain: the post-TRIPS patent bargain and the TRIPS-plus bargain.

### *2.6.2 The patent bargain and its relationship with access to medicines*

The patent system of rights is statute-conferred in nature and has been viewed as a bargain between inventors and the state. In the words of Abraham Lincoln:

> The patent system ... secured to the inventor, for a limited time, the exclusive use of his invention; and thereby added the fuel of interest to the fire of genius, in the discovery and production of new and useful things.[79]

[77] See www.forbes.com/sites/aroy/2012/04/24/how-the-fda-stifles-new-cures-part-i-the-rising-cost-of-clinical-trials/; Phase III trials are reportedly the biggest contributor of rising costs of innovation.

[78] Kali Murray, *A Politics of Patent Law – Crafting the Participatory Patent Bargain* (Routledge, 2013) .

[79] *2nd Lecture on Discoveries and Inventions*; reproduced as Chap. 32 in ed. David Vaver, *Intellectual Property Rights: Critical Concepts in Law*, Vol. 3 (Routledge, 2006), p. 36. For a useful account of the origins of many aspects of the patent system and its "bargain", see generally, Christopher May, *The Venetian Moment – New Technologies, Legal Innovation and Institutional Origins of Intellectual Property*, Chap. 31 in the same book.

In exchange for the state's recognition and protection of the inventor's rights under the patent granted for his/her invention, the inventor is expected to set out his/her claim and specifications clearly – certainly clearly enough for another skilled in the art to employ the invention. This second aspect is the "price" or *quid pro quo* for the statutory rights under the patent. As such a patent grants a legal monopoly (albeit for a limited duration) the boundaries of this monopoly are critical. As Lord Hoffman explained in *Kirin-Amgen*:

> The need to set clear limits upon the monopoly is not only . . . in the interests of others who need to know the area "within which they will be trespassers" but also in the interests of the patentee, who needs to be able to make it clear that he lays no claim to prior art or insufficiently enabled products or processes which would invalidate the patent.[80]

He further explained the nature of the "contract" in a patent:

> An invention is a practical product or process, not information about the natural world. That seems to me to accord with the *social contract* between the state and the inventor which underlies patent law. The state gives the inventor a monopoly in return for an immediate disclosure of all the information necessary to enable performance of the invention. That disclosure is not only to enable other people to perform the invention after the patent has expired. If that were all, the inventor might as well be allowed to keep it secret during the life of the patent. It is also to enable anyone to make immediate use of the information for any purpose which does not infringe the claims. *The specifications of valid and subsisting patents are an important source of information for further research*, as is abundantly shown by a reading of the sources cited in the specification for the patent in suit. Of course a patentee may in some cases be able to frame his claim to a product or process *so broadly* that in practice it will be impossible to use the information he has disclosed, even to develop important improvements, in a way which does not infringe. But *it cannot be right to give him a monopoly of the use of the information as such.*[81]
>
> [Italic emphasis added]

The bargain in its classic sense requires sufficient disclosure to enable future innovation by others. Lord Neuberger stated the terms of the "bargain" as follows, in *Human Genome Sciences Inc v Eli Lilly*:[82]

> . . . it is worth remembering the purpose of the patent system, namely to provide a temporary monopoly as an incentive to innovation, while at the

[80] *Kirin-Amgen Inc v Hoechst Marion Roussel Limited* [2004] UKHL 46, in para. 21, citing Lord Russell in *Electric and Musical Industries Ltd v Lissen Ltd* (1938) 56 RPC 23, at p. 39.
[81] *Ibid.*, para. 77. [82] [2011] UKSC 51, para. 99.

> same time facilitating the early dissemination of any such innovation through an early application for a patent, and its subsequent publication. Although this is true in any sector, it has particular force in the pharmaceutical field, where the pharmaceutical field, where even many of those who are sceptical about the value of intellectual property rights accept that there is a public interest in, and a commercial need for, patent protection.

In *Brenner v Manson*, the US Supreme Court explained the *quid pro quo by* framing the public benefit as a central plank of the system:

> The basic quid pro quo contemplated by the Constitution and by Congress for granting a patent monopoly is the benefit derived by the public from an invention with substantial utility. Unless and until a process is refined and developed to this point – where specific benefit exists in currently available form – there is insufficient justification for permitting an applicant to engross what may be a broad field.[83]

In *Teva Canada v Pfizer Canada*, the Canadian Supreme Court explained the patent bargain as follows[84]:

> The patent system is based on a "bargain", or *quid pro quo*: the inventor is granted exclusive rights in a new and useful invention for a limited period in exchange for disclosure of the invention so that society can benefit from this knowledge. This is the basic policy rationale underlying the Act. The patent bargain encourages innovation and advances science and technology. Binnie J. explained the *quid pro quo* as follows in *AZT*, at para. 37:
>
> A patent, as has been said many times, is not intended as an accolade or civic award for ingenuity. It is a method by which inventive solutions to practical problems are coaxed into the public domain by the promise of a limited monopoly for a limited time. Disclosure is the *quid pro quo* for valuable proprietary rights to exclusivity which are entirely the statutory creature of the *Patent Act*.
>
> (Italics original.)

The Court went on to explain that the patent "bargain theory" was reflected in sections 27(1)–(3) of the Canadian Patents Act.[85]

In *First Currency Choice Pte Ltd v Main-Line Corporate Holdings Ltd*, the Singapore Court of Appeal explained the objectives of the patent system, being of the view that the following quotation as an "excellent conspectus of the congenital difficulties in regulating a patent system":

[83] 383 U.S. 519 (1966), pp. 534–535.

[84] [2012] 3 R.C.S. 625, at p. 639. The Court went on to briefly discuss the argument of "sound prediction" (pp. 640–642).

[85] *Ibid.*, p. 643.

> An efficient patent system is essential for the promotion of innovation. Patents give inventors a temporary monopoly on a new idea in return for disclosing how it works, so that others can subsequently build upon it. But if a patent is granted for something that is not novel (people are already doing it), or is obvious (any Tom, Dick, or Harry in the field could think it up), it can hamper innovation by turning a widely used invention or process into one person's monopoly. The trouble is that examiners cannot always tell when a patent is unwarranted. . . .[86]

The Court observed:

> These insightful observations neatly encapsulate the pragmatic rationale underpinning as well as the intrinsic tensions in the modern approach towards administering a patent protection scheme – namely, striking the *right balance* between stimulating the creative energies of inventors, while promoting the free flow of ideas and encouraging entrepreneurship. Striking the appropriate balance is certainly not a prosaic task . . .
>
> [Italic emphasis added]

The patent bargain may be explained by the grant of a statutory, time-limited monopoly of a suite of rights in exchange for disclosure of certain information by the grantee to teach society about the invention. As stated by the English Court of Appeal in *Halliburton v Smith International*, in the context of the European Patent Convention:

> . . . the basic requirement, that a patent should teach the practice of the invention sufficiently well for the notional addressee to perform the invention is fundamental to the patent system: the teaching is the price the patentee pays for his time-limited monopoly.[87]

There has been much discussion about how the debate on IPRs and access to medicines might be "reframed", especially in light of the

[86] [2008] 1 SLR(R) 335 para. 1, citing *The Economist Technology Quarterly* (8 September 2007) at 23–24, "A patent improvement". The rest of the quotation (omitted here) relates to the question of examining "prior art" to determine what is patentable. It should be noted that patents confer a *legal* monopoly to exclude others' use but not necessarily an *economic* monopoly. However, several features of modern patent law, as exacerbated by the manner in which FTAs try to expand the *legal* monopoly, seem to be directed at attaining or protecting an *economic* monopoly. Whilst ownership of a patent does not guarantee an economic monopoly, the means of legal influence through FTAs could significantly boost the patent owner's chances of gaining an economic monopoly, primarily by exclusion of competitors from either marketing their products for a longer time than the domestic statutory duration of a patent, or by delaying regulatory approval (and therefore market entry, again) by not allowing the regulator to use research data pertinent to approval but arising from the patent owner's own earlier application.

[87] [2006] EWCA Civ. 1715, para. 12.

trade-based focus in treatment of IPRs in TRIPS and FTAs. Owing to the emergence of TRIPS-plus regimes in certain FTA signatory countries, it is pertinent to consider whether there is a treaty-influenced or "treatified" patent bargain in such states. Depending on the nature of the TRIPS-plus commitments, the patent bargain in such systems may have been modified from its classical form stated above. The modifications alter the calculus of the basic patent bargain in its classic form, so that it is no longer adequate to view it simply as a *quid pro quo* of monopoly rights in exchange for sufficient disclosure.

### *2.6.3 The patent bargain and "treatification": the TRIPS and TRIPS-plus era*

The patent bargain has been explained in terms of social contract theory and rewards theory.[88] The original patent bargain notion consisted of the compact between the state and the innovator, where the state statutorily and deliberately permits a monopoly of limited duration to the innovator for his/her invention, in exchange for adequate disclosure so that others may further engage in innovative activity based on the disclosed information. Apart from such statutory protection, the laws of the state may also provide for legitimate business secrets of such an innovator company not required to be disclosed for the patent grant. In exchange for the valuable patent grant, the patent holder enjoys certain rights. According to the Panel in *Canada – Pharmaceuticals*:

> In theory, the rights of the patent owner are generally viewed as a right to prevent competitive commercial activity by others, and manufacturing for commercial sale is a quintessential competitive commercial activity, whose character is not altered by a mere delay in the commercial reward.[89]

The patent bargain theory has been the subject of debate against the background of new technologies, new methods of business and commercial practices, and it calls for a rebalancing of rights and obligations.[90]

[88] See, e.g., Vincenzo Denicolò and Luigi Alberto Franzoni, The Contract Theory of Patents, *Int. Rev. L. & Econs.*, Vol. 23 (2004), 365–380. For alternative formulations, see also Dan L. Burk and Mark A. Lemley, *The Patent Crisis and How the Courts Can Solve It* (University of Chicago, 2009) esp. Chapters 6 and 7.

[89] See para. 7.35, note 47 above.

[90] See, e.g., Shubha Ghosh, *Patents and the Regulatory State: Rethinking the Patent Bargain Metaphor After Eldred*, http://papers.ssrn.com/sol3/papers.cfm?abstract_id=574141 (accessed 6 September 2014).

In the context of pharmaceutical patents, one author has characterized the division in opinions of the patent as running the spectrum from the "privilege view", with a patent being "a mere privilege granted by a nation and are inherently subject to limitations to accommodate other societal goals, such as access to medicine", to a patent being an "uber-right", that is, "a type of super-property right that should seldom, if ever, be subject to exception".[91]

The TRIPS Agreement affected this view of the bargain in that it introduced a number of treaty protections beyond what had existed in international law and in many national systems. Aspects of the patent bargain as expressed in the TRIPS itself may be found in Arts. 28 and 29. As the negotiating history of TRIPS shows the discussions were often divided as to the extent the new agreement should impose new patent (and other IP) rights. During the negotiations, the submission of draft language by a group of developing countries attempted to include *explicit obligations on the part of patent owners* themselves, as part of TRIPS.[92] This was ultimately not included in the TRIPS Agreement. This interesting idea, however, may be useful when considering FTA/ bilateral investment treaty negotiations (see Chapter 4).

The TRIPS-plus phenomenon arises primarily from negotiation of bilateral or regional FTAs. Thus, a country which enters into a treaty with such commitments and which has to amend its patent laws to reflect them no longer necessarily reflects the traditional patent bargain as

91 Cynthia Ho, Unveiling Competing Patent Perspectives, *Hous. L. R.*, Vol. 46 (2009–2010), 1047, at p. 1050.

92 Communication from Argentina, Brazil, Chile, China, Colombia, Cuba, Egypt, India, Nigeria, Peru, Tanzania and Uruguay, MTM.GNG/NG11/W/71, 14 May 1990. Draft Article 5(2) provided as follows:

> The owner of the patent shall have the following obligations:
>
> (i) to disclose the invention in a clear and complete manner to permit a person versed in the technical field to put the invention into practice and in particular to indicate the best mode for carrying out the invention;
> (ii) to give information concerning corresponding foreign application and grants;
> (iii) to work the patented invention in the territory of the Party granting it within the time limits fixed by national legislation and subject to the sanctions provided for in Chapter VI;
> (iv) in respect of licence contracts and contracts assigning patents, to refrain from engaging in abusive or anticompetitive practices adversely affecting the transfer of technology subject to the sanctions provided for in Chapter VI and VII.

described earlier. An FTA may alter this bargain by requiring the state to (a) extend the monopoly duration; (b) provide exclusive protection for the innovator's data; and (c) require the state's medicine licensing authority to act as a watchdog for patent holders in case a generic manufacturer that makes an application for approval to market a medicine like that of the patent holder. In addition, such FTAs may enlarge the class of subject matter that is to be patentable in the state, with the result that more classes of subject matter become eligible for the patent monopoly.

Besides enhanced rights pertaining to patents, TRIPS and several other bilateral agreements include obligations on protection of undisclosed information. Article 39 of TRIPS addresses the protection of undisclosed data.[93] There is some disagreement over the operation of this provision,

[93] Art. 39 provides:

1. In the course of ensuring effective protection against unfair competition as provided in Article 10*bis* of the Paris Convention (1967), Members shall protect undisclosed information in accordance with paragraph 2 and data submitted to governments or governmental agencies in accordance with paragraph 3.
2. Natural and legal persons shall have the possibility of preventing information lawfully within their control from being disclosed to, acquired by, or used by others without their consent in a manner contrary to honest commercial practices[10] so long as such information:
   (a) is secret in the sense that it is not, as a body or in the precise configuration and assembly of its components, generally known among or readily accessible to persons within the circles that normally deal with the kind of information in question;
   (b) has commercial value because it is secret; and
   (c) has been subject to reasonable steps under the circumstances, by the person lawfully in control of the information, to keep it secret.
3. Members, when requiring, as a condition of approving the marketing of pharmaceutical or of agricultural chemical products which utilize new chemical entities, the submission of undisclosed test or other data, the origination of which involves a considerable effort, shall protect such data against unfair commercial use. In addition, Members shall protect such data against disclosure, except where necessary to protect the public, or unless steps are taken to ensure that the data are protected against unfair commercial use.

[10] For the purpose of this provision, "a manner contrary to honest commercial practices" shall mean at least practices such as breach of contract, breach of confidence and inducement to breach, and includes the acquisition of undisclosed information by third parties who knew, or were grossly negligent in failing to know, that such practices were involved in the acquisition.

See also Art. 10*bis* of the Paris Convention, which deals with unfair competition, referring to "any act contrary to honest practices in industrial or commercial matters".

as to how WTO members are to provide protection against "unfair commercial use".[94] One view is that it provides for two separate and distinct obligations on WTO members in relation to such data, namely non-*disclosure* of data by government authorities and non-*use* (such as governmental reliance on it for approval of other pharmaceutical approval applications).[95] Others view Art. 39 as allowing member governments to use such data (such as in reliance of it for marketing approval of other similar products).[96]

Given that the disagreements and the fact that there is so far no WTO case law interpreting the provision, the following has been an attempt to summarize the position thus far (subject to the prerequisites in Art. 39):

> (i) the flexibilities and pro-public health interpretation in the Doha Declaration cover the TRIPs Agreement as a whole and therefore apply to test data protection under Article 39.3; (ii) there is no explicit TRIPs requirement to provide data exclusivity, but some form of protection against unfair commercial use is required; and (iii) the fact that two forms of protection are to be provided under Article 39.3 of the TRIPs Agreement highlights that protection against unfair commercial use must involve more than merely not disclosing the data.[97]

For a discussion of the negotiating history of Art. 39, see Daniel Gervais, *The TRIPS Agreement:, Drafting History and Analysis*, 4th ed. (Sweet & Maxwell, 2012) at pp. 532–546, G. Lee Skillington and Eric M. Solovy, The Protection of Test and Other Data Required by Article 39.3 of the TRIPS Agreement, *Nw. J. Int'l L. & Bus.*, Vol. 24, No. 1 (2003–2004), at pp. 15–20 and Carlos Correa, *Protection of Data Submitted for the Registration of Pharmaceuticals: Implementing the Standards of the TRIPS Agreement* (2002), pp. 53–55, at www.who.int/medicines/areas/policy/protection_of_data.pdf (accessed 12 March 2015).

[94] See WTO, WIPO and WHO, *Promoting Access to Medical Technologies and Innovation – Intersections between Public Health, Intellectual Property and Trade* (2012), pp. 65–66, at p. 65. See also generally, WHO, *Public Health – Innovation and Intellectual Property Rights*, Report of the Commission on Intellectual Property Rights, Innovation and Public Health (CIPR), (2006), at pp. 124–126.

[95] See EU paper, *Legal Issues related to Compulsory Licensing under the TRIPS Agreement and Questions on TRIPS and Data Exclusivity*, 2001, trade.ec.europa.eu/doclib/html/122031.htm, at pp. 18–19, 21 (accessed 11 March 2015).

[96] See WTO, WIPO and WHO, *Promoting Access to Medical Technologies and Innovation – Intersections between Public Health, Intellectual Property and Trade* (2012), pp. 65–66. Some in Brazil, e.g., have expressed the view that Art. 39 excludes test data exclusivity: see *Brazil's Patent Reform – Innovation Towards National Competition* (Center for Strategic Studies and Debate2013), at pp. 158–160.

[97] *Ibid.*, pp. 64–65; the study also states a number of conditions at the same page.

LDC members of the WTO – such as Cambodia, Lao PDR and Myanmar – are not obliged to provide the protection under Article 39 until 1 January 2016.[98]

Data exclusivity clauses in FTAs may prevent not only disclosure but also reliance on data for regulatory approval of similar generic products.[99] This has the practical result of either requiring a generic drug manufacturer to conduct its own clinical tests or await expiry of the period of data exclusivity stipulated in the FTA. Given that generic manufacturers may not be able to incur the expense of such trials the practical reality is that market entry of such products will be delayed. At a policy level, requiring repetitive testing for similar drugs has been identified as a possible matter for concern in the context of the EU's regulations.[100]

In some cases, interpretative guidance has been provided in the FTAs on test data exclusivity provisions.[101]

Such FTA-based requirements are TRIPS-plus but some (such as data exclusivity protection) are in fact "patent-plus", as they are not

[98] Decision of the TRIPS Council, 29 November 2005, IP/C/40, and Decision of the TRIPS Council, 1 July 2002, IP/C/25. At the time of writing, discussions are ongoing as to a possible extension of indefinite duration for LDCs; see www.ip-watch.org/weblog/wp-content/uploads/2015/02/LCD-pharma-waiver-proposal-Feb-2015.pdf (accessed 9 October 2015) and http://europa.eu/rapid/press-release_IP-15-5620_en.htm (accessed 9 October 2015).

[99] See, e.g., the Korea–US FTA (Art. 18.9) and the US–Singapore FTA (Art. 16.8).

[100] See, e.g., Directive of the Council of 6 November 2001 on the Community code relating to medicinal products for human use, as amended by Directive 2004/27/EC of the European Parliament and of the Council of 31 March 2004, which states, while recognizing the need for rules to safeguard human health through standards and protocols for tests and trials on medicines, "(10) However, there are reasons of public policy for not conducting repetitive tests on humans or animals without over-riding cause." To this end, it also requires "(26) In order to facilitate the movement of medicinal products and *to prevent the controls carried out in one Member State from being repeated in another, minimum requirements* should be laid down for manufacture and imports coming from third countries and for the grant of the authorization relating thereto." [Italic emphasis added]

[101] See, e.g., Art. 11.30.1, the EU–Singapore FTA (signed but not yet in force at the time of writing), which provides:

The Parties recognise the importance of the *Declaration on the TRIPS Agreement and Public Health*, adopted on 14 November 2001 by the Ministerial Conference of the WTO, at Doha. In interpreting and implementing the rights and obligations under Sub-Section E (Patents) and Sub-Section F (Protection of Test Data Submitted to Obtain an Administrative Marketing Approval to put a Pharmaceutical Product on the Market), the Parties shall ensure consistency with this Declaration.

dependant on whether the product is patented or not.[102] Given these "treatified" rights which accrue to the benefit of patent holders, and in the case of data exclusivity, rights attaching to post-patent products, signatory states are in fact guaranteeing a series of rights exceeding those known in the traditional patent bargain, as explained earlier. In other words, this new FTA-driven construct alters the patent bargain with respect to pharmaceutical patents for these countries.

This alteration of the bargain however often does not produce a balanced "bargain" in that the provisions mentioned may not be adequately be balanced by the inclusion of provisions which safeguard the other part of the bargain: the public interest in further innovation and competition, and therefore competitive pricing.

### 2.6.4 *The patent bargain, revisited*

As early as in the patent system of fifteenth century Venice, credited as being the system of origin of elements of the modern patent system, the importance of a balance between private and public interests was already recognized.[103]

The TRIPS-mandate minimum standards for national patent "bargains" also recognizes such a balance, as can be seen from its various provisions that provide members with "flexibilities" in rule- and policy-making. The balance under TRIPS has, however, been the subject of debate and criticism.[104]

The introduction of TRIPS-plus treaty provisions (such as in FTAs) influences the overall balance of national patent bargains in the national systems of treaty signatories. This has arguably resulted in a modified

[102] See EU paper, *Ibid.*, at pp. 18–19, 21, and WTO, WIPO and WHO, *Promoting Access to Medical Technologies and Innovation – Intersections between Public Health, Intellectual Property and Trade* (2012), at p. 65.

[103] Christopher May, *The Venetian Moment*, at p. 13, note 79 above.

[104] See, generally, Elizabeth Ng Siew Kuan, Balancing Patents and Access to Medicine, *SAcLJ*, Vol. 21 (2009),457, Barton and Emanuel, The Patents-Based Pharmaceutical Development Process: Rationale, Problems, and Potential Reforms, *JAMA*, Vol. 294, No. 16 (26 October 2005 ) 2075–2082, Elizabeth Ng Siew Kuan, Chapter 4 – "Global Health and Development: Patents and Public Interest", in *Incentives for Global Public Health: Patent Law and Access to Essential Medicines*, ed. T Pogge, M Rimmer, K and Rubenstein (United States of America: Cambridge University Press 2010), and *Global Health & Development: Patent & Public Interest in Connecting International and Public Law* (Cambridge University Press, 2009).

patent bargain for such countries, goes beyond what the classical patent bargain provides for and to the extent it is TRIPS-plus, also differs from the minimum-plus-flexibilities "bargain" mandated under TRIPS.

If one were to accept this as the new patent bargain agreed upon in such countries, what is its new *quid pro quo* and how does it impact access to medicines? Perhaps, the more appropriate term is "patent-plus" rights which innovators enjoy under TRIPS-plus treaty provisions. Such rights comprise a suite of enhanced rights, ranging from more hospitable patentability thresholds to the concomitant availability in many cases of investor-State dispute mechanisms to enforce these rights. This last aspect may seem to be peripheral to the debate on pharmaceuticals and access, but in fact forms an important, new aspect of TRIPS-plus, patent-plus rights in the new 'treatified' bargain. This aspect will be discussed further in Chapters 4 and 7. For now, an ongoing dispute serves as an apt illustration. As explained in Chapter 4, national judicial interpretation of the patent bargain has recently been challenged in international investor-dispute arbitration. In *Eli Lilly v Government of Canada*, the claimant Eli Lilly Company argues as follows in relation to what its two patents granted by the Canadian government entails:

> A patent constitutes a commitment to the patentee that it will have exclusive rights to make, use, and sell its invention until the expiry of the patent. By granting exclusivity to Lilly, Canada was upholding its end of the bargain between the inventor and the State that undergirds the entire Canadian Patent Act. . . . In other words, the . . . patents were *more* than a mere representation to Lilly from the government of Canada; they were a bundle of legally enforceable rights.[105]

This case, which remains pending at the time of writing, is immense importance in that an arbitral tribunal is being called upon to adjudicate upon Canadian judicial authorities on patentability criteria under Canadian law, in the context of alleged violations of NAFTA provisions. It not only entails a study of the patent bargain under Canadian domestic law, but also represents an attack on the Canadian judicial case law process through the lens of investment treaty law.

[105] Claimant's Memorial, para. 286; www.international.gc.ca/trade-agreements-accords-commerciaux/topics-domaines/disp-diff/eli.aspx?lang=eng (accessed 6 March 2015).

### *2.6.5 Managing the new patent bargain: the modern state's balancing act*

The balancing of innovator rights and incentives, the state's interest in promoting innovative activity and its protection of the public interest (for present purposes, particularly in areas such as public health) has become increasingly complex. Patents are but one facet of this balancing act, as there are a number of associated interests which innovators of medicines are interested in protecting, such as the test data which shows whether their products meet safety and efficacy requirements. In the pharmaceutical context, in particular, patents have both been praised as well as vilified but neither position alone will lead to a satisfactory and sustainable patent bargain in today's world.[106]

As will be seen in Chapters 4 and 5, patents and such associated interest areas have been the subject of treaty negotiations, resulting in some cases in protection exceeding those in the TRIPS Agreement.

#### 2.6.1.1 State interests and trade-offs in negotiations

Within state interests, trade and investment technocrats may form a powerful subgroup which seeks aggressive economic growth through means such as entering new trade and investment agreements. Trade partners in negotiating such agreements may pressure a state to consider and accept trade-offs, by way of health commitments, for the "overall package" that have significant implications for healthcare measures of the state, whether immediate or over the long term.[107]

[106] In the context of discussing patents in biomedical innovation it has been said:

> One way of thinking about biomedical research and treatment is as an ecology in which a variety of factors and players affect whether, how, and what biomedical research is conducted and whether, how, and which treatments are delivered and to whom. Different forces play niche roles in this ecology, and patents are one such force. Yet the actual and the ideal roles of patents in this ecology are contested. Some commentators argue that patents are inappropriate, ineffective, or even harmful to innovation and access, while others argue that patents are vital to achieving these goals.

Josephine Johnston and Angela A. Wasunna, Patents, Biomedical Research and Treatments – Examining Concerns, Canvassing Solutions, Hastings Center Special Report, Jan–Feb 2007.

[107] An example of a longer-term implication is that risk of exposure to investor-state disputes being brought against the state for alleged violations of the treaty when a state acts in regulating public health. For a discussion of this aspect, see Chapters 4 and 5.

#### 2.6.1.2 Treaty protection and commercial practices

Exacerbating this natural tension are efforts of such private pharmaceutical patent holders to corral patent provisions in the FTAs for the further protection of their rights, beyond what TRIPS requires. TRIPS-plus provisions include those requiring protection for data exclusivity and patent linkages. FTA provisions of this nature are "patent-plus" requirements – and may, in many cases, promote "innovation-status quo" monopoly rights – allowing patent holders to continue to enjoy additional means of protection beyond their patents, without necessarily showing any innovation or improvement of existing products or technology.

Secondly, quite apart from treaty support, patent owners *themselves* often engage in commercial practices that in effect give them extended enjoyment of their patent rights. As the European Patent Office observed:

> What was once the preserve of a small legal and technical department now often forms an integral part of boardroom strategy. Patents are no longer simply a defensive shield, but a key weapon of corporate strategy.[108]

For example, some companies enter into agreements with generic drug producers, in which the former pays the latter to delay bringing their generic products to the market. Such an agreement – known as a "pay-for-delay" agreement, effectively allows patent owners to maintain their monopoly for a longer time, and allow pricing in a manner that is not challenged by competitor products being in the market. (A variant is where the patent holder arranges to pay a competitor not to pursue a lawsuit to challenge its patent, in a "reverse-payment" agreement.)

Another practice is a refusal to license the use of the patent. Other practices are "evergreening" (which is application for a new patent for relatively minor improvements of an existing patented product), patent thickets/clusters and royalty stacking.[109] In some cases, regulatory authorities have found patent holders to have obtained patents through misrepresentation or mislisting.[110] In other cases, pharmaceutical companies simply acquire generic competitors.[111]

[108] European Patent Office, *The Changing Economic Role of Patents*, in *Scenarios for the Future* (2007), p. 17. For an overview of such practices, see the EU Report on the Inquiry on the Pharmaceutical Sector, note 33 above.

[109] *Ibid.*, p. 4, note 5.

[110] In the US, examples include the following cases: Dippin' Dots, Inc v Mosey, 476 F.3d 1337 (Fed.Cir.) and In the Matter of Biovail Corp., Dkt. No. C-4060, FTC Decision and Order, 2 October 2002. For other types of patentee conduct, see generally, Alfonso, Foster and Reingold, US: IP and Antitrust, *The Antitrust Review of the Americas* (2013).

[111] See, for instance, the discussions in http://ec.europa.eu/competition/publications/cpn/2009_2_15.pdf.

## 2.7 A very contentious debate

TRIPS-plus provisions have come under fire from many quarters, but the demands for such provisions are continuing in, for example, the TPP negotiations. Given the criticisms and more importantly, the grave impact of such provisions on pricing of medicines in FTA members, some rethinking is certainly necessary. Some have ventured so far as to suggest scrapping the patent system altogether while others have suggested less drastic approaches.[112]

Writing in 2004, about a decade after the WTO system's coming into existence, Maskus and Reichman identified, eloquently, a striking irony of globalization, as follows:

> We find it ironic that, as tariffs, quotas, and other formal barriers to trade are dismantled, there has been a strong push to re-regulate world technology markets. Although the ratcheting up of global IPRs could adversely affect the growth prospects of developing countries, these nations have so far exerted little influence on standard-setting exercises. Indeed, the progressive re-regulation of world markets for knowledge goods is not driven by a broad consensus of economic agents in the developed world. Rather, pressures to elevate IP norms are exerted by powerful private interests whose lobbying activities hold sway in legislative and regulatory initiatives in rich countries and international forums.
>
> *These efforts are largely detached from the traditional goal of domestic IP systems to strike a balance between commercial profitability and public-interest concerns.*"[113]
>
> [Italic emphasis added]

In 2013, the Chief Executive Officer of Bayer, manufacturer of a drug for treating cancer known as Nexavar (see Chapter 5 on a related lawsuit in India), explained candidly that the drug was intended for the Western market which could afford it, and not for the Indian

[112] See, e.g., WIPO, *Multilateral Agreements and a TRIPS-Plus World*, 2003; Quaker International Affairs Programme, *Bilateral agreements in a TRIPS-Plus World: The Chile-USA Free Trade Agreement*; Ellen 't Hoen, *Report of the Commission on Intellectual Property Rights, Innovation and Public Health: A Call to Governments*, May 2006, Bulletin of the WHO; and Suthersanen, Dutfield and Chow, eds., *Innovation without Patents: Harnessing the Creative Spirit in a Diverse World*, Chap 9, *The ASEAN States*, Edward Elgar Publishing 2007 (SMU online access).

[113] Keith E. Maskus and Jerome H. Reichman, The Globalization of Private Knowledge Goods and the Privatization of Global Public Goods, *JIEL*, Vol. 7, No. 2 (2004), 279–320, at pp. 282–283. The authors also refer to some developed countries having a "tendency to espouse a protectionist IP agenda" as a result of "policy capture" by certain interest groups or industries, at p. 286.

market.[114] This led to a slew of highly critical and even vitriolic responses, thanks in part to an initial misquote by a reporting agency of his/her exact language.[115] More measured remarks explaining the pharmaceutical industry's stance exist, of course.[116]

In other cases, the language of protectionism has been raised in the debate, both by pharmaceutical exporting countries and importing countries. In 2013, for example, a former United States Trade Representative (USTR) referred to "innovation mercantilism" and "free riding" on others' technology, criticising, among other things, India's use of compulsory licensing.[117]

Those who argue on behalf of developing countries have employed the protectionism argument as well:

> High-protectionist visions of intellectual property law have become a kind of latter-day religion promoted by the special interests that have long dominated the political scene in the United States, the European Union (EU), and Japan.[118]

While rhetoric will not solve the problems of access, constructive dialogue, comprehensive understanding of the various elements of the debate and practical problems, and creativity in crafting solutions can.

[114] See reports at *Bloomberg*, www.bloomberg.com/news/2014-01-21/merck-to-bristol-myers-face-more-threats-on-india-patents.html and www.businessweek.com/news/2014-01-21/merck-to-bristol-myers-face-more-threats-on-india-drug-patents#p2.

[115] See, e.g., reports at IP Watch, at www.ip-watch.org/2014/02/18/battles-over-patents-is-india-changing-the-rules-of-the-game/ and Columbia Journalism Review, at www.cjr.org/the_audit/bloombergs_viral_misquote_1.php (accessed 14 April 2015).

[116] See, e.g., Statement of dissent by representatives of the research-based pharmaceutical industry, 6 December 2004, Appendix 2, *Millennium Development Goals Working Group Report 2005* – Appendix 2, at www.unmillenniumproject.org/documents/TF5-medicines-Complete.pdf (accessed 14 April 2015).

[117] Charlene Barshevsky, *The New Protectionist Threats*, 8 May 2013, Wall Street Journal.

[118] Jerome H. Reichman, Intellectual Property in the 21st Century: Will Developing Countries Lead or Follow? *Hous. L. R.*, Vol. 46 (2009–2010) 1115, at p. 1120. After explaining the weaknesses of the US patent system, Professor Reichman lamented as follows:

> ... one may also ask why certain Asian patent offices blindly supported these same proposals for a further upward ratcheting of international patent norms. It was as if their governments were asking the other OECD countries, "Please give us all your insoluble problems and contradictions as soon as possible, so we can undermine our own national systems of innovation, too."
>
> (pp. 1122–1123)

## 2.8 Other relevant elements of the debate

Before moving on to more specific systems of Asia in the next chapter, this discussion would not be complete without mentioning a number of further matters which can affect pricing of, and access to, pharmaceuticals.

### *2.8.1 Tariffs applicable to pharmaceuticals in Asia*

Tariffs levied on imported pharmaceutical products add to the final cost of such products to the users, and are therefore a component of the overall debate on access, as many countries maintain these for pharmaceutical goods.

Most ASEAN countries maintain tariff bindings with regard to pharmaceutical products (see Table 2.1), with Singapore and Japan having the lowest average bound duties, and India, Indonesia and Myanmar having the highest.

Four ASEAN countries maintain zero applied tariffs to pharmaceutical products (Brunei Darussalam, Cambodia, Malaysia and Singapore), while India and Thailand maintain the highest applied levels (see Table 2.2).

Note that *among* ASEAN members, the ASEAN economic integration plan provides an intra-ASEAN preferential tariff treatment. The ASEAN Trade in Goods Agreement sets out tariff elimination and reduction obligations of the member countries.[119] The majority of products – including pharmaceutical products – are, by early 2015, subject to zero tariffs. There are, however, some significant exceptions in the pharmaceutical sector. For instance, Indonesia has committed to MFN tariffs but not to elimination or reduction of tariffs in respect of "medicaments for the treatment of cancer, HIV/AIDS or other intractable diseases".[120]

[119] ASEAN members' individual ATIGA tariff liberalization schedules (including Schedule A items) are available at: www.asean.org/news/item/annex-2-tariff-schedules (accessed 7 May 2015).

[120] See ATIGA tariff schedule of Indonesia, *Ibid.*, at item 2599. Article 19(2)(h), ATIGA permits such exclusions to be listed in a "Schedule H", by reference to the reasons in Article 8 on general exceptions (which resembles Article XX of the General Agreement on Tariffs and Trade). Note that Indonesia also maintains local production requirements: see Ministry of Health Decree No. 1010/MENKES/PER/XI/200, 3 November 2008; see also the WHO, www.who.int/medicines/areas/coordination/Indonesia_PSCPQuestionnaire_21072011.pdf (accessed 7 May 2015).

Table 2.1. *Bound tariffs for pharmaceutical products of ASEAN and selected Asian countries*

| | Bound duty | | | | |
|---|---|---|---|---|---|
| Country | Binding status (B/P/U)[a] | Average duty | Minimum | Maximum | Duty free TL (%) |
| Brunei Darussalam | B | 20.0 | 20 | 20 | 0.0 |
| Cambodia | B | 9.9 | 7 | 10 | 0.0 |
| India | P | 38.1 | 25 | 40 | 0.0 |
| Indonesia | B | 34.9 | 30 | 40 | 0.0 |
| Lao PDR | B | 15.5 | 5 | 20 | 0.0 |
| Malaysia | P | 16.0 | 5 | 30 | 0.0 |
| Myanmar | P | 30.0 | 30 | 30 | 0.0 |
| Philippines | P | 19.5 | 10 | 20 | 0.0 |
| Singapore | B | 0.3 | 0 | 6.5 | 95.2 |
| Thailand | P | 24.6 | 10 | 30 | 0.0 |
| Vietnam | B | 2.5 | 0 | 15 | 51.5 |
| China | B | 4.4 | 0 | 6 | 3.4 |
| Japan | B | 0.1 | 0 | 2.6 | 96.8 |
| Republic of Korea | P | 0.5 | 0 | 6.5 | 91 |

*Source:* WTO Integrated Database, as at April 2014.[b]

[a] B: Fully bound; P: Partially bound; U: Unbound. (Based on Harmonized System Chapter 30 (Pharmaceutical Products).)

[b] The *WTO Integrated Database* is available at: http://tariffdata.wto.org/; the author acknowledges with gratitude the kind advice and assistance of Ms Adelina Mendoza of the WTO Economics & Statistics Division on the WTO *Database*.

Vietnam has a scheduled 5% tariff level on certain medicines such as amoxicillin, ampicillin and penicillin,[121] and has *excluded* commitments for "medicaments for the treatment of cancer, HIV/AIDS or other intractable diseases".[122]

[121] ATIGA tariff schedule of Vietnam, note 119 above; see items 3004.10.10, 3004.10.20, and 3004.10.15–16, scheduled as "Priority Integration Sectors negative list" items, excluded from priority-sector integration liberalization (see ATIGA Art. 19(2)(a)(i)).

[122] See *Ibid.*, item 3006.92.10, excluded as an "Schedule H" item under ATIGA Art. 19(2)(h); see explanation in note 120 above.

Table 2.2. *Applied MFN tariffs for pharmaceutical products of ASEAN and selected Asian countries*

| | Latest applied MFN duty (as at April 2014) | | | | |
|---|---|---|---|---|---|
| Country | Year | Average duty | Minimum | Maximum | Duty free TL (%) |
| Brunei Darussalam | 2011 | 0.0 | 0 | 0 | 100.0 |
| Cambodia | 2012 | 0.0 | 0 | 0 | 100.0 |
| India | 2013 | 9.7 | 0 | 10 | 3.2 |
| Indonesia | 2013 | 3.5 | 0 | 15 | 37.4 |
| Lao PDR | 2008 | 7.3 | 5 | 10 | 0.0 |
| Malaysia | 2013 | 0.0 | 0 | 0 | 100.0 |
| Myanmar | 2013 | 1.3 | 0 | 1.5 | 10.5 |
| Philippines | 2012 | 3.2 | 0 | 20 | 0.5 |
| Singapore | 2013 | 0.0 | 0 | 0 | 100.0 |
| Thailand | 2011 | 7.9 | 0 | 30 | 17.8 |
| Vietnam | 2013 | 1.7 | 0 | 14 | 77.3 |
| China | 2011 | 4.5 | 0 | 10 | 11.0 |
| Japan | 2013 | 0.3 | 0 | 5.3 | 91.6 |
| Republic of Korea | 2013 | 4.3 | 0 | 8 | 45.1 |

*Source:* WTO Integrated Database.[a]

[a] The *WTO Integrated Database* is available at: http://tariffdata.wto.org/; the author acknowledges with gratitude the kind advice and assistance of Ms Adelina Mendoza of the WTO Economics & Statistics Division on the WTO *Database.*

### 2.8.2 *National pricing, internal taxes and charges – some examples*

Internal taxes, such as value-added taxes and goods and services taxes (GSTs), add to the costs of medicines. This would be the result of domestic policy and it is open to the state to remove such taxes for medicines – at least "essential" ones, determined by reference, for example, to the state's formulary list.

#### 2.8.2.1 GST in Singapore and Malaysia

In Singapore, for example, while the country operates a virtually tariff-free system, the internal GST of 7% applies to pharmaceutical products. For pharmaceuticals included in the standard drugs list (SDL) government subsidies are available.[123] For qualifying households, certain forms of GST relief are also available.[124]

Procurement of medicines for public hospitals is undertaken by the Singapore Health Services Pte Ltd (SingHealth) Group Procurement Office (GPO).[125] The GPO was established in 2000 to streamline procurement activities in public healthcare facilities and to achieve strategic purchasing, with SingHealth being set up as a not-for-profit establishment.[126] Previously, such procurement was conducted by the Ministry of Health's Pharmaceutical

[123] See Ministry of Health, *Drug Subsidies* (showing the list of drugs subsidized at public healthcare institutions as of 1 April 2015), www.moh.gov.sg/content/moh_web/home/costs_and_financing/schemes_subsidies/drug_subsidies.html (accessed 15 April 2015). Other cases of drug subsidies are available; see generally, www.moh.gov.sg.

[124] See, e.g., the GST Voucher scheme at: www.gstvoucher.gov.sg/ (accessed 30 June 2014).

[125] See Singapore's Annex 1 and Annex 3, and the SingHealth website at: www.singhealth.com.sg/AboutSingHealth/GroupProcurement/AboutUs/Overview/Pages/Home.aspx and http://gpo.nhg.singhealth.com.sg/Tender/GPO/pub/index.html (accessed 25 April 2014).

[126] The then Minister for Health stated in Parliament on 10 March 2001 (italic emphasis added):

> . . . profits has never been the main objective of our restructuring exercise. We restructure so that we can be more efficient, more productive and more responsive. In this way, we bring the costs down and offer a better service to Singaporeans. I want to make our position on this explicit. SingHealth and NHG would therefore be repositioned as not-for-profit entities. As not-for-profit entities, their main objective is not to maximise profits, neither are they under pressure to pay dividends to their shareholders. To underscore this "not-for-profit status", the Minister for Finance has agreed to exempt the two clusters from paying corporate tax. Instead, the tax savings and any surpluses generated as a result of their efficiency gains will be ploughed back for medical research, training and patient care. While the concept of not-for-profit is relatively new in Singapore, it is well established in the US. In the US, some of the very well-known medical centres, such as the Mayo Clinic Foundation in Minnesota and Johns Hopkins Medicine in Maryland are operated as not-for-profit organisations. By modelling themselves as not-for-profit organisations, our restructured hospitals will have a different organisational culture from the private hospitals. Our restructured hospitals will continue to preserve their sense of public service. *Let me clarify that not-for-profit does not mean that our restructured hospitals do not have to worry about costs. The not-for-profit organisation must still exercise financial prudence, cost their services accurately, control their costs and improve productivity.*

(Singapore embarked on restructuring of its public hospitals to promote greater efficiency and economy from the late 1980s; see statement by the then Minister of State for Health and later, Acting Minister for Health in Parliament on 17 March 1986 and 13 March 1987.) On 20 July 2004, the then Minister of State for Health further explained in Parliament that cost savings by the GPO system are "passed directly" on to patients.

Department.[127] The GPO uses a system of open tendering for pharmaceuticals.[128]

Administrative costs of procuring and dispensing add to the overall costs of pharmaceuticals.[129] In 2012, the then Minister for Health explained the system of handling such costs for medicines in public hospitals[130] and the procurement system of medicines in Singapore.[131]

[127] Statement by the Minister for Health in Parliament on 1 June 1998.

[128] See SingHealth website at: http://gpo.nhg.singhealth.com.sg/Tender/GPO/pub/index.html (accessed 25 April 2014).

[129] See statement in Parliament by the then Minister of State for Health, on 3 December 1993; Hansard Singapore.

[130] The following is an excerpt from the then Minister's statements in Parliament, 14 February 2012; Hansard, Singapore:

As the public hospitals operate on a non-profit basis, *any revenue from sale of drugs, consultation fees and other services help to defray their overall operating costs.* In FY2010, the public healthcare institutions incurred operating deficits of over $2 billion, before subvention from the government . . . The margins for drugs sold in the public hospitals (excluding overheads and before subsidy) varies and *they typically range between 5% and 30%, on a per drug basis.* This helps the public hospitals to *cover overhead costs of procuring and storing each drug, preparing, packing and dispensing each prescription, and running the pharmacy and inventory. Any surplus is used to cover other operating expenses* of the healthcare institution. Various schemes such as the Standard Drug List (SDL) and the Medication Assistance Fund (MAF) help keep these drugs affordable for lower income Singaporeans who are able to access these drugs often at significantly below cost. For needy patients, they can also apply to Medifund for financial assistance. [Italic emphasis added]

[131] The following are excerpts from the then Minister's statements in Parliament, 7 March 2012; Hansard, Singapore:

> Currently, our clusters purchase drugs through a centralised Group Procurement Office (GPO) where possible. However, some drugs are better purchased at the individual cluster or hospital-level for various reasons, for example, where quantities are small, or if the drugs are specific to that institution, it is better for them to purchase the drugs directly. Price differences could still arise, even for drugs purchased via the GPO. For example, same drugs purchased in different batches may have different prices because different tenders carry different pricing. So some hospitals may be using drugs from an earlier batch or a later batch so their base price will be different, their costs will be different and also institutions have different cost structures and therefore we have to take into account giving institutions some flexibility so that they can take into account their different clinical setups and cost structures, but they are all subject to general subsidy guidelines. Some drugs are sold at a higher margin in order to cross-subsidise other drugs which may be very expensive and therefore needs a much higher discount to keep it affordable. In this way, we ensure that overall drugs are affordable for patients.

More generally, the reduction of costs of such medicines finds support from a government budget point of view where they are either paid for or subsidized with tax dollars. Apart from individuals who pay for their own medicines or co-pay for them, companies with health insurance plans and their insurers would also benefit from such a reduction in costs.

In April 2015, Malaysia introduced a new GST at 6%. The government has announced that pharmaceuticals on the National Essential Medicines List (NEML) and approved by the Minister of Finance will be exempt from the GST.[132]

### 2.8.3 *Corruption and government procurement of medicines*

Corruption and other unethical conduct in the supply chain of medicines contribute to problems of access.[133] Countries in Asia, like those in other regions, face issues of corruption in this context.[134] In China, a large pharmaceutical company has faced anti-corruption enforcement action

> I would like to stress that contrary to what [*name of member of Parliament omitted*] has said, our healthcare institutions are not-for-profit organisations. Whatever revenue that is generated from operations, be it from the pharmacy or be it from drugs, or charges from consultations, is taken in as overall revenue to the institution and used to partially defray the costs of operations, including overheads. These revenues are, however, insufficient to fully cover their costs. In FY2010, the public healthcare institutions incurred operating deficits of over $2 billion, before subvention from the government . . .

[132] See Royal Malaysian Customs, *Guide on Healthcare Services*, 19 November 2014, p. 7; text available at: http://gst.customs.gov.my/en/rg/SiteAssets/industry_guides_pdf/Revised/Guide%20on%20HEALTHCARE%20SERVICES%2019112014.pdf (accessed 5 March 2015). The Malaysian NEML, 4th ed., 2014, is available at: www.pharmacy.gov.my/v2/sites/default/files/document-upload/neml-4th-edition-29092014.pdf (accessed 5 March 2015). See also Malaysian Community Pharmacy Guild, *GST for Medicines*, 14 October 2014, at www.mcpg.org.my/index.php?root=MjIy&detail=MTI3.

[133] See generally, Jerome P. Kassirer, *The Corrupting Influence of Money on Medicine, in Global Corruption Report* (Pluto Press, Transparency International, 2006), at pp. 85–90, and *Report of the Special Rapporteur on the right of everyone to the enjoyment of the highest attainable standard of physical and mental health, Anand Grover, on access to medicines*, para. 60, note 64 above.

[134] See, e.g., Stuart Cameron, *Corruption in the Ministry of Public Health, Thailand*, and *Malpractice in the Office of the Drug Controller in Karnataka, India, in Global Corruption Report 2006, ibid.*, at pp. 100 and 101 respectively. E.g. of the types of conduct (involving monetary bribes and other acts) observed in the pharmaceutical sector, see Part Two on *Country Reports*, which includes reports on a number of Asian countries.

in recent times.[135] In Vietnam, one study has attributed higher drug costs predominantly to corruption.[136] To compound the problem, it has been observed that corrupt officials may also shield counterfeiters from exposure and legal action.[137]

Given that medicines are publicly procured by state authorities, procurement processes and the costs they impose on medicines procured form a relevant part of the debate on access to medicines. According to the WHO, 10–25% of public procurement spending (including on pharmaceuticals) is lost to corrupt practices, with unethical practices occurring at various points along the medicines chain from R&D to dispensing and promotion.[138] In discussions on access to medicines in the right to health context, procurement of medicines has also been found to be an area which states should take responsibility to address.[139]

The WTO's Government Procurement Agreement (GPA), which establishes a variety of legal commitments in procurement measures, is a plurilateral agreement, rather than a multilateral one, in that WTO members may choose not to participate in it. Indeed, to date, only 15 members (counting the EU as one) – or 42 (counting EU constituent members) – are signatories to the GPA.[140] Of these, Hong Kong, Japan, Korea, Singapore and Chinese Taipei are the only Asian signatories. China is in the process of accession, while Indonesia, India, Malaysia, Mongolia, Sri Lanka and Vietnam are observers. The objectives of the GPA include requiring greater transparency and competition in

[135] See, e.g., Patti Waldmeir, Corruption A Symption of Healthcare Ills in China, *Financial Times*, 3 November 2014 (accessed 13 April 2015).

[136] See Taryn Vian, Derick W. Brinkerhoff, Frank G. Feeley, Matthieu Salomon and Nguyen Thi Kieu Vien, *Confronting Corruption in the Health Sector in Vietnam: Patterns and Prospects*, Health and Development Discussion Paper No. 14, October 2011, Center for Global Health and Development, Boston University School of Public Health, and Ben Hirschler, *Vietnam Study Finds Bribes Dominate Medicine Prices*, 28 February 2012, *Reuters*.

[137] *Global Corruption Report 2006*, note 133 above, pp. 97–98.

[138] WHO, *Medicines: Corruption and Pharmaceuticals*, Fact Sheet No. 335 (December 2009).

[139] See, e.g., *Resolution* adopted by the UN Human Rights Council, on *The Right of Everyone to the Enjoyment of the Highest Attainable Standard of Physical and Mental Health*, para. 7(f): "to ensure that procurement practices and procedures for medicines are transparent, fair and competitive", 10 June 2011, A/HRC/17/L.16, and Report of the Special Rapporteur, note 64 above, at para. 74.

[140] See WTO website at: www.wto.org/english/tratop_e/gproc_e/memobs_e.htm (accessed 21 January 2015). New commitments were negotiated and accepted on 15 December 2011, to further these commitments. See generally also WHO, *The World Medicines Situation 2011 – Procurement of Medicines*.

signatory members in procurement activities for goods and services. The extent of each signatory's commitments is determined by their Annexes to the GPA.

### 2.8.4 National price control systems

Drug price control systems exist in various Asian countries and differ in their application.[141] Specifically, the regulation of price markups (or lack thereof) also has a significant impact on pharmaceutical prices.[142] Some countries make use of external reference pricing systems, in which pharmaceutical prices of other countries (of, for example, similar *per capita* income level) are used as benchmarks for domestic pricing.[143] In some countries, the government-controlled prices must be either displayed on packaging or made available to the public regularly.[144] The following provides an overview of such policies in Asia. Distribution markup costs, which can add significantly to the price of medicines, are often unregulated.[145]

At its accession to the WTO, Cambodia committed that any price controls on goods would be WTO-consistent.[146]

China's national drug price policy falls under the charge of the National Development and Reform Commission (NDRC).[147] The NDRC maintains maximum price controls on pharmaceuticals, allowing

[141] On pharmaceutical price controls generally, see ed. Schoonveld,*The Price of Global Health – Drug Pricing Strategies to Balance Patient Access and the Funding of Innovation*, Ashgate, 2nd ed. (Gower Publishing Limited, 2015).

[142] See mark-ups in low- and middle-income countries, WHO, *Guideline on Country Pharmaceutical Pricing Policies*, Table ES1.1., pp. 49–60. For a more detailed discussion on regulation on price mark-ups, see generally, Health Action International, *Regulation of Mark-ups in the Pharmaceutical Supply Chain*).

[143] See Health Action International, *External Price Referencing*, Policy Brief No. 1, February 2014.

[144] See WHO, *Public-Private Roles in the Pharmaceutical Sector – Implications for Equitable Access and Rational Drug Use – Health Economics and Drugs Series*, No. 005 (1997), para. 6.3.

[145] See UNHRC rapporteur report, which observes that such costs can add as much as 40% of pharmaceutical costs; note 64 above, at p.10.

[146] Paras. 26–7, Cambodia, WTO Accession Working Party Report, WTO/ACC/KHM/21.

[147] See CFDA website: http://eng.sfda.gov.cn/WS03/CL0767/61640.html (accessed 20 July 2014). See also, generally, Sarah L. Barber, Baobin Huang, Budiono Santoso, Richard Laing, Valerie Paris and Chunfu Wu, The Reform of the Essential Medicines System in China: A Comprehensive Approach to Universal Coverage, *J. Glob. H.*, Vol. 3, No. 1 (June 2013).

for production costs and a profit margin.[148] Until recently, it had not relaxed such price controls. In April 2014, China removed the price controls over a limited number of pharmaceuticals, and it has been reported that the government was planning to further remove controls on the prices of a larger range of drugs.[149] In May 2015, China announced that it would remove the remaining price controls by 1 June 2015.[150]

China's system also provides for special review and approval processes for applications for, *inter alia*, "new drugs for the treatment of diseases such as AIDS, malignant tumors and rare diseases, etc. with significant clinical advantage" and "new drugs for the treatment of diseases, for which effective therapeutic methods are not available".[151] Other domestic policies, such as the role of public healthcare providers in prescribing medicines and the incentive system for such providers which links the profit margin with procurement prices, also affect pharmaceutical prices in China.[152]

India has a national drug price control system to keep costs of medicines low. The system is operated by the National Pharmaceutical Pricing Authority (NPPA).[153] The roots of the drug price control policies

[148] See Art. 48–55 of the *Regulations for Implementation of the Drug Administration Law of the People's Republic of China* (Decree of the State Council No. 360). On China's central drug procurement bidding system under the Ministry of Health rules. See Yu Fang, Chapter 3, Pharmaceutical Pricing Policies in China, in, *Pharmaceutical Prices in the 21st Century*, ed. Zaheer-Ud-Din Babar (2015) Springer International Publishing, and Chen Yang and Lei Li, Drug Procurement Bidding, *China Business Review*, 1 January 2010.

[149] See Li Hui and Adam Jourdan, *China Backs Away from Price Controls on Basic Drugs*, *Reuters*, 17 April 2014, www.reuters.com/article/2014/04/17/china-drugs-idUSL3N0MF1XB20140417 and *China to Scrap Drug Price Caps Next Year – State Newspaper*, *Reuters*, 26 November 2014, www.reuters.com/article/2014/11/26/china-pharmaceuticals-idUSL3N0TG2LU20141126 (accessed 22 February 2015).

[150] *China Lifts Price Controls on Most Drugs*, *Xinhua*, 5 May 2015, http://news.xinhuanet.com/english/2015-05/05/c_134211498.htm (accessed 7 May 2015). See earlier reports, such as the China government website reporting the NRDC's statement, at http://english.gov.cn/state_council/ministries/2015/03/06/content_281475066855492.htm and Laurie Burkitt, Drug Companies Face Pressure Despite China Price Pledge, *Wall Street Journal*, 5 March 2015, www.wsj.com/articles/china-plans-to-lift-price-controls-on-drugs-1425520528 (accessed 20 April 2015).

[151] Article 45, Provisions for Drug Registration, SFDA Order No. 28, Normative Document, text available at: http://eng.sfda.gov.cn/WS03/CL0768/61645.html (accessed 20 July 2014).

[152] See Qiang Sun, Michael A. Santoro, Qingyue Meng, Caitlin Liu and Karen Eggleston, Pharmaceutical Policy in China, *Health Affairs*, Vol. 27, No. 4 (July/August 2008), 1042–1050, available at: hphci.org/content/pdf/Pharmaceutical%20Policy%20In%20China.pdf (accessed 22 February 2015), especially at pp. 1044–1047.

[153] See NPPA website at: www.nppaindia.nic.in/index1.html (accessed 12 June 2014). India publishes an annual *Drug (Price Control) Order* (DPCO). The DPCO 2013, e.g., includes

are found in the report of a committee set up by the Indian government in 1974, known as the *Hathi Committee*, which issued an important report in April 1975.[154] Not long after, the national Drug Policy was announced in 1978, and the Drugs Prices (Control) Order 1979 was issued in 1979, replacing the prior Drugs Prices (Control) Order 1970.[155] In 1986, the Indian government issued its revised Drug Policy, aiming to improve the price control system and emphasizing a technological push to upgrade the domestic pharmaceutical industry.[156] Until recently, India had planned price controls for off-patent generic medicines.[157] On 10 July 2014, the NPPA issued a notice to control prices of 108 drugs used for treating diabetes, cardiovascular disease, HIV and malaria.[158]

Korea maintains price ceilings for medicines for purposes of reimbursement under its National Health Insurance Programme.[159] The prices may be reviewed by the Ministry of Health & Welfare. In 2010, the Ministry sought to reduce the price of Glivec – a drug marketed by Novartis – by 14% but the decision was overturned by a Korean court on the ground that the existing price was not

over 600 medicines. See generally, Sudip Chaudhuri, Chapter 7, *Pharmaceutical Prices in India*, in *Pharmaceutical Prices in the 21st Century*, note 148 above.

154 Named after its chair, Shri Jaisukhlal Hathi.

155 See *Report of the Task Force to Explore Options other than Price Control for Achieving the Objective of Making Available Life-saving Drugs at Reasonable Prices*, 20 September 2005, pp. 10–11, http://pharmaceuticals.gov.in/drpronabreport.pdf (accessed 11 August 2014). For general overview and further developments on drug prices controls in India up to 2005, see generally pp. 9–18.

156 Drug Policy of 1986, http://nppaindia.nic.in/drug_pol86/txt1.html (accessed 5 March 2015). For an example of the domestic litigation arising in relation to implementation of price controls, see *GlaxoSmithKline Pharmaceuticals Limited (Formerly known as SmithKline Beecham Pharmaceuticals (India) Limited v Union of India)*, Supreme Court judgment, December 2013, judgment available at: http://judis.nic.in/supremecourt/imgs1.aspx?filename=41053 in which the Supreme Court upheld a 15-day adjustment period for new price controls.

157 See Amy Kazmin, *India Scraps Generic Drug Price Controls*, *Financial Times*, (23 September 2014), www.ft.com/cms/s/0/41bdec2c-4319-11e4-9a58-00144feabdc0.html#axzz3TU09iCrE (accessed 5 March 2015).

158 NPPA website: www.nppaindia.nic.in/index1.html. See also http://businesstoday.intoday.in/story/govt-caps-prices-of-108-more-medicines-drugmakers-to-be-hit/1/208165.html (accessed 20 July 2014).

159 For an overview of the Korean system, see Annex, *Pharmaceutical Price Controls in OECD Countries*, US Department of Commerce, pp. 75–76. For developments in 2010, see Kim and Chang, *IP Newsletter*, Fall 2010, 1–3, www.kimchang.com/UserFiles/files/KandC_IP_Fall_2010[3].pdf.

unreasonable.[160] A compulsory licence in respect of the drug had been refused in 2003.[161]

Indonesia controls the price of pharmaceuticals, requiring for instance that profit margins may not exceed 50%.[162] The Indonesian government has introduced policies to promote the use of less expensive generic pharmaceutical products in place of brand-name products.[163] However, it appears that the use of such generic products remains suboptimal due to various reasons.[164]

Myanmar makes use of an NEML.[165] Drug pricing is determined to a large extent by the Myanmar Pharmaceutical Medical Products and Entrepreneur Association in collaboration with the Ministry of Commerce, with markups on prices of 3–5% being used.[166]

Between 2008 and 2009, the Philippines introduced a medicines price control mechanism via the Cheaper Medicines Act 2008 and the Presidential Executive Order No. 821, 2009.[167] The Maximum Drug

[160] *South Korean Court Oks Pricing of Novartis' Anti-Leukemia Drug, First Word Pharma*, 22 January 2010, www.firstwordpharma.com/node/490057#axzz3WJjEc3Ad (accessed 4 April 2015).

[161] Bong-min Yang and Hye-Young Kwan, *TRIPS and New Challenges for the Pharmaceutical Sector in South Korea, in The New Political Economy of Pharmaceuticals*, ed. Hans Löfgren and Owain David Williams (Palgrave Macmillan, 2013) at pp. 209–210.

[162] See WHO, *Pharmaceuticals: Why Reform is Needed*, Indonesia Health Sector Review, Policy Note Series, March 2009.

[163] See Yusi Anggriani, Mohamed Izham M. Ibrahim, Sri Suryawati and Asrul A. Shafie, The Impact of Indonesian Generic Medicine Pricing Policy on Medicine Prices, *J. Gen. Meds.*, Vol. 10, No. 3–4 (2013) 219–229, and Health Action International, *Indonesia – Medicine Prices, Availability and Affordability.*

[164] *Ibid.*

[165] See NEML 2010, at Myanmar government website, www.moh.gov.mm/file/NLEM.pdf (accessed 5 March 2015). The Central Medical Store Depot (CMSD), Department of Health, Division of Medical Care, Ministry of Health, supplies all public healthcare facilities but it has been reported as at 2011, that due to budget constraints, the CMSD is able to supply *less than half* the drugs (92 out of 341) in the NEML: see pp. 6 and 11, available at: www.searo.who.int/entity/medicines/myanmar_situational_analysis.pdf?ua=1 (accessed 5 March 2015). See also generally, Thida Aye and James Finch, Pharmaceuticals in Myanmar – Law and Procedure, *SJICL*, Vol. 4 (2000) 115–147.

[166] WHO, Kathleen A. Holloway, *Myanmar Mission Report 19–26 October 2011*, 30 December 2011, at pp. 27–28, available at: www.searo.who.int/entity/medicines/myanmar_situational_analysis.pdf?ua=1 (accessed 5 March 2015).

[167] The text of the Executive Order is available at: www.gov.ph/2009/07/27/executive-order-no-821-s-2009/ (accessed 11 September 2014). See generally, James Hookway, *Philippines President Arroyo Orders Drug Price Controls*, 28 July 2009, http://online.wsj.com/articles/SB124879732286587201. See also, generally, Douglas Ball and Klara

Retail Price in this country covers 38 drug molecules (72 drugs).[168] The Philippines also makes use of an index to promote transparency in government procurement of medicines, through its Drug Price Reference Index (DPRI).[169] The DPRI includes acquisition costs including landed cost, packaging, drug content, quality assurance, manufacturing overheads and Food and Drug Administration (FDA) fees but excludes other costs such as value-added taxes and storage fees.[170] The DPRI system provides for external price referencing for innovator drugs approved for inclusion in the Philippine National Formulary; in such cases, prices may be set with reference to markets particularly those with similar income status such as Thailand and India.[171]

Between 2003 and 2005, Vietnam took steps to introduce pricing measures for pharmaceuticals.[172] Under Vietnam's Law on Pharmacy No. 34/2005/QH11 passed by the National Assembly on 14 June 2005, pharmaceutical prices are set by trading entities themselves according to market forces.[173] Given rising pharmaceutical costs, Vietnam has reportedly been considering the introduction of a price control law and to use international reference pricing.[174]

Tisocki, *Medicines Price Components in the Philippines*, HAI Global, and Douglas B. Batangan and Noel Juban, *Philippines Pharmaceutical Situation 2009*, WHO Health Facility Survey on Medicines.

168 See Republic of Philippines Department of Health website, www.doh.gov.ph/node/1332.html (accessed 5 March 2015).

169 See Department of Health, *The 2013 Drug Price Reference Index*, 2nd ed., 13 October 2014, at www.doh.gov.ph/sites/default/files/DPR2013.pdf (accessed 5 March 2015).

170 *Ibid.*, p. 2. The VAT is 12%. 171 *Ibid.*, p. 3.

172 See the excellent overview in Anh Tuan Nguyen, Rosemary Knight, Andrea Mant, Quang Minh Cao and Geoffrey Brooks, Medicine Pricing Policies: Lessons from Vietnam, *Southern Med. Rev.*, Vol. 3, No. 2 (October 2010), 12–19. See also *Report of the Special Rapporteur on the right of everyone to the enjoyment of the highest attainable standard of physical and mental health*, Anand Grover, UN General Assembly, Twentieth Session, Human Rights Council, 4 June 2012, Addendum, A/HRC/20/15/Add.2, para. 30.

173 Text available at: www.moj.gov.vn/vbpq/en/Lists/Vn%20bn%20php%20lut/View_Detail.aspx?ItemID=6838 (accessed 20 July 2014).

174 See A/HRC/20/15/Add.2, paras. 30–31, www.ohchr.org/Documents/HRBodies/. . ./A-HRC-20–15-Add2_en.pdf (accessed 22 February 2015), www.thanhniennews.com/health/vietnam-seeks-better-control-of-drug-prices-24844.html and http://en.vietnamplus.vn/Home/Cabinet-debates-medicine-price-controls/20143/47803.vnplus (accessed 20 July 2014).

### 2.8.5 Generic substitution policies

Regulations which actively promote or require substitution of brand-name pharmaceutical products with generic products exist in some countries in Asia.[175]

The policies are not always clearly articulated or firmly implemented. In several countries, generic substitution is promoted/permitted but not mandated.

Among newer ASEAN member countries, Lao PDR has an express policy for the government to promote generic substitution.[176] The Philippines and Indonesia also have articulated generic substitution policies.[177] There is presently no clear or coordinated regional approach within ASEAN countries (which are aiming at greater economic, social and political integration) in the area of generic substitution policy.

Outside the ASEAN region, Japan and Korea have generic substitution policies although they are non-mandatory in nature.[178]

[175] See Mohamed Azmi Hassali, Jayabalan Thambyappa, Fahad Saleem, Noman ul Haq and Hisham Aljadhey, Generic Substitution in Malaysia: Recommendations from a Systematic Review, *J. App. Pharm. Sci.*, Vol. 2, No. 8 (2012) 159–164. For policies in the EU, see P. Dylst, A. Vulto and S. Simoens, Demand-Side Policies to Encourage the Use of Generic Medicines: An Overview, *Expert Rev. Pharmacoecon. Outcomes Res.*, Vol. 13, No. 1 (February 2013), pp. 59–72, S. M. Ess, S. Schneeweiss and T. D. Szucs, European Healthcare Policies for Controlling Drug Expenditure, *Pharmacoeconomics*, Vol. 21, No. 2 (2003), 89–103 and McKinsey & Company, note 11 above (from Chapter 8), pp. 47–48.

[176] See para. 3.7, of the Lao PDR *Revised National Medicine Policy* (Ministry of Health Revised Lao Trade Portal, 19 August 2003) www.laotradeportal.gov.la/index.php?r=site/display&id=252 (accessed 5 March 2015).

[177] For the Philippines, see the Generics Act of 1988, Republic Act No. 6775; text available at www.fda.gov.ph/issuances/29035-repubic-act-no-6675 and the Universally Accessible Cheaper and Quality Medicines Act of 2008, Republic Act No. 9502/2008 (commonly known as the "Cheaper Medicines Act"). For Indonesia, Malaysia and Thailand, see Mohamed Azmi Hassali, Alian A. Alrasheedy, Andrew McLachlan, Tuan Anh Nguyen, Saleh Karamah AL-Tamimi, Mohamed Izham Mohamed Ibrahim and Hisham Aljadhey, The Experiences of Implementing Generic Medicine Policy in Eight Countries: A Review and Recommendations for a Successful Promotion of Generic Medicine Use, *Saudi Pharm. J.*, Vol. 22 (2014) 491–503.

[178] See Tomasz Mroczkowski, *The New Players in Life Science Innovation: Best Practices in R & D from Around the World*, (Pearson Education Ltd, 2012) at pp. 82–83. For Japan, see also Eriko Kobayashi, Hiroshi Karigome, Tomoya Sakurada, Nobunori Satoh and Shiro Ueda, Patients' Attitudes towards Generic Drug Substitution in Japan, *Health Policy*, Vol. 99 (2011) pp. 60–65. See also generally, World Bank, *A Generic Drug Policy as a Cornerstone to Essential Medicines in China*, China Health Policy Notes (2010).

### 2.8.6 *User attitudes*

Barriers which may limit the use of lower priced generic products may also arise from the user. For example, low prices may be perceived as representing lower quality pharmaceutical products.[179] Another possible scenario is familiarity with an existing (brand-name) pharmaceutical and a reluctance to switch to another, lower priced product.[180] Another issue may be low awareness among consumers of generic substitution policies aimed at lowering medicine prices.[181]

### 2.8.7 *Marketing approval time*

Marketing approval times vary within Asia, and long delays can also contribute as an obstacle to access. In Singapore, approval times are

[179] See Anggriani *et al.*, note 163 above, p. 227.

[180] On attitudes in Malaysia, see Omotayo Fatokun, Mohamed IM Ibrahim and Mohammed A. Hassali, Factors Determining the Post-Patent Entry of Generic Medicines in Malaysia: A Survey of the Malaysian Generic Pharmaceutical Industry, *J. Gen. Meds.*, Vol. 10, No. 1 (2013), 22–33, at p. 31. On Singapore's situation, the then Minister of State for Health stated on 20 July 2004:

> With the introduction of electronic prescription systems, doctors are now kept informed about the costs of drugs and available alternatives. Doctors are encouraged, where possible, to use generic instead of patented drugs.
>
> To succeed in our efforts to keep healthcare costs affordable, we need the support of both the doctors and the public. Medical care in Singapore can remain affordable if we ensure that we spend our healthcare dollars in a sensible manner. Many older medicines, which are generic, are effective and have a well proven track record of curing or controlling a disease. *It does not always mean that a newer patented drug is better than an older generic drug. Often, the latest medical treatments only provide a theoretical benefit but at a significantly higher cost. The long-term benefits and risks of these new treatments are often unknown. Doctors therefore need to look at medical treatments objectively and not be swayed by the subjective influences of advertising and marketing efforts. The public should also not think that expensive or newer medicine is always better.*
>
> [Italic emphasis added]

[181] See *Low Consumer Awareness Hampers Adoption of Cheaper Medicines Law and Generics Act in PH*, Philippine Institute for Development Studies, www.pids.gov.ph/index2.php?pr=152 (accessed 5 March 2015), based on the study by the Philippine Institute for Development Studies (PIDS) and the Department of Health (DOH), Eleanora de Guzman and Maria Adoracion Fausto, *The Impact of the Cheaper Medicines Act on Households in Metro Manila: A Qualitative Study*, Discussion Paper Series No. 2014–20, March 2014.

generally viewed positively as being quite efficient.[182] In some other countries, there is a relatively long lag between testing of an innovative product and its regulatory approval.[183]

### 2.8.8 *Counterfeit pharmaceuticals*

One problem which presents both safety and competition dimensions is that of counterfeit pharmaceuticals, which is a problem that has been growing.[184]

According to the USTR:

> Reports indicate that anywhere from 10–40 percent of drugs sold in Indian markets are counterfeit and could represent a serious threat to patient health and safety.[185]

It also reports as follows, with regard to China:

> ... in China, some domestic chemical manufacturers that produce API have avoided regulatory oversight by failing to declare that bulk chemicals are intended for use in pharmaceutical products. This contributes to China being a major source country for APIs used in counterfeit pharmaceutical products.[186]

According to a survey of the pharmaceutical industry by the US International Trade Administration in 2010, China and India were the largest sources of active pharmaceutical ingredients (APIs) and generics for the US market, respectively, while China, India and Italy are the world's largest producers of APIs.[187]

[182] See, e.g., observations in *Re Novartis AG of Alcon Inc* [2010] SGCCS 2, para. 53 and *Re GSK Trading Services Limited and UCB SA* [2009] SGCCS 1, para. 26.

[183] See, e.g., Harriet Wileman and Arun Mishra, Drug Lag and Key Regulatory Barriers in the Emerging Markets, *Perspect. Clin. Res.*, Vol. 1, No. 2 (April–June 2010), 51–56.

[184] See WHO Bulletin, *Growing Threat from Counterfeit Medicines*, www.who.int/bulletin/volumes/88/4/10-020410/en/ (accessed 5 March 2015), and Merri C. Moken, Fake Pharmaceuticals: How they and Relevant Legislation or Lack thereof Contribute to Consistently High and Increasing Drug Prices, *Am. J. L. & Med.* (2003), 525–542. On counterfeit medicines and Asia, see Philip Stevens and Helmy Haja Mydin, *Fake Medicines in Asia*, EMHN Briefing No. 1, 1–9.

[185] USTR Special 301 Report 2014, p. 20. See also Melanie Lee and Ben Hirschler, China's "Wild East" Drug Stores, *Reuters*, 28 August 2012, www.reuters.com/article/2012/08/28/us-china-pharmaceuticals-idUSBRE87R0OD20120828 (accessed 25 July 2014).

[186] *Ibid.*

[187] *Pharmaceutical Industry Profile*, 2010, pp. 6–7; www.ita.doc.gov/td/health/PharmaceuticalIndustryProfile2010.pdf (accessed 23 July 2014).

In 2006, the WHO established the International Products Anti-Counterfeiting Taskforce (IMPACT), an international initiative which brings together relevant stakeholders (such as regulatory and enforcement authorities, pharmaceutical associations and NGOs) to address the sale and production of counterfeit medicines.[188] Separately, the WHO Working Group on Substandard/Spurious/Falsely-Labelled/Falsified/Counterfeit (SSFFC) Medical Products works to identify activities in relation to counterfeit and substandard pharmaceuticals, through a Member State Mechanism (MSM).[189] The MSM's activities include studying the issues of accessibility and affordability of medicines and their impact on counterfeit medicines.[190]

In 2011, the EU issued a Directive against falsified medicines entering the supply chain.[191] In 2013–2014, the US FDA started a 2-year pilot programme to secure the supply chain for imported pharmaceuticals.[192] The European Council has also established the MEDICRIME Convention, with the aim of promoting criminalization and punishment of drug counterfeiting among signatory states.[193]

#### 2.8.8.1 Distinguishing the issues of counterfeit and generic medicines

While counterfeit medicines are a genuine source of concern in terms of public safety, warnings have been issued against *conflation* of this issue with that of *generic* medicines, which are legitimate and genuine goods on the market.[194] Caution is needed so that anti-counterfeiting measures are not used as protectionist ones which result in driving up the cost

[188] WHO, www.who.int/impact/about/en/ (accessed 4 April 2015). See generally, Beverley D. Glass, Counterfeit Drugs and Medical Devices in Developing Countries, *Res. Rep. Trop. Med.*, Vol. 5 (2014) 11–22.

[189] WHO, http://apps.who.int/gb/ssffc/ (accessed 4 April 2015). The MSM was established in 2012 by the decision of the World Health Assembly; see WHA65.19, 26 May 2012, *Resolutions and Decisions, Annexes*, 65th World Health Assembly, at pp. 28–32.

[190] *Report of the Third Meeting of the MSM*, para. 8, A/MSM/3/3, October 2014.

[191] Directive 2011/62/EU.

[192] See FDA News Release, 18 February 2014, www.fda.gov/NewsEvents/Newsroom/PressAnnouncements/ucm386275.htm?source=govdelivery&utm_medium=email&utm_source=govdelivery (accessed 25 July 2014).

[193] See Council of Europe website: www.coe.int/t/dghl/standardsetting/medicrime/medricrime_ENG_BAT_web_v2.pdf (accessed 5 March 2015), and statement of Russian delegation, WIPO Standing Committee on the Law of Patents, SCP/18/12).

[194] See, e.g., *Report of the Special Rapporteur on the right of everyone to the enjoyment of the highest attainable standard of physical and mental health, Anand Grover, on access to medicines*, note 64 above, para. 66.

of medicines through elimination of legitimate competitor goods. Developing countries such as India, which have a strong generic medicines industry, are developing ways to counteract allegations of counterfeit, substandard medicines from their industries.[195]

## 2.9 Conclusion

The debate on access to medicines is clearly a multifaceted one, with many stakeholder interests.[196] A variety of factors contribute to the overall costs of pharmaceutical products, and these include patenting policy and activity, regulatory systems, national infrastructure, tariffs and other non-tariff measures.[197] In addition, users are not simply patients/consumers, as intermediary parties – such as doctors, healthcare institutions, government agencies – are frequently the actual actors who purchase, select and/or negotiate prices of pharmaceutical products or set policies affecting these prices. Apart from costs, access to pharmaceuticals is also affected by physical access to facilities to obtain pharmaceuticals, a country's infrastructural obstacles, level of corruption and private sector practices. Therefore, to overcome the obstacles to access, whether price-related or otherwise, concerted efforts from a variety of actors are required.

[195] See India, *Foreign Trade Policy Statement, 2015–2020*, at p. 41; http://dgft.gov.in/exim/2000/FTPstatement2015.pdf (accessed 21 April 2015), announcing the first "track and trace" system with effect from 1 April 2015.

[196] A number of factors contributing to cost and hence access, issues, for pharmaceuticals are identified below:

> . . . Despite drastic reductions in drug prices during the past decade, the costs of antiretroviral therapy programmes continue to rise. Although drugs must continue to be made more affordable—including first- and second-line regimens—*potential gains are highest in reducing the non-drug-related costs* of providing treatment, such as diagnostics, hospitalization, monitoring treatment and out-of-pocket expenses. *These costs are currently twice the cost of the drugs themselves.*
>
> [Italic emphasis added]

UNAIDS, *Getting to Zero: 2011–2015 Strategy*, Joint United Nations Programme on HIV/AIDS (UNAIDS), 2010, pp. 39 and 40.

[197] See also the views in Calvin W.L. Ho and Klaus M. Leisinger, Intellectual Property and Access to Essential Medicines: A Tenuous Link?, *Asian Bioethics Rev.*, Vol. 5, No. 4 (December 2013), 376–382.

Further, from a safety point of view, counterfeit pharmaceutical products are certainly a concern, but these should not be confused with generic products in the overall discussion.

The next chapter will examine the legal and policy construct in Asia, focusing on some of the key issues that have received attention through either FTA or case law activity and which could have a potential impact on access in Asia. The discussion aims to point to opportunities that exist for governments to provide clearer laws and practices in these areas, particularly in ASEAN, where members are seeking greater integration, and generally in Asia, where several countries are, or will be, participating in significant international trade negotiations.

# 3

# Trade, investment and medicines

## The existing legal and policy construct

This chapter aims to provide an explanation of the legal and policy framework, highlighting points of ambiguity and exploring the national legal and policy construct existing within patent and marketing approval systems in some major systems in Asia, as the region embarks on various FTA negotiations which may have implications for the patent bargain in them. The analysis focuses on the patenting of medicines in Asia and regulatory steps by which medicines reach final users, taking into account the roles of the above components of the system. These comprise the laws, regulations and directives on the twin tracks of patent protection and drug registration in selected Asian countries and the impact that FTAs may have on them. The next chapter will then examine the legal risks that may arise from such ambiguity and from entering into such FTAs which may extend standards of protection beyond the TRIPs system of requirements as well raise issues of ISDS.

### 3.1 The existing legal and policy construct

As discussed in Chapter 2, the regulation of delivery of medicines occurs not only in the context of national laws; international, regional and bilateral treaties impose a separate layer of legal regulation which addresses a wide variety of issues, ranging from patentability to licensing of medical technology. The legal and policy construct exists at the national and international levels, and each conditions the formulation of the other. National systems embody their priority interests as well as the international obligations to which they have committed. Conversely, international legal and policy systems are shaped through inter-state negotiations, which in turn are conditioned by national agendas of those negotiating the systems.

Regional and bilateral agreements generally contain provisions requiring the elimination or reduction of tariffs and non-tariff barriers that may impede cross-border trade and investment in medicines, which can generally lead to freer trade in them. However, as has been mentioned, such agreements may also contain intellectual property provisions that affect the patenting, licensing and pricing of medicines, and the data associated with the creation and testing of new medicines, so as to have a potentially adverse impact on the affordability of medicines in the signatory countries. Such provisions generally favour pharmaceutical patent holders and pose obstacles to generic producers' entry into a market; exceptionally, however, one bilateral agreement – the CECA – includes obligations that aim to provide speedier reviews for *generic* medicines approvals.[1] Some countries in Asia (e.g., the Philippines) have also enacted specific legislation to address the availability of generic medicines.

The tension that regulators face in balancing the interests of patent holders, generic medicine manufacturers, consumers and the state itself (as a supplier of public health services and payer for medicines) is present in many systems. The legal and policy constructs have to be reviewed continually to strike this balance, as the demographic, disease profile, economic capacity of citizens and other relevant factors change. While legal frameworks and their laws underpin and determine a great deal of the functioning of the system, the policy aspect, which encompasses considerations beyond the law, such as social policy, healthcare costs and citizenry welfare, are equally important as they drive and shape laws on medicines. A third component of the system comprises the major stakeholders, including policy-makers, international institutions, private, public or other entities involved in the research and development of medicines and their delivery to the public, health regulators and healthcare advocates.

At the same time, a country's regime governing the creation, testing and delivery of medicines to final users and the provision of incentives for innovative medical efforts comprises a complex ecosystem of laws, policies and international instruments, all of which play different yet sometimes overlapping roles. As has been pointed out in Chapter 2, these include not only intellectual property and marketing approval aspects, but also tariff and non-tariff barriers which affect delivery of medicines, and the enforcement of corruption and competition laws. In this regulatory "ecosystem", it is evident that the patent system is but one component.

[1] Comprehensive Economic Cooperation Agreement between India and Singapore, signed on 29 June 2005 and entered into force on 1 August 2005. India is a leading global producer and exporter of generic medicines.

## 3.2 ASEAN – the need for greater clarity in the collective legal and policy construct

As ASEAN aims to create a single production and investment base in Asia, its collective stance in the area of regulation of medicines and how FTA activity should be handled in this area will be important. Presently, within ASEAN, there does not yet appear to be such a collective stance. There are three reasons for this: first, the medicine regulatory work focuses on harmonization. Second, IP and competition conversations and initiatives have not been clearly linked to the regulation of medicines in the FTA (or global context). Third, health and affordability of medicines is handled under the ASCC pillar while IP, competition and other commercial or trade aspects fall under the AEC. Within countries, competition law and policy which could be a check on IPR abuses are not harmonized – competition law is relatively new to many ASEAN members, and for some, indeed, the laws are currently being prepared. As will be seen below, too, the legal thresholds for patentability and national policies on new technologies are not uniform, and in some cases, even remain unclear.

Through the *Bali Concord III*, ASEAN expressed its aspirations to speak as one in the global community. In particular, it has adopted "an ASEAN common platform on global issues" and has, as a community, further resolved "at the global level to [e]nsure access to adequate and affordable healthcare medical services as well as accessibility to safe, non-counterfeit, affordable, and effective medication".[2] Yet, ASEAN does not reflect a clear, unified voice on global IPR, innovation and medicine policy. Singapore, a signatory to the USSFTA, is a strong proponent of patents and associated rights, while other, newer ASEAN members are still working towards implementing their IP laws.[3]

## 3.3 Key elements

Under the TRIPs Agreement, WTO members are given a degree of flexibility to determine what may be patentable under their national laws, subject to certain minimum requirements. States make use of this

[2] Preamble and para. C.3(a), *Bali Declaration on ASEAN Community in a Global Community of Nations, "Bali Concord III"*, signed in Bali, Indonesia, 17 November 2011.

[3] See, e.g., Singapore WTO TRIPS Council non Paper 2000, JOB(00)/7853. On IP in ASEAN member states, see, for example, ASEAN, *Intellectual Property and the ASEAN Economic Community (AEC) in 2015*, especially p. 3.

flexibility to carve out national policies on what is to be excluded from patents and the level of strictness of criteria for patentability. Emerging areas which may involve moral or public interest dimensions (such as human genes and sequences of such genes) raise difficult questions of patentability and can lead to polarized views. The law lags behind science, and part of the challenge – particularly for developing countries in Asia with relatively new patent laws –involves the analysis and evaluation of new scientific developments under the rubric of existing legal concepts.

### *3.3.1 Patentability*

National systems set out the scope and limits of patentable material. Patentability forms a critical "gateway" or threshold for obtaining a patent for an invention. Art. 27 of TRIPs imposes minimum requirements which WTO members must adhere to in providing for their national patent systems. Patentability criteria which WTO members should include are requirement of an inventive step (for which there are differing standards across national systems), novelty and industrial application. Art. 27 also permits a degree of flexibility as to what a WTO member may exclude. Within the sphere of flexibility of what may be excluded, national systems often differ in what they exclude.

Patentability is generally determined first at the national IP office level, usually with the possibility of appeals being brought to the judicial system. On appeal the level of deference given to the decisions of the IP office, or of a lower trial court where primary evidence and witnesses are examined, may be strong.[4] States – usually beginning with their patent offices – therefore act as "gatekeepers", providing the statutory monopoly of rights in a patent. Patentability is a "gatekeeping" tool as it requires a determination of whether an invention is eligible for consideration to be granted the patent monopoly. This gatekeeping function allows the authority to exclude weak claims, and in this regard, Professor Peter Drahos' reference to the old adage – "prevention is better than cure" – is most apposite, particularly in respect of developing countries where expensive litigation after the fact is not a viable policy tool.[5] In *Biogen*

[4] See, e.g., Lord Hoffman's observations in *Biogen Inc v Medeva plc* [1997] RPC 1, at p. 45., echoed in *Human Genome Sciences, Inc v Eli Lilly* [2011] UKSC, at e.g. paras. 166 and 169 (although in this case, the UK Supreme Court found sufficient cause to reverse the lower court's decision).

[5] One author has observed that " . . . a guiding principle for developing countries should be prevention rather than cure ie it is much better to find ways to prevent pharmaceutical

*Inc v Medeva plc*, Hoffman LJ cautioned against over-inclusive patents, which may tilt the balance towards overprotection, and therefore arrest further innovative activity:

> It is inevitable in a young science, like electricity in the early nineteenth century or flying at the turn of the last century or recombinant DNA technology in the 1970s, that dramatically new things will be done for the first time. But care is needed not to stifle further research and healthy competition by allowing the first person who has found a way of achieving an obviously desirable goal to monopolise every other way of doing so.[6]

National patent law principles such as requirements of disclosure in the patent application, determining justifiability of the breadth of the claim and examining the prior art are some of the tools to execute the "gatekeeping" function. Naturally, the manner in which this function is executed affects the quality of the resulting patents that make the grade. In some systems, such as those in China and Japan, there is an additional "layer" of checks on the acceptance of patents, in the form of a re-examination or review procedure.[7]

patents of doubtful validity getting on to the patent register rather than trying to remove them through costly litigation." – Peter Drahos, *"Trust me": Patent Offices in Developing Countries*, Working Paper, Centre for Governance of Knowledge and Development, November 2007, p. 26.

[6] [1997] RPC 1, at p. 52. See also generally, *Protection of Creators and the Innovation Incentive*, ed. Frederick M. Abbott, Carlos M. Correa and Peter Drahos, *Emerging Markets and the World Patent Order* (Edward Elgar Publishing Ltd., 2013) One author has argued that over-wide patent applications, particularly in the context of biotechnology patents, are "effective 'submarine' applications" and that a "submarine patent is written as widely as possible and is intended to lurk in the patent system to snare later unsuspecting innovators with a claim that the advance they have made is actually covered by the prior patent"; Christopher May, *On the Border: Biotechnology, the Scope of Intellectual Property and the Dissemination of Scientific Benefits, in The Role of Intellectual Property Rights in Biotechnology Innovation*, ed. David Castle (Edward Elgar Publishing Ltd., 2009) p. 266. But see Lord Hoffman's caution against such over-wide claims; Chapter 2, text to note 72. See generally, also, Jerome H. Reichman, Intellectual Property in the 21st Century: Will Developing Countries Lead or Follow? *Hous. L. R.*, Vol. 46 (2009–2010), 1115, at p. 1133, pointing out that even the US is ratcheting up requirements domestically, at p. 1134, note 87.

[7] See PRC government website, at www.chinaipr.gov.cn/direrdcusarticle/directions/administrative/appatent/appdirections/200907/284944_1.html, and Japanese Patent Office website, at www.jpo.go.jp/tetuzuki_e/t_tokkyo_e/1312-002_e.htm (accessed 6 April 2015), and Yoshinari Kishimoto, *How to Challenge Patent Validity*, May 2005, Managing IP, at www.sughrue.com/files/Publication/4e04d0af-555d-4af9-8828-20983bb31da3/Presentation/PublicationAttachment/42915bdd-6b16-47ac-9357-2a17030b9b3d/YoshiMIPArticle%281%29.pdf (accessed 6 April 2015).

### 3.3.2 *Biotechnological developments and patentability uncertainties*

The issue of patentability as applied to new technologies – such as manipulation of genes and other biological material – is a difficult one, and one in which states have not yet found a consensus on where to draw the line. Art. 27 of the TRIPs Agreement does not expressly deal with the patentability of genetic material.[8]

National policies on this issue often differ, resulting in different legislative stances, or different interpretations of patentability requirements for claims over such material. In advocating flexibility in determining exclusions from patentability 1989, India argued as follows during the Uruguay Round negotiations (to establish the WTO and its agreements such as TRIPs):

[8] The relevant part of Art. 27(b) provides that the following subject matter may be excluded from patentability, without defining its terms (such as "micro-organisms"):

> (b) plants and animals other than micro-organisms, and essentially biological processes for the production of plants or animals other than non-biological and microbiological processes. . . .

In a paper to the TRIPS Council in 2000, IP/C/W/209, the US took the following view of this term (at pp. 1–2):

> The principles of international law regarding the interpretation of treaties and international agreements should be used in determining what is meant by the term micro-organism in Article 27.3(b) of the TRIPS Agreement. Articles 31 and 32 of the Vienna Convention on the Law of Treaties require, *inter alia*, that treaties be interpreted "in good faith in accordance with the ordinary meaning to be given to the terms of the treaty in their context and in light of its object and purpose". The Concise *Oxford Dictionary of Current English* defines "micro-organism" as "an organism not visible to the naked eye, *e.g.*, bacterium or virus". That definition should be sufficient to distinguish plants and animals generally from micro-organisms for purposes of the discussion in the TRIPS Council. . . . To be patentable, a micro-organism cannot be as it exists in nature . . . *The key to what should be patentable as a micro-organism is not what name is given the biological material on which the invention is based, but is the subject matter claimed; is that subject-matter new, does it involve an inventive step, and is it capable of industrial application. If the subject-matter meets those criteria, Article 27.3(b) requires that it be patentable.*
>
> (Italic emphasis added.)

In its report, *The Ethics of Patenting DNA* (2002), the UK's Nuffield Council on Bioethics took a different view: "The question of whether DNA sequences are eligible for patenting is *distinct from* the question of whether they meet these legal criteria" [i.e. that they must be novel, inventive and having utility or being capable of industrial application]; at para. 3.18 (italic emphasis added). The text of the report is available at: http://nuffieldbioethics.org/project/patenting-dna/ (accessed 14 March 2015).

> There are . . . a whole range of moral, ethical, environmental and other issues involved in the patenting of living things and genetically engineered micro-organisms. The full dimensions of scientific and technological development in these areas are yet to be comprehended. *Even in the industrialised countries*, the legal and other implications involved in the granting of patents in areas such as bio-technology and genetic engineering are *in a state of flux*, and the wisdom of granting product patents in bio-technology and for higher forms of life is being subject to serious scrutiny.[9]
>
> [Italic emphasis added]

A group of developing countries also presented the following view at the TRIPs Council:

> 4.1 The grounds for concerns over the patentability of life forms, obtaining in societies far and wide, are ethical, religious and cultural. The co-modification and marketing of life structures violate the cultural principles of quite a number of societies.
>
> 4.2 The concerns also arise from the fact that patents over research material may restrict further research, that discoveries do no [*sic*] amount to inventions, that the cost of medicines keeps escalating due to the required systems of patent protection, and that research increasingly targets products for the affluent rather than for the general public health.[10]

The regulatory dilemma in deciding what ought to be patentable can be particularly acute in such "new" areas, and even developed country systems may not be *ad idem*. As Lord Neuberger acknowledged in *Human Genome Sciences, Inc. v Eli Lilly and Co*:[11]

> Just as it would be undesirable to let someone have a monopoly over a particular biological molecule too early, because it risks closing down

[9] Para. 19, Communication from India, *Standards and Principles Concerning the Availability Scope and Use of Trade-Related Intellectual Property Rights*, Negotiating Group on Trade-Related Aspects of Intellectual Property Rights, including Trade in Counterfeit Goods, Multilateral Trade Negotiations, The Uruguay Round, MTN.GNG/NG11/W/37, 10 July 1989.

[10] Communication of Mauritius on behalf of the African Group, IP/C/W/206, submitted for TRIPs Council Meeting, 20 September 2000. Brazil's communication to the same meeting discusses ethical issues of patenting life forms, arguing that: "Whenever a Member considers that some specific technologies related to patents on technology are contrary to ethical, cultural or religious standards, Article 27.3(b) should consider the possibility of providing flexibility for Members to limit or to deny intellectual property rights over such technologies, in order to prevent that their development is encouraged"; see IP/C/W/228, part IV. On Japan's position, see Communication by Japan for the same meeting, IP/C/W/236, at pp. 4–6.

[11] [2011] UKSC 51, para. 130.

> competition, so it would be wrong to set the hurdle for patentability too high ... *Quite where the line should be drawn in the light of commercial reality and the public interest can no doubt be a matter of different opinions and debate.*
>
> [Italic emphasis added]

It was further observed in the same decision that in interpreting industrial applicability or usefulness under UK patent law, Kitchin J had referred to the US jurisprudence on "utility", a term found in the US patent legislation.[12] On appeal, the UK Supreme Court held that, while "the analyses in the US cases deserve great respect", there are "obvious risks in relying on US jurisprudence when considering the precise nature of the requirements of Article 57 in relation to a claim for a patent for biological material under the EPC". It concluded:

> particularly when it comes to a nice question such as the precise delineation of boundaries between patentability and unpatentability on the ground of industrial application, it would be *unsurprising if the law was not identical under the two jurisdictions.*[13]
>
> [Italic emphasis added]

Within the Asian context, several countries have a general exception against patentability of subject matter where patenting would raise questions of morality. These include India[14] and Singapore,[15] but there is no explicit statutory provision as to patentability (or not) of genes or isolated genetic materials. Even in developed countries, as will be seen below, there is disagreement on this.

As several countries in Asia aspire to promote innovation and development of biotechnology products – including, possible, biopharmaceuticals and gene therapy – the divergences in national laws in the developed world are cause for pause and study. The following section examines the flexibility with regard to patentability in national

[12] Para. 219, *Eli Lilly and Co v Human Genome Sciences, Inc.* [2008] EWHC 1903 (Pat).

[13] Judgment of Lord Neuberger, in paras. 39–41, *Human Genome Sciences, Inc.* v *Eli Lilly and Co* [2011] UKSC 51.

[14] Section 3(b) of the Patents Act provides that the following are not inventions: "(b) an invention the primary or intended use or commercial exploitation of which could be contrary public order or morality or which causes serious prejudice to human, animal or plant life or health or to the environment". See also Biotechnology patent Examination guidelines (March 2013), pp. 12–13).

[15] Section 13(2) of the Singapore Patents Act provides: "(2) An invention the publication or exploitation of which would be generally expected to encourage offensive, immoral or anti-social behaviour is not a patentable invention." But see the *Examination Guidelines for Patent Applications at IPOS*, 14 February 2014, at pp. 250–251. See also note 64 below.

laws, and how some divergences have recently emerged in developed country jurisdictions.

#### 3.3.2.1 Genes and isolated gene sequences

The issues surrounding patentability of human genetic material are too complex to be fully set out here. The present objective is to provide an overview of some of the significant recent litigation in the US, EU and Australia, to highlight the uncertainties in this area, and to provide Asian policy-makers a starting point for shaping (or reshaping) their laws and treaty negotiating considerations.

#### 3.3.2.2 The breast cancer (BRCA) patent claims

A number of contentious patent proceedings have arisen in at least three continents relating to the BRCA1 and BRCA1 genes used in breast and ovarian cancer screening tests.[16] The most recent decisions come from the US Supreme Court and the Australian Federal Court, as will be mentioned below. The US decision, in particular, has raised debates as to its scope and its effect on patents for naturally occurring genetic material in general.[17]

[16] The BRCA1 and BRCA2 genes are useful as persons with mutations in them may be at a higher risk of developing certain cancers (such as breast and ovarian cancers): see Gert Matthijs, *DNA Diagnostics in Practice, in Gene Patents and Public Health*, ed. Geertrui Van Overwalle (Establissements Emile Bruyant, 2007), at p. 35. Different researchers worked to discover BRCA1 and BRCA2 but Myriad Genetics Inc took each forward research result with further steps, and obtained patents for these genes in various jurisdictions; see Matthijs, at pp. 35–38. The BRCA litigation in the US, EU and Australia, as well as celebrity news on the BRCA gene (see e.g. Angelina Jolie, *My Medical Choice*, 14 May 2013, New York Times, at www.nytimes.com/2013/05/14/opinion/my-medical-choice.html?_r=0) have helped to focus global attention on this issue. See also, generally, WIPO, Bioethics and Patent Law: The Case of Myriad, WIPO Magazine, August 2006, and Esther Van Zimmeran, Dianne Nicol, Richard Gold, Julia Carbone, Subhashini Chandrasekharan, A. Lane Baldiwin and Robert Cook-Deegan, *The BRCA Patent Controversies: An International Review of Patent Disputes, Chapter 8 in Breast Cancer Gene Research and Medical Practices: Transnational Perspectives in the Time of BRCA*, ed. Sahra Gibbon, Galen Joseph, Jessica Mozersky, Andrea zur Nieden and Sonja Palfner (Routledge, 2014). For an overview of the EU litigation, see Gert Matthijs, Isabelle Huys, Geertrui Van Overwalle and Dominique Stoppa-Lyonnet, The European BRCA Patent Oppositions and Appeals: Coloring inside the Lines, *Nat. Biotechnol.*, Vol. 31, No. 8 (2013) 704–710.

[17] For an example of the kind of discussion arising from the US decision, see Mark Ratner, Myriad Decision Aftershocks Ripple Through Biotech, *Nat. Biotechnol.*, Vol. 31, No. 7 (2013), 663–665.

#### 3.3.2.3 US Supreme Court: from *Diamond* to two recent cases

In *Diamond v Chakrabarty*,[18] the US Supreme Court held that genetically modified organisms may be patented, albeit with limitations.[19] In March 2012, two cases were decided in quick succession by the US Supreme Court within days of each other. These cases provide important directions and clues as to the scope of patentable material. While not wishing to stifle innovation, a clear line in the sand was drawn between these two cases, to distinguish what is not patentable.

In *Mayo Collaborative Services and Mayo Medical Laboratories et al. v Prometheus Inc*,[20] the US Supreme Court held that relationships between concentrations of certain metabolites in the blood and the likelihood that a particular drug dosage will prove ineffective or cause harm were not themselves patentable.

In *Association for Molecular Pathology et al. v Myriad Genetics Inc et al.*,[21] the validity of a number of Myriad's patents for the BRCA1 and BRCA2 genes was put in question. Myriad had obtained these patents after discovering the location and sequence of the genes, and this was important for detecting mutations of these genes in individuals, as they could indicate risks of breast and ovarian cancer. The US Supreme Court issued an order remanding the matter in light of the *Mayo* decision, and vacated the judgment of the lower court that had permitted Myriad's patents.[22] The Court held that naturally occurring deoxyribonucleic acid (DNA) sequences extracted from human cells were *not* patentable, distinguishing such sequences from synthetically created strands of exons (called complementary DNA or cDNA) which could be patented.[23]

[18] 447 U.S. 303, 206 USPQ 193 (1980).

[19] See judgment at: https://supreme.justia.com/cases/federal/us/447/303/ (accessed 10 December 2014). See generally, USPTO at: www.uspto.gov/web/offices/pac/mpep/s2105.html (accessed 10 December 2014).

[20] Decision of the US Supreme Court, 30 March 2012, 566 U.S. __ (2012). For a summary of this case, see Jonathan Moss, Prometheus Fails to Steal Fire from the Gods, *J. Intell. Prop. L. & Prac.*, Vol. 7, No. 8 (2012), 550, and John R. Thomas, *Mayo v Prometheus: Implications for Patents, Biotechnology, and Personalized Medicine*, 6 November 2012.

[21] Order of the US Supreme Court, at p. 2 www.supremecourt.gov/orders/courtorders/032612zor.pdf.

[22] See James Vicini, *Supreme Court Sets Aside Myriad Gene Patent Ruling*, 26 March 2012, *Reuters*, at: www.reuters.com/article/2012/03/26/us-myriadgenetics-patent-idUSBRE82P0N220120326. In a prior decision the US Supreme Court had held that "works of nature" are not patentable: *Funk Brothers Seed Co v Kalo Inoculant Co*, 333 US 127 (1948) at 130. See also *Diamond v Chakrabarty*, 447 US 303 (1980).

[23] Exons are nucleotides in DNA which code for amino acids, which are in turn used to build proteins for the human body. Interestingly, Dr Chakrabarty, the patent holder in

As will be seen below, Myriad's patents in the EU and Australia have also been subjected to legal challenge. While this was not a case relating to pharmaceuticals, it has important implications for the patentability of naturally occurring genetic material (even if isolated), and whether a company like Myriad would be able to control the location and isolation process, and to prevent others from developing genetic testing based on it. If the patents had been upheld, Myriad would have had the *exclusive* right to isolate the genes (for testing) in individuals.

#### 3.3.2.4 The EU

The EU Directive of 1998, incorporated into implementing rules of the European Patent Convention, permits patenting of biotechnological inventions.[24]

In 2004, in the face of oppositions to patents held by Myriad Genetics Inc. for BRCA1 and BRCA2 genes, the European Patent Office (EPO) revoked a patent in respect of the "BRCA1" gene but upheld in a limited fashion (limited to testing for gene mutations specifically in Ashkenazi Jewish women) its patent of the "BRCA2" gene.[25]

#### 3.3.2.5 Australia

There has been a great deal of debate in Australia on the patentability of human genetic material.[26] The issue recently came to the fore

*Chakrabarty* case (above), recently expressed his disagreement with this ruling; see his statements reported at a conference held in Singapore in August 2014, *Bacterium Creator's Patent Opens Legal Can of Worms*, http://news.asiaone.com/news/science-and-tech/bacterium-creators-patent-opens-legal-can-worms (accessed 9 October 2014). For post-*Myriad* USPTO views, see e.g. guidance on patentability, USPTO Memorandum, 4 March 2014, www.uspto.gov/patents/law/exam/myriad-mayo_guidance.pdf (accessed 13 March 2015). See generally, See Julia Carbone, E. Richard Gold, Bhavan Sampat, Subhashini Chandrasekharan, Lori Knowles, Misha Angrist and Robert Cook-Deegan, DNA Patents and Diagnostics: Not a Pretty Picture, *Nat. Biotech.*, Vol. 28, No. 8 (August 2010), pp. 784–791.

[24] EU Directive 98/44/EC, adopted 6 July 1998. See also the EU Parliament's related Resolution, 2012/2623(RSP), 10 May 2012.

[25] See Matthijs G, The European Opposition against the BRCA Gene Patents, *Fam. Cancer*, Vol. 5, No. 1 (2006), 95–102, and Lori B Andrews, Patents: The Need for Bioethics Scrutiny and Legal Change, *Yale Journal of Health Policy, Law, and Ethics*, Vol.5 (2005), 403 (with J. Paradise). See also Gert Matthijs, Isabelle Huys, Geertrui Van Overwalle, Dominique Stoppa-Lyonnet, The European BRCA Patent Oppositions and Appeals: Coloring Inside the Lines, *Nat. Biotechnol.*, Vol. 31, No. 8 (August 2013), p. 704.

[26] The Patent Amendment (Human Genes and Biological Materials) Bill 2010 [No. 2] that would have excluded patentability of genetic and biological material was proposed but

through a lawsuit brought by an individual against Myriad Genetics Inc in Australia courts to challenge claims over patenting of isolated human genetic material. The case reached the Australian Federal Court in 2014. By contrast to the US Supreme Court's views, the Australian full Federal Court decided in that isolated genetic sequences or nucleic acid *were patentable subject matter* under the Australian Patents Act 1990.[27] As isolated nucleic acid may be used in gene therapy and genetic testing (such as for susceptibility to breast cancer in individuals), a decision as to whether such material is patentable (so that it is the subject to a patent-holder's monopoly rights) or not (so that it is freely available for use by other researchers and innovators) has significant and broad public health implications. The Court considered the text of the statute and its legislative history and concluded that the gene material in question was patentable, arguing as follows:

> This case is not about the wisdom of the patent system. It is about the application of Australian patent law, as set out in the Act and as developed by the courts since the *Statute of Monopolies*.
>
> It is not about whether, for policy or moral or social reasons, patents for gene sequences should be excluded from patentability. That has been considered by the ALRC and by Parliament and has not occurred. It is not a matter for the court, but for Parliament to decide. Parliament has considered the question of the patentability of gene sequences and has chosen not to exclude them but to make amendments to the Act to address, in part, the balance between the benefits of the patent system and the incentive thereby created, and the restriction on, for example, subsequent research.
>
> . . .
>
> In Australia, there is no statutory or jurisprudential limitation of patentability to exclude "products of nature". To the contrary, the High Court has specifically rejected such an approach. A mere discovery is not

rejected in Sep 2011 by the Senate Legal and Constitutional Affairs Legislation Committee for being too broad in scope;

see Report of the Committee and generally, the discussions in *Australian Government Response to Senate Community Affairs References Committee Gene Patents Report*, November 2011, and in *Australian Law Reform Commission*. The subsequent Intellectual Property Laws Amendment (Raising the Bar) Act 2011 provides an experimental use exemption.

27 *D'Arcy v Myriad Genetics Inc* [2014] FCAFC 115 (5 September 2014). See overview of country approaches in *Genes and Ingenuity*, ALRC report no. 99, and generally, *Genes and Ingenuity: Genes Patenting and Human Health*, Australian Law Reform Committee (ALRC) report no. 99, Chap. 6 *Patentability of Genetic Materials and Technologies.*

> patentable and an idea is not patentable, but a "manner of manufacture", as that term has been developed, is.[28]

The Federal Court found the US case, *Myriad*, to be distinguishable.[29]

The plaintiff filed a request to appeal this decision to the Australian High Court, which granted special leave to the plaintiff to appeal the case.[30] On 7 October 2015, in a much-anticipated and unanimous judgment, the High Court of Australia allowed the appeal and *set aside* the decision of the Federal Court.[31] The High Court held that, having regard to the relevant factors,[32] an isolated nucleic acid – that is, genetic coding for the BRCA1 protein – with specified variations, was not a manner of manufacture within section 18(1)(a) of the Patents Act 1990, and was therefore *not* patentable.[33] It held that

> The term "manner of manufacture" and the concept it embodies, was and is no more pregnant with rules and applications awaiting discovery, than is the common law.

The High Court referred to the laws and practices of other jurisdictions – China, Japan, Korea, India and Singapore – on the patentability of isolated nucleic acids.[34] Noting that such jurisdictions might permit patenting of isolated nucleic acids, it nonetheless went on to cite with approval the following sentiments of Lord Neuberger in *Human Genome Sciences*[35]:

> "There have been moves over the past fifty years (and more) to harmonise patent law across jurisdictions (the EPC and TRIPS . . . being two important examples), and it is a laudable aim to seek to ensure that all aspects of the law of patents are identical throughout the world. However, the achievement of such an aim is plainly not currently practicable, and,

[28] *Ibid.*, paras. 203, 204 and 207. On the legislative history and reform discussions, see paras. 156–161.

[29] See note 27 above, paras. 130–138.

[30] See report at www.iplawwatch.com/tag/yvonne-darcy-v-myriad-genetics-inc-anor/ (accessed 13 March 2015).

[31] See http://eresources.hcourt.gov.au/showCase/2015/HCA/35 (accessed 8 October 2015).

[32] See paras. 28–30 of the judgment, *ibid.*

[33] Section 18(1)(a) of the Act provides:
"Subject to subsection (2), a patentable invention is an invention that, so far as claimed in any claim:
(a) is a manner of manufacture within the meaning of section 6 of the Statute of Monopolies."

[34] Para. 34 of the judgment. [35] Note 11, above.

> although they have a great deal in common, there are significant and fairly fundamental differences . . ."
>
> "Accordingly, particularly when it comes to a nice question such as the precise delineation of boundaries between patentability and unpatentability on the ground of industrial application, it would be unsurprising if the law was not identical under the two jurisdictions".[36]

Importantly, the High Court preferred to look at the *substance* of the invention and considerations flowing from it, rather than to view the invention as being simply information embodied in a product created by human action.[37] It also made the following important observations:

> When proper regard is paid to their emphasis on genetic information, the subject matter of the claims lies at the boundaries of the concept of "manner of manufacture". That it does lie at the boundaries is further evidenced by the odd consequence that if the claims are properly the subject of a patent, the patent could be infringed without the infringer being aware of that fact. That consequence coupled with the very large, indeed unquantified size of the relevant class of isolated nucleic acids, all of which bear the requisite information, raises the risk of a chilling effect upon legitimate innovative activity outside the formal boundaries of the monopoly and risks creating a penumbral de facto monopoly impeding the activities of legitimate improvers and inventors.[38]

The decision has clarified firstly that under the Australian patent legislation, human intervention and activity to isolate human genetic material is not indicative that there is a patentable invention, and secondly, the High Court is prepared to evaluate the evidence, the legislative words as well as broader societal implications of its decision. It took care to express its mindfulness of the limits of judicial law-making in this context:

> The proposition that a broad statutory concept applies to a new class of case on the boundaries of existing judicial development of that concept requires consideration of the limits of judicial lawmaking inherent in common law methodology. Where an affirmative application of the

[36] At para. 35 of the judgment. [37] At para. 94 of the judgment.

[38] At para. 93 of the judgment (original footnote omitted). Notably, the High Court drew a distinction between claiming a patent over "*isolated fragments* of naturally occurring DNA as comprise the BRCA1 gene as are found upon examination to contain the (naturally occurring) specified mutations and polymorphisms" and over "a *process or method* of detecting the increased likelihood of certain kinds of malignancy by isolating the BRCA1 gene and examining it for the presence of any of the specified mutations and polymorphisms"; para. 168 of the judgment (italics added).

> concept is likely to result in the creation of important rights as against the world, to involve far-reaching questions of public policy and to affect the balance of important conflicting interests, the question must be asked whether that application is best left for legislative determination. The patentability of nucleotide sequences derived from human DNA is in that category. The inherent patentability of the invention as claimed would powerfully imply patentability of any claim for an isolated nucleic acid coding for a specified polypeptide.[39]

### 3.3.2.6 UK

In 2002, the Nuffield Council on Bioethics had issued a report – *The Ethics of DNA Patenting*.[40] The report, while acknowledging the economic importance of patents, set out a number of important recommendations cautioning *against* ease of obtaining patents on genes or DNA material.

In 2011, in *Human Genome Sciences Inc v Eli Lilly and Company*,[41] the UK Supreme Court was faced with the issue of patentability of a gene sequence for the protein, "neutrokine-α". The Court took a relatively generous view of the requirement that material had to be "capable of industrial application". (The Court did this by holding that the test was one of "plausibility" of use, as opposed to use as shown by doing research.) Lord Walker, one of the Justices who gave judgment in the case, referred to two strong policy arguments in this holding, the first being "to reduce the risk of a chilling effect on investment in bioscience" (the second being a further alignment of UK's interpretation of the European Patent Convention with those of other countries).[42]

One might distinguish the US and UK cases on the basis that they deal with different subject matters and with different criteria for patentability. However, the above cases reflect two points clearly: the *boundaries* of patentability with respect to genetic and biological material are still being charted even by relatively well-developed intellectual property law systems, such as those in the US and UK, as the above discussions demonstrate.

[39] At para. 7 of the judgment.

[40] The report is available at: http://nuffieldbioethics.org/project/patenting-dna/ (accessed 14 March 2015).

[41] [2011] UKSC 51; judgment (2 November 2011).

[42] Para. 171, *ibid.* For a recent discussion of stem cell patents in the EU context, see Scott Parker and Paul England, "Where now for stems [*sic*] cell patents?", *Journal of Intellectual Property Law and Practice*, Vol. 7, No. 10 (2012), 738. See also the UK decision, *Amgen v TKT*, discussed in *Gert Matthijs*, note 16 above, at pp. 56–58.

#### 3.3.2.7 Canada

Canada has, as a matter of practice, permitted patents for genetic material.[43] This has not, however, been without its detractors and critics. In November 2014 litigation was initiated by a Canadian hospital to challenge certain patents held by the defendants in respect of genetic material in the form of isolated nucleic acids.[44] In *Canada (Commissioner of Patents) v Harvard College*, the Canadian Supreme Court held that higher life forms (such as transgenic animals) are excluded from patentability by virtue of their not being either "manufactures" or "compositions" of matter within the meaning of the definition of "invention" (found in Section 2 of the Canadian Patent Act, read with Section 28.3).[45] Reflecting the discomfort of a judiciary in pronouncing on matters of patentability of such higher life forms in applying the Patent Act, the Court held candidly as follows:

> The [preceding] concerns ... illustrate the fact that the Patent Act in its current form is not well suited to address the unique characteristics possessed by higher life forms. It is not an appropriate judicial function of the courts to create an exception from patentability for human life given that such an exception requires one to consider both what is human and which aspects of human life should be excluded.
>
> The patenting of higher life forms raises special concerns that do not arise in respect of non-living inventions. Unlike other inventions, biologically based inventions are living and self-replicating. ... the products of biotechnology are incredibly complex, incapable of full description, and can contain important characteristics that have nothing to do with the invention ...
>
> The lack of direction currently in the Act to deal with issues that might reasonably arise signals a legislative intent that higher life forms are currently not patentable ... this Court does not possess the institutional competence to deal with issues of this complexity, which presumably will require Parliament to engage in public debate, a balancing of competing social interests, and intricate legislative drafting.[46]

The Court was of the view that while lower life forms may be patentable under Canadian law, this did not mean it was so for higher life

[43] See, Canadian Intellectual Property Office (CIPO), *Manual of Patent Office Practice*, Chapter 17.

[44] See in particular paras. 50–59, Statement of Claim, *Children's Hospital of Eastern Ontario v University of Utah Research Foundation, Genzyme Genetics and Yale University*, at www.cheo.on.ca/en/challenge-gene-patents (accessed 13 March 2015).

[45] 2002 SCC 76; see also Canadian Intellectual Property Office website at: www.cipo.ic.gc.ca/eic/site/cipointernet-internetopic.nsf/eng/wr03153.html#n12_01 (accessed 6 July 2014).

[46] Paras. 167, 181 and 183, *ibid.*

forms. In drawing this distinction, it drew support from trade agreements, stating that specific exception for plants and animals in trade agreements such as TRIPs and NAFTA demonstrates that "the distinction between higher and lower life forms is widely accepted as valid".[47] The Supreme Court also opined against the patentability of human life and its "commodification".[48]

### 3.3.3 *Divergent approaches to new technologies in developed countries: a question for Asian countries*

The above discussion illustrates the complexity and divergences in legal approaches to patentability with regard to genetic material, even in developed countries with advanced R&D policies and debates. Policies on patentability of genetic material are not equally clear in all Asian countries; some jurisdictions have expressed statutory or guideline language on this, while others have not. The relevance of such policy to access to medicines lies in the access to use of genetic material in development of medicines, diagnostic tests and tools, and medical R&D in general.[49] Where patents are granted for such subject matter it will, as in other areas, create a legal monopoly in favour of the rights holder for a fixed duration. This will allow the rights holder to impose "transaction costs" on parties wishing to use such material for further innovation and R&D, possibly adding to the price of any innovative product.[50] In this regard, some have argued that such subject matter should form part of the public "commons", so that the patenting of such matter would have the effect of forming an "anti-commons", that is, reducing or curtailing

[47] Para. 205.

[48] Para. 175. For examples of criticisms to this decision, see Timothy Caulfield, Biotechnology Patents, Public Trust and Patent Pools: the Need for Governance?, in 358–360, and Jey Kumarasamy, *Are Human Genes Patentable? (Pt II – Gene Patentability in Canada)*, 21 March 2014, McGill J. L. & H. Online, at http://mjlh.mcgill.ca/blog.php?blog_id=116 (accessed 13 March 2015). *Pt. I* is available at: http://mjlh.mcgill.ca/blog.php?blog_id=106.

[49] See also the discussion in Koichi Sumikura, *Intellectual Property Rights Policy for Gene-Related Inventions – Toward Optimum Balance between Public and Private Ownership*, in *The Role of Intellectual Property Rights in Biotechnology Innovation*, note 6 above, at pp. 74–80, and Thomas Baldwin, *Ethics and Patents for Genetic Diagnostic Tests*, in *Gene Patents and Public Health*, note 16 above.

[50] On transaction costs in the biotechnology context, see e.g. the discussion in F. Scott Kieff, *On the Comparative Institutional Economics of Intellectual Property in Biotechnology*, in *The Role of Intellectual Property Rights in Biotechnology Innovation*, note 6 above, at pp. 286–290.

further innovative research.[51] For a further discussion of patentability of gene-related material in Asian countries, see Section 3.3.3.1.[52]

Apart from human gene-related subject matter, the regulation of other relatively new subject matter will also require the careful attention of Asian regulators. An example would be biosimilar pharmaceuticals, which aim to compete with biologic pharmaceutical products (products using materials of living organisms, instead of purely synthetic chemicals). The analogy has been drawn with generic products as substitutes for chemical pharmaceuticals.[53] The US FDA recently granted marketing approval for the first biosimilar product.[54] Even in countries which aim to be magnets for bioscientific R&D and discovery and production of biopharmaceutical products, the approach remains a cautious one, with boundaries of patentability in new areas not entirely defined (perhaps for good reason). Foreseeable regulatory issues for Asia include patentability of products of bioscience and protection of related test data.[55]

#### 3.3.3.1 Patentability – approaches to biotechnology in Asia

Patentability in newer areas, such as that of living matter or their constituent components, differs across ASEAN.[56] As will be seen below a number of Asian countries have recently issued patent examination guidelines to provide guidance to on patentability of new technologies such as genes and genetic materials.

#### 3.3.3.2 Singapore

In 2012, Singapore amended its Patents Act to reform the examination system, and these changes are in the implementation stage. Under the system existing prior to the amendments, patents were granted without examination by the national agency the Intellectual Property Office of

[51] See, e.g., Christopher May, note 6 above, at pp. 231–232, and Timothy Caulfield, Robert M. Cook-Deegan, F. Scott Kieff and John P. Walsh, Evidence and Anecdotes: An Analysis of Human Gene Patenting Controversies, *Nat. Biotechnol.*, Vol. 24, No. 9 (September 2006), 1091–1094.

[52] For a convenient collection of links to Asian intellectual property office sites and materials, see European Patent Office website, www.epo.org/searching/asian/links.html (accessed 9 October 2014).

[53] See Charley Grant, *Pharma's Biosimilar Threat*, 24 March 2015, The Wall Street Journal, p. 28.

[54] See FDA announcement, www.fda.gov/NewsEvents/Newsroom/PressAnnouncements/ucm436648.htm (accessed 25 March 2015).

[55] The US, for example, has enacted legislation on competition of biologic drugs; see the Biologics Price Competition and Innovation Act of 2009, Title VII, H.R.3590–686.

[56] See generally, WHO, *Genetics, Genomics and the Patenting of DNA*.

Singapore (IPOS), but rather by "self-assessment".[57] Singapore adopted a new "positive grant" examination system replacing the prior "self-assessment" system with effect from 14 February 2014.[58] This was to introduce greater rigor at the patent application stage, to ensure that only high quality patents are granted.[59] *Sans* a proper, rigorous examination, weak or unmeritorious patents may be granted, and such patents give the holders the full benefit of the generous protection afforded by the Patents Act. Unless another party incurs the time and cost to challenge such a patent, it remains as effective and powerful as any patent, blocking would-be competitors from using the product or process to, say, develop improvements. The self-assessment system therefore provides an important opportunity and mechanism for Singapore to guard against granting

[57] See IPOS consultation letter, 21 March 2012, and Annexes, available at: www.ipos.gov.sg/AboutIP/IPLegislation/PublicConsults.aspx, and speech by the Minister for Law, note 2 above.

[58] The Patents (Amendment) Bill, No. 13 of 2012, introduced in Parliament on 14 May 2012, passed in July 2012.

[59] The rationale was explained in Parliament by the Minister for Law as follows:

> We will enhance our patent regime. At present, we adopt a self-assessment system, where patents can be granted even if objections are raised in the examination report. To enhance it, we should move to a positive grant system where patents will be granted **only if patentability requirements are fully met first**. This will increase confidence in the quality of Singapore patents.
>
> (Bold emphasis added.)

Parliamentary Report, 6 March 2012, available at www.lawnet.com.sg.

At the 2nd reading of the Patents (Amendment) Act in Parliament, the Senior Parliamentary Secretary to the Minister for Law added:

> Amendments to
>
> replace our current self-assessment patent system with a new positive grant patent system. All applications for patents must undergo a search and examination process. Under our existing self- assessment patent system, patent applications need not, however, fully fulfil Singapore's patentability criteria in order to be granted. This means that patents may be granted without a fully positive examination report. . . . **The move to the positive grant patent system will minimise the occurrence of weak patents being granted in future, enhance the standing of Singapore granted patents, and give increased assurance to companies and investors of the quality of Singapore-granted patents.** . . .
>
> (Bold emphasis added.)

Parliamentary Report, 10 July 2012. See also see www.mlaw.gov.sg/news/parliamentary-speeches-and-responses/second-reading-speech-by-senior-parliamentary-secretary-for-education-and-law-ms-sim-ann-on-the.html (accessed 19 March 2015).

weak patents, which would instead of spurring the innovative spirit – what Singapore is seeking to encourage through various policies – would impede it.[60]

In Singapore, Section 13 of the Patents Act, which governs patentability, contains no express exclusion of life forms, animals or plants.[61]

However, it is also not clear if newer technology claims may still be excludable under Section 13(2), as being "immoral". In a "non-paper" submitted to the TRIPs Council in 2000, Singapore had argued in favour of extending patent protection to plants and animals.[62] In the same

[60] It should be noted, that under s. 80(1)(a) of the Patents Act, a patent may be revoked if it is shown that the invention was not a patentable invention. This revocation process allows an applicant to raise and challenge the *patentability* of an invention. S. 80 was raised in *Dien Ghin Electronic (S) Pte Ltd v Khek Tai Ting (trading as Soon Heng Digitax)* [2011] 3 SLR 227. For an earlier discussion of patent law considerations in relation life-sciences and genetic engineering, see George Wei, Life-Sciences, Genetic Engineering and Patent Rights: A Singapore Perspective, *BSLR*, Vol. 10, No. 4 (2009), 135–146.

[61] Note that in 1995, Singapore had deleted s. 13(2) and (5) relating to certain exclusions from patentability, vide Act No. 40 of 1995, explained in the Bill Supplement thus: "so as to enable Singapore to keep up with the continuing evolution and changes in science and technology." Prior to repeal, section 13(2) read:

(2) It is hereby declared that the following (among other things) are not inventions for the purposes of this Act, that is to say, anything which consists of —

(a) a discovery, scientific theory or mathematical method;
(b) a literary, dramatic, musical or artistic work or any other aesthetic creation whatsoever;
(c) a scheme, rule or method for performing a mental act, playing a game or doing business, or a program for a computer; or
(d) the presentation of information,

but the foregoing provisions shall prevent anything from being treated as an invention for the purposes of this Act only to the extent that a patent or an application for a patent relates to that thing as such.

Prior to repeal, section 13(5) read:

> (5) The Minister may, by order published in the *Gazette*, vary the provisions of subsection (2) for the purposes of maintaining them in conformity with developments in science and technology.

See also generally, the following excellent works: George Wei, Chapter 2, *Fitting Biological Products within the Intellectual Property Framework: Challenges Facing the Policy Makers*, in *Intellectual Property and Biological Resources*, Marshall Cavendish International (Singapore) Private Ltd (2004);*Life Sciences, Genetic Engineering and Patent Rights: A Singapore Perspective*, (2009) Vol. 10 Issue 4 *Bio-Science Law Review* 1-12; Chapter 4, *An Introduction to Genetic Engineering, Life Sciences and the Law*, Singapore University Press (2002); and *Industrial Design Law in Singapore*, (2012) Academy Publishing, Singapore Academy of Law, *Addendum p. xv.*

[62] Non-paper by Singapore – Art. 27.3(b), TRIPS Council, 11 December 2000, JOB(00)7853.

submission, Singapore stated that "[e]xcept for reasons of public order or morality, Singapore believes that biological inventions deserve patent protection in the same manner as that accorded to other areas of science and industry", while recognizing that "biotechnology is a new field and it is therefore not surprising that various difficulties are encountered over the award of patents". As Singapore aims to be a biomedical and life science hub *as well as* an intellectual property law "hub",[63] a clear position as to patentability in newer types of claims arising from advancements in the life sciences is important as this will provide certainty not only to the inventors and potential patent applicants, but also to the health research community at large. While the IPOS has issued a very detailed examination guide for patents including patentability criteria under Singapore legislation and case law, newer technologies such as genes, sequencing and manipulations of genes, and their use in therapies have not been discussed in a specific and detailed manner as might be expected.[64] While there are some explicit statements in the Guide, they appear to be relatively brief and somewhat tentative, especially given the silence of the patent legislation on certain subject matter. For example, the IPOS has also expressed the view that the use of human embryos for commercial or industrial purposes is not governed by any express provision under the Singapore patent law, and "such inventions will considered *on a case by case basis*". [Italic emphasis added]

In relation to patentability of genes, the IPOS Guidelines state that:

> "The issue of morality under section 13(2) should be considered when assessing inventions relating to genes."[65] . . .

[63] See *Intellectual Property (IP) Hub Master Plan*, April 2013, IP Steering Committee, Singapore Ministry of Law.

[64] See *Examination Guidelines for Patent Applications at IPOS*, 14 February 2014, note 15 above, generally Part 4 on "Inventive Step"; Part 4, pp. 175–176 which provides "*Biotechnology Examples*" in the context of *Unity of Inventions*; and Part 8 on *Patentable Subject Matter and Industrial Applicability*, especially Part D. on *Morality*, at pp. 249–251, on *Human Cloning* and *Patentability of Genes, Genetic Material and Embryos*. The 2014 Guidelines are available at:

[65] *Examination Guidelines* (2014), *ibid.*, at p. 250. Section 13(2) states: "(2) An invention the publication or exploitation of which would be generally expected to encourage offensive, immoral or anti-social behaviour is not a patentable invention". See George Wei, Chapter 2, *Fitting Biological Products within the Intellectual Property Framework: Challenges Facing the Policy Makers*, in *Intellectual Property and Biological Resources*, note 61 above, at pp. 51–60.

This clearly flags the possibility of a Section 13(2) morality objection to a gene-related patent application, although the types of circumstances in which it will indeed be raised are far from clear.

Nonetheless, drawing from the European Patent Office's stand, the IPOS Guidelines further state:

> In *Howard Florey Institute [Relaxin], V8/94 Relaxin, OJ EPO 6/1995*, the invention is directed to mRNA isolated from tissue taken from pregnant women with consent. The Opposition Division of the EPO held that DNA is not "life", but a chemical substance which carries genetic information and can be used as an intermediate in the production of proteins which may be medically useful. The patenting of a single human gene therefore has nothing to do with the patenting of human life – even if every gene in the human genome was cloned (and possibly patented), it would be impossible to reconstitute a human being from the sum of its genes. *Thus where the subject matter involves genes* per se, *the Examiner should proceed to assess whether it meets the section 13(1) requirements.*[66]
>
> [Italic emphasis added]

Taken together, the IPOS therefore appears to take the stance that while patenting of a human gene is not prohibited *a priori*, such a patent claim must fulfil the usual patentability requirements of Section 13(1), *and* would be subjected to morality considerations under Section 13(2).[67] In this regard, the *Examination Guidelines* provide the following caveat:

> Only in the clearest cases should section 13(2) be invoked if the general public would regard the invention as abhorrent or from which the public need protection.[68]

As for biosequences (such as nucleotides and peptides), these are discussed in the context of Unity of Inventions and not under the morality section.[69]

The application of these Guidelines provide a degree of flexibility and the general instruction to consider gene patent claims in connection

66 *Examination Guidelines* (2014), note 64 above, ain para. 8.98.

67 *Ibid.*, pp. 250–251. Contrast this position with that of IP Australia, which states that human genes in the human body are not patentable but DNA or gene sequences that have been isolated may be patentable; see: www.ipaustralia.gov.au/get-the-right-ip/patents/about-patents/what-can-be-patented/patents-for-biological-inventions/ (accessed 3 May 2014).

68 *Ibid.*, p. 251.

69 *Ibid.*, pp. 175–176. The guidelines state at p. 175: "Biosequences can generally be considered using the same principles as used for chemical inventions . . . "

with Section 13(2) in effect leaves a great deal to be decided by the examiner and, in the case of appeal, by the Singapore courts.

Section 13 of the Malaysian Patents Act excludes *inter alia* "(b) plant or animal varieties or essentially biological processes for the production of plants or animals, other than man-made living micro-organisms, micro-biological processes and the products of such micro-organism processes . . . " Notably, paragraph (b)above excludes "processes" rather than the living matter itself.

The Examination Guidelines of the Malaysian Intellectual Property Office state as follows[70]:

> To find a substance or micro-organism freely occurring in nature is also merely [*sic*] discovery and therefore unpatentable. If, however, as will generally be the case, it were *necessary to extract and isolate* it then *a process* developed for this purpose, and *also the material* when obtained by this process, could both be patentable. A gene which is discovered to exist in nature may be patentable if a technical effect is revealed.
>
> [Italic emphasis added]

For microorganisms, the Guidelines state as follows:

> The question whether a process is "essentially biological" is one of degree depending on the extent to which there is technical intervention by man in the process; if such intervention plays a significant part in determining or controlling the result it is desired to achieve, the process would not be "essentially biological" and hence not excluded.
>
> . . . Man-made living micro-organisms, micro-biological processes and the products of such micro-organism processes are specifically exempted from the exclusion. The term microbiological process is to be interpreted as covering not only industrial processes using micro-organisms, but also processes for producing new micro-organisms, e.g. by genetic engineering. The product of a microbiological process may also be patentable *per se* (product claim). Propagation of the micro-organism itself is considered to be a microbiological process; consequently, the micro-organism can be protected per se as a product of that process. The term micro-organism also covers plasmids and viruses.[71]

On the exceptions to patentability under the Act, the Guidelines state as follows:

> Section 31(1) states that the grant of a patent shall not be refused and a patent shall not be invalidated on the ground that the performance of any

[70] Para. 3.1, Malaysia Intellectual Property Office, *Guidelines for Patent Examination.*

[71] Para. 3.3, *ibid.*

> act in respect of the claimed invention is prohibited by any law or Regulation, except when the performance of that act would be contrary to public order or morality. The purpose of this is to exclude from protection inventions likely to induce riot or public disorder, or to lead to criminal or other generally offensive or immoral behaviour; one obvious example of subject-matter which should be excluded under this provision is a letter-bomb. This provision is likely to be invoked only in rare and extreme cases. A fair test to apply is to consider whether it is probable that the public in general would regard the invention as so abhorrent that the grant of patent rights would be inconceivable. If it is clear that this is the case, objection should be raised by the examiner otherwise not.[72]

In Thailand, Article 9 of the Patents Act expressly excludes certain categories of subject matter from being patented. Article 9 excludes from patentability, *inter alia*, the following: "naturally occurring microorganisms and their components, animals, plants or extracts from animals or plants"; "methods of diagnosis, treatment or cure of human and animal diseases"; and "inventions contrary to public order, morality, health or welfare".

Newer technologies that lead to claims of inventions with regard to genes and gene sequences are possibly covered under Art. 9(1) or even 9(5). However, it appears that methods of producing transgenic animals, plants and micro-organisms are patentable.[73]

Under the Indonesian Patent Law of 2001, Article 7 excludes, *inter alia*: "all living creatures, except micro-organism" and "any biological process which is essential in producing plant or animal, except non-biological process or microbiological process".

Sections 21 and 22 of the Philippines IP Code, as amended by the Universally Accessible Cheaper and Quality Medicines Act of 2008, govern patentability.[74] Section 22.4 excludes from patentability: "plant varieties or animal breeds or essentially biological process for the production of plants or animals" but not "micro-organisms and non-biological and microbiological processes."

[72] Para. 4.1.

[73] See Fabrice Mattei, IP and Other Legal Issues Relevant to Biotech Entrepreneurs in Thailand, Biotechonology in Malaysia and Thailand, *Asia-Pacific Biotech News*, Vol. 11, No. 8 (2007), 498–505.

[74] Republic Act No. 8293.

See also Rules 201 and 202, *The Revised Implementing Rules and Regulations for Patents, Utility Models and Industrial Designs Implementing Rules and Regulations of the Republic Act No. 9502, otherwise known as the "Universally Accessible Cheaper and Quality Medicines Act of 2008".*

In Vietnam, Article 59 of the Law on Intellectual Property of 2005 excludes from patentability, *inter alia*: "plant varieties, animal breeds"; "processes of plant or animal production which are principally of biological nature other than microbiological ones"; and "human and animal disease prevention, diagnostic and treatment methods".

Under the Patent Law of the People's Republic of China,[75] patenting of life forms is not permitted, but the patenting of genes is permitted.[76] The Patent Guidelines provide further guidance, and allows for patenting of genes and sequences.[77]

India's Patent Act governs patentability, with the key provision being Section 3.[78] The section does not expressly deal with patentability of genes or biological material but does contain a prohibition against patenting of:

[75] See http://english.sipo.gov.cn/laws/lawsregulations/201101/t20110119_566244.html.

[76] Article 5 provides:

> Patent rights shall not be granted for invention-creations that violate the law or social ethics, or harm public interests. Patent rights shall not be granted for inventions that are accomplished by relying on genetic resources which are obtained or used in violation of the provisions of laws and administrative regulations.

Article 25 provides:

> Patent rights shall not be granted for any of the following:
>
> (1) scientific discoveries;
> (2) rules and methods for intellectual activities;
> (3) methods for the diagnosis or treatment of diseases;
> (4) animal or plant varieties;
> (5) substances obtained by means of nuclear transformation; and
> (6) designs that are mainly used for marking the pattern, color or the combination of the two of prints.
>
> The patent right may, in accordance with the provisions of this Law, be granted for the production methods of the products specified in Subparagraph (4) of the preceding paragraph.

[77] State Intellectual Property Office, PRC, *Guidelines for Patent Examination*, 2010; see especially Part II, Chapter 10, part 9. See also generally, WHO, note 56 above; according to the WHO:

> ... in 2001, Shanghai Joint Gene Technology Co. Ltd, the largest gene technology company in China, applied for more than 3700 gene patents, including patents for genes dealing with cancer, obesity, high blood pressure and senile dementia, which are expected to be of high value for clinical diagnosis and the development of new medicines.

[78] See www.ipindia.nic.in/.

(b) an invention the primary or intended use or commercial exploitation of which could be contrary public order or morality or which causes serious prejudice to human, animal or plant life or health or to the environment;
(c) the mere discovery of a scientific principle or the formulation of an abstract theory or discovery of any living thing or non-living substance occurring in nature . . .

In 2013, the Indian government issued more specific guidelines.[79] Under these guidelines, gene sequences may form a patentable subject matter. In particular, according to the guidelines:

> A claim to a specific fragment of polynucleotide may be considered to be novel, but subject to fulfilment of the inventive step and non-patentability under relevant clauses of Section 3 of the Act.[80]

The Japan Patent Office's Examination Guidelines includes the following statement on non-patentable subject matter:[81]

> (2) Mere discoveries and not creations
>
> One of the requirements for a statutory invention is to be a "creation", and thus, mere discoveries, such as discoveries of natural things like an ore or natural phenomena, for which an inventor does not consciously create any technical idea, are not considered to be a statutory invention.
>
> However, *if things in nature such as chemical substances or microorganisms have been isolated artificially from their surroundings, then those are creations and considered to be a statutory invention.*
>
> [Italic emphasis added]

The Korean Patent Act permits exclusion from patentability, certain applications on the grounds of public order or morality or public health concerns.[82] Article 32 provides as follows:

> **Article 32 (Unpatentable Inventions)**
>
> Inventions that are feared to have risks to contravene public order or morality or to injure public health shall not be patentable, notwithstanding Article 29(1) and (2).

[79] *Guidelines for Examination of Biotechnology Applications for Patent*, Office of the Controller General of Patents Designs and Trade Marks, India, March 2013.
[80] *Ibid.*, para. 7.2. See also paras. 8 and 9.
[81] Japan Patent Office, *Examination Guidelines for Patent and Utility Model in Japan*, Part II, Chapter 1, Industrially Applicable Inventions. See also explanation of Japan's position, see Communication by Japan for the TRIPS Council meeting, 11 December 2000, IP/C/W/236, at pp. 3–4.
[82] See www.kipo.go.kr/upload/en/download/PatentAct.pdf.

Under the Patent Examination Guideline for Biotechnological Inventions 1998, it was clarified that gene sequences which satisfy other requirements of novelty, inventiveness and industrial applicability may be patentable subject matter.[83]

### *3.3.4 New uses*

#### 3.3.4.1 Patentability of new uses of known substances or compounds (such as existing medicines)

Under patent law, upon expiry of the patent, anyone is free to exploit the invention for commercial profit. The TRIPs Agreement does not change this. Commercial strategies on the part of pharmaceutical patent holders have, however, been designed to deal with this. One such strategy is the so-called "evergreening" approach – the term itself is controversial and has no universal definition – in which the patent holder, who is faced with impending patent expiry, applies for a fresh patent on the invention on the basis of a variant of the drug, allegedly having no significant new or increased therapeutic benefit (compared to that of the existing drug).[84] Permitting a patent of such a variant would in effect extend the patent protection of the drug. Patentability requirements in an FTA can potentially facilitate the possibility of "evergreening", such as where a trade partner may have low or lax threshold standards of patentability in its patent law.[85] With many "blockbuster" prescription medicine patents reaching expiry in recent years, one of the strategies for rights holders is to apply existing medicines to new (second or third) uses, as this can be significantly cheaper than inventing a new medicine "from scratch".[86]

[83] Korea Patent Office, *Examination Guidelines for Biotechnology Inventions*, 1998. See also, generally, Jay Young-June Yang and Mon Sohn, Korean Examination Guidelines for Biotechnology Inventions, *JWIP*, Vol. 2, No. 5, (1999) 845–850. The more general examination guidelines issued in 2013 are available at www.kipo.go.kr/kpo/user.tdf?a=user.english.html.HtmlApp&c=92006&catmenu=ek03_06_01 (accessed 9 October 2014). On unpatentable inventions, see Chapter 6 (p. 307).

[84] See, e.g., *Medecins Sans Frontiéres*' criticisms at www.msfaccess.org/content/evergreening-drugs-attack-access-medicines (accessed 11 April 2015). The matter of whether a new drug merits patent protection where its improvement of therapeutic benefit is called into question was examined by the Indian Supreme Court recently, in *Novartis AG v Union of India*; see note in Table 3.1, note e below and the accompanying text.

[85] Chapter 4 provides examples of FTAs which explicitly require that the signatories recognise that new uses of inventions may be patentable subject matter.

[86] A.R., *Teaching Old Pills New Tricks*, The Economist, 13 February 2013; the author calls this "repurposing" of drugs. See also Ted Agres, *New Life for Old Drugs, Drug Discovery*

A question which yields diverse answers in different jurisdictions is whether a second (or subsequent), new medical use of a known substance (such as an existing approved medicine) may be the subject of a patent. The two ends of the spectrum in answering this question have been pointed out in comparing the *Andean Community Decision 486*, the common IP law of the Andean Community states, with the European approach after EPA Enlarged Board of Appeal Decision 2/08 discussed below.[87] In the case of the Andean Community Decision, such new use cannot be the subject of a patent.[88]

In the EU approach, such use may be patentable if that new use does not form part of the state of the art. At the same time, the US approach allows such use to be patented. Within ASEAN, some countries, such as Lao PDR and Thailand, take the approach of excluding such new uses from patentability (see Table 3.1).

New uses for known products are explicitly required to be patentable in certain FTAs.[89] Article 15.9(2) of the US–Morocco FTA, for example, expressly creates obligations beyond those required under TRIPs, with respect to ensuring patentability of new uses of known products:

> "2. Each Party shall make patents available for the following inventions:
>
> (a) plants, and
> (b) animals.
>
> In addition, the Parties confirm that *patents shall be available for any new uses or methods of using a known product, including new uses of a known product for the treatment of humans and animals.*"
>
> [Italic emphasis added]

The question of whether a new drug is a significant improvement over existing drugs (so as to merit a fresh patent and all its attendant rights) may also arise, apart from the patent context, in at least two other important but separate contexts in regulatory systems: in determining whether to fund or

*and Development*, 29 July 2011, at: www.dddmag.com/articles/2011/07/new-life-old-drugs (accessed 13 June 2014).

87 See trilateral study by the WTO, WIPO and WHO, *Promoting Access to Medical Technologies and Innovation* (2012), at pp. 128–130.

88 See text of the *Decision* at: www.comunidadandina.org/ingles/normativa/D486e.htm (accessed 1 May 2014). Art. 21 provides: "**Article 21**.- Products or processes already patented and included in the state of the art within the meaning of Article 16 of this Decision may not be the subject of new patents on the sole ground of having been put to a use different from that originally contemplated by the initial patent."

89 See, e.g., the US–Oman FTA (signed 2006, in force 2009) and US–Morocco FTA (signed 2004, in force 2006).

subsidize a new drug,[90] in determining whether to give marketing approval protection period extension.[91] In these two latter contexts authorities undertake an evaluation of the improvements and new attributes of the new drug in order to make their decisions. Given that time and cost are incurred in developing new drugs, state agencies may actively promote finding new uses (or re-purposing) of existing drugs.[92] Interestingly, the existing debate does not, however, appear to make any connection between the patent application evaluation of such new uses and these other evaluations. Table 3.1 summarizes the position on patentability of new uses of known products in various Asian countries.

#### 3.3.4.2 India

In 2013, the Supreme Court of India ruled on Section 3(d) of the Patents Act in the landmark decision, *Novartis AG v Union of India*, interpreting the section to impost a strict requirement of improvement in therapeutic efficacy before a new use could qualify as an "invention" under the Act.[93] Section 3(d) provides as follows:

[90] E.g., Singapore's Ministry of Health uses the Standard Drugs List (SDL) which is modelled after the WHO's Essential Drugs List, and one of the determining factors for whether a new drug is included in the SDL (for subsidisation to patients) is "whether the drug offers a major improvement in terms of efficacy and effectiveness, as compared to existing standard drugs"; see Ministry of Health, www.moh.gov.sg/content/moh_web/home/pressRoom/Parliamentary_QA/2011/Standard_Drug_List_SDL.html (accessed 11 April 2015). See also Keng Ho Pwee, Health Technology Assessment in Singapore, *International Journal of Technology Assessment in Health Care*, Vol. 25, Supplement (2009), 234–240, at pp. 236–237, and Van der Erf S, Lim JYF and Chow WL, *The Impact of Pharmacoeconomic Assessment of National Formulary Decisions: A Comparative Analysis of Three Asia-Pacific Countries and Three European Countries*, www.davejunia.com/chsr/files/Conf_The%20Impact%20of%20Pharmacoeconomic%20Assessment%20on%20National%20Formulary%20Decisions%20A%20Comparative%20Analysis%20of%20Three%20Asia%20Countries%20and%20Three%20European%20Countries.pdf (accessed 10 April 2015).

[91] The EU evaluates whether there is a "significant clinical benefit in comparison with existing therapies"; see Art. 10, Directive 2001/83/EC of the European Parliament and of the Council of 6 November 2001 on the Community code relating to medicinal products for human use, as amended by Directive 2004/27/EC of the European Parliament and of the Council of 31 March 2004.

[92] See, e.g., www.ncats.nih.gov/research/reengineering/rescue-repurpose/rescue-repurpose.html (accessed 11 April 2015). On ways of discovering drugs and their uses, see Chapters 1 and 2, John P. Griffin, John Posner and Geoffrey R. Barker, eds., *The Textbook of Pharmaceutical Medicines*, 7th ed. (Wiley-Blackwell, BMJ Books, 2013),.

[93] MANU/SC/0281/2013. See also, generally, Frederick M. Abbott, *Inside Views: The Judgment In* Novartis v. India*: What The Supreme Court Of India Said*, 4 April 2013, at IP Watch website: www.ip-watch.org/2013/04/04/the-judgment-in-novartis-v-india-what-the-supreme-court-of-india-said/ (accessed 2 April 2014). See also Gopakumar Nair, Andreya Fernandes and Karthika Nair, Landmark Pharma Patent Jurisprudence in India, *JIPR*, Vol. 19, (2014), 79–88.

Table 3.1. *Patentability of second/subsequent uses in ASEAN and selected non-ASEAN countries*

| ASEAN country | Patentability of new uses for known substances or products | Provision/remarks |
|---|---|---|
| Brunei Darussalam | Yes, in limited circumstances | Section 14(7) Patents Order 2011 |
| Cambodia | No<br>Note that Art. 4 read with Art.136 permits Cambodia to exclude pharmaceuticals from patentability until 1 January 2016 in accordance with the Doha Agreement on Trade and Public Health | |
| Indonesia | No explicit provision (but Swiss-type claims allowed?); see esp. Art. 3 Patent Law 14/2001 | Incremental improvements: Article 2, paras. 3–4 read with Article 7 of Indonesian Patent Law No. 14/2001[a] |
| Lao PDR | No | Art. 13 Law on Intellectual Property[b] |
| Malaysia | Yes | Section 14(4) Patents Act[c] |
| Myanmar | Intellectual property law being prepared[d] | |
| Philippines | No, unless enhanced efficacy is shown | S 22 of the Intellectual Property Code Republic Act No. 8293 inserted by the Universally Accessible Cheaper and Quality Medicines Act of 2008 and Implementing Rules[e] |
| Singapore | Yes | S 14(7) Patents Act[f] (see also text elaborating on the Singapore position, below) |
| Thailand | Appears not based on case law[g] | Sections 6–9 Patent Act B.E. 2522 (1979)[h] |
| Vietnam | No | Article 60, Intellectual Property Law[i]<br>Section 25.6 of Circular 01/2007/ TT-BKHCN, and Rule 23 of Guideline for examination of patent applications[j] |

Table 3.1. (*cont.*)

| ASEAN country | Patentability of new uses for known substances or products | Provision/remarks |
|---|---|---|
| **Non-ASEAN country** | | |
| China | No explicit provision in Patent Law of the People's Republic of China (as amended up to 2008)[k] | On incremental improvements, see Decision of Standing Committee of the NPC 2008 amending Art. 22 of the Patent Law of the PRC[l] |
| Hong Kong | No | |
| India | No, unless proven improvement in therapeutic efficacy | S. 3(d) Patents Act and *Novartis v Government of India*, Supreme Court decision (2013); and 2014 Patent Examination Guidelines para. 12[m] |
| Korea | | Art. 96(2) Patent Act[n] |

[a] Indonesia questionnaire, AIPPI, 2011, at www.aippi.org/download/commitees/217/GR217indonesia.pdf. The answer to para. 19 states, with regard to assessment of the inventive step, the relevance of "advantageous effects" of a claimed invention (improvements over prior art): "The advantageous effect must be real enough for the people skilled in the art after reading it as such advantageous technical effects must be capable of being applied industrially".

[b] See Art. 13, Law No. 01/NA, Law on Intellectual Property (as amended, 2011).

[c] Act 291 of 1983 as amended; see also para. 3.5 of the Examination Guidelines (note 70, above) which permits "Swiss-type" claims.

[d] See WTO Trade Policy Review, Myanmar, Report of the Secretariat, February 2014, available at:

[e] Implementing Rules and Regulations of the Republic Act No. 9502, otherwise known as the "Universally Accessible Cheaper and Quality Medicines Act of 2008". Article 26.2 of the Intellectual Property Code provides:

> In the case of drugs and medicines, there is no inventive step if the invention results from the mere discovery of a new form or new property of a known substance which does not result in the *enhancement of the known efficacy* of that substance, or the mere discovery of any new property or new use for a known substance, or the mere use of a known process unless such known process results in a new product that employs at least one new reactant.
>
> (Italic emphasis added.)

[f] The invitation to provide feedback and the related background information on this aspect of the Public Consultation Paper, comprising a letter and its Annex A, are found at pp. 32–34, www.ipos.gov.sg/Portals/0/IP%20legislation/LetterAnnexA.pdf (accessed 29 April 2014). This information was based on the EPO practice of allowing "Swiss-type claims" prior to its more recent decision in 2010 to exclude such claims.

Table 3.1. (*cont.*)

[g] See Srila Thongklang and Suebsiri Taweepon, "Patentabiity of 'New Use' Patents in Thailand", at www.dancham.or.th/images/upload/files/informed_counsel_vol3_no2%20_2.pdf (accessed 25 September 2015) and Mirandah, "Thailand Supreme Court Limits Patent Protection for New Use Claims", www.mirandah.com/pressroom/item/87-thailand-supreme-court-limits-patent-protection-for-new-use-claims (accessed 25 September 2015).

[h] Text available at WIPO, www.wipo.int/wipolex/en/details.jsp?id=3807 (accessed 10 May 2015).

[i] Law No. 50/2005/QH11 of November 29, 2005, on Intellectual Property.

[j] Text available at WIPO, www.wipo.int/wipolex/en/details.jsp?id=5013 (accessed 10 May 2015).

[k] Text available at: www.wipo.int/wipolex/en/details.jsp?id=5484 (accessed 1 May 2014).

[l] The amended para. 3 reads: "Creativity means that, compared with the existing technologies, the invention possesses *prominent substantive features* and indicates *remarkable advancements*, and the utility model possesses substantive features and indicates advancements". (Italic emphasis added.)

[m] See www.ipindia.nic.in/whats_new/biotech_Guidelines_25March2013.pdf (accessed 15 May 2014). Note that in August 2014, the Indian Patent Office issued new examination guidelines for pharmaceutical industry products; at the time of writing these were still in draft form – see www.ipindia.nic.in/iponew/draft_Pharma_Guidelines_12August2014.pdf (accessed 21 October 2014).

[n] Text of the Act available in English at the Korea Intellectual Property Office website: www.kipo.go.kr/upload/en/download/PatentAct.pdf (accessed 6 March 2014).

> **3. What are not inventions.** – The following are not inventions within the meaning of this Act,—
>
> . . .
>
> (d) the *mere discovery of a new form* of a known substance *which does not result in the enhancement of the known efficacy of that substance* or the mere discovery of any new property *or new use for a known substance* or of the mere use of a known process, machine or apparatus unless such known process results in a new product or employs at least one new reactant. [Italic emphasis added]

Its explanatory statement further provides:

> *Explanation.*—For the purposes of this clause, salts, esters, ethers, polymorphs, metabolites, pure form, particle size, isomers, mixtures of isomers, complexes, combinations and other derivatives of known

> substance shall be considered to be the same substance, *unless they differ significantly in properties with regard to efficacy.*
>
> [Italic emphasis added]

Importantly, in interpreting the meaning of "efficacy" in Section 3(d), the Indian Supreme Court held that:

> in the case of a medicine that claims to cure a disease, the test of efficacy *can only be* "therapeutic efficacy".[94]
>
> [Italic emphasis added]

The Supreme Court was, however, careful to expressly observe that the Indian patent legislation did not exclude patentability of incremental inventions.[95]

Soon after this decision, the USTR issued the following criticism:

> The United States is concerned that section 3(d), as interpreted, may have the effect of limiting the patentability of potentially beneficial innovations. Such innovations would include drugs with fewer side effects, decreased toxicity, improved delivery systems, or temperature or storage stability. In practice, this standard has already been applied to deny patent protections to potentially beneficial innovations, some of which enjoy patent protection in multiple other jurisdictions.[96]

In October 2014, revised pharmaceutical patent examination guidelines were issued by the Indian authorities. These guidelines cover a great deal of ground but two major observations are provided here. First, they state the claims filed as follows would not be considered as inventions:

1) Use of compounds in the treatment of ———————
2) A product or a substance (which is known) for the treatment of new disease (which is nothing but use/application claim)[97]

[94] *Novartis* judgment, *ibid.*, para. 180.

[95] *Novartis* judgment, note in Table 3.1, note e above, para. 191.

[96] USTR Special 301 Report 2014, p. 40. India is listed in the USTR's "Priority Watch List" for 2014. The Report also criticizes India's patent system for providing both a pre-grant and a post-grant opposition system (p. 40).

[97] Para. 6.1, *Guidelines for Examination of Patent Applications in the Field of Pharmaceuticals*, 29 October 2014. The accompanying comment in the Guidelines is:

> The above two categories of claims are not to be considered as inventions, since the claimed subject matter neither pertains to product or process. Further, an objection with regard to Section 3(i) and 3(d) would be invoked.

The Guidelines also refer to the case of *Fumapharm AG v The Controller of Patents and Designs*; OA/6/2009/PT/KOL, at para. 6.2, Guidelines, ibid. OA/6/2009/PT/KOL. Stakeholder comments on an earlier draft of the Guidelines are available at: http://ipindia.nic.in/iponew/comments_PharmaGuidelines/FeedBack_Pharmaceuticals.htm. See also generally, the

Secondly, in relation to second medical uses of known pharmaceuticals, the *Guidelines* incorporate the Supreme Court's approach in *Novartis*, with heavy emphasis on the test focused on *therapeutic* efficacy.[98] During the public consultations on an earlier draft of the *Guidelines*, concerns had been raised over this emphasis.[99] The Director of Patents has in two subsequent rulings applied Section 3(d) against two other foreign pharmaceutical companies to deny a patent application in one case, and to revoke an existing patent in the other.[100]

#### 3.3.4.3 Thailand

Under Thailand's most recent patent examination guidelines, it is provided that in examination of the inventive step for a claim of pharmaceutical invention, enhancement of known efficacy is a key factor that will be taken into consideration.[101] It further states that efficacy in this context means "non-obvious" efficacy or "other non-obvious enhancement in any properties". In assessing changes of properties, clear test results and a substantial increase in quality are to be taken into consideration. For pharmaceutical compositions and formulations of a known active compound, determination of novelty and inventive step involves consideration of whether the composition provides "surprising effects in solving technical solution in comparison to the prior art not necessarily therapeutic effects".

#### 3.3.4.4 Examples of pharmaceutical patent litigation in Thailand

With regard to new uses of known pharmaceutical products, it appears that case law has not examined this yet.[102] However, the access to

Indian *Guidelines for Examination of Biotechnology Applications for Patents*, 25 March 2013 www.ipindia.nic.in/whats_new/biotech_Guidelines_25March2013.pdf and *Guidelines for Processing of Patent Applications Relating to Traditional Knowledge and Biological Material.*

98 See paras. 10.4–10.9.

99 See, e.g., the comments of the US-India Business Council, at http://ipindia.nic.in/ipo new/comments_PharmaGuidelines/FeedBack_Pharmaceuticals.htm.

100 See Michael Mezher, *Bayer, Boehringer Latest to be Hit by Indian Patent Office Decisions*, 11 March 2015, Regulatory Affairs Professionals Society, at www.raps.org/Regulatory-Focus/News/2015/03/11/21696/Bayer-Boehringer-Latest-to-be-Hit-by-Indian-Patent-Office-Decisions/ (accessed 13 April 2015).

101 *Examination of Chemical and Pharmaceutical Patent for Invention and Petty Patent Applications*, 2013.

102 The Thai Supreme Court addressed the "new use" issue in a *non*-pharmaceutical context; see *the Chidlom Marine Services* litigation (2012). See www.tilleke.com/resources/patent ability-%E2%80%9Cnew-use%E2%80%9D-patents-thailand, and www.mondaq.com/404.asp?404; www.mondaq.com:80/x/178230/Patent/Patentability±of±New±Use ±Patents±in.

medicine issue has surfaced in Thai courts in the form of challenges to the scope of a pharmaceutical patent. In 2002, in a suit brought by AIDS Access Foundation, a Thai court ordered that the patent of Bristol-Myers Squibb (BMS) be amended to restore a dosage in its original claim, that is, 5–100 mg per unit of the antiretroviral drug known as didanosine (ddl).[103] (BMS had, some years after its patent, applied to remove this dosage so that its patent scope would have been expanded, and the Thai Department of Intellectual Property had permitted that removal.) This court decision allowed others to manufacture the drug in other dosages e.g. 125 or 200 mg dosages, that is, in effect reduced the scope of BMS' patent by restoring it to the pre-amendment, more limited formulation.

#### 3.3.4.5 China

Chinese patent law permits patenting of incremental improvements to existing inventions but this is subject to a high threshold requirement, namely, that "the invention possesses *prominent substantive features* and indicates *remarkable advancements*".[104]

#### 3.3.4.6 Singapore: patentability of second/subsequent medical uses of known substances

The Singapore Patents Act contains two key provisions relevant to patentability of second medical use of known substances or pharmaceuticals. These are Sections 16(2) and 14(7).

Sections 16(2)–(3) provide as follows, in relation to the requirement of industrial application:

> (2) An *invention of a method of treatment* of the human or animal body by surgery or therapy or of diagnosis practised on the human or animal body *shall not* be taken to be capable of industrial application.
>
> (3) Subsection (2) shall not prevent a product consisting of a substance or composition from being treated as capable of industrial application merely because it is invented for use in any such method.
>
> [Italic emphasis added]

[103] Khabir Ahmad, Thailand Court Forces Reversal of Drug Firm Antiretroviral Patent, *The Lancet*, Vol. 360, No. 9341 (2002), p. 1231.

[104] Note in Table 3.1, note b above.

Section 14(7), on the other hand, provides as follows:

> (7) In the case of an invention consisting of a substance or composition for use in a method of treatment of the human or animal body by surgery or therapy or of diagnosis practised on the human or animal body, the fact that the substance or composition forms part of the state of the art *shall not prevent the invention from being taken to be new* if the use of the substance or composition in any such method does not form part of the state of the art.
>
> [Italic emphasis added]

In 2009, the IPOS conducted a public consultation on possible amendments to the Patents Act, which included consideration of second or subsequent medical use claims. The bases of consultation on this point were twofold: first, Section 14(7) mirrored the equivalent UK patent provision, and secondly, at that time, the European Patent Office (EPO) had permitted what were known as "Swiss-type claims" for second medical uses framed in a particular manner in the patent claim. The claim would, for example, take the form of "Use of X *for the manufacture* of a medicament to treat Y", where X is the known medical substance and Y is the medical condition to be treated with X (italic emphasis added to show focus on manufacturing process in the "Swiss-type" claim). The claims were so named as the Swiss federal patent authorities had first accepted such patent claims in 1984. The IPOS was therefore interested to gather comments as to whether to follow this European practice then prevailing, follow the US approach (which permits as inventions treatment methods), the *status quo* (presumably not permitting "Swiss-type" claims under section 14(7)), or to completely disallow patenting for second or subsequent medical uses (to prevent "evergreening").

Meanwhile, in 2010, the EPO's Enlarged Board of Appeal (EBA) ruled against such claims, and the practice in the UK – a party to the European Patent Convention 2000 (EPC 2000) – has seen changed to reflect this.[105] "Swiss-type" claims are no longer accepted, and instead, a statutory, more direct claim has been put in place.[106]

[105] EPO's EBA Decision 2/08. For the current UK guidelines, see *Examination Guidelines for Patent Applications relating to Medical Inventions in the Intellectual Property Office*, May 2013, especially pp. 25–43.

[106] See section 4A(4), UK Patents Act, and *Examination Guidelines for Patent Applications relating to Medical Inventions in the Intellectual Property Office* (UK), May 2013, at paras. 92–107. Section 4A(4) of the UK Act states:

> "(4) In the case of an invention consisting of a substance or composition for a specific use in any such method, the fact that the substance or composition forms part of the state of the art shall not prevent the invention from

In its *Guide on Patentability Issues arising during Search & Examination* (2010) the IPOS indicated that the "Swiss-type" claim was acceptable under section 14(7).[107] In a Consultation Paper, the IPOS did recognize that the approach of leaving it to the courts to decide whether second or subsequent new medical uses could be patentable, could lead to uncertainty.[108]

Nonetheless, post-2010 amendments to the Singapore Patents Act have left Section 14(7) unchanged, which seems to indicate that the IPOS has preferred to retain the *status quo*, at least for the time being. Under the present IPOS Guidelines issued in February 2014, second medical uses of known substances or compounds are discussed and "Swiss-type" claims (and certain variants) remain acceptable.[109] Singapore practice therefore now differs from with EU and UK practice in this regard. Acceptance of such claims still appears to occur elsewhere in the Commonwealth as well such as in New Zealand.[110]

### 3.3.5 *Patent linkage*

While patentability addresses inventiveness, marketing approval addresses the safety and efficacy of medicines. Therefore, a patented product may not be granted marketing approval if, for instance, it is not safe; conversely, a product generally need not be patented to be granted marketing approval. The key to marketing approval is providing

> being taken to be new *if that specific use does not form part of the state of the art.*"
>
> (Italic emphasis added.)

The new formulation under the UK Guidelines no longer refers to manufacture, and takes the following form: "Substance X for the use in treatment of medical condition Y".

107 See pp. 35–36 of the 20 September 2010 Guide; text of the 2010 Guide is available at: (accessed 3 May 2014). Section 14(7) provides a departure from section 16(2); an *invention of a known substance or composition* for use in treatment otherwise excluded by section 16(2) *can* be treated as new (and therefore be patentable) if *its use* in any such method does not form part of the prior art.

108 Annex A, p. 31.

109 *Examination Guidelines for Patent Applications at IPOS*, 14 February 2014, pp. 234–235.

110 See New Zealand Intellectual Property Office website at: www.iponz.govt.nz/cms/patents/patent-topic-guidelines/current-practice-guidelines/5.-examination-of-patent-applications/5.2-guidelines-for-the-examination-of-swiss-type-claims (accessed 30 April 2014). Under the Australian patent law system, it appears that new uses of known substances are treated as being patentable: see section 2.9.2.17.1 of the Examiners Manual, at www.ipaustralia.gov.au/pdfs/patentsmanual/WebHelp/Patent_Examiners_Manual.htm (accessed 30 April 2014). Malaysia also permits "Swiss-type" claims; see note 86, above.

evidence of safety through testing of the product. In the case of medicines, this is carried out through clinical trials on human subjects. Should a patented product be proven safe it is allowed to be marketed. While this may be the general position (permitted by TRIPs) certain systems no longer operate this way. Through FTA provisions requiring "patent linkage" a party applying for approval of a medicine which is the subject of an existing patent must now face the patent holder and challenge his patent in order to advance his own marketing approval application. Essentially, such commitments make the marketing approval authorities a watchdog of sorts for the patent holders, as the authorities are obliged to trigger certain processes if a patent exists for the product (or a similar product). This kind of linkage greatly facilitates the work of pharmaceutical patent holders as the authorities serve as a notification node for obtaining alerts and information on potential competitors' products when such applications are filed.[111] Tables 3.2 and 3.3 summarize the position on patent linkage in ASEAN and a number of non-ASEAN countries.

Table 3.2. *ASEAN pharmaceutical approval systems*[a]

| ASEAN country | Whether approval is linked to existing patents | Relevant legislation |
|---|---|---|
| Brunei Darussalam | No | Medicines Order, 2007[b] |
| Cambodia | No | Kram (Law) on the Management of Pharmaceuticals (as amended in 2008)[c] |
| Indonesia | No | Agency: NA-DFC[d] |
| Lao PDR | No | Regulation Governing Drug Registration[e] |
| Malaysia | No | Control of Drugs and Cosmetics Regulations[f] |
| Myanmar | – | National Drug Law, 1992 |

[111] See generally, Benjamin P. Liu, *The Missing Link: US-China Patent Linkage Experience and Its Echo around the Pacific*, at http://citation.allacademic.com/meta/p_mla_apa_research_citation/5/5/9/9/5/p559959_index.html?phpsessid=med4688rl4en0khj9igska0617, and Sandeep K Rathod, Patent Linkage and Data Exclusivity: A Look at Some Developments in India, *Journal of Generic Medicines*, Vol. 8 (July 2011), 140–149.

Table 3.2. (*cont.*)

| ASEAN country | Whether approval is linked to existing patents | Relevant legislation |
|---|---|---|
| Philippines | No | Food, Drug and Cosmetics Act, and Administrative Order No. 2005–0001, Revised Policies and Guidelines Governing Patent and Trade Secret Rights in relation to the Registration of Pharmaceutical Products[g] |
| Singapore | Yes | Medicines Act |
| Thailand | No | Drug Act[h] |
| Vietnam | No | Pharmacy Law 2005[i] |

[a] For a fuller comparative study of pharmaceutical regulatory differences in Asia, see generally, *Analysis Report: Identification and Clarification of the Differences in Regulatory Requirements between Asian Economies*, Asia Partnership Conference of Pharmaceutical Associations (APAC), Regulations and Approvals Expert Working Group, 12 April 2013 Tokyo, Japan Member, www.jpma.or.jp/english/apac/pdf/130722_05.pdf (accessed 5 February 2014).

[b] Text available at WHO website, http://apps.who.int/medicinedocs/documents/s18037en/s18037en.pdf.

[c] June 17, 1996.

[d] See www.pom.go.id/index.php/home/en.

[e] 1441/MOH, 2003, www.laotradeportal.gov.la/index.php?r=site/display&id=41.

[f] See www.pharmacy.gov.my/v2/en/documents/sale-drugs-act-1952-and-regulations.html.

[g] See respectively, www.fda.gov.ph/industry-corner/downloadables/217-requirements-for-drug-registration/95567-actd-atr-requirements and www.wipo.int/wipolex/en/text.jsp?file_id=313013 (accessed 4 October 2014). The 2005 Administrative Order aims to de-link the work of the patent rights authority (IPO) and of the authority in charge of approvals of pharmaceutical applications (BFAD). It also provides that where the IPO or a competent Court finds that an applicant "decide, with finality that the applicant has no intellectual property right involving, or attached to, the pharmaceutical product" then the certificate of marketing approval would be "automatically deemed cancelled and/or revoked".

[h] B.E. 2510 (1967) http://thailaws.com/law/t_laws/tlaw0071_1.pdf.

Table 3.2. (*cont.*)

[i] Chapter III, No.34/2005/QH11 passed by the National Assembly on 14 June 2005. For Vietnam's IP laws, see generally, www.noip.gov.vn/web/noip/home/en?proxyUrl=/noip/cms_en.nsf/%28agntDisplayContent%29?OpenAgent&UNID=A3257F48CA99547A4725773100292BFB; but see Article 14, Ministry of Health Circular, Circular on Registration of Drugs, 22/2009/TT-BYT, which permits the Ministry to refuse registration of a drug if an IP-related dispute arises during an application. However the responsibility is not on the Ministry to enquire on patent status when receiving a drug application; on the contrary, the provision places the responsibility on "drug registration establishments" (see Articles 4 and 14(1)). I am grateful to Professor Tu T. Nguyen for drawing my attention to this Circular.

Table 3.3. *Pharmaceutical approval systems in selected Asian countries (non-ASEAN)*

| Country | Whether approval is linked to existing patents | Relevant legislation |
|---|---|---|
| China | Yes[a] | Drug Administration Law of the People's Republic of China[b] and implementing Regulations Trial Measure 2002, which in 2005 became the Measure for the Administration of Drug Registration[c] |
| Hong Kong* | | Pharmacy and Poisons Ordinance[d] |
| Macau | | |
| India | No | Pharmacy Act 1948, Drug and Cosmetics Act 1940 |
| Japan | No | Pharmaceutical Affairs Law[e] |
| Korea | Yes | Pharmaceutical Affairs Act |

[a] Art. 18, Provisions for Drug Registration, SFDA Order No. 28.

[b] Order of the President of the People's Republic of China, Order No. 45, 2001. See also Provisions for Drug Registration, *ibid.* See Benjamin P. Liu, Fighting Poison with Poison? The Chinese Experience with Pharmaceutical Patent Linkage, *J. Marshall Rev. Intell. Prop. L.*, Vol. 11 (2012), 623.

[c] Liu, *ibid.*, pp. 630.

[d] Cap. 138.

[e] See www.jpma.or.jp/english/parj/pdf/2012.pdf or See apps.who.int/medicine docs/documents/s18577en/s18577en.pdf.

#### 3.3.5.1 Judicial decisions of interest on patent linkage and infringement actions against generic competitors

In Malaysia, it does not appear that the courts have examined patent linkage issues specifically. However, in the context of an application for an injunction against infringement of a patent by a generic manufacturer, and in considering the issue of adequacy of granting damages in lieu of such an injunction, there has been a general recognition that a patent holder is given a period of exclusivity to allow an innovator to recoup his investment in R&D,[112] and an injunction may be granted to prevent "premature erosion of pricing" by a generic competitor by a party that has "incurred virtually no up-front investment relating to research and development".[113]

Singapore introduced a statutory system of patent linkage following its signing of the USSFTA, and the implementing provisions are in section 12A of the Medicines Act and related regulations.[114] The Health Sciences Authority (HSA) is the regulator having oversight of approval and marketing of medicines in Singapore, and administers the section 12A system. Although quite new, the provisions have already been the subject of a small number of cases.

In what appears to be the first case under Section 12A, *AstraZeneca AB (SE) v Sanofi-Aventis Singapore Pte Ltd*,[115] the patent linkage mechanism was triggered by the marketing approval applications by the defendant in the action, Sanofi-Aventis, for its *Rosucard* Film-coated tablets in 10 mg, 20 mg and 40 mg dosage forms. It duly notified the plaintiff, AstraZeneca AB, under section 12A(3)(a), of its applications. Subsequently, the plaintiff commenced the action in question seeking a declaration of infringement by the defendant. Under the section 12A system, this action triggered a 30-day moratorium on processing of the defendant's applications. The plaintiff requested further and better particulars (FBP) from the defendant, which were furnished. The present case involved a countering request by the defendants for FBP from the plaintiff on, inter alia, how the defendant would infringe its patent. The plaintiff resisted this request. In its judgment the court applied principles used in relation to considering further and better particular requests in general (i.e. non-section 12A-related) patent infringement lawsuits.

[112] See para. 24, *Aventis Farma Sa (M) Sdn Bhd & Anor v Rohibul Sabri bin Abas @ Megat (t/a as Dabur Enterprise) & Anor* [2008] *MLJ*, Vol. 3, p. 451.

[113] *Ibid.*

[114] See Chapter 4 for an explanation of the section 12A system of patent linkage.

[115] [2013] SGHCR 7.

Upon examining the English position, the Singapore court rightly observed that there were "competing considerations" underlying the tension between what it considered the "traditional" position (leaning against the grant of FBP) and the "modern position" (leaning in favour of granting FBP).[116] In the event, the court found the "modern" approach to be not only pertinent but "highly persuasive" in this case, with the result that it ordered the plaintiff to furnish FBP. While this result is unremarkable, certain observations in the process of reasoning are instructive.

First, the court pointed to the fact that in such cases, infringement might not be readily observable without costly experiments, so that "an absence of particularization at an early stage may result in the wasted costs of misdirected and inadequate experiments" (presumably by the defendants in this context). Secondly, the court was mindful that the present request for FBP had arisen in the context of section 12A, under which all that was required to trigger a 30-day moratorium on the defendants' applications was an action for prospective infringement.[117] Thirdly, the mechanism of Section 12A is based on prospective infringement, and in the present case, the plaintiff's own request to the defendants of the FBP had been sought "to know how the Defendant could be infringing the Plaintiff's Patent".[118] This feature of the system clearly

[116] See paras. 43–48 of the judgment. The "traditional" view was represented by the views such as those in *Lux Traffic Controls Limited v The Staffordshire Public Works Company* [1991] 4RPC 73, while the "modern" view was represented by views such as those in *Novartis AG, Ciba Vision AB v Johnson & Johnson Medical Limited* [2008] EWHC 293 (Pat).

[117] The court further observed:

> Although the precise scheme for patent actions differs from that regarding actions for breaches of confidence, the same concerns and considerations should apply. Indeed, such concerns are recognised, in part, in the form of remedies against groundless threats of patent infringement. In my view, these concerns apply *a fortiori* to actions taken out under the Medicines Act, because such actions trigger a stay on the processing of product licences for medicinal products for a period of 30 months from the date of the commencement of the action, based simply on *prospective* infringement .... The availability of the moratorium, whilst possibly intended to encourage settlement of patent infringement claims before the allegedly infringing products enter the market, can be used to delay the manufacture, importation and sale of medicinal products by competitors of the patentee, and may further have the effect of hindering public access to the competitor's products. Accordingly, the modern position as embodied in *Novartis* is, for the reasons alluded to throughout this judgment ..., patently applicable to cases such as the present.

Para. 48 of the judgment.

[118] See para. 9 of the judgment.

illustrates how it aids patent holders both in being alerted to and seeking information on potential infringements of their patents. This is what makes it an attractive tool to request of other countries in FTA negotiations, as it will create a valuable, previously absent system of alerts and information-gathering to facilitate patent infringement actions.

In a related earlier action, the court had considered the relationship between patent infringement actions under the Patents Act and those initiated pursuant to Section 12A.[119] The court concluded that Section 12A established a separate and independent cause of action.[120]

In *Astrazeneca AB v Ranbaxy (Malaysia) Sdn Bhd*,[121] the plaintiff, AstraZeneca AB, brought an action for infringement under the Patents Act against the defendant, who had applied for marketing approval for its products. In so doing the plaintiff also sought certain particulars from the defendant, who resisted the request on the ground that it had furnished the plaintiff with a Section 12A notice with certain information. The court found that such a notice was not sufficient to meet the requirements under the Patents Act (Section 78, in particular) requiring the furnishing of information.[122]

### 3.3.6 *Data exclusivity*

As regulatory authorities require evidence of safety and efficacy of medicines in order to grant marketing approvals, pharmaceutical companies need to conduct several clinical tests for their products. The testing, as mentioned, comprises a series of steps that can be very costly. Hence the testing data is seen to be of commercial value to these companies who invest in them. As a result there is a demand for legal protection of such valuable data. In the US, these were treated as "trade secrets" and were protected as such in early FTAs such as the North American Free Trade Agreement (NAFTA).[123]

Art. 10*bis* of the Paris Convention, which preceded the TRIPs Agreement, requires protection against "unfair competition"[124]; it does

[119] *AstraZeneca AB (SE) v Sanofi-Aventis Singapore Pte Ltd* [2012] SGHC 16.
[120] Paras. 32 and 35 of the judgment. [121] [2012] SGHC 7.
[122] See paras. 23–27 of the judgment, *ibid.*
[123] Text available at USTR, https://ustr.gov/trade-agreements/free-trade-agreements/north-american-free-trade-agreement-nafta (accessed 8 May 2015).
[124] Article 10(2) provides:

> (2) Any act of competition contrary to honest practices in industrial or commercial matters constitutes an act of unfair competition.

not specifically address pharmaceutical test data. In the decade prior to establishment of the WTO, the US and EU had each enacted legislation which protected such test data.[125] The TRIPs Agreement brought about multilateral provisions on "undisclosed information".[126] Article 39 requires protection of such information (as described in it), but importantly, *permits disclosure* "where necessary to protect the public" *or* where steps have been taken to protect such data against "unfair commercial use".[127] The use of the disjunctive "or" is significant, as TRIPs envisages that either situation provides enough justification for disclosure. Since

See Daniel Gervais, *The TRIPS Agreement: Drafting History and Analysis*, 4th ed. (Sweet & Maxwell/Thomson Reuters, 2012), at p. 542.

[125] See G. Lee Skillington and Eric M. Solovy, The Protection of Test and Other Data Required by Art. 39.3 of the TRIPS Agreement, *Nw. J. Int'l L. & Bus.*, Vol. 24 (2003–2004), 1, esp. at pp. 8–12.

[126] See earlier discussion in Chapter 2.

[127] Article 39 of TRIPS provides as follows:

> **"SECTION 7: PROTECTION OF UNDISCLOSED INFORMATION**
> **Article 39**
>
> 1. In the course of ensuring effective protection against unfair competition as provided in Article 10*bis* of the Paris Convention (1967), Members shall protect undisclosed information in accordance with paragraph 2 and data submitted to governments or governmental agencies in accordance with paragraph 3.
> 2. Natural and legal persons shall have the possibility of preventing information lawfully within their control from being disclosed to, acquired by, or used by others without their consent in a manner contrary to honest commercial practices 10 so long as such information:
>
> (a) is secret in the sense that it is not, as a body or in the precise configuration and assembly of its components, generally known among or readily accessible to persons within the circles that normally deal with the kind of information in question;
> (b) has commercial value because it is secret; and
> (c) has been subject to reasonable steps under the circumstances, by the person lawfully in control of the information, to keep it secret.
>
> 3. Members, when requiring, as a condition of approving the marketing of pharmaceutical or of agricultural chemical products which utilize new chemical entities, the submission of undisclosed test or other data, the origination of which involves a considerable effort, shall protect such data against unfair commercial use. In addition, *Members shall protect such data against disclosure, except where necessary to protect the public, or unless steps are taken to ensure that the data are protected against unfair commercial use.*"
>
> (italic emphasis added)

TRIPs, several bilateral FTAs have required not only that such data be protected, but in terms stronger than those of this provision in TRIPs, as will be seen below.

Four main observations may be made of Article 39(3). First, the protection required under it is to protect against "unfair commercial use". This is not defined, which leaves WTO members with a fair amount of leeway to apply such protection. Views differ on the interpretation of this phrase.[128]

Secondly, no period of protection is established under the provision. Thirdly, the protection is for "undisclosed test or other data, the origination of which involves a considerable effort". This provides conditions for the kind of data subject to such protection, ensuring that a requisite degree of effort – which may involve innovative, intellectual, financial or research effort – has been involved in its generation. As will be seen below, a number of Asian countries have begun to include the requirement of "considerable effort" in their data protection laws. Fourth, Art. 39(3) permits two exceptions: "except where necessary to protect the public, or unless steps are taken to ensure that the data are protected against unfair commercial use".[129]

These observations show that the data protection obligation is neither an absolute one, nor is it to be applied only according to demands of data owners. On the contrary, the language and intent provide WTO members flexibility in implementation within their national systems. Moreover, if one reads Article 39(3) in light of Article 7, the protection under Article 39(3) "should contribute to the promotion of technological innovation and to the transfer and dissemination of technology, to the *mutual advantage of producers and users* of technological knowledge and *in a manner conducive to social and economic welfare*, and *to a balance of rights and obligations*" (italic emphasis added).

Notably, Article 39(3) provides neither for "exclusivity" nor for a fixed (or any) duration of protection. Indeed, the negotiating history of the provision indicates that there was at least one proposal that the

[128] See the excellent, detailed discussion in Nuno Pires de Carvalho, *The TRIPS Regime of Antitrust and Undisclosed Information*, (Kluwer Law International, 2008), Section 7; Gervais, note 111 above, at pp. 541–546; and Carlos Correa, *Protection of Data Submitted for the Registration of Pharmaceuticals: Implementing the Standards of the Trips Agreement* (South Centre, 2002), Part VIII, pp. 24–33, available at http://apps.who.int/medicinedocs/en/d/Jh3009ae/12.html#Jh3009ae.12 (accessed 12 September 2014).

[129] See also Chapter 8. Regarding the necessity of disclosure, Gervais, note 111 above, opines that it should be read consistently with Art. XX(b) of GATT 1994 – see p. 546.

protection should, if named, be for a "reasonable period".[130] Another was for the period to bear some relation to the degree of *efforts* in origination of the data.

As has been explained elsewhere, data exclusivity protection has a number of attractive features.[131] Certain TRIPs-plus FTA provisions have not only surpassed the requirements of Art. 39(3), but have also provided for minimum data protection periods (e.g. 5 years, and if certain US lobbyists for the TPP Agreement had their way, 12 years for biologic medicines) which is not necessarily always reflective of the degree of effort expended by data-producers. Such FTA provisions are also not necessarily accompanied by balancing provisions, and may not give due regard to wider "social and economic welfare" in signatory states.

The WTO dispute settlement system has not had occasion to interpret the phrase "unfair commercial use".

#### 3.3.6.1 Data exclusivity and Asia

The following discussion examines the legal regime in selected Asian countries which relate to protection of pharmaceutical data protection. Tables 3.4 and 3.5 provides a summary of the relevant laws and regulations, and a number of specific provisions from among these are selected for more detailed discussion below.

It can be seen from the above that not every country which provides protection includes the requirement in TRIPs Art. 39 that protected data should have been obtained through "considerable effort". The argument has been made that the protection of such data does not in fact protect the innovation incentive, or does so indirectly, but rather, that it protects the investment made by the innovator in obtaining the data.[132]

#### 3.3.6.2 Data exclusivity in Malaysia since 2011

Under Malaysia's data exclusivity system, data exclusivity is governed by the Directive on Data Exclusivity of 28 February 2011. Data exclusivity is

[130] See discussion of negotiating history, Correa, note in Table 3.2, note d above.

[131] See Charles Clift, *Data Protection and Data Exclusivity in Pharmaceuticals and Agrochemicals*, Chap. 4.9, *Intellectual Property Management in Health and Agricultural Innovation: A Handbook of Best Practices*, p. 433; www.iphandbook.org/handbook/orders/.

[132] See, e.g., Christian A. Fackelman, *Clinical Data, Data Exclusivity and Private Investment Protection, in Pharmaceutical Innovation, Competition and Patent Law*, ed. Josef Drexel and Nari Lee, (Edward Elgar Publishing, 2013), and Correa, note in Table 3.2, note d above, at p. 18.

Table 3.4. *ASEAN pharmaceutical laws and data protection*[a]

| ASEAN country | Whether undisclosed data for pharmaceuticals is protected by legislation | Duration/other remarks | Relevant legislation |
|---|---|---|---|
| Brunei Darussalam | No legislation but common law protection exists for trade secrets | – | Patents Order, 2011 and Medicines Order, 2007[b] |
| Cambodia | Limited to protection in relation to producing evidence in court | – | Art. 128, Law on the Patents, Utility Model Certificates and Industrial Design 2002[c]<br>Note: A new Law on the Protection of Undisclosed Information and Trade Secret is under consideration.[d] See also WTO Working Party Report.[e] |
| Indonesia | No express provision for data exclusivity for pharmaceuticals data although a general trade secrets law exists.[f] | – | Law No. 30 of December 20, 2000 regarding Trade Secret[g]<br>Agency: NA-DFC[h] |
| Lao PDR | Yes | 5 years, for undisclosed test or data the "origination of which involves a considerable effort" | Art. 61 (revised), Law of Intellectual Property 2011[i]<br>See also Law on Drugs and Medical Products (2011)[j] |

Table 3.4. (*cont.*)

| ASEAN country | Whether undisclosed data for pharmaceuticals is protected by legislation | Duration/other remarks | Relevant legislation |
|---|---|---|---|
| Malaysia | Yes | 5 years for a drug containing a New Chemical Entity and 3 years for second indication of a registered drug product; protection for undisclosed data the "origination of which involves a considerable effort" subject to application and grant procedure, appeal available (see main text below for more details of this system) | Directive on Data Exclusivity[k] |
| Myanmar | Laws under construction | – | – |
| Philippines | Subject to s. 72.4 of the Cheaper Medicines and Rule 9, Implementing Rules[l] | Undisclosed information submitted by patent holder subject to protection if origination of the data involved "considerable efforts"[m] | Food, Drug and Cosmetics Act[n] |
| Singapore | Yes | 5 years No reference to "considerable effort" | Medicines Act[o] ("protected period" under s.19A(2)) |
| Thailand | No express protection for data exclusivity for pharmaceuticals data, although general protection for "trade secrets" exists[p] | – | Trade Secrets Act 2002[q] |

| | | | |
|---|---|---|---|
| Vietnam | Yes | 5 years, where information was obtained by "investment of considerable efforts" | Art. 84, Law of Intellectual Property[r]<br>Health Minister's Decision No. 30/2006/QD-BYT of September 30, 2006, Promulgating the Regulation on Data Security of Drug Registration Records<br>and<br>Regulation on Data Security of Drug Registration Records[s]<br>Ministry of Health Circular, *Guiding the Keeping of Confidentiality of Trial Data in Medicine Registration* and Art. 128 of the Intellectual Property Law of 2005[t]<br>Ministry of Health Circular on Registration of Drugs, Circular No. 22/2009/TT-BYT dated 24 November 2009 (Article 16) |

*Source:* Compiled by author, as of September 2014.

[a] For an overview of data exclusivity laws around the world as at 2004, see *A Review of Existing Data Exclusivity Legislation in Selected Countries* (2004) at www.who.int/intellectualproperty/topics/ip/en/Data.exclusivity.review.doc.

[b] S79/2007; text available at WHO website, http://apps.who.int/medicinedocs/documents/s18037en/s18037en.pdf. See also See WTO Trade Review Policy, Brunei Darussalam, Report of the Secretariat, February 2015, WT/TPR/S/309, at p. 52; the report is available at: www.wto.org/english/tratop_e/tpr_e/tp409_e.htm (accessed 6 April 2015).

[c] Royal Decree NS/RKM/0103/005, (unofficial) English text available at WIPO, www.wipo.int/edocs/lexdocs/laws/en/kh/kh002en.pdf (accessed 8 May 2015).

[d] See www.cambodiainvestment.gov.kh/investors-information/intellectual-property-rights.html (accessed 20 July 2014).

[e] Para. 191, Accession Working Party Report, note in Table 3.5, note i.

[f] See European Commission Market Access Database, at: http://madb.europa.eu/madb/barriers_details.htm?barrier_id=095258&version=2; but see Article 14(3) of Health Ministry Regulation of R.I. no. 10101 MENKES/PER/XI/2008 provides that a "registration document is a confidential document which can only be used for evaluation by the relevant officer." (source: http://bricwallblog.wordpress.com/tag/data-exclusivity/). See also generally, UNCTAD, *Development Dimensions of Intellectual Property in Indonesia: Access to Medicines, Transfer of Technology and Competition*, pp. 12 (2011), and Articles 23 and 29, Law of the Republic of Indonesia No. 5 of 1999 Concerning the Ban on Monopolistic Practices and Unfair Business Competition.

[g] Text available at: www.wipo.int/wipolex/en/text.jsp?file_id=182062 (accessed 3 April 2014).

[h] See www.pom.go.id/index.php/home/en.

[i] Law No. 01/NA. Text in English available at: www.wipo.int/wipolex/en/text.jsp?file_id=310926. See also para. 210–211, WTO Accession Working Party Report, note in Table 3.5, note i.

[j] Law No. O7/NA; Art. 29 provides:

> The State protects and keeps the secret information of those persons who have duly registered their drugs and medical products in accordance with this Law and the Law on Intellectual Property, Agreements and Treaties of which the Lao PDR is a Party.

[k] Ministry of Health, entry into force on 1 March 2011: see http://portal.bpfk.gov.my/index.cfm?&menuid=105&parentid=82. The Directive is administered by the Director of Pharmaceutical Services.

[l] Art. 72.4 provides:

> 72.4. In the case of drugs and medicines, where the act includes testing, using, making or selling the invention including any data related thereto, solely for purposes reasonably related to the development and submission of information and issuance of approvals by government regulatory agencies required under any law of the Philippines or of another country that regulates the manufacture, construction, use or sale of any product: *Provided*, That, in order to protect the data submitted by the original patent holder from unfair commercial use provided m Article 39.3 of the Agreement on Trade-Related Aspects of Intellectual Property Rights (TRIPS Agreement), the Intellectual Property Office, in consultation with the appropriate government agencies, shall issue the appropriate rules and regulations necessary therein not later than one hundred twenty (120) days after the enactment of this law.

Rule 9(iv), Implementing Rules and Regulations of the Republic Act No. 9502, otherwise known as the "Universally Accessible Cheaper and Quality Medicines Act of 2008", mirrors the language of TRIPS Art. 39.3, and provides for data protection against "unfair commercial use".

[m] Rule 9, (iv.a) and (iv.b) of the Implementing Rules.

[n] See www.fda.gov.ph/industry-corner/downloadables/217-requirements-for-drug-registration/95567-actd-atr-requirements.

[o] See sections 19A and 19D of the Act. These provisions, introduced after Singapore entered into the USSFTA, are discussed in greater detail in the next chapter. Section 19A refers to "confidential supporting information" which may be protected, explained as follows:

> "confidential information" includes —
>
> (a) trade secrets; and
> (b) information that has commercial value that would be, or would be likely to be, diminished by disclosure;
>
> "confidential supporting information" means confidential information given —
>
> (a) in, or in relation to, an innovative medicinal product application; and
> (b) about the medicinal product that is or was, as the case may be, the subject of that application.

[p] The explanation in para. 165, Thailand, Trade Policy Review, WTO Secretariat Report, 8 February 2012, WT/TPR/S/255/Rev.1, is as follows:

> "Under the Trade Secrets Act B.E. 2545 (2002), undisclosed information is only protected if it is not generally known or readily accessible within the business, if its secrecy makes it valuable, and if it is in the control of the information owners who have taken steps to keep it secret. Generally, the Government has to keep secret any confidential information it receives subject to limited exceptions. The protection lasts as long as the information remains confidential. *Test data for pharmaceutical and agricultural chemicals products qualify, but Thai law does not give data 'exclusivity'; the law does not prevent the data from being used for the approval of generic versions.*"
>
> (Italic emphases added.)

[q] B.E. 2545 (2002); text in English available at: www.wipo.int/wipolex/en/text.jsp?file_id=129785. The Act defines "trade secrets" as follows:

> "*Trade Secrets*" means trade information not yet publicly known or not yet accessible by persons who are normally connected with the information. The commercial values of which derive from its secrecy and that the controller of the trade secrets has taken appropriate measures to maintain the secrecy.

Notified to the WTO TRIPS Council, IP/N/1/THA/U/1, 28 March 2003.

[r] See WTO Trade Policy Review, Vietnam, Report of the Secretariat, WT/TPR/S/287/Rev.1, at para. 3.4.10.2.2.7.

[s] Notified to the WTO under Art. 63.2 of the TRIPS Agreement, 12 February 2008, under document: IP/N/1/VNM/U/1.

[t] Circular no. 05/2010/TT-BYT, 1 March 2010, text available at: www.lawfirm.vn/?a=doc&id=695. By Art. 17.2 of this Circular, its provisions replace the Minister of Health's Decision No. 30/2006/QD-BYT of 30 September, 2006, promulgating the *Regulation on Keeping Confidentiality of Data in Medicine Registration Dossiers*. The 2005 Law is available at: www.noip.gov.vn/noip/resource.nsf/vwSelectImageResourceUrl/6A7273451828D79C472576720020482D/$FILE/Intelletual%20Property%20Law%20of%20Viet%20Nam%202005.pdf. See also paras. 437–438, Vietnam, WTO Accession Working Party Report, WT/ACC/VNM/48.

Table 3.5. *Data exclusivity protection in selected Asian (non-ASEAN) countries*

| Country | Whether undisclosed data for pharmaceuticals is protected by legislation | Duration/other remarks | Relevant legislation |
| --- | --- | --- | --- |
| China | Yes[a] | 6 years | Drug Administration Law of the People's Republic of China and implementing Regulations[b]<br>Trial Measure 2002, which in 2005 became the Measure for the Administration of Drug Registration[c]<br>See also China's WTO Accession Working Party Report[d] |
| Hong Kong | No | | Pharmacy and Poisons Ordinance[e] |
| India | No | | Pharmacy Act 1948, Drug and Cosmetics Act 1940 |
| Japan | Not expressly but during the re-examination process effective data protection is available[f] | | Pharmaceutical Affairs Law[g] |
| Korea | Yes | 5 years (post-KORUS) | Pharmaceutical Affairs Act[h] |
| Taiwan | Yes | 5 years from issue of license for a "new drug of new molecular entity" | Pharmaceutical Affairs Act[i] |

[a] Art. 18, Provisions for Drug Registration, SFDA Order No. 28.

[b] Order of the President of the People's Republic of China, Order No. 45, 2001, http://eng.sfda.gov.cn/WS03/CL0755/ and Articles 35 and 72, Regulations for Implementation of the Drug Administration Law of the People's Republic of China (Decree of the State Council No. 360). See generally, Benjamin P. Liu, *Fighting Poison with Poison? The Chinese Experience with Pharmaceutical Patent Linkage, J. Marshall Rev. Intell. Prop. L.*, Vol. 11 (2012), 623.

[c] Liu, *ibid.*, pp. 630.

[d] Para. 284, WTO/ACC/CHN/49.

[e] Cap. 138. See also the *Position Paper of the Hong Kong Association of the Pharmaceutical Industry on the Review of the Patent System in Hong Kong*, at: www.hkapi.hk/images/newsletter/Patent%20System%20in%20Hong%20Kong_30012012_161233.pdf (accessed 21 July 2014) and *Hong Kong Pharmaceutical Patents: Breaking the Ranks*? October 2004, AsiaLaw.

[f] See www.lexology.com/library/detail.aspx?g=0a295b70-e577-461b-99e5-48e0eeeac512.

[g] Art. 40 of the Act and Rule 14 of the Implementing Rules of the Pharmaceutical Affairs Act. See also generally *Pharmaceutical Administration and Regulations in Japan* (2012), Japan Pharmaceutical Manufacturers Association: www.jpma.or.jp/english/parj/pdf/2012.pdf (accessed 31 March 2014). www.aippi.org/download/helsinki13/presentations/Pres_Pharma_4_YOkumura_300813.pdf.

[h] Art. 88 of the Act. See also IFPMA 2011, at p. 76, www.ifpma.org/fileadmin/content/Publication/IFPMA_2011_Data_Exclusivity__En_Web.pdf.

[i] Art. 40–2 of the Act; see www.fda.gov.tw/EN/law.aspx (accessed 31 March 2014).

subjected to a formal application process and must be applied for from the Director of Pharmaceutical Services. The Directive governs such applications and the grounds upon which they may be granted or refused. The Directive applies both to new drugs containing new chemical entities (NCE) and to second indications of a registered drug product.

If an application is successful, the grant will be notified in a Register of Data Exclusivity maintained and published by the government.[133] As of February 2014, six notifications have been published.

The prerequisites for a grant of application for a NCE under the Directive include:

- Submission of undisclosed, unpublished and non-public domain pharmaceutical test data that has been obtained with "considerable effort";
- Application is made within 18 months from the date the product is first registered or granted marketing authorization; and granted data exclusivity / test data protection in the country of origin or in any country, recognized and deemed appropriate by the Director of Pharmaceutical Services.

The grant of data exclusivity to an applicant does not *a priori* prevent granting of registration for other drug products with the same active moiety. This is provided if either:

- The later applicant provides undisclosed, unpublished and non-public domain pharmaceutical test data, the origin of which requires considerable effort to demonstrate the quality, safety and efficacy of the drug product submitted for registration (i.e. does not rely on the earlier registrant's test data but produces its own), or
- The later applicant has obtained consent in writing for right of reference or use of the test data from a person authorized by the earlier registered NCE drug product.

Malaysia's Directive also expressly subjects its data exclusivity protection to the Doha TRIPs and Public Health Declaration and to public health.[134]

[133] See Malaysia's National Pharmaceutical Control Bureau website at: http://portal.bpfk.gov.my/index.cfm?&menuid=105&parentid=82 (accessed 29 April 2014).

[134] Para. 5 of the Directive states: "NON-APPLICATION OF DATA EXCLUSIVITY

5. Nothing in the Data Exclusivity shall:

(i) apply to situations where compulsory licenses have been issued or the implementation of any other measures consistent with the need to protect public health and ensure access to medicines for all; or

### 3.3.6.3 The Philippines

The Philippines Universally Accessible Cheaper and Quality Medicines Act of 2008 provides for protection of undisclosed data through amendments to Section 72 of the Intellectual Property Code, in the following terms:

> SEC. 7. Section 72 of Republic Act No. 8293, otherwise known as the Intellectual Property Code of the Philippines, is hereby amended to read as follows:
>
> ...
>
> "SEC. 72. *Limitations of Patent Rights.* -The owner of a patent has no right to prevent third parties from performing, without his authorization, the acts referred to m Section 71 hereof in the following circumstances: 72.4. In the case of drugs and medicines, where the act includes testing, using, making or selling the invention including any data related thereto, solely for purposes reasonably related to the development and submission of information and issuance of approvals by government regulatory agencies required under any law of the Philippines or of another country that regulates the manufacture, construction, use or sale of any product: *Provided*, That, in order to protect the data submitted by the original patent holder from unfair commercial use provided m Article 39.3 of the Agreement on Trade-Related Aspects of Intellectual Property Rights (TRIPs Agreement), the Intellectual Property Office, in consultation with the appropriate government agencies, shall issue the appropriate rules and regulations necessary therein not later than one hundred twenty (120) days after the enactment of this law. . . . "

Rule 9, (iv.a) and (iv.b) of the Implementing Rules provide as follows:

> (iv.a) **Use of Data by BFAD**. The BFAD shall not be precluded from using all data, including, but not limited to, pre-clinical and clinical trials, of an applicant when evaluating other applications.
>
> (iv.b) **Data Protection from Unfair Commercial Use**. Data submitted by the original patent holder shall be protected against unfair commercial use as provided in Article 39.3 of the TRIPS.
>
> (72.4)
>
> When required as a condition of approving the marketing of drugs and medicines which utilize new chemical entities, any submitted undisclosed test or other data, the origination of which involves a considerable effort, shall be protected against unfair commercial use. In addition, such data shall be protected against disclosure, except where necessary to protect the

(ii) prevent the government from taking any necessary action to protect public health, national security, non-commercial public use, national emergency, public health crisis, or other extremely urgent circumstances declared by the government.

public or unless steps are taken to ensure that the data are protected against unfair commercial use.

(39.3)

The BFAD referred to above is the Bureau of Food and Drugs, now known as the Food and Drug Administration (FDA), the entity under the Department of Health entrusted with approvals of pharmaceutical products.[135] Rule 9(iv.a) merits attention in particular as it treats the use *by the FDA* of undisclosed data (by its inclusive reference, "all data") in evaluating *other* applications, as being excluded from the "unfair commercial use" prohibition covered in paragraph (iv.b). It is therefore arguable that the protection provided in the sentence "When required …" is distinguished from other forms of use, or use by persons other than the FDA, for the purpose of the "unfair commercial use" prohibition. In effect, it appears that the Philippines has adopted a *narrow* construction of the protection required under Article 39.3 of TRIPs. Further, since it has carved out from its legislative protection FDA use for evaluation of other applications, this appears to imply that the "unfair commercial use" protection extends only to *third-party* use. To date, this provision has not been challenged at the WTO as being non-compliant with Art. 39.3.

### 3.3.7 *Patent term extensions*

Under the TRIPs Agreement, patent terms are to be of at least 20 years duration. Whether a WTO member decides to extend such a period is a matter of national discretion. However, since there is often a syncopation between the time of grant of a patent and time of grant of marketing approval (sometimes due to a deliberate delay by a party with regard to applying for the latter), it has been argued that the effecting marketing duration of a patented medicines is less than 20 years. As a result there have been calls for *patent term extensions* to make up for any periods of delay in the marketing approval process that leads to the above kind of situation.[136]

[135] For information on the Administration and the legislation it administers, see www.fda.gov.ph/about-food-and-drug-administration (accessed 12 September 2014). The change was brought about by The Food and Drug Administration (FDA) Act of 2009, A971; see www.who.int/medical_devices/02_agnette_peralta.pdf.

[136] Jorn Sonderholm, Wild card patent extensions: see article, Wild-Card Patent Extensions as a Means to Incentivize Research and Development of Antibiotics, *J. L. Med. & Ethics* (2009), 240; note: this issue contains other articles on innovation and pharmaceutical products.

#### 3.3.7.1 Patent term extensions in Asia

Under the minimum requirements of the TRIPs Agreement, member states are obliged to provide of patent terms of at least 20 years in duration. In some countries, however, patent term extensions are permitted – sometimes as a result of FTA obligations. Tables 3.6A and 3.6B set out the patent terms of each of these countries and whether the laws permit any extensions.

Table 3.6A. *Legislative patent periods in ASEAN*[a] *and other selected Asian countries*

| ASEAN country | Patent period/whether available | Relevant legislation |
|---|---|---|
| Brunei Darussalam | 20 years from filing<br>Extension available[b] | Patents Order, 2011[c] |
| Cambodia | 20 years from filing | Law on Patents, Utility Models and Industrial Designs[d] |
| Indonesia | 20 years from filing | Law No. 14 of 1 August 2001 regarding Patents[e] |
| Lao PDR | 20 years from filing | Law No. 01/NA of 20 December 2011, on Intellectual Property (as amended)[f] |
| Malaysia | 20 years from filing | Patents Act 1983[g] |
| Myanmar | – | [A new intellectual property law has been submitted to the Myanmar Parliament and is awaiting approval at time of writing][h] |
| Philippines | 20 years from filing | Intellectual Property Code of the Philippines[i] |
| Singapore | 20 years from filing<br>Extension available[j] | Patents Act[k] |
| Thailand | 20 years from filing | Patent Act[l] |
| Vietnam | 20 years from filing | Law on Intellectual Property[m] |

[a] See generally, Paul Wedel, ed., *Comparative Assessment Study of Patent and Trademark Offices in Southeast Asia*, www.academia.edu/3883157/Comparative_Assessment_Study_of_Patent_and_Trademark_Offices_in_Southeast_Asia (accessed 5 February 2014).

[b] Section 36, S57/2011; text available at Brunei Registry Office website, www.bruipo.com.bn/files/Patents_Order_2011.pdf (accessed 5 February 2014).

[c] Section 35, *ibid.*

[d] NS/RKM/0103/005, Art. 45.

Table 3.6A. (*cont.*)

[e] Art. 8. For other IP Laws of Indonesia, see generally: www.wipo.int/wipolex/en/profile.jsp?code=ID (accessed 3 April 2014).
[f] Art. 48.
[g] Section 35, Act 291 of 1983 as amended by A1264 of 2006, as amended up to Act No. A1264 of 2006.
[h] See generally information at WIPO website, www.wipo.int/wipolex/en/profile.jsp?code=MM. See also reports from the WTO Trade Policy Review of Myanmar, 11 and 13 March 2014, at: www.wto.org/english/tratop_e/tpr_e/tp393_e.htm (accessed 31 May 2014). This was the first such WTO review for Myanmar.
[i] Section 54, Republic Act No. 8293.
[j] Section 36A, Patents Act, Cap. 221, 2005 Rev. Ed.
[k] Section 36, Patents Act, Cap. 221, 2005 Rev. Ed.
[l] Patent Act (3rd issue) B.E.2542 (A.D.1999).
[m] Art. 93.2, Law No. 50/2005/QH11 of November 29, 2005, on Intellectual Property (promulgated by the Order No. 28/2005/L-CTN of 12 December 2005, of the President of the Socialist Republic of Vietnam).

Table 3.6B. *Legislative patent periods in selected Asian countries (non-ASEAN)*

| Country | Patent period/whether extensions available | Relevant legislation |
|---|---|---|
| China | 20 years from filing | Patent Law of the People's Republic of China (as amended)[a] |
| Hong Kong | 20 years from filing | Patents Ordinance[b] |
| Macau | 20 years from filing | Industrial Property Code[c] |
| India | 20 years from filing | Patents Act, 1970 (as amended in 2005)[d] |
| Japan | 20 years from filing<br>Extension available[e] | Patent Act (Act No. 121 of April 13, 1959, as amended up to Act No. 63 of 2011)[f] |
| Korea | 20 years from filing<br>Extension available | Patent Act (as amended in 2012)[g] |
| Taiwan | 20 years from filing<br>Extension available for pharmaceutical or agrichemical patents[h] | Patent Act 2013[i] |

[a] Art. 42, Law as amended up to the Decision of the Standing Committee of the National People's Congress on Amending the Patent Law of the People's Republic of China (adopted at the 6th Meeting of the Standing Committee of the Eleventh National People's Congress on 27 December 2008.

Table 3.6B. (*cont.*)

[b] Section 39, Cap. 514.

[c] Art. 103, "Legal Regime of Industrial Property" approved by Decree-Law No. 97/99/M of 13 December 1999, as amended by Law No. 11/2001 of 6 August 2001, on the Establishment of the Customs Services of the Macao SAR, P.R.C.

[d] Section 53.

[e] Art. 67(2), 67–2 and 68–2, Patent Act, as amended in 2005.

[f] Art. 67(1), Patent Act, *ibid.*

[g] Patent Act, Act of 1946, as amended up to Act No. 11654 of 22 March 2013. See also Pharmaceutical Affairs Act, as amended in 2012, at www.wipo.int/edocs/lexdocs/laws/en/kr/kr094en.pdf, requiring patent-drug approval linkage.

[h] Art. 53, *ibid.*

[i] Art. 52; Taiwan Intellectual Property Office website: www.tipo.gov.tw/ct.asp?xItem=479550&ctNode=6817&mp=2 (accessed 5 February 2014).

## 3.4 Exhaustion of rights and parallel imports

Article 6 of the TRIPs Agreement and paragraph 5(d) of the Doha Declaration on the TRIPs Agreement and Public Health provide WTO members with discretion as to whether to permit, in their national laws, parallel imports – subject only to Article 3 (National Treatment) and Article 4 (MFN).[137]

It is important to note that the references here are to parallel imports of genuine goods, not counterfeit ones. Parallel imports are a means for a country to enable lower-priced medicines (and other goods) to be

[137] A pithy illustration by the WTO explains this principle well in the context of pharmaceuticals:

> For example, suppose company A has a drug patented in the Republic of Belladonna and the Kingdom of Calamine, which it sells at a lower price in Calamine. If a second company buys the drug in Calamine and imports it into Belladonna at a price that is lower than company A's price, that would be a parallel or grey import.
>
> The legal principle here is "exhaustion", the idea that once company A has sold a batch of its product (in this case, in Calamine), its patent rights are exhausted on that batch and it no longer has any rights over what happens to that batch.

WTO website, www.wto.org/english/tratop_e/trips_e/factsheet_pharm02_e.htm#dohadecl5d (accessed 3 October 2014). On the concepts of national and international exhaustion of rights, see WIPO, www.wipo.int/sme/en/ip_business/export/international_exhaustion.htm (accessed 3 October 2014).

imported for the use of its people.[138] Importantly, parallel imports of medicines differ from those of other products in that countries usually regulate the *import and marketing* of medicines under marketing approval laws separate from IP laws. Under Singapore law, for example, a license is still required for the parallel import of medicinal products.[139] The use of parallel importation for medicines must therefore be viewed in this light.

Parallel importation laws in Asian countries vary and not all permit such imports.[140]

Article 69 of China's Patent Law appears to adopt an international exhaustion of patent rights.[141]

Japan has not explicitly provided for exhaustion but it is thought that it follows a system of international exhaustion.[142]

India's Patent Act was amended in 2002 to introduce a provision relating to parallel imports, and a further amendment was enacted in 2005 which has led to the present provision,[143] section 107A(b). This is a provision that exempts certain acts from being patent infringements, and provides as follows:

> **107A. Certain acts not to be considered as infringement.**—For the purposes of this Act,—
>
> (a) any act of making, constructing, using, selling or importing a patented invention solely for uses reasonably related to the development and submission of information required under any law for the time being in force, in India, or in a country other than India, that regulates the manufacture, construction, use, sale or import of any product;
> (b) importation of patented products by any person from a person who is duly authorised under the law to produce and sell or distribute the product, shall not be considered as an infringement of patent rights.

[138] As will be seen in Chapter 7, however, such price arbitraging may affect the willingness of innovator companies to use differential pricing for medicines, since such pricing assumes no leakage of goods due to price arbitrage.

[139] Section 5(2), Medicines Act, Cap. 176.

[140] See WIPO Standing Committee on the Law of Patents, Twenty-first Session, 3–7 November 2014, Geneva, *Exceptions and Limitations to Patent Rights: Exhaustion of Patent Rights*, SCP/21/7.

[141] See para. 21, *Exceptions and Limitations to Patent Rights: Exhaustion of Patent Rights*, note 113 above.

[142] Cédric Julien Poget, *Parallel Imports of Pharmaceuticals: Evidence from Scandinavia and Policy Proposals for Switzerland* (Birkhäuser, 2008), p. 128.

[143] See Shamnad Basheer and Mrilini Kochupillai, "Exhausting" Patent Rights in India: Parallel Imports and TRIPS Compliance, *JIPR*, Vol. 13 (2008), 486–497, at p. 489.

While this does not explicitly state whether it is national or international exhaustion which applies, one view is that it is the latter.[144]

### 3.4.1 ASEAN

While ASEAN does not appear to have a common policy on exhaustion of patent rights, member states in effect permit parallel imports of pharmaceuticals on the basis of *international* exhaustion, in one form of legislative provision or another.

Countries which provide for international exhaustion include Cambodia,[145] Malaysia,[146] Thailand,[147] and Vietnam.[148] Parallel imports appear to be permitted in Brunei Darussalam as a matter of practice.[149] In Indonesia, Article 135(a) of the Patent Act provides for international exhaustion but only insofar as it is an exemption from criminal provisions.[150] The Philippines' system on exhaustion is classified as a

[144] Note 116, above.

[145] See Art. 44(i), Law on the Patents, Utility Model Certificates and Industrial Designs, Royal Decree NS/RKM/0103/005. See Hor Peng, Kong Phallack, Jörg Menzel (eds.), *Introduction to Cambodian Law* (Konrad Adenauer Stiftung, 2012), at p. 275.

[146] Section 58A, Patents Act. See also generally, Salmah Bahari, Mohamed Izham and Mohamed Ibrahim, *National Medicine Policy – A Malaysian Perspective*, (Penerbit Universiti Sains Malaysia, 2011).

[147] Section 36 exception (7), Patent Act B.E. 2522 (1979), as amended. See Jakkrit Kuanpoth, *Appropriate Patent Rules in Developing Countries – Some Deliberations Based on Thai Legislation*, citing section 36(7) of the Thai Patent Act; available at http://ro.uow.edu.au/cgi/viewcontent.cgi?article=1350&context=lawpapers (accessed 3 October 2014).

[148] Ministry of Health Decision 1906/2004/QD-BYT dated May 28, 2004 Promulgating the Regulation on Parallel Import of Medicines for Prevention and Cure of Human Diseases – text of the Decision in English is available at: www.lawfirm.vn/?a=doc&id=885 (accessed 3 October 2014); see Dr Nguyen Nhu Quynh, Ministry of Science and Technology, *Parallel Trade of Pharmaceuticals: A Discussion from Developing Country Perspective*, http://iprenforcement.most.gov.vn/cac-bai-nghien-cuu-shtt/parallel-trade-of-patented-pharmaceuticals-a-discussion-from-developing-country-perspective (accessed 3 October 2014) and Tu Thanh Nguyen, *Technology Transfer and Competition Law: Options for Developing Countries*, in *Sustainable Technology Transfer, A Guide to Global Aid and Trade Development*, ed. Hans Henrik Lidgard, Jeffery Atik and Tu Thanh Nguyen (Kluwer Law International, 2012). Note also the US-Vietnam Bilateral Trade Agreement 1997 which contains provisions on IPRs affecting pharmaceutical inventions (Art. 7 on patents and Art. 9 on trade secrets) www.usvtc.org/trade/bta/text/chapter2.htm.

[149] See WTO Trade Review Policy, Brunei Darussalam, Report of the Secretariat, February 2015, WT/TPR/S/309, at p. 51.

[150] UNCTAD, *Development Dimensions of Intellectual Property in Indonesia: Access to Medicines, Transfer of Technology and Competition*, p. 18.

mixed one.[151] Section 72.1 of the IP Code (as amended by the Universally Acceptable Cheaper and Quality Medicines Act of 2008) and Rule 15, Section 1(iv) of the QUAMA Implementing Rules explicitly permit the parallel implementation of medicines on the basis of international exhaustion.[152] While Singapore generally permits parallel imports under international exhaustion, such imports into Singapore are not permitted where they are the subject of compulsory licensing and intended for another country.[153]

## 3.5 Conclusion

The national patent bargain, as it was understood under national law, has been profoundly affected by the TRIPs Agreement obligations and post-TRIPs bilateral and regional agreements. The mandatory minimum standards stipulated for patent systems under the Agreement have helped to reshape many national laws, and in some cases, required the introduction of new mechanisms and levels of protection. For some WTO members, in addition to these minimum obligations, additional, stricter commitments have been undertaken through the participation in extra-WTO legal treaties, such as in FTAs. Secondly, the flexibilities in the TRIPs Agreement, such as in the areas of patentability and protection of undisclosed data, have provided member countries with a degree of freedom to determine the type and level of protection beyond the minimum standards. Patentability, with its critical gatekeeping function, is an example of where the countries have a certain freedom for regulation. However, important debates are taking place even in developed countries, such as the US, those in the EU and Australia, as to patentability of the subject matter of biotechnology, such as genes, gene sequences, biologic and biosimilar products, and protection of test data for the latter, because of the potential impact on access to researchers or users, the resultant access to medicines, and in some cases, implications in terms of morality. In Asia, many patent laws contain provisions prohibiting patentability where it would be a breach of public morals, and it is not clear how this link to morality will be given effect in the

[151] See para. 28, *Exceptions and Limitations to Patent Rights: Exhaustion of Patent Rights*, note 113 above.

[152] See also s. 147 and s. 159.4 of the Code.

[153] Section 66, Patents Act. See also related statements by the then Senior Minister of State for Law, Singapore Parliamentary debate at the second reading of the Patents (Amendment) Bill, 25 August 2008.

context of such subject matter. Deep divisions of opinion also exist among countries as to the data protection provision in Article 39(3) of the TRIPS Agreement and how it should be interpreted.

For countries in Asia which are participating or looking to participate in FTA negotiations – and particularly for ASEAN, which is aiming for a deeper level of economic, social and political integration – a number of important legal questions need to be examined domestically and answered clearly. As has been shown in the chapter, there is a divergence in approaches to some of these newer areas of technology and newer issues, and in some cases, a lacuna in the policy areas in the first place. The existing legal and policy constructs in this area may be said to be in a state of flux. It is critical for the policy matters raised to be properly debated and decided internally, not only to better prepare for global and regional debate and for FTA negotiations, but also to provide clarity to investors who may wish to devote significant funds to the field of biotechnology and biopharmaceutical products in Asian countries.

# 4

# Treaties in Asia

## Embedded risks and lessons from disputes elsewhere

### 4.1 Legal obligations in TRIPS

As discussed in Chapter 2, World Trade Organization (WTO) members are legally bound by the Trade-Related Aspects of Intellectual Property Rights (TRIPS) Agreement to observe the requirements for IP systems, and implementation (and non-implementation) of their obligations is subject to challenge and scrutiny under the WTO's dispute settlement system. In addition, Asian countries which have acceded to the WTO in the last 15 years have had to make commitments during the accession process which exceed those in existing WTO multilateral agreements. This is due to the fact that such countries had to undergo a demanding negotiation process with WTO members in order to gain their agreement to admit them. In acceding to the organization, such countries are typically bound by commitments recorded in two key documents: the Working Party Report and the Protocol of Accession. Within ASEAN, Cambodia, Lao PDR and Vietnam submitted to this process, with Protocols of Accession containing commitments on patents/pharmaceuticals.[1] In 2001, China acceded to the WTO, with the Working Party Report and Protocol of Accession addressing such commitments as well. The implementation of these commitments is also subject to WTO dispute settlement scrutiny.

One notable matter in the area of dispute settlement related to the TRIPS Agreement, however, is that so far WTO members have suspended the application of non-violation complaints under the Agreement.[2]

[1] See Karin Timmermans, Monopolizing Clinical Trial Data: Implications and Trends, 13 February 2007, *PLOS Medicine*, especially the text to footnotes 36–45.

[2] The non-violation complaint stems from Art. XXIII(b) of GATT 1994; it is of limited application in general GATT and WTO jurisprudence. See WTO website, www.wto.org/english/tratop_e/dda_e/status_e/nonviolation_e.htm (accessed 8 May 2015).

## 4.2 Examples of TRIPS disputes in the WTO system

Given that there has been a great deal of highly visible and publicized debate over the TRIPS Agreement and its flexibilities, a relatively small number of WTO disputes are directly related to its provisions. Though few in number, these cases provide useful and valuable lessons on how panels understand the Agreement and its exceptions. They illustrate not only the type of challenges brought but also the use of important TRIPS exceptions (such as Article 30). Such lessons can be applied not only in domestic law and policy-making but also when negotiating new bilateral or regional trade and investment agreements (as will be examined in the next chapter).

In an early complaint under the TRIPS Agreement in the WTO dispute settlement system, India's "mailbox" system was challenged by the US.[3]

In *Canada – Pharmaceutical Patents*, the WTO panel accepted Canada's argument that its Bolar provision was justified under Article 30 of the TRIPS Agreement.[4]

In *Brazil – Patents*, Brazil's local working requirement was the subject of a challenge by the US. However, as the case was settled there is no decision on the merits of the claim.[5]

Between 1999 and 2000, the US brought two WTO complaints against Argentina. These related to Argentina's pharmaceutical patents protection and test data protection.[6] Both complaints were eventually settled by the parties.

Two complaints have been brought against the EU and its member state (the Netherlands) for alleged seizures of generic medicines while in transit, in ports and airports in the Netherlands. These complaints have been brought by India and Brazil.[7] The requests for consultations were made in 2010 and the disputes remain pending at the time of writing.

[3] *India – Patent Protection for Pharmaceutical and Agricultural Chemical Products*, Appellate Body decision WT/DS50/AB/R, adopted 16 January 1998.

[4] See discussion in Chapter 2.

[5] *Brazil – Measures Affecting Patent Protection*, DS/199, mutually agreed solution reached in July 2001; see WTO website, www.wto.org/english/tratop_e/dispu_e/cases_e/ds199_e.htm (accessed 8 May 2015). See generally, Ellen F. M. 't Hoen, TRIPS, Pharmaceutical Patents and Access to Essential Medicines: Seattle, *Doha and Beyond*, pp. 44–46, www.who.int/intellectualproperty/topics/ip/tHoen.pdf (accessed 12 August 2014). See also Chapter 5.

[6] DS171, *Argentina – Patent Protection for Pharmaceuticals and Test Data Protection for Agricultural Chemicals*, and DS196, *Argentina – Certain Measures on the Protection of Patents and Test Data.*

[7] WTO, DS408, *European Union and a Member State – Seizure of Generic Drugs in Transit*, www.wto.org/english/tratop_e/dispu_e/cases_e/ds409_e.htm and DS409, *European Union and a Member State – Seizure of Generic Drugs in Transit.*

A number of countries requested to join these consultations, among them is notably, China, a large supplier of pharmaceutical ingredients and products, is most notable among them. With respect to India's complaint, the EU subsequently proposed amending its border regulation in question. In 2011, India announced it had reached an "Understanding", by way of an "informal settlement", with the EU on this matter.[8] The key features in this Understanding have been described by India as follows:

> India has taken note of the commitments offered by the EU. India has reiterated the core principle of the Understanding that the mere fact that medicines are in transit through EU territory, and that there is a patent title applicable to such medicines in the EU territory, does not in itself constitute enough grounds for customs authorities in any Member State to suspect that the medicines at stake infringe patent rights. However, a situation in which medicines are in transit through EU territory and there is adequate evidence that satisfies the customs authorities that there is a substantial likelihood of diversion of such medicines on to the EU market may constitute enough grounds for customs authorities to suspect that the medicines at stake infringe patent rights in the EU.[9]

As part of the Understanding, India assured the EU it would not proceed with the WTO complaint as long as the EU authorities adhered to the Understanding.[10]

Clarity of customs rules with respect to pharmaceutical products in transit is critical. For procurers of generic medicines (such as MSF whose procurement centres are based in Europe and redistribute procured medicines to beneficiary countries such as those in Asia),[11] such

[8] See *India EU Reach an Understanding on Issue of Seizure of Indian Generic Drugs in Transit*, 28 July 2011, India Ministry of Commerce and Industry website, http://pib.nic.in/newsite/erelease.aspx?relid=73554 (accessed 12 August 2014).

[9] *Ibid.* For views expressing disquiet over the subsequent proposal and amendment by the EU, see Brook K. Baker, *Settlement of India/EU WTO Dispute re Seizure of In-transit Medicines: Why the Proposed EU Border Regulation Isn't Good Enough* (2012), http://digitalcommons.wcl.american.edu/cgi/viewcontent.cgi?article=1026&context=research and *New EU Customs Regulation Might Allow Wrongful Seizures Of Generic Drugs In Transit, NGOs Say*, IP Watch, 17 October 2013, at www.ip-watch.org/2013/10/17/new-eu-custom-regulation-might-allow-wrongful-seizures-of-generic-drugs-in-transit-ngos-say/ (accessed 12 August 2014).

[10] In its announcement of the Understanding, India also stated: "India's options to revive the dispute remain intact in case the EU does not abide by the core principles agreed to in the Understanding." (*ibid.*)

[11] MSF has three procurement European centres, located in Belgium, France and the Netherlands; see www.unicef.org/supply/files/6_General_presentation_MSF_Meeting_UNICEF_Oct_06_final_version.pdf (accessed 12 August 2014).

seizures can affect their supply chain to needy patients. In order to ensure clarity, it is necessary to distinguish between counterfeit and generic pharmaceutical products. This point of distinction was emphasized by Brazil, China and India during a TRIPS Council meeting in 2012.[12]

WTO complaints have also been brought against the US and the EU. In *United States – US Patents Code*, Brazil has complained that the US Patents Code is inconsistent with TRIPS Articles 27 and 28.[13] In *European Communities – Patent Protection for Pharmaceutical and Agricultural Chemical Products*, Canada brought a complaint against the EU with regard to regulations in which a patent term extension scheme allegedly applied only to pharmaceutical and agricultural patents.[14] At the time of writing both matters remained pending.

Finally, WTO complaints relating to tobacco control measures in the form of "plain packaging" laws and regulations also raise TRIPS issues, and will offer valuable lessons.

## 4.3 TRIPS-like and TRIPS-plus commitments in Asia FTAs

Much of recent literature discusses the implications of provisions in FTAs that contain "TRIPS-plus" provisions, that is, bilateral treaty provisions that provide for commitments beyond and stronger than those that have been made in the WTO under the TRIPS Agreement. The trend of such TRIPS-plus provisions has been examined in the context of enhancement and increased protection of rights of patent holders, as compared with the wider public concern of ease of access to medicines that may be subject to such patents.[15]

China, a relatively recent entrant in the FTA "race", has signed ten bilateral FTAs, some of which include TRIPS-plus provisions.[16] Some other FTA signatories, such as Singapore and Korea, have also chosen to include such provisions. TRIPS-plus provisions require relinquishment

[12] See *India, Brazil & China Defend Generic Drugs at WTO*, 52 June 2012, The Economic Times, http://articles.economictimes.indiatimes.com/2012-06-25/news/32409062_1_counterfeit-medicines-fake-drugs-generic-drugs (accessed 12 August 2014).

[13] DS224, *United States – US Patents Code.* [14] DS153.

[15] For instance, for a discussion of Jordan's position as an FTA partner of the United States and its TRIPS-plus commitments in relation to pharmaceutical trade under the US–Jordan FTA, see Bashar H Malkawi, *Asian J of WTO & Int'l Health Law and Policy*, Vol. 4, No. 1 (March 2009), 95–128.

[16] See China's Ministry of Commerce website containing texts of China's FTAs, at: http://fta.mofcom.gov.cn/english/fta_qianshu.shtml. An example of an FTA with TRIPS-plus commitments is the China–Switzerland FTA.

of additional regulatory freedom, providing patent holders better protection than what is obligatory under TRIPS (such as increased duration of patent rights and special protection of data exclusivity in respect of data from research and development of a drug). The stronger the protection, the more the obstacles are for competitors, especially those who manufacture generic drugs for similar diseases.[17] Pharmaceutical companies holding patents have been known, for example, to engage in "evergreening". Although not a formal legal concept, "[e]vergreening is a term referring to the numerous ways in which patent owners of pharmaceutical products use the patent laws to extend their monopoly privileges beyond periods that are normally allowed by law, particularly over high-revenue-earning drugs".[18] "Evergreening" comprises various ways of expanding or extending protection by patent holders of their rights, by essentially building on the existing patent, and some FTAs contain provisions which may in effect facilitate evergreening.

In addition to FTA commitments, countries newly acceding to the WTO system have also been required to include TRIPS-plus commitments.[19]

## 4.4 FTAs and legal risks arising from negotiated language

### *4.4.1 FTAs are not just about "free trade"*

The scope of many FTAs is very broad, often encompassing commitments in trade in goods, services, investment, movement of natural persons, intellectual property rights, transparency and dispute settlement. FTAs are permissible as regional trade agreements (RTAs) under GATT 1994 and GATS. While the WTO has a notification requirement and a committee which examines notified RTAs, there is no specific mechanism in the WTO to regulate FTA provisions, and therefore nothing to prevent TRIPS-plus provisions even though these in effect erode TRIPS flexibilities. Some recent FTAs have included such

[17] For a study of practices of pharmaceutical companies to block or delay the market entry of generic drug makers, see Chapter 3 of the EU's *Final Report on the Pharmaceutical Sector*, 2009, at: http://ec.europa.eu/competition/sectors/pharmaceuticals/inquiry/communication_en.pdf. On the relationship between competition law and regulation of abuse of patents, and other issues relating to generic manufacturers, see generally: http://ec.europa.eu/competition/sectors/pharmaceuticals/inquiry/index.html.

[18] Gaurav Dwivedi, Sharanabasava Hallihosur, Latha Rangan, Evergreening: A Deceptive Device in Patent Rights, *Technology in Society*, Vol. 32 (2010), 324–330.

[19] See, e.g., China, WTO Working Party Report, WT/ACC/CHN/49, at para. 284.

provisions which have significant implications for access to medicines in the signatory countries.[20]

### 4.4.2 *Balancing trade, investment and public health interests in FTA negotiations*

Given the *realpolitik* involved in trade negotiations, a negotiating party (or parties where it is plurilateral rather than bilateral) has to balance trade and non-trade considerations (such as geopolitical ones) to decide how much to push back in the face of demands across the table. Agreements often comprise a series of trade-offs as well as domestic negotiations.[21]

The following excerpt from the parliamentary speech of the Singapore Minister for Trade & Industry (MTI) sums up the position of the government on balancing trade negotiations and healthcare affordability:

> [The member of Parliament's name] is concerned that Singapore's interests may not be adequately protected as we negotiate with other countries participating in the Trans-Pacific Partnership. Let me assure [the member of Parliament's name] that Singapore and Singaporeans are at the heart of our FTA negotiations. We proactively consult other government agencies as well as relevant stakeholders to ensure that Singaporeans benefit from our FTAs with other countries. Our approach to the TPP is no different.
>
> More specifically, [the member of Parliament's name] also asked about the trade-off between a strong Intellectual Property (IP) regime versus healthcare cost through generic drugs. A strong IP regime is a critical enabler for Research and Development, and in fact our strong IP regime is an important factor why we have succeeded in building up a significant biomedical sciences sector over the last 15 years.
>
> *At the same time, keeping drugs affordable for Singaporeans is a very important consideration. Therefore, MTI will work with MOH and other government agencies to make sure we are able to strike the right balance in our TPP negotiations.*[22]
>
> [Italic emphasis added]

[20] See, e.g., Ruth Lopert and Deborah Gleeson, The High Price of "Free Trade": U.S. Trade Agreements and Access to Medicines, *J. L., Med. & Ethics* (Spring 2013), 199–223.

[21] See, e.g., Johanna von Braun, *The Domestic Politics of Negotiating International Trade – Intellectual Property Rights in US-Colombia and US-Peru Free Trade Agreements* (Routledge, 2012).

[22] Parliamentary speech during Committee of Supply debate, 6 March 2014, available in Hansard.

As Singapore's Chief Negotiator of the US–Singapore FTA (USSFTA) has candidly explained, FTAs are not just about economic considerations:

> Singapore wants an FTA with the U.S. for a combination of economic and strategic reasons. The U.S. is Singapore's largest foreign investor and second largest trading partner. The U.S. is also Singapore's most important source of technology and management know-how. *Singapore's interest in the U.S., however, transcends business and economics. Singapore wishes to entrench the presence of the U.S. in the region because it underpins the security of the whole Asia-Pacific region. Singapore regards the U.S.-Singapore FTA as a symbol of continued U.S. commitment to the region. Therefore, for Singapore, the USSFTA is not just about securing tariff-free entry for Singapore's exports to the U.S. market. It is not just about attracting more Foreign Direct Investment (FDI) to Singapore. It is also about enhancing the prospects of peace and stability in the region.*[23]
>
> [Italic emphases added]

Another member of the negotiating team further explained:

> For example, in negotiating the USSFTA, Singapore was mindful that there were some difficult areas, including the area of intellectual property rights where we had to consider committing to TRIPS-plus standards. While this was seen as an opportunity to speed up the review of Singapore's then IP system, clearly there were trade-offs that needed to be taken into consideration when accepting the overall 'package' of obligations.
>
> . . . With respect to pharmaceutical products, for example, the negotiations over parallel imports apparently remained a sticking point and this was resolved only at the last day of negotiations.[24]

Given the "new normal" regarding FTAs and ISDS discussed above, it is imperative that negotiating states understand well the legal risks involved in agreeing to FTA provisions, particularly those which can have a long-term impact on health-policy-making.

[23] Tommy T.B. Koh, Chapter 1, *The USSFTA: A Personal Perspective, in The United States Singapore Free Trade Agreement – Highlights and Insights* (Institute of Policy Studies World Scientific Publishing Co. Pte. Ltd., 2004) See also related remarks of the then Ambassador of Singapore to the US, Amb. Chan Heng Chee, before the US International Trade Commission hearing on the USSFTA on 24 April 2003, at: www.mfa.gov.sg/content/mfa/overseasmission/washington/newsroom/press_statements/2003/200304/press_200304_01.html (accessed 23 July 2014).

[24] Liew Woon Yin, the then Director-General, Intellectual Property Office of Singapore (IPOS) and USSFTA negotiator, in Chapter 12, *ibid.*, at p. 126.

In a recent Trade Policy Review report on Malaysia, the WTO Secretariat Report stated the following:

> The Trans-Pacific Partnership negotiations (Section 2.4.2), in which Malaysia is actively involved, aim, *inter alia*, at a high-level Intellectual Property chapter, which would extend copyright protection, impose high-level enforcement standards, and *deal with pharmaceuticals and access to medicines.*[25]
>
> (Italics emphasis added)

In some of its earlier FTA negotiations, the US required some of the highest TRIPS-plus standards in the protection of IP and control of marketing approvals for medicines.[26] 2007 however marked a change in the US FTA policy.[27]

The USTR states its current view of treatment of IP and access to medicines in negotiating FTAs [such as the Trans-Pacific Partnership (TPP) and the Transatlantic Trade and Investment Partnership (TTIP)] as follows:

> **Providing Access to Medicines for the World's Poor While Promoting the Development of New Life-Saving Drugs.** In TPP, we are proposing a "differentiated approach" for pharmaceutical IP protections taking into account countries' individual levels of development while ensuring that the incentives remain in place for pharmaceutical innovation.[28]

It has also described its approach to access to medicines in the TPP as a "flexible" one.[29] However, given that no negotiating text is publicly

[25] Para. 3.118, WTO Secretariat Report, Malaysia, Trade Policy Review, WT/TPR/S/292/Rev.2, 8 April 2014. Trade policy reviews are conducted periodically for all WTO members, and reports are made by both the national government and the WTO Secretariat to provide detailed updates on all trade laws and policies of the country under review.

[26] See, e.g., the US' FTAs with Morocco, and with Latin American countries such as Peru and Colombia.

[27] See Maria Fabiana Jorge, New US Trade Policy: A Turning Point? *Journal of Generic Medicines*, Vol. 5 (2007), 5–8. For an overview of the position prior to the change in policy, see same author, Trade Agreements and Public Health: Are US Trade Negotiators Building an Intellectual Property Platform Against the Generic Industry? Are they Raising the Standards to Go Beyond the US Law? *Journal of Generic Medicines*, Vol. 4 (2007), 169 – 179.

[28] USTR Fact Sheet, *Values Driving US Trade Policy*, February 2014.

[29] USTR statement, 29 November 2013, www.ustr.gov/about-us/press-office/blog/2013/November/stakeholder-input-sharpens-focuses-us-work-on-pharmaceutical-IP-in-TPP and related commentary in IP Watch, www.ip-watch.org/2013/11/28/ustr-says-its-tpp-proposal-on-ip-and-public-health-shows-flexibility/ (accessed 23 July 2014). The statement also states that the US is taking a "more flexible" approach under which

available, it is difficult to understand at this moment how the said approaches of "differentiation" and "flexibility" are being promoted.

In light of the possibility of IP chapters being seen as trade-off chapters, FTA language may be a "template" language with little time for analysis in the new treaty context, or may comprise compromise language which is less precise than might be hoped for. This brings legal risks as such language potentially leaves a wide margin for interpretation by a dispute settlement tribunal, should a dispute subsequently arise.[30]

## 4.5 Non-FTA mechanisms with a bearing on national laws

Pressure to amend national IP laws may come from sources other than treaty commitments. For example, the US publishes an annual list of "priority watch" countries under its "Special 301" process.

Recently, India announced that it would initiate a regular process to review the trade obstacles faced by the Indian exporters in partner countries.[31]

## 4.6 Types of TRIPS-plus FTA provisions

### *4.6.1 Patentability*

While TRIPS leaves members to determine the scope of patentability – subject to some basic recognized requirements[32] – some FTAs have limited this flexibility by requiring explicit inclusion of subject matter which might otherwise been within the signatory countries' area of flexibility. Such provisions are therefore TRIPS-plus.

trade partners could retain "reasonable" pre-grant opposition proceedings. See also *Development-Based Transition Mechanism For IP Rules Seen as Losing Ground*, on the US' and Mexico's respective proposals with transitional, "phase-in" mechanisms for adherence to particular IP obligations, and *TPP IP Negotiators Wrestle Over Scope of 'Bolar' Exemption for Drugs*, both in *Inside US Trade – World Trade Online*, 24 May 2015.

[30] See discussion by this author in Investor-State Dispute Settlement Reform – Examining the Formative Aspect of Investment Treaty Commitments: Lessons from Commercial Law and Trade Law, *Transnational Dispute Management*, Vol. 11, No. 1 (January 2014), as well as generally, other articles in the issue.

[31] See text at: www.miti.gov.my/cms/documentstorage/com.tms.cms.document.Document_b3c9637d-c0a8156f-5df15df1-d75dff43/MICECA%20-%20edit%20as%20at%2003052011.pdf (accessed 23 July 2014).

[32] See Article 27, TRIPS Agreement.

#### 4.6.1.1 Asian FTAs: TRIPS-plus patentability provisions

Not all FTAs signed by Asian countries contain TRIPS-plus provisions for patents or for pharmaceutical products. Examples of agreements that do not contain such provisions are the Korea–Chile FTA[33] and a number of China's FTAs.

The EU has been negotiating an FTA with India. However, the access to medicines aspect has been a source of concern especially for India. In an effort to address these concerns, the EU issued a "fact sheet" on its approach to access to medicines in negotiating this FTA.[34] The fact sheet states, *inter alia*, that:

- The FTA will in no way affect access to medicines in India, nor in other developing countries.
- The FTA will not require India to introduce any kind of data exclusivity provisions.
- The FTA will not require India to introduce any kind of patent term extension provisions, especially since preliminary information indicates that market authorisations in India are handled in an expeditious manner. While this issue was initially suggested for discussion in the FTA negotiations, the EU has decided not to pursue it further.

As for TRIPS-plus provisions, the EU explained:

> The TRIPS Agreement dates back more than 15 years. In the meantime, new challenges have emerged (e.g. biodiversity-related issues protection of traditional knowledge) and technical development has overtaken the agreement (e.g. digital environment, internet). There is also a change as regards the economic importance of certain rights. There is for example a growing interest in designs and geographical indications. This is why the EU, India and many other countries have progressively introduced, in their domestic legislation, additional rules which are more accurate, or intended to be more effective, than what TRIPS provided for.
>
> Where India and the EU have a shared interest in issues that complement or clarify the TRIPS Agreement, therefore, these will be included in the FTA. This means that certain provisions going beyond TRIPS, which are already present in Indian legislation, may be reflected in the FTA.

The USSFTA is an example of an FTA which reduces the scope of non-patentable matter. Article 16.7 provides for the Parties to exclude such inventions from patentability as are provided for in Arts. 27.2

[33] See Chapter 16, Korea–Chile FTA.

[34] See the EU website, http://trade.ec.europa.eu/doclib/docs/2013/april/tradoc_150989.pdf (accessed 27 July 2014).

and 27.3(a) of the TRIPS Agreement.[35] This therefore disallows the parties from excluding those inventions that are mentioned in Article 27.3(*b*) of the TRIPS Agreement, namely:

> (b) plants and animals other than micro-organisms, and essentially biological processes for the production of plants or animals other than non-biological and microbiological processes. However, Members shall provide for the protection of plant varieties either by patents or by an effective *sui generis* system or by any combination thereof. The provisions of this subparagraph shall be reviewed four years after the date of entry into force of the WTO Agreement.[36]

Article 18.8(1) of the KORUS agreement expressly requires patentability to be available for "new uses or methods of using a known product".

By contrast, the Japan–Switzerland Agreement preserves a version of Article 27.3(b).[37]

Patentability and generic producers' rights of course raise a wider question than what is found in FTAs. Significantly, a few recent US Supreme Court cases show that there is a need to balance the rights of generic drug manufacturers with those of drug patent holders under the law.[38]

#### 4.6.1.2 FTAs and new uses of known pharmaceutical substances

As part of patent holders' efforts to obtain further protection, new uses or new methods of using existing patented products, such as patented pharmaceutical products, are sometimes claimed as new inventions, meriting a new patent. While patents formerly claimed mainly rights over the original product or process, now increasingly pharmaceutical companies are claiming patents for such new uses or methods. We have seen in Chapter 2 that Asian countries take different approaches to whether they permit patenting of new uses for known pharmaceuticals

[35] See generally, the CRS Congressional Research Report 2004 on the USSFTA and its IPR provisions: http://congressionalresearch.com/RL31789/document.php?study=The±U.S.-Singapore±Free±Trade±Agreement.

[36] See Art. 16.7.1, USSFTA. In 2004, Singapore enacted the Plant Varieties Protection Act 2004, Cap. 232A, 2006 Rev. Ed.

[37] See Art. 117.3 of the Agreement of Free Trade and Economic Partnership between Japan and the Swiss Confederation, footnote 42, *infra*, and accompanying text.

[38] See, e.g., http://newsandinsight.thomsonreuters.com/Legal/News/2012/03_-_March/Supreme_Court_sets_aside_Myriad_gene_patent_ruling/.

substances or medicines. Where they do permit such matter to be patented, there may be a further requirement that the new use be shown to have a demonstrable increase in efficacy of use, apart from the conventional requirements of an inventive step, novelty and industrial application. For some countries in Asia, FTA commitments explicitly require patentability for such new uses.

Article 18.8(1) of the Korea–United States (KORUS) FTA, for example, provides a specific inclusion of new uses of a known product:

> ... each Party confirms that patents shall be available for *any new uses or methods of using a known product.*
>
> [Italic emphasis added]

This provision, as drafted, may however be a double-edged sword. While it *does not stipulate any degree* of effort or of "newness", or criteria for assessment, that must be shown before such new uses or methods are patentable, such silence may also give the signatory state a degree of flexibility to determine and define such a degree of effort or "newness" and criteria in its national law, to control the scope of this obligation and the potential of evergreening for marginal improvements. Section 3(d) of India's Patents Act is an example of such a national control mechanism applied to pharmaceutical patents.[39]

The Canada–Korea FTA provides for patentability of new uses of known products but explicitly subjects them to basic patentability requirements:

> **Article 16.12: Patents**
>
> 1. Each Party shall make patents available for any invention, whether a product or process, in all fields of technology, provided that the invention is new, involves an inventive step, and is capable of industrial

[39] See the new Section 3(d) introduced by the Indian Patents (Amendment) Act 2005:

> *(d)* the mere discovery of a new form of a known substance which does not result in the enhancement of the known efficacy of that substance or the mere discovery of any new property or new use for a known substance or of the mere use of a known process, machine or apparatus unless such known process results in a new product or employs at least one new reactant.
>
> *Explanation*—For the purposes of this clause, salts, esters, ethers, polymorphs, metabolites, pure form, particle size, isomers, mixtures of isomers, complexes, combinations and other derivatives of known substance shall be considered to be the same substance, unless they differ significantly in properties with regard to efficacy.

See also earlier discussion in Chapter 3.

> application[13]. In addition, each Party confirms that *patents shall be available for any new uses or methods of using a known product, provided that the invention is new, involves an inventive step, and is capable of industrial application.*[40]
>
> [Italic emphasis added]
>
> [13] For the purposes of this Article, a Party may treat the term "inventive step" as synonymous with "non-obvious" and the term "capable of industrial application" as synonymous with "useful".

Article 17.9 of the US–Australia FTA, Article 16.12 of the Canada–Korea FTA, and Article 13.8 of the KAFTA also ensure that new uses of known products are patentable subject matters.

By contrast, Article 16.7 of the USSFTA does not contain any explicit patentability requirement regarding new uses.

It should also be noted that not all recent FTAs entered into by Asian states contain TRIPS-plus patent requirements. For example, the Korea–EU FTA[41] and the signed text of the EUSFTA (awaiting ratification/implementation) do not contain any TRIPS-plus patentability requirements, while the Canada–Korea FTA (signed in September 2014) contains no commitments respect of *pharmaceutical* patents (such as patent extensions, patent linkage or data exclusivity).[42] The Korea–India FTA contains no TRIPS-plus commitments and even excludes the intellectual property (IP) provisions from treaty dispute settlement.[43]

### *4.6.2 Patent term extensions*

Some FTAs require an extension of patent terms for delays in marketing approvals for pharmaceutical products. Such provisions are TRIPS-plus as the TRIPS Agreement does not oblige a WTO member to provide such extensions beyond the mandatory minimum period of 20 years from the filing date under Article 33.

[40] Text available at: www.international.gc.ca/trade-agreements-accords-commerciaux/agr-acc/korea-coree/16.aspx (accessed 21 July 2014).

[41] Text available at www.customs.go.kr/kcshome/main/content/ContentView.do?contentId=CONTENT_ID_000002363&layoutMenuNo=23271 (accessed 11 April 2015).

[42] See http://international.gc.ca/trade-agreements-accords-commerciaux/agr-acc/korea-coree/overview-apercu.aspx?lang=eng#five (accessed 11 April 2015).

[43] Chapter 12, Comprehensive Economic Partnership Agreement between the Republic of Korea and the Republic of India.

An example of such an FTA provision is Article 10.35 of the EU–Korea FTA, which provides for extensions of the duration of rights conferred by patents "to compensate the patent owner for the reduction in the effective patent life as a result of the first authorization to place the product on their respective markets".[44] The duration required under such provisions, which usually runs from the granting of marketing approval to the first applicant, varies. Under Article 10.35, it is 5 years.

In other FTAs, longer periods of protection may be required.

For example, Article 3(b)(i) of the EFTA–Singapore FTA provides:

> The Parties shall ensure in their national laws at least the following:
>
> ...
>
> with respect to pharmaceutical and plant protection products that are subject to patent protection, an extension of the patent term to compensate the patentee for curtailment of the patent term as a result of the marketing approval process[3], subject to a maximum of 5 years. The overall term of patent protection, including any extension, shall not exceed 25 years from the filing date of the patent application. This compensatory extension will only apply when the marketing approval process takes more than 5 years ...
>
> [3] Marketing approval process is understood to encompass the period between the filing date of the patent application and the date of the first market authorisation of the product in the territory of the Party granting this authorisation. Singapore reserves the right to depart from the definition of the marketing approval process if it is found to be inconsistent with well established and recognised national or regional standards.

The EUSFTA negotiated text provides for patent term extensions (Article 11.31).

#### 4.6.2.1 Patent term extensions: Singapore's implementation of the USSFTA

Singapore stands out as the only country which has incorporated highly detailed patent extension conditions which have been included as a result of the FTA activity. Upon entering the USSFTA, Singapore introduced a new section (Section 36A) to its Patents Act in 2004.[45] Section 36A(1) permits extensions on three main grounds:

[44] Art.10.35.2, EU–Korea FTA.

[45] The text of section 36A is available at: http://statutes.agc.gov.sg/aol/search/display/view.w3p;page=0;query=CompId%3A0553033d-30d8-44e1-8bd1-b5fa715e1dfb;rec=0;resUrl=http%3A%2F%2Fstatutes.agc.gov.sg%2Faol%2Fbrowse%2FtitleResults.w3p%3Bletter%3DP%3Btype%3DactsAll (accessed 11 September 2014).

a) Where there was an unreasonable delay by the Registrar of Patents in granting the patent;
b) Where the patent was granted on the basis of information relating to a corresponding application filed in a prescribed patent office under Section 29(2)(c)(ii) and there was an unreasonably delay in granting of the patent office, and that patent office has granted an extension;
c) Where the subject of the patent includes any substance which is an active ingredient of any pharmaceutical product, that
   i There was an unreasonable curtailment of the opportunity to exploit the patent caused by the process of obtaining marketing approval for a pharmaceutical product, being the first pharmaceutical product to obtain marketing approval which uses the substance to obtain marketing approval which uses the substance as an active ingredient; and
   ii The term of the patent has not previously been extended on this ground.

Ground (c) above is elaborated upon in section 36A(5), as follows:

> (5) A curtailment of the opportunity to exploit a patent, the subject of which includes a substance which is an active ingredient of any pharmaceutical product, caused by the process of obtaining marketing approval for a pharmaceutical product, being the first pharmaceutical product to obtain marketing approval which uses the substance as an active ingredient, shall not be treated as an unreasonable curtailment under subsection (1)(*c*) unless such requirements as may be prescribed are satisfied.

Rule 51A of the Patents Rules further sets out the implementation of these provisions.[46]

[46] Under this Rule, paragraph (7) provides as follows:

> (7) For the purposes of Section 36A(5), a curtailment of the opportunity to exploit a patent, the subject of which includes a substance which is an active ingredient of any pharmaceutical product, caused by the process of obtaining marketing approval for a pharmaceutical product, being the first pharmaceutical product to obtain marketing approval which uses the substance as an active ingredient, shall not be treated as an unreasonable curtailment under Section 36A(1)(*c*) unless —
>
> (a) the marketing approval was obtained after the date of issue of the certificate of grant; and
> (b) the interval between the date the application for marketing approval was filed and the date marketing approval was obtained, excluding any period attributable to an act or omission of the applicant for marketing approval, exceeds 2 years.

The requirement for patent extension in Article 16.8.4 of the USSFTA is brief, leaving signatories a degree of flexibility to develop details of how such extensions are to be implemented:

> 4. With respect to any pharmaceutical product that is subject to a patent:
>
> (a) each Party shall make available an extension of the patent term to compensate the patent owner for unreasonable curtailment of the patent term as a result of the marketing approval process; . . .

The Singapore provisions explained earlier in this chapter therefore provide useful lessons in the face of the above commitment, by requiring limiting the situations eligible for an extension, and in the determination of the duration of an extension. They represent a fairly good effort at attempting to balance the extension commitment with ensuring that extensions are targeted to address the type of reason for their grant.

The text of the Canada–EU FTA (CETA) released in September 2014 includes a "sui generis" period of additional protection for eligible patent holders, within which system exceptions may be allowed.[47]

For what may constitute such "attributable" period, see Rule 51(11). Under paragraph (8) of the Rule, the extension period for a grant on ground 36A(1)(c) is determined as being the shortest of the following:

(a) a period equivalent to the interval between the date of issue of the certificate of grant (of the patent) and the date marketing approval was obtained;
(b) the period by which the interval referred to in paragraph (7)(*b*) (above) exceeds 2 years;
(c) a period of 5 years.

The hierarchy of possible extension periods in paragraph (8) is important in two respects: first, it requires a selection of the shortest possible extension of the three periods, so that no undue extension of the patent (and its monopoly rights) beyond the "curtailed opportunity to exploit" period is granted; and secondly, even if (in the unusual event that) the durations in (a) and (b) exceed 5 years, the *maximum* extension in effect will be for *5 years* under (c) where that is the shortest. In addition, where paragraph (7)(b) applies (with respect to the interval between the application for marketing approval and the actual grant of such approval), there is a minimum requirement that the interval must *exceed 2 years*; any interval less than that disqualifies the application of (7)(b), and therefore, Section 36A(5) of the Act, for want of an "unreasonable curtailment".

47 Article 9, *Canada–European Union Comprehensive Economic and Trade Agreement* (CETA). At the time of writing the CETA is awaiting ratification and implementation. Art. 9.5 in particular provides the possibility of exceptions to permit making and sale of products during such a period of "sui generis" protection, opening the door to the *export* of generic products during such periods; see the Canadian government's explanation of this point in *Technical Summary of Final Negotiated Outcomes – Canada–European Union Comprehensive Economic and Trade Agreement, Agreement-in-Principle*, October 2013, at p. 19.

### 4.6.3 *Provisions limiting patent opposition, revocation and renewals*

Countries which permit a pre-grant opposition procedure include Brazil, Egypt, India and Thailand.[48] Some countries have, on the other hand, entered into FTA commitments that exclude such procedures.

The USSFTA is an example of an agreement which removes the right of a state to provide for pre-grant opposition proceedings.[49] It has been argued that such proceedings provide an important threshold check on unmeritorious or weak patent applications and their removal makes it more difficult for a third party to challenge such patents. There are two reasons for this.

Some FTA provisions may also seek to limit the grounds for revocation once a patent is granted. Article 18.4 of the KORUS FTA, for example, limits the grounds for revocation of a patent to the grounds for refusing a grant of patent. Article 18.4 also prevents the state parties to the KORUS FTA from providing for opposition to grant of patent before a grant is made. This is a hurdle to those who may wish to block the granting of a patent without merit.

KORUS FTA strengthens the protection of registered patents in several ways. For one, the FTA makes it a challenge both to refuse registration and to revoke a patent. Article 18.8(3) permits limited exceptions subject to certain conditions, while Article 18.8(4) limits the grounds for revocation of patents.

### 4.6.4 *Provisions limiting parallel imports*

Most countries in Asia permit parallel imports, with several adopting the approach of international exhaustion.

#### 4.6.4.1 Singapore

The Patents Act permits parallel importation of medicines, subject to specific constraints. The relevant provisions are Section 66(2)(g), read with Sections 66(3) and 66(5A). Section 66(2)(g), read with 66(3), permits parallel importation of a patented medicinal product if the product has already been sold or distributed in Singapore by the patentee or his licensee, who is authorized to sell or distribute it in Singapore. In

[48] See generally, WIPO Secretariat, *Opposition Systems*, 11 December 2009, SCP/14/5, and WIPO Secretariat, *Opposition Systems*, SCP/17/9.

[49] See generally, WIPO, paper on opposition systems: www.wipo.int/edocs/mdocs/scp/en/scp_19/scp_19_ref_scp_18_4.pdf.

cases where such previous sale or distribution is absent, importing the product would amount to infringement where such import amounts to a breach of the patentee's licence agreement with his licensee who is authorized to distribute the product outside Singapore, and the importer is shown to have actual or constructive knowledge of this.

In explaining this provision, the government said:

> This is a delicate balance we have sought to preserve between the interests of the patent owner and the interests of users of pharmaceutical products. Essentially, the patent owner has a "first mover advantage" in the Singapore market, but once he is in, will have to compete with the parallel importers.
>
> . . . in drafting this amendment, we have given special consideration to ensure that no Singaporean would be denied adequate medical care as a result of this new right.[50]

While this arrangement imposes a limitation on parallel importation, it has been argued to be a relatively limited restraint.[51] Importantly, it still permits parallel importation where the patentee or his licensee has already introduced the medicine in the Singapore market. Prior to amendment in 2004, the provision permitted parallel importation whether or not there had been previous sale or distribution in Singapore.[52] The amendment occurred as a result of the commitment in Article 16.7(2) – read with footnote 16–10 – of the USSFTA.[53]

[50] Parliamentary speech by the then Senior Minister of State for Law on the 2004 amendments to the Patents Act, 15 June 2004.

[51] See Ng-Loy Wee Loon, Chapter 6, *Parallel Imports of Pharmaceuticals: Doha versus Free Trade Agreements*, p. 168, in *Intellectual Property and Free Trade Agreements*, ed. Christopher Heath and Anselm Kamperman Sanders (Hart Publishing, 2007).

[52] Section 66(2)(g) previously provided that an act would not be infringement if: "it consists of the import, use, disposal or offer to dispose of, of any patented product, or of any product obtained by means of a patented process or to which a patented process has been applied, *which is produced by or with the consent (conditional or otherwise) of the proprietor of the patent or any person licensed by him*, and for this purpose "patent" includes a patent granted in any country outside Singapore in respect of the same or substantially the same invention as that for which a patent is granted under this Act and "patented product", "patented process" and "licensed" shall be construed accordingly." (Italics added.) The present section 66(3) containing the qualifying conditions was added in 2004.

[53] Art. 16.7(2) and footnote 16–10 respectively provide as follows –

Art. 16.7.2.:

> "Each Party shall provide that patent owners shall also have the right to assign, or transfer by succession, a patent and to conclude licensing contracts. Each Party shall provide a cause of action to prevent or redress the procurement of a patented pharmaceutical product, without the authorization of the patent owner, by a party who knows or has reason to know that such product is or

### *4.6.5 FTA provisions affecting national formulary and reimbursement systems*

The TRIPS Agreement does not explicitly regulate national pricing policies and formulary systems in respect of pharmaceutical products. Member states therefore have a degree of flexibility in policy-making in these areas. Some FTAs, however, have added some constraints to this flexibility, such as in the requirements of formulary listing and reimbursement systems and practices.

While the USSFTA does not include any pricing or form of formulary control, the KORUS and US–Australia FTAs do. Provisions place price reimbursement systems under surveillance and subject measures under them to treaty challenge by investors.[54]

The KORUS FTA provides as follows, introducing treaty disciplines over a number of aspects of the Korean formulary listing and reimbursement system:

> **Article 5.2: Access to Innovation**
>
> To the extent that healthcare authorities at a Party's central level of government operate or maintain procedures for listing pharmaceutical products, medical devices, or indications for reimbursement, or setting the amount of reimbursement for pharmaceutical products or medical devices, under healthcare programs operated by its central level of government,1 the Party shall:
>
> (a) ensure that the *procedures, rules, criteria, and guidelines that apply to the listing of pharmaceutical products*, medical devices, or indications for reimbursement, or setting the amount of reimbursement for pharmaceutical products or medical devices are *fair, reasonable, and non-discriminatory*;
>
> (b) ensure that the Party's determination, if any, of the reimbursement amount for a pharmaceutical product or medical device, once approved by the appropriate regulatory authority as safe and effective, is *based on competitive market-derived prices; or* if its

> has been distributed in breach of a contract between the right holder and a licensee, regardless of whether such breach occurs in or outside its territory. 16–10 Each Party shall provide that in such a cause of action, notice shall constitute constructive knowledge."

Footnote 16–10: "Party may limit such cause of action to cases where the product has been sold or distributed only outside the Party's territory before its procurement inside the Party's territory."

[54] See, e.g., KORUS, and the Australia–US FTA, Annex 2-C, and side letter on the Australian Pharmaceutical Benefits Scheme.

determination is not based on competitive market-derived prices, then that Party shall:

(i) *appropriately recognize* the value of the patented pharmaceutical product or medical device in the amount of reimbursement it provides;
(ii) permit a manufacturer of the pharmaceutical product or medical device to apply, based on evidence of safety or efficacy, *for an increased amount of reimbursement over that provided for comparator products*, if any, used to determine the amount of reimbursement; and
(iii) permit a manufacturer of the pharmaceutical product or medical device, after a decision on a reimbursement amount is made, to *apply for an increased amount of reimbursement* for the product based on evidence the manufacturer provides on the product's safety or efficacy; and . . . [Italic emphases added]

It can be seen that these requirements not only address qualitative scrutiny for reasons of transparency and non-discrimination, but also stipulate significant benchmarks for determining reimbursement amounts. While paragraphs (i)–(iii) address situations where the national authority may depart from using "market-derived prices", they nonetheless encumber it with the need to apply an ambiguous standard of "appropriate recognition" of the value of the patented medicine. This ambiguity alone is fertile ground for disagreement and provides potential for investor claims by patent holders. It should also be noted that the provision does not explicitly provide for factors that a national authority may take into account in its reimbursement policy while determining the "value of the patented pharmaceutical product", such as health technology assessment, and pharmacoeconomic factors, such as therapeutic value and cost-effectiveness.[55]

The Australia–US FTA expressly refers, on the other hand, to "the need to recognize the value of innovative pharmaceuticals through the operation of competitive markets or by adopting or maintaining procedures that appropriately value the objectively demonstrated therapeutic significance of a pharmaceutical".[56]

[55] See e.g. Jacoline Bouvy and Sabine Vogler, *Pricing and Reimbursement Policies: Impacts on Innovation*, Background Paper 8.3, *Priority Medicines for Europe and the World – "A Public Health Approach to Innovation"* (2013), and József Bodrogi and Zoltán Kálo, Principles of Pharmacoeconomics and their Impact on Strategic Imperatives of Pharmaceutical Research and Development, *Br. J. Pharmacol.*, Vol. 159, No. 7 (April 2010), 1367–1373.

[56] Annex 2-C, Agreed Principle para. 1(d); see related concerns expressed in respect of New Zealand in relation to the TPP negotiations, in Deborah Gleeson, Ruth Lopert and Papaarangi Reid, How the Trans Pacific Partnership Agreement Could Undermine

A question that may be asked at this point is whether such provisions will appear in a TPP agreement, and if so, whether they have been sufficiently adjusted to take into consideration the level of development of negotiating countries from Asia.

#### 4.6.5.1 EUSFTA

Under the EUSFTA whose negotiations were largely completed in September 2013, there are provisions that have a significant impact on Singapore's pharmaceutical procurement processes. Unlike the USSFTA which carries a large number of commitments in its IP chapter, the EUSFTA provisions are found in an Annex to the goods chapter.[57] In particular, the provisions under para. 3 of Annex 2-C of the FTA affect measures on listing, pricing and/or reimbursement of pharmaceutical products.[58] These provisions aim to increase transparency in how the Singapore systems of pharmaceutical procurement and subsidization

PHARMAC and Threaten Access to Affordable Medicines and Health Equity in New Zealand, *Health Policy*, Vol. 112, No. 3 (October 2013), pp. 227–233.

[57] Annex 2-C, Chapter 2, National Treatment and Market Access for Goods, EUSFTA, released text of chapters available at: http://trade.ec.europa.eu/doclib/press/index.cfm?id=961 (accessed 29 April 2014).

[58] Para. 3 states as follows:

3. To the extent that a Party's healthcare authorities introduce or operate procedures for the listing, pricing and/or reimbursement of pharmaceutical products, the Party shall:

(a) ensure that the criteria, rules, procedures, and any guidelines, where relevant, that apply to the listing, pricing and/or reimbursement of pharmaceutical products, are objective, fair, reasonable and non-discriminatory, and are available upon request to interested persons;
(b) ensure that decisions on all applications for the pricing or approval of pharmaceutical products for reimbursement are adopted and communicated to the applicant within a reasonable and specified period from the date of the receipt of the application. If the information submitted by the applicant is deemed inadequate or insufficient and the procedure is suspended as a result, the Party's competent authorities shall notify the applicant of what additional information is required and resume the original decision-making process upon receipt of this additional information;
(c) provide applicants with appropriate opportunities to provide comments at relevant points in the pricing and reimbursement decision-making processes without prejudice to the applicable domestic law on confidentiality;
(d) in case of a negative decision on listing, pricing and/or reimbursement, provide the applicant with a statement of reasons that is sufficiently detailed to understand the basis of the decision, including the criteria applied and, if appropriate, any expert opinions or recommendations on which the decision is based. In addition, the applicant shall be informed of any remedies available under domestic law and of the time limits for applying for such remedies.

operate.[59] They do not require any particular pricing level; neither do they require any decision to take special account of whether a pharmaceutical product is patented or not. Apart from requiring reasoned decision-making and an opportunity for comment, para. 3(a) does impose a substantive standard for decisions on listing, pricing and/or reimbursement, namely, that healthcare authorities are to:

> ensure that the criteria, rules, procedures, and any guidelines, where relevant, that apply to the listing, pricing and/or reimbursement of pharmaceutical products, are *objective, fair, reasonable and non-discriminatory*, and are available upon request to interested persons.

Unlike the KORUS provisions above, however, they do not directly regulate the reimbursement *value* of medicines as determined by the national authority or link them to market prices.

Under EU law, member states have a degree of leeway with respect to pricing and reimbursement for medicines. For pricing, members are permitted to retain discretion to include *price increases, price freezes, price reductions and profitability controls* in their individual laws. The EU Transparency Directive however aims to regulate the *transparency* with which these national laws and regulations are applied.[60] The EUSFTA provisions are therefore broadly consistent with the EU philosophy behind this Directive. The Directive however uses the standard of "objective and verifiable criteria" for decisions made,[61] not that in para. 3(a) of the EUSFTA as stated above. Significantly, Article 14 of the Directive *de-links* such pricing, reimbursement and health insurance inclusion decisions from intellectual property rights.[62]

[59] See Chapter 3.

[60] *Directive of the European Parliament and of the Council relating to the transparency of measures regulating the prices of medicinal products for human use and their inclusion in the scope of public health insurance systems*, repealing Council Directive 89/105/EEC was adopted by the Commission on the 1 March 2012, especially Arts. 3–6 and 7–9.

[61] See, e.g., Arts. 3.9, 4.6, 5.1 and 7.7 of the Directive.

[62] Art.14 provides as follows:

> **Non-interference of intellectual property rights**
>
> 1. Applications, decision-making procedures and decisions to regulate the prices of medicinal products in accordance with Article 3 or to determine their inclusion within the scope of public health insurance systems in accordance with Articles 7 and 9 shall be considered by Member States as administrative procedures which, as such, are *independent from the enforcement of intellectual property rights.*

Comparing the KORUS and EUSFTA regimes on pricing and reimbursement, it can be seen that the KORUS provisions go further, in areas such as ensuring transparency and due process.

### 4.6.6 *Data exclusivity provisions*

#### 4.6.6.1 The Forerunner of TRIPS: NAFTA

NAFTA Article 1711, the precursor to more recent "data exclusivity" FTA provisions, provides for the protection of "trade secrets". Specifically, Arts. 1711.5–1711.7 provide as follows:

5. If a Party requires, as a condition for approving the marketing of pharmaceutical or agricultural chemical products that utilize new chemical entities, the submission of undisclosed test or other data necessary to determine whether the use of such products is safe and effective, the Party shall protect against disclosure of the data of persons making such submissions, where the origination of such data involves considerable effort, except where the disclosure is necessary to protect the public or unless steps are taken to ensure that the data is protected against unfair commercial use.
6. Each Party shall provide that for data subject to paragraph 5 that are submitted to the Party after the date of entry into force of this Agreement, no person other than the person that submitted them may, without the latter's permission, rely on such data in support of an application for product approval during a reasonable period of time after their submission. For this purpose, a reasonable period shall normally mean not less than five years from the date on which the Party granted approval to the person that produced the data for approval to market its product, taking account of the nature of the data and the person's efforts and expenditures in producing them. Subject to this provision, there shall be no limitation on any Party to implement abbreviated approval procedures for such products on the basis of bioequivalence and bioavailability studies.

2. The *protection of intellectual property rights shall not be a valid ground to refuse, suspend or revoke decisions* relating to the price of a medicinal product or its inclusion within the public health insurance system.
3. Paragraphs 1 and 2 shall apply without prejudice to the Union and national legislation relating to the protection of intellectual property.

[Italic emphases added]

7. Where a Party relies on a marketing approval granted by another Party, the reasonable period of exclusive use of the data submitted in connection with obtaining the approval relied on shall begin with the date of the first marketing approval relied on.

Three points are noteworthy here.

First, the period of protection of submitted data is "a reasonable period of time after their submission" which is "normally" not less than 5 years. This suggests that there may be circumstances where the period is less. This differs from the set time limits prevalent in more recent FTAs, usually periods of 5 years.

Second, the period is linked to "the nature of the data and the person's efforts and expenditures in producing them". Such a link is absent in more recent FTA provisions.

Third, the protection afforded under paragraph 5 is "against disclosure of the data of persons making such submissions, *where the origination of such data involves considerable effort, except where the disclosure is necessary to protect the public or unless steps are taken to ensure that the data is protected against unfair commercial use*" (italic emphasis added). Variants of this language found its way into Article 39 of the TRIPS Agreement but not the limiting language of paragraphs 6 and 7 mentioned above.

In the WTO context, during negotiations of the TRIPS Agreement there were proposals to include data exclusivity in Article 39. Members eventually agreed to Article 39.3 in its present, more general form, protecting against "unfair commercial use".[63] The prerequisite of "considerable effort" in the provision is not defined, which leaves its interpretation open.

#### 4.6.6.2 US law

Data exclusivity exists under US law and varying periods apply to different types of drug approval applications. The Hatch–Waxman Act provides the following periods of protection for the stated categories of application:

- New chemical entities (NCEs): 3 years
- New uses: 5 years
- Orphan drugs: 7 years[64]

[63] Trilateral study by WTO, WIPO and WHO, note, p. 64.

[64] See FDA website, www.fda.gov/Drugs/DevelopmentApprovalProcess/ucm079031.htm#How long is exclusivity granted for? and www.fda.gov/Drugs/DevelopmentApprovalProcess/

For paediatric drugs, 6 months may be added to the 3- and 5-year periods.[65] For biologic pharmaceuticals, the protection is the most extensive, namely 12 years.[66]

Notably, for NCEs, US domestic law only provides a data exclusivity period of 3 years.

#### 4.6.6.3 EU law

EU law provides for data exclusivity protection with regard to confidential data submitted to regulatory authorities for the purpose of obtaining pharmaceutical marketing approval. The period of exclusivity was initially 6–10 years. This has been extended to 8–11 years.[67]

#### 4.6.6.4 Data exclusivity and FTAs with Asian parties

The extended periods of 5–6 years for NCEs in FTAs, such as in the USSFTA, surpass domestic US law protection, that is, are "US-law-plus".[68] In the case of the USSFTA, as well as the KORUS FTA, the duration also begins from the *date of grant* of marketing approval and not from the (earlier) date of the application.[69]

In the case of the KORUS FTA, Article 18.8(5) requires that the use of "subject matter" of a patent should be for "generating information" to support an application for marketing approval, and any sale/export of

SmallBusinessAssistance/ucm069962.htm; relevant legislation: Federal Food, Drug, and Cosmetic Act under Section 505(c)(3)(E) and 505(j)(5)(F) and 21 C.F.R. § 314.108. The information on orphan drug data exclusivity is at: www.fda.gov/ForIndustry/DevelopingProductsforRareDiseasesConditions/HowtoapplyforOrphanProductDesignation/default.htm (accessed 24 July 2014). See generally www.fda.gov/Drugs/DevelopmentApprovalProcess/ucm090410.htm (accessed 23 July 2014). See also generally, *Knowledge Ecology International* (KEI) website at: http://keionline.org/testdata. (accessed 23 July 2014).

65 Section 505A, *Food and Drug Administration Modernization Act* of 1997 (the Modernization Act).

66 Biologics Price Competition and Innovation Act 2009. See Linfong Tzeng, Follow-on Biologics, Data Exclusivity, and the FDA, *Berk. Tech. L. J.*, Vol. 25 (2010), 135–158. Other issues relating to biologic pharmaceutical products are discussed further in Chapter 8.

67 See footnote 95 below and accompanying text; see also, generally, EMA website for legislative details: www.ema.europa.eu/ema/index.jsp?curl=pages/regulation/document_listing/document_listing_000211.jsp&mid=WC0b01ac0580031b0a (accessed 23 July 2014).

68 For a discussion of concerns on data exclusivity and developing countries in the context of a possible US–Malaysia FTA, see Robert Galantucci, Data Protection in a U.S.-Malaysia Free Trade Agreement: New Barriers to Market Access for Generic Drug Manufacturers, *Fordham I.P., Media and Ent. L. J.*, Vol. 17, No. 4 (2007) 1083–1123.

69 Art. 16.8.1, USSFTA.

goods using such "subject matter" should similarly only be for the purpose of supporting such an application.

Article 18.9 of KORUS FTA applies in particular to "certain regulated products" and their patents, namely pharmaceutical products and agricultural chemical products. The provision essentially provides a number of extensions of periods of blockage of grant of marketing approval to other drug manufacturers. Besides such blockage periods, it is noteworthy that the protection terms and periods in Article 18.9 KORUS FTA exceed those of Arts. 16.8.1–16.8.2 of USSFTA.[70] By comparison, Article 10.36 of the EU–Korea FTA imposes a period of 5 years for the protection of data with respect to pharmaceutical products,[71] while the Japan–Switzerland Agreement, however, imposes a period of 6 years from time of approval of the original application.[72]

The EFTA–Korea FTA provides an interesting variant on the data exclusivity obligation. While it requires the signatory Parties in the first instance to *prevent reliance* on undisclosed test or other data submitted by a first applicant for marketing approval of its pharmaceutical product, it also states that they may "instead" allow in their national legislation other applicants to rely on data for marketing approval, *if* the first applicant is "adequately compensated".[73] The treaty does not explain the meaning of this last phrase but it provides an important alternative. Whether it is an illusory alternative would depend on what is meant by "adequate".

The EU–Singapore FTA draft text[74] shows provisions on patent extensions and data exclusivity protection provisions have been agreed upon. The period of protection proposed under this FTA is 5 years, but provides for a future review with a view to increasing this duration.[75] Article

[70] The USSFTA commitments by Singapore on data protection can be seen in s. 19A of the Medicines Act. See also saving clause in footnote 16–14 USSFTA. Note that Chapter 5 of the KORUS FTA contains separate provisions on the sale and marketing of Pharmaceutical Products and Medical Devices. One of the commitments in the chapter deals with pricing of covered products and devices; see, e.g., Art. 5.2(b) which requires ensuring "competitive market-derived prices" or "reimbursement" based on product safety or efficacy.

[71] The full text of the EU–Korea FTA is available at: http://ec.europa.eu/trade/creating-opportunities/bilateral-relations/countries/korea/.

[72] Art. 121, Agreement of Free Trade and Economic Partnership between Japan and the Swiss Confederation, footnote 47, *infra*.

[73] Article 3, Free Trade Agreement between the EFTA States and the Republic of Korea.

[74] Text available at: http://trade.ec.europa.eu/doclib/press/index.cfm?id=961 (accessed 6 February 2014).

[75] Art. 11.33, footnote 29: "The Parties shall, five years after the entry into force of this Agreement, initiate without prejudice discussions on the possible extension of the

11.30 refers to the WTO Declaration on TRIPS and Public Health as well as the "paragraph 6" documents. Interestingly, Chapter 16 provides for a "Mediation Mechanism".

Under the EFTA–Hong Kong FTA, the duration of protection for pharmaceuticals data is at least 8 years from granting of marketing approval to the first applicant.[76]

The Canada–Korea FTA does not contain any TRIPS-plus data exclusivity commitments.[77]

In the ongoing TPP negotiations, the US has demanded of its trade partners, a similar 12-year period of protection for biologics. According to the USTR,

> ... Biologic drugs need data protection because those drugs require enormous amounts of time and money to develop. Before entrepreneurs (in the United States and across the world) are willing to make the investment in new therapies, they want to know that they will have rights to their own research for a certain period of time in order to see a return on their investments.
>
> In the TPP negotiations, opinions vary on the best term of patent protection for biologics. Standards also vary across the TPP regions. Some TPP countries currently have no data protection for biologic drugs. Some have 5 years. Others have 8 years. *Traditionally, the U.S. approach to trade negotiations has been to base proposals on existing U.S. law, where the current standard is 12 years.*[78] [Italic emphasis added]

However, more generally, on access to medicines, the USTR stance with regard to TPP negotiations is as follows:

> **Access to Medicines**
>
> The United States is a leading voice both for strong IPR protections and for access to medicines for the world's poor, including in developing country TPP partners. We believe the best approach to pharmaceutical IPR issues in the TPP would be one that offers countries flexibility based on their individual circumstances. That's why we've begun to work with TPP partners to gauge their interest in a "differential approach," and to

duration of the protection of test data submitted to obtain an administrative marketing approval to put a pharmaceutical product on the market." The duration of protection for agricultural chemical products is 10 years: Art. 11.34.

76 See Art. 4, EFTA–Hong Kong FTA.

77 See www.international.gc.ca/trade-agreements-accords-commerciaux/agr-acc/korea-coree/toc-tdm.aspx?lang=eng (accessed 21 July 2014).

78 USTR statement of November 2013, USTR website: https://ustr.gov/about-us/policy-offices/press-office/blog/2013/November/stakeholder-input-sharpens-focuses-us-work-on-pharmaceutical-IP-in-TPP (accessed 25 September 2015).

identify ways to tailor potential flexibilities based on countries' existing laws and international obligations.

This flexible approach is based on precedent: Previous U.S. trade agreements covered by the May 10, 2007 bipartisan agreement. Under May 10, developing free trade agreement partners (like Peru) were *offered greater flexibility* relative to more developed trade agreement partners (like Korea). *In TPP, we are seeking to pursue a similar idea, using previous agreements – like those with Peru, Australia, Chile, Korea, and Singapore – as benchmarks, but keeping an open mind as to how these standards can be tailored to reflect the situations of individual partners.*[79]

[Italic emphasis added]

#### 4.6.6.5 Israel: example of test data law amended after agreement with the US

In the case of Israel, recent amendments to its data exclusivity period arose not as a direct result of an FTA. In 2010, the US and Israel reached an agreement regarding certain amendments to be made by Israel to its intellectual property laws.[80] In July 2011, Israel amended its Pharmacists' Ordinance to extend the duration of data exclusivity from 5 to 6 years *after* marketing approval is granted in Israel where the first worldwide application occurs in Israel, *or* 6.5 years (instead of 5.5 years) *after* receiving final marketing approval from a recognized country.[81] This unique formulation permits, in the first situation, an Israeli company to have a slightly shorter wait before it may rely on protected data.

[79] *Ibid.*

[80] See para. 108, WTO Trade Policy Review, Report of the WTO Secretariat, referring to the USTR-Government of Israel Agreement of 18 February 2010. See also USTR announcement of removal of Israel from the Special 301 Report "Watch" list, February 2014, at: www.ustr.gov/about-us/press-office/press-releases/2014/February/Israel-removed-from-Special-301-Report. Israel had signed an FTA in 1985; see http://tcc.export.gov/Trade_Agreements/All_Trade_Agreements/exp_005439.asp (accessed 23 July 2014). Prior to the 2010 agreement between the US and Israel the US had initiated in 2009, a Special 301 "out-of-cycle review" (OCR) of Israel. This is a process permitting USTR examination of trade policies and negotiations with a trade partner, outside the annual Special 301 "reporting cycle"; see www.ustr.gov/about-us/press-office/press-releases/2010/february/deputy-ustr-sapiro-announces-united-states-and-is. The author is grateful to Mr Wolf Meier-Ewert, TRIPS Division, the WTO, for the helpful information regarding these developments. For information on 2005 amendments to the Israeli Pharmaceutical Ordinance to refine the pharmaceutical data provisions and related case law, see Eran Liss and Dan Adin, *Intellectual Property Law and Practice in Israel* (Oxford University Press, 2012) at pp. 586–588.

[81] See para. 63(3), Israel, WTO Trade Policy Review, Report of the Government of Israel; see commentary at KEI: http://keionline.org/node/1576 (accessed 23 July 2014).

It is noteworthy that in Israel, attempts by pharmaceutical companies to sue for use of pharmaceutical data on the basis of unjust enrichment and unauthorized use of trade secrets have been made in the courts.[82]

The Australia–Japan FTA employs what might be seen as a minimalist provision, referring to and reiterating the obligation in TRIPS with respect to undisclosed information.[83] The FTA also exceptionally does not include ISDS. Instead the signatory parties retain a choice of whether to utilize the FTA options (e.g. state-to-state arbitration under Chapter 19) or "dispute settlement procedures available under any other international agreement to which both Parties are party, including the WTO Agreement".[84]

In 2011, India expressed the view – in the context of the EU–India FTA negotiations – that it will not accept TRIPS-plus data exclusivity provisions.[85]

#### 4.6.6.6 Singapore: pharmaceutical data protection after the USSFTA

In the case of Singapore's experience, after the USSFTA came into force on 1 January 2004, Singapore introduced amendments to its system of approval and licensing of sale of medicines under the existing Medicines

[82] See Eran Liss and Dan Adin, *Intellectual Property Law and Practice in Israel*, note 80 above, at pp. 586–587.

[83] Art. 16.13, *Agreement between Australia and Japan for an Economic Partnership*, entered into force 15 January 2015. Art. 16.13 provides:

> Each Party shall protect undisclosed information in accordance with Article 39 of the TRIPS Agreement.

Definitions in Art. 14.2(f) read with Art. 16.2 of the FTA render undisclosed information as investments that may be covered under the FTA.

[84] Art. 19.3(1), *ibid.*

[85] The then Minister of Commerce & Industry has been reported as saying the following at a meeting of the consultative committee of the Indian Parliament on challenges in Intellectual Property Rights-international and domestic:

> "While data protection is something which must be respected, data exclusivity is well beyond the provisions of Article 39.3 of TRIPS agreement. India does not provide data exclusivity for pharmaceuticals and agrochemicals which is in the paramount interest of our generic pharmaceutical industry as grant of data exclusivity would have considerable impact in delaying the entry into the market of cheaper generic drugs."

See India government press release, 29 March 2011, http://pib.nic.in/newsite/erelease.aspx?relid=71341 (accessed 21 July 2014). See also *India against Inclusion of Data Exclusivity in any FTA* (6 April 2011), Economic Times; http://articles.economictimes.indiatimes.com/2011-04-06/news/29388653_1_data-exclusivity-drug-seizure-issue-data-protection (accessed 21 July 2014).

Act to implement its treaty obligations.[86] One such amendment of some significance is Section 12A which establishes patent linkage. As a result, the Singapore Health Sciences Authority (HSA), the licensing authority for medicinal products, now licenses pharmaceutical products subject to patent information regarding relevant, similar products being provided. However, it should be noted that this provision also includes an important inroad into the expanded protection provided to originator–patent holders. This is because the section additionally provides a means for an applicant for a licence to serve a notice on a patent proprietor, requiring him to obtain a court order blocking the licensing for infringement. Failure to obtain such a court order after service of the notice would clear the way for the HSA to grant the licence.[87] This mimics – in a highly abbreviated form – a part of the thinking in the US patent system, commonly known as the "Hatch-Waxman" amendments to the US Federal Food, Drug and Cosmetics Act.[88] These US amendments were introduced to facilitate generic producers' entry into the market by allowing generic drug manufacturers to make certain "certifications" as well as challenge the validity of the proprietorship of the patent. The Singapore provisions are not as extensive as the US legislative provisions, but still provide an important opportunity for generic manufacturers to obtain licensing (or under the FTA, marketing approval) even where there is an existing patent, subject to the conditions in the provision. This is an important and unique provision in the USSFTA that is often overlooked, which could be of relevance in negotiating other FTAs in which patent linkage is requested.

Under the Section 12A system, where an originator submits confidential information in support of an application for a licence for a relevant "innovative medicinal product", the originator enjoys protection of such data by the licensing authority in two ways. The protection comprises firstly an obligation on the part of the authority that it will "take reasonable steps" to keep as confidential, the applicant's

[86] Cap. 176. The amendments were introduced via the Medicines (Amendment) Act 2004 (No 26 of 2004). Concomitantly, amendments were made to the Patents Act (Cap. 221) via the Patents (Amendment) Act 2004 (No 19 of 2004). See generally the HSA's "Guidance on Medicinal Product Registration in Singapore" (2011).

[87] Section 12A(3)-(6), Medicines Act, *ibid.*

[88] Drug Price Competition and Patent Term Restoration Act of 1984; see explanations of the amendments by the US Food and Drug Administration (FDA) in www.fda.gov/downloads/Drugs/GuidanceComplianceRegulatoryInformation/Guidances/ucm079342.pdf and www.fda.gov/newsevents/testimony/ucm115033.htm.

"confidential supporting information"[89] (e.g. clinical trials data), and secondly, that it will not use such information in determining whether to grant another application.[90] Both provisions operate during the "protected period", defined as follows:

> "protected period", in relation to confidential supporting information relating to an innovative medicinal product application received by the licensing authority, means a period of 5 years from the date the innovative medicinal product application is or was, as the case may be, received by the licensing authority.

This means that in effect, the non-originator applicant is unable to obtain access via the licensing authority to (and therefore rely on) test data of the originator for the said period, with the result that it may have to conduct its own tests or trials. It also means that such an applicant may not benefit during such period from the licensing authority using the protected information for his application.

However, certain specific exceptions which allow disclosure or use of such "confidential supporting information" is permitted under Sections 19B and 19C of the Patents Act. Section 19B(1)(a) permits disclosure or use of information with the applicant's consent, or, more significantly, "if that disclosure or use is, in the opinion of the licensing authority, necessary to protect the health and safety of members of the public".

Section 19D of Singapore's Medicines Act – also a reflection of USSFTA commitments – further prohibits the licensing authority from granting a licence to a non-originator applicant for a period of 5 years in respect of the medicinal product for which information had been submitted or for a "similar medicinal product" *on the basis of an earlier-granted licence* (to an originator) and of safety or efficacy information supporting that earlier-granted licence. This has the effect of delaying entry into the market of the non-originator's medicine.[91] This has to be read in conjunction with the further conditions under Section 12A (explained below);

[89] Defined in section 19A as . . .

[90] Section 19A, Medicines Act.

[91] Section 19D provides:

> **19D.**—(1) Where —
>
> (a) information has been provided by an applicant for a product licence to the licensing authority relating to the safety or efficacy of a medicinal product in support of such application; and
>
> (b) a product licence has been granted in respect of that medicinal product (referred to in this section as the earlier licence),

Date of grant of patent (by patent authority)

↓

Date of application for grant of marketing licence (to licensing authority)
(submission of confidential information
protected under s. 19D)

↓

5 years of data protection by licensing authority under s. 19A and 19D

↓

Final 18 months before expiry of relevant patent

↓ Date of application of later applicant

Expiry of relevant patent ↓

Grant of marketing licence to later applicant (*after* expiry of relevant patent)

**Figure 4.1** Example of an "A3" sequence

specifically, if the later applicant is applying for a licence but is not asserting that the patent in question is valid (under the "A3" procedure of Section 12A),[92] such an applicant has to wait until expiry of the patent before the licensing authority may grant him the licence in any case, and may not make such an application for grant earlier than 18 months before expiry of the patent.[93] (Where he does assert that the patent is invalid, the separate procedure "B", explained below, would apply.) The sequence and time limitations in situation "A3", in combination with that in section 19D, may take the course shown in Figure 4.1.

Sections 19A and 19D therefore place statutory obligations on the *regulatory authority* (the Singapore HSA) to firstly protect the confidential information, secondly not use or rely on such information should a subsequent application for the medicine or a similar medicine be received, and thirdly, to refrain from granting a licence to such a later applicant on the basis of an earlier-granted licence.[94]

the licensing authority may not, for a period of 5 years from the date of such grant, grant a product licence to another person in respect of that or a similar medicinal product on the basis of the grant of the earlier licence unless the holder of the earlier licence has given his consent to the grant on that basis.

92 See note 109 below and the accompanying text.

93 Medicines (Licensing, Standard Provisions and Fees) (Amendment) Regulations 2004, Sub. Leg. No. S. 384/2004.

94 The wording of Art. 19D(1)(b) ("on the basis of the grant of the earlier licence") is a little curious, as one might have expected the prohibition to have been based on *reliance on*

Under the Medicines Act regime therefore, a patent holder has the benefit of a 5-year period of protection for data submitted for its marketing approval application (Sections 19A and 19D) and a period of protection in which a competitor may not obtain grant of marketing approval which lasts till the patent expiry date (Section 12A read with its regulations, where the competitor does not assert that there is an invalid patent). By contrast, under the EU system, there is an 8-year data protection period, and a 10-year marketing approval protection period (which may be extended).[95]

### 4.6.6.7 The EUSFTA

In late 2012, the EU and Singapore concluded negotiations for the EUSFTA.[96] Chapter 11, which deals with intellectual property rights, contains data exclusivity provisions. For pharmaceutical products, the duration of protection is 5 years, but interestingly for Singapore this is from the date of *receipt of the first application*, while for the EU this is from the date of approval of the first application. Singapore has therefore managed to negotiate for a shorter protection duration than that which applies to the EU; there is however also provision for a review 5 years after entry into force of the FTA of this duration, with a view to a possible extension.

> **Protection of Test Data**
>
> **Article 11.33**
>
> **Protection of Test Data Submitted to Obtain an Administrative Marketing Approval to put a Pharmaceutical Product on the Market**
>
> When a Party requires the submission of test data or studies concerning the safety and efficacy of a pharmaceutical product prior to granting

*confidential information* submitted for the grant of the earlier licence, rather than on the earlier licence itself.

[95] This is the so-called "8+2+1" system under Art. 10 of EU Directive 2001/83 and Art. 14 of Regulation 726/2004. (Art. 10(1) of the Directive and Art. 14(11) of the Regulation permit an extension up to 11 years for the marketing approval protection. Extensions depend on whether there is "significant clinical benefit" shown.) See Christian R. Fackelman, Clinical Data, Data Exclusivity and Private Investment Protection in Europe, in *Pharmaceutical Innovation, Competition and Patent Law*, ed. Josef Drexel and Nari Lee (Edward Elgar Publishing, 2013), especially at pp. 149–164.

[96] The FTA – save for the Investment chapter –was initialled on 20 September 2013 and awaits ratification; the EU suspended the conclusion of investment provisions pending a public consultation initiated in early 2014 (arising mainly from its negotiations of the TTIP with the US). On the public consultation on ISDS which ended on 13 July 2014, see http://trade.ec.europa.eu/consultations/index.cfm?consul_id=179 and http://trade.ec.europa.eu/doclib/docs/2014/july/tradoc_152693.pdf (accessed 25 July 2014).

approval for the marketing of such product, the Party shall not, for a period of at least five years from the date of approval in that Party, permit third parties to market the same or a similar product, on the basis of the marketing approval granted to the party which had provided the test data or studies, unless the party which had provided the test data or studies has given its consent.[28, 29, 30]

[28] The conditions and procedures for the provision of the protection contemplated under this Article shall be determined by the respective legislations of the Parties.

[29] The Parties shall, five years after the entry into force of this Agreement, initiate without prejudice discussions on the possible extension of the duration of the protection of test data submitted to obtain an administrative marketing approval to put a pharmaceutical product on the market.

[30] In the context of this Article, it is understood that such confidential supporting test data or studies shall not be used for the purposes of determining whether to grant any other application for a period of at least five years:

(a) in the case of Singapore, from the date that the first application is received;

(b) in the case of the Union, from the date of approval of the first application,

unless the party which had provided the test data or studies has given his consent.

### 4.6.7 *Patent linkage FTA provisions*

As explained by one author:

> "'Patent linkage' is a practice of linking regulatory approval for a generic medicine to the patent status of the originator product."[97]

While patent linkage exists in the patent legislation of some countries (such as Singapore and the US), other countries, such as the EU, do not make such linkages.[98]

#### 4.6.7.1 Patent linkage – KORUS FTA

KORUS FTA imposes patent linkage obligations under Arts. 18.8(5) and 18.9. Article 18.9(4) further extends the period of protection against marketing approval in Article 18.9(1)–(2) *beyond* the life of the patent

[97] Hiroko Yamane, *Understanding TRIPS – Globalisation of Intellectual Property Rights and Access to Medicines* (Hart Publishing, 2011), at p. 445.

[98] See EU Directive 2001/83/EC.

in question. This means that upon the expiry of a patent during the period under those provisions, a generic manufacturer must still wait till the end of the protection period before a grant of approval can be obtained. Such treaty provisions give a significant *de facto* extension of protection to the patent holder by continuing to block marketing of the generic good for a period, despite expiry of the patent.

The KORUS FTA is accompanied by a side letter on patent linkage dispute settlement.[99] The side letter signals agreement between the US and Korea that neither will resort to Article 22.4 (which sets out the grounds for triggering dispute settlement process beginning in Article 22.7) for the first 18 months of the entry into force of the agreement but that each would instead resort to consultations under Article 22.3 to arrive at a mutually satisfactory solution.

An issue under the KORUS patent linkage provisions which came to light in the early 2015 is whether the patent linkage requirements in the KORUS FTA include (beyond small molecule drugs) biological pharmaceuticals.[100] The US recently confirmed through a letter that it would, and that it would also seek such coverage in the TPP agreement which is under negotiation.[101]

#### 4.6.7.2 Patent linkage – Singapore's new section 12A procedure

Until 2004, the regime for licensing medicines in Singapore was distinct from the patenting regime. Previously, the relevant law containing the licensing regime, the Medicines Act, permitted the authority to license a drug based on considerations of safety and efficacy.

The USSFTA came into force on 1 January 2004 and soon after, Singapore introduced amendments to its system of approval and licensing of sale of medicines under the existing Medicines Act.[102] One such amendment of some significance is section 12A which

[99] See www.ustr.gov/trade-agreements/free-trade-agreements/korus-fta/final-text (accessed 23 July 2014).

[100] See *Korea Accuses U.S. of Failing to Provide 'Patent Linkage' for Biologics*, 19 March 2015, *Inside US Trade*, and.

[101] *Ibid.*, and see letter from US Ambassador Mark W. Lippert to Minister Chung Seung, Ministry of Food and Drug Safety, Republic of Korea, dated 17 February 2015, at InfoJustice website, http://infojustice.org/archives/34087 (accessed 8 May 2015).

[102] Cap. 176. The amendments were introduced via the Medicines (Amendment) Act 2004 (No 26 of 2004). Concomitantly, amendments were made to the Patents Act (Cap. 221) via the Patents (Amendment) Act 2004 (No 19 of 2004).

establishes patent linkage; licensing of a medicine would be "linked" to the issue of whether a patent exists "in respect of any medicinal product to which the application relates".[103] An applicant for a licence would need to state if there are any relevant patent(s) for the medicine that the licence would pertain to. Following such a statement the applicant would need to . . .

As a result, the Singapore HSA, the licensing authority for medicinal products, now has the task of making the above inquiry as to existence of a relevant patent before granting a licence.

However, Section 12A of the Medicines Act provides a means for an applicant for a licence to serve a notice on a patent proprietor requiring him to obtain a court order blocking the licensing for infringement. Failure to obtain such a court order after service of the notice would clear the way for the HSA to grant the licence.[104] This mimics in a highly abbreviated form a part of the thinking in the US patent system, commonly known as the "Hatch-Waxman" amendments to the US Federal Food, Drug and Cosmetics Act.[105] The US amendments were introduced to facilitate generic producers' entry into the market by allowing generic drug manufacturers to make certain "certifications" as well as challenge the validity of the proprietorship of the patent. The Singapore provisions are not as extensive as the US amendments but still provide an important opportunity for generic manufacturers to obtain licensing (or under the FTA, marketing approval) even where there is an existing patent, subject to the conditions in the provision.

When an originator wishes to obtain approval to market his product in Singapore, he files a new drug application (NDA). A non-originator applicant for a licence for a generic drug files a generic drug application (GDA). The HSA defines a generic drug as follows:

> A generic product is *essentially similar* to a currently registered product in Singapore (known as the "Singapore reference product") but excludes biologics. *Essentially similar* is defined as having the same qualitative and quantitative composition in terms of active substances, having the same pharmaceutical form and being bioequivalent. By extension, the concept

[103] Section 12A(1). [104] Section 12A(3)-(6), Medicines Act.

[105] Drug Price Competition and Patent Term Restoration Act of 1984; see explanations of the amendments by the US Food and Drug Administration (FDA) in www.fda.gov/downloads/Drugs/GuidanceComplianceRegulatoryInformation/Guidances/ucm079342.pdf and www.fda.gov/newsevents/testimony/ucm115033.htm.

of essentially similarity also applies to different conventional immediate release oral dosage forms (i.e. tablets and capsules) which contain the same active ingredient(s).[106]

HSA maintains an online database listing reference products. A GDA may proceed to an "abridged" evaluation route, or a "verification" evaluation route. The former applies where the product has been approved by at least one regulatory agency. The latter – the faster process of the two – applies where the product has been approved by at least two of the agencies listed by the HSA.[107] Where neither route applies, a full evaluation has to be undertaken.[108]

Under the amended law, a non-originator applicant for a licence is given an opportunity to test a weak or bad patent under a procedure provided for under section 12A of the Medicines Act.[109] This provision permits the applicant to declare that grant of a licence to his drug will either not be an infringement of the originator's patent, or that the patent is invalid. Following such a declaration the patentee is given a period of 45 days in which he must file an action in court for infringement, failing which the licence may be granted to the applicant. This mechanism is illustrated in Figure 4.2.

[106] HSA Guidelines (2011), p. 45.

[107] The five listed agencies are the US Federal Drug Administration, Health Canada, Australian Therapeutics Goods Administration, European Medicines Agency and the UK Medicines and Health Products Regulatory Agency.

[108] See HSA Guidelines (2011), pp. 26, 42–43 and 72. For the differences in duration of evaluation routes, see Appendix 1 of the Guidelines. See also, generally, Liu, *et al.*, *Evidence-Based Decision-Making on Medical Technologies in China, Japan, and Singapore, Value in Health*, Vol. 12, Suppl 3 (2009), S12–S17.

[109] Section 12A permits four categories of applications for non-originator drug applicants: A1, A2, A3 and B, as follows:

A1: where the applicant declares that there is no patent for the product for which a licence is being applied for;

A2: where there is a patent but the applicant declares he is the owner or that the patentee consents or acquiesces to the grant of the licence;

A3: where there is a patent and the applicant declares that he is not applying for grant of a licence before the expiry of the patent, and the application is being made no earlier than 18 months before such expiry; and

B: there is a patent but the applicant declares it is invalid, or that the act for which the licence is sought will not infringe the patent.

A small number of cases relating to section 12A has been observed so far; see *AstraZeneca AB (SE) v Sanofi-Aventis Singapore Pte Ltd* [2013] SGHCR 7 and *AstraZeneca AB (SE) v Sanofi-Aventis Singapore Pte Ltd* [2012] SGHC 16.

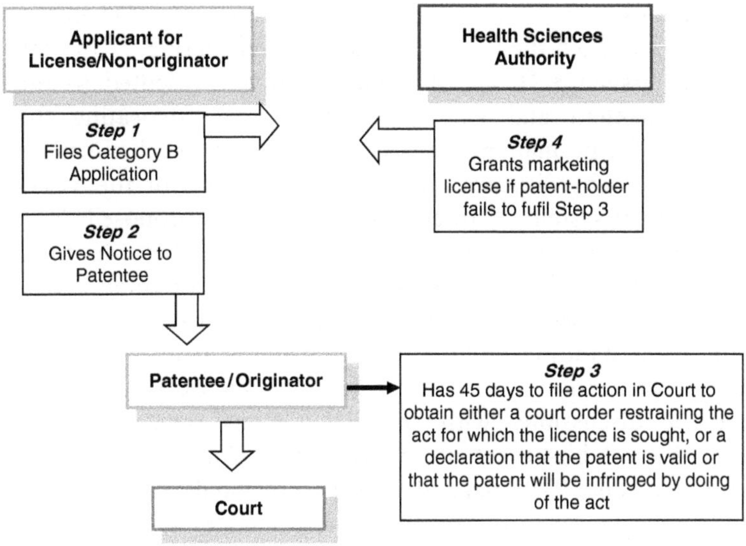


**Figure 4.2** Section 12A "Category B" application mechanism

## 4.7 The treaty-based ISDS system – legal risks and concerns

### *4.7.1 The new frontiers of investor-state disputes and the nexus with IPR-health provisions*

In recent years, given the rise in number of bilateral investment treaty commitments and investor-state disputes, concern has grown over various aspects of the handling of such disputes. Criticisms have been raised at the process, the outcomes and the players.[110] Expressed concerns span a variety of aspects, including a lack of consistency in legal reasoning among tribunals, the lack of an appellate mechanism, issues relating to selection and choice of arbitrators, costs and the fact that tribunals can make determinations affecting matters of public interest.

A small number of Latin American countries (Bolivia, Ecuador and Venezuela) have chosen to denounce the International Centre for Settlement of Investment Disputes (ICSID) Convention to begin the process of exiting its ISDS system.[111] In the negotiations for the TTIP between the EU and the US which are ongoing at the time of writing,

[110] See, e.g., Michael Waibel, *The Backlash Against Investment Arbitration: The Perceptions and the Reality* (Wolters Kluwer L. & Bus., 2010).

[111] See UNCTAD, http://unctad.org/en/Docs/webdiaeia20106_en.pdf and *Venezuela-ICSID: Leaving is Easier Said than Done*, Financial Times, 17 January 2012.

strong objections have been raised to the inclusion of ISDS in the trade agreement, which the EU Commission is considering.[112] In May 2015, the EU issued a "Concept Paper" – to be presented to the European Parliament and EU trade ministers – advocating the establishment of a permanent court (with an appeal mechanism) to deal with investment disputes, drawing from aspects of the WTO dispute settlement system.[113] In addition, the EU has proposed, *inter alia*, to include stronger provisions to safeguard member countries' "right to regulate" under the terms of negotiated free trade agreements.[114] The proposed court would be "multilateralized", either by setting up a self-standing body or by "embedding it in an existing multilateral organization".[115] In September 2015, the EU Commission proposed for discussions with the EU Council and European Parliament, the establishment of an Investment Court system comprising a first instance Tribunal and an Appeal tribunal.[116] The Appeal Tribunal "would be operating on the same principles as the WTO's Appellate Body".[117] The proposal neatly illustrates the discontent – at least in the EU Commission's view – with the existing system of *ad hoc* investor-state arbitration tribunals.

Some Asian countries – such as Indonesia and India – are also taking various actions to address their concerns. Indonesia announced recently

[112] See *France and Germany to Form United Front against ISDS*, EurActiv France, 15 January 2015, www.euractiv.com/sections/trade-society/france-and-germany-form-united-front-against-isds-311267 (accessed 23 January 2015). See also the EU's consultation process over the ISDS question, at EU website, http://trade.ec.europa.eu/consultations/index.cfm?consul_id=179 (accessed 23 January 2015). The consultation report, which benefited from a reportedly 150,000 replies, was published on 13 January 2015. In a further related development, it has been reported that France and Germany are even requesting the re-opening of discussions of the Canada–EU FTA (CETA), to amend the ISDS provisions in it: *Paris and Berlin call for review of EU–Canada trade deal*, Euractiv report, 27 January 2015 (accessed 30 January 2015).

[113] EU Concept Paper, *Investment in TTIP and Beyond – the Path for Reform, Enhancing the right to regulate and moving from current ad hoc arbitration towards an Investment Court.*

[114] Paras. I.1 and I.3, EU Concept Paper, *ibid.* See also generally, www.euractiv.com/sections/trade-society/eu-floats-isds-reform-ahead-trade-talks-us-314344 (accessed 6 May 2015).

[115] Para. V, EU Concept Paper, note 113, above.

[116] See http://europa.eu/rapid/press-release_IP-15-5651_en.htm (accessed 7 October 2015) and text at http://trade.ec.europa.eu/doclib/docs/2015/september/tradoc_153807.pdf (accessed 7 October 2015).

[117] *Ibid.* At the time of writing, this proposal had not yet been put to the US as a negotiating proposal.

that it would terminate its bilateral investment treaties.[118] India, having recently faced (and lost) an investor-state arbitration by an Australian investor, has undertaken a review of its model bilateral investment agreement and proposed a new draft model agreement.[119]

Apart from investment treaties some Asian countries such as Myanmar are also undertaking reviews of their foreign investment statutes. Myanmar enacted a Foreign Investment Law in 2012, which is expected to be replaced imminently by a new law of 2015 which will deal both with international and domestic investments.[120]

In the context of public health issues, two recent examples of disputes have received a great deal of attention. The first is in the area of tobacco regulation, involving Australia's "plain packaging" legislation introduced in 2012; disputes are ongoing in international arbitration and at the WTO.[121] The second is in the pharmaceutical field, in which Eli Lilly, a US-based pharmaceutical manufacturer, has an ongoing ISDS dispute against the government of Canada.[122] This case will be discussed further below.

Apart from the general concerns mentioned above, a number of specific matters in ISDS are of particular interest in the context of investors who are in the field of pharmaceuticals and related health

[118] See Ben Bland and Shawn Donnan, *Indonesia to terminate more than 60 bilateral investment treaties*, Financial Times, 6 March 2014. India is reviewing its Bilateral Investment Protection Agreements (BIPA); see *BIPA Talks Put on Hold*, 22 January 2013, The Hindu. It is planning to introduce a new Bilateral Investment Treaty (BIT) model; see *India to replace BIPA with a new pact to protect investments*, Economic Times, 19 November 2014, http://articles.economictimes.indiatimes.com/2014-11-19/news/56265699_1_bipa-investment-protection-model-text (accessed 22 January 2015). Also see the next note.

[119] *White Industries Ltd v Republic of India*, UNCITRAL, Final Award (30 November 2011). See also generally, Hsu, Asian Treaty Makers and Investment Arbitration: Negotiating with a Wary Eye, *Contemp. Asia Arb. J.*, Vol. 5, No. 2 (2012), 243–263. On the new draft model agreement, see text at India government website, https://mygov.in/sites/default/files/master_image/Model%20Text%20for%20the%20Indian%20Bilateral%20Investment%20Treaty.pdf (accessed 6 May 2015).

[120] The 2015 Act will consolidate and replace the Myanmar Citizens Investment Law (Pyidaungsu Htluttaw Law No. 18 of 29 July 2013) and the Foreign Investment Law (Pyidaungsu Htluttaw Law No. 21, 2012, 2 November 2012).

[121] On the investor-state arbitration, see the Australian government website, at www.ag.gov.au/tobaccoplainpackaging (accessed 22 January 2015). On the WTO disputes, see DS434, initiated by Ukraine; DS435, initiated by Honduras; DS441, initiated by the Dominican Republic; DS458 initiated by Cuba; and DS467, initiated by Indonesia.

[122] See Government of Canada website, at www.international.gc.ca/trade-agreements-accords-commerciaux/topics-domaines/disp-diff/eli.aspx?lang=eng (accessed 23 January 2015).

products. The following part links common ISDS issues with this more specific context.

#### 4.7.1.1 Broad scope of "investments" – IPRs and other intangible rights

For example, Article15.1.13 of the USSFTA includes "licences" conferred pursuant to applicable domestic law as protected "investments". A licence for marketing approval of a drug granted under the Medicines Act could therefore be a covered "investment". This would subject the "investment" (and its treatment by Singapore) to the commitments in Chapter 18 of the FTA, including those relating to National Treatment, Most-Favoured Nation, the minimum standard of treatment and expropriation. As will be explained later in this chapter, various uncertainties surround such provisions and their application, so that the treaty investment commitments not only lead to potential liability for measures, but also subject the interpretation of such commitments to these wider uncertainties.

#### 4.7.1.2 Broad "treatment" investment provisions – FET

Investment protection provisions include guarantees of certain minimum treatment or standards of protection for foreign investors. Among these, the "fair and equitable" treatment standard is one which has been interpreted by numerous investment arbitral tribunals and the subject of much academic discussion. While a number of recurrent factors have begun to emerge from various arbitral tribunal decisions, and there has even been an attempt by the EU to capture these in its evolving FTA negotiations for greater certainty,[123] it remains an undefined concept which is subjected to interpretation by tribunals which are not bound by precedent, even if they do refer to interpretations in prior awards. As each tribunal is required to undertake the interpretative exercise in accordance with recognized international rules on treaty interpretation,[124] the precise provision, context and other matters in each case may differ as well. In addition, a tribunal in the NAFTA context, after a small number of NAFTA decisions raised the question of whether such treatment went beyond the standard under customary international law, the NAFTA Free Trade Commission issued an interpretative note in 2001 to clarify that it did not exceed the customary

[123] See notes 34 above and note 153 below.

[124] See, e.g. Arts. 31 and 32 of the Vienna Convention on the Law of Treaties.

international law standard.[125] Other FTAs have since begun to include such a similar clarification on the scope of this standard of treatment. Examples include: USSFTA Article 15.5, exchange of letters (following NAFTA Free Trade Commission interpretative note), and the ASEAN-Korea Investment Agreement (Article 5). Despite these interpretative and treaty developments, the precise boundaries of FET remain debatable and subject to the interpretation of a tribunal seised of a dispute should one arise under such a provision in an FTA.

In the case of the EU, it has aimed to achieve greater certainty in dealing with FET in its recent FTA negotiations, such as those with Canada and with Singapore. The EU approach has been to delineate the confines of the FET obligation by making express reference in its treaty texts to certain factors that have been applied in various arbitral awards, such as the legitimate expectations of the investor, a denial of justice, breach of due process and manifestly arbitrary conduct by the state.[126]

Myanmar's proposed new Foreign Investment Law contains a FET provision which attempts to provide some guidance and limitation on the scope of the FET obligation.[127]

[125] See NAFTA Free Trade Commission, *Notes of Interpretation of Certain Chapter 11 Provisions*, 31 July 2001.

[126] See negotiated texts of CETA (Art. X.9) at http://trade.ec.europa.eu/doclib/docs/2014/september/tradoc_152806.pdf and of the EUSFTA (Art. 9.4) at http://trade.ec.europa.eu/doclib/docs/2014/october/tradoc_152844.pdf (accessed 24 October 2014). The EUSFTA is awaiting ratification by the EU and implementation by Singapore at the time of writing.

[127] On this proposed law, see note 118 above. The proposed provision on FET provides as follows:

**Section 11: Regulatory Transparency and Fair and Equitable Treatment**

(1) The Union guarantees to all Investors fair and equitable treatment. Fair and equitable treatment includes the obligation not to deny justice in criminal, civil, or administrative adjudicatory proceedings in accordance with the principle of due process.

(2) Fair and equitable treatment as stated in this sub-section shall include, but is not limited to:

  (a) The right to obtain information on any action or decision which has an effect on any Investor and its Direct Investment;

  (b) The right to due process and right to appeal including, but not limited to, any changes to the terms of approval, permission or licence granted to by the Union Government to the Investors and their Direct Investments; and

  (c) The due compliance of the Union Government and Government Entities with contracts, if any, that it may have entered into with any Investor.

(3) All laws, regulations, procedures, administrative rulings of general application, and adjudicatory decisions respecting any matter covered by this Law are to be promptly published or otherwise made publicly available.

India, which has recently embarked on a review of its model BIT – the Bilateral Investment Promotion and Protection Agreement (BIPA) – has released a draft model treaty for public consultation.[128] The draft excludes any reference to "fair and equitable" treatment, and instead provides as follows:

> **Article. 3 Standard of Treatment**
>
> 3.1 Each Party shall not subject Investments of Investors of the other Party to Measures which constitute:
>
> (i) Denial of justice under customary international law[1];
> (ii) Un-remedied and egregious violations of due process; or
> (iii) Manifestly abusive treatment involving continuous, unjustified and outrageous coercion or harassment.
>
> 3.2 A determination that there has been a breach of another provision of this Treaty, or of a separate international agreement, does not establish that there has been a breach of this Article.
>
> [1] For greater certainty, it is clarified that "customary international law" only results from a general and consistent practice of States that they follow from a sense of legal obligation.

The draft also subjects investors to a number of positive obligations, such as not participating in corruption of local officials (Article 9) and disclosure obligations (Article 10).

#### 4.7.1.3 Expropriation, indirect, if undefined

There is no single definition of "expropriation". While early arbitral case law addressed primarily instances of direct expropriation, allegations of "indirect" expropriation have increased. In response to this, recent treaties have begun to spell out more detailed explanations of this term to try to clarify its scope and meaning. The US Model BIT, for example, includes an explanatory Annex which attempts to better set out the scope and boundaries of "indirect" expropriation.[129] Examples of such actual provisions can be found in the Singapore-India FTA ("CECA"), Annex 3 and, Singapore-Peru FTA Annex 10A.

[128] The text of the draft Model BIT is available at: https://mygov.in/sites/default/files/master_image/Model%20Text%20for%20the%20Indian%20Bilateral%20Investment%20Treaty.pdf (accessed 5 April 2015).

[129] The United States Model Bilateral Investment Treaty 2004 and the most recent 2012 Model Treaty each contain an Annex with such language (in both cases, named "Annex B").

This is clearly to permit for regulatory flexibility and to discourage challenges unless certain minimum conditions of expropriation are considered and met. Examples of such language can be found in Annex 11-B of the KORUS FTA.

Similarly, the ACIA, which came into force on 29 March 2012, similarly contains some exclusionary language with regard to expropriation.[130] Apart from setting out factors to determine if expropriation has occurred, it also provides specifically as follows:

"Non-discriminatory measures of a Member State that are designed and applied to protect legitimate public welfare objectives, such as public health, safety and the environment, do not constitute expropriation of the type referred to in sub-paragraph 2(b)."[131]

National investment laws may also contain provisions regulating expropriation and compensation.[132]

[130] ACIA, Annex 2. [131] *Ibid.*, para. 4.

[132] Myanmar's proposed new foreign investment law (see note 118 above) contains such a provision, which provides as follows:

**Section 15: Protection against Expropriation**

(1) The Union Government shall neither expropriate nor nationalize nor impose any measures that are in effect tantamount to expropriation (otherwise known as indirect expropriation), of any investments except:

14

(a) for a public purpose; and
(b) in a non-discriminatory manner; and
(c) against payment of prompt, fair, adequate and effective compensation; and
(d) in accordance with due process of law.

(2) Expropriation shall not include non-discriminatory measures of general application

which governments normally take for the purposes of regulating economic or social activity in their jurisdiction, including without limitation the measures enumerated in General Exception in Section 22.

(3) For a measure or series of measures to be indirect expropriation as in sub-section (1) above, there must be a case-by-case, fact-based inquiry that considers at least the following cumulative factors:

(a) whether the action creates an adverse effect on the economic value of an Investment;
(b) whether the action breaches the government's prior binding written commitment to the Investor (by contract, licence or other legal document); and
(c) whether the government action, including its objective, is disproportionate to the public purpose.

#### 4.7.1.4 Investor-state disputes and pharmaceutical patents – some recent developments

In *Eli Lilly v Government of Canada*, an ongoing investor-state arbitration, a Canadian common law doctrine has come under challenge in an arbitration initiated by a US investor under the NAFTA.[133] The investor Eli Lilly is challenging Canada's "promise utility" doctrine in determining patentability under Canadian law. The application of this doctrine by Canadian courts – allegedly from 2005 – had led to invalidation of two pharmaceutical patents belonging to Eli Lilly, namely *Strattera* and *Zyprexa*. Eli Lilly argues that the following NAFTA provisions have been violated by the Canadian government[134]:

> Article 1701(1) "provide in its territory to the nationals of another Party adequate and effective protection and enforcement of intellectual property rights, while ensuring that measures to enforce intellectual property rights do not themselves become barriers to legitimate trade."
>
> Article 1709(1) Canada is required to grant patents for inventions that "are new, result from an inventive step and are capable of industrial application."
>
> Article 1709(7) "patents shall be available and patent rights enjoyable without discrimination as to field of technology"
>
> Article 1709(8) Canada may only revoke a patent on grounds that would have justified a refusal to grant the patent in the first instance.

It further argues that by not observing the above obligations the Canadian government has violated Articles 1110 (unlawful expropriation) and 1105 (minimum standard of treatment) of the NAFTA.[135] In

[133] Arbitration initiated under UNCITRAL Rules (1976) and the NAFTA; Notice of Arbitration dated 12 September 2013.

[134] Eli Lilly argues in para. 9 of its Notice of Arbitration (*ibid.*): "Specifically, the judiciary in Canada has created *a new doctrine* to assess whether an invention meets the condition of being 'useful' or 'capable of industrial application' " (italic emphasis added). It further argues that this was not the standard that applied when the *Strattera* and *Zyprexa* patents were applied for or when they were examined by the Canadian Intellectual Property Office (para. 10). Canadian Federal Court cases that have applied the doctrine include *Apotex Inc. v. Wellcome Foundation Ltd.*, 2002 SCC 77, [2002] 4 SCR 153, *Apotex* v *Pfizer Canada Inc.*, 2011 FCA 236, *Eli Lilly Canada Inc. v. Apotex Inc.*, 2009 FCA 97, 78 C.P.R. (4th) 388 (F.C.A.), *Sanofi-Aventis v. Apotex Inc.*, 2013 FCA 186, *Eurocopter v. Bell Helicopter Textron Canada Ltd.*, 2013 FCA 219 and *Pharmascience Inc. v. Canada (Health)*, 2014 FCA 133. The *Eurocopter* case does not relate to a pharmaceutical invention, and the FCA opined that the doctrine may be applied to mechanical inventions as well (at para. 146).

[135] See Claimant's Notice of Arbitration and Memorial, available at: www.international.gc.ca/trade-agreements-accords-commerciaux/topics-domaines/disp-diff/eli.aspx?lang=eng (accessed 6 March 2015).

challenging the "utility promise" doctrine Eli Lilly is arguing that the Canadian standard for utility as defined by this doctrine as being out of step and an "outlier" when compared with those of other jurisdictions.[136] Eli Lilly also argues that the utility requirement was the subject of "substantial harmonization", which the government of Canada disagrees with.[137]

The significance of this arbitration is at least threefold. First, in the field of intellectual property, it challenges the "promise" doctrine as a patentability test in the Canadian system which is the subject matter of dispute. Second, the challenge squarely places domestic judicial decision-making – as opposed to executive action, for example, at the heart of the dispute. In this connection, the Canadian government argues in its defence that the NAFTA Tribunal:

> cannot act as yet another court of appeal. *International law provides no basis to second-guess the reasoning of the Canadian courts on questions of Canadian law in the absence of a denial of justice.* Claimant uses inflammatory language such as "arbitrary," "illogical and absurd" and "discriminatory" to describe the Federal Court's interpretation of Canadian law. These are mere labels which cannot mask the lack of legal foundation of Claimant's position.[138]
>
> [Italic emphasis added]

Third, a question that arises is whether such a challenge might not have a chilling effect[139] on judicial decision-making, by linking the

[136] Para. 9, Notice of Arbitration, note 133 above; see also Marc André-Séguin, *The $500-million Doctrine*, September–October 2013, The National Magazine, www.nationalmagazine.ca/Articles/Sept-Oct-2013/The-$500-million-doctrine.aspx (accessed 6 July 2014). The doctrine springs from a principle aiming to make a "sound prediction" of what a patent application "promises"; see e.g. Canadian Supreme Court case: *Apotex Inc. v. Wellcome Foundation Ltd.* [2002] 4 S.C.R. 153, 2002 SCC 77), and the Canadian Manual of Patent Office Practices takes both into account in assessing "utility" of an invention: particularly paras. 12.08.01, 12.08.03 and 12.08.004. This case remains pending at the time of writing.

[137] See respectively, para. 276 of the Claimant's Memorial, and para. 361–2, Counter-memorial of the government of Canada, available at: www.international.gc.ca/trade-agreements-accords-commerciaux/topics-domaines/disp-diff/eli.aspx?lang=eng (accessed 6 March 2015).

[138] Para. 215, Counter-memorial of the government of Canada, available at: www.international.gc.ca/trade-agreements-accords-commerciaux/topics-domaines/disp-diff/eli.aspx?lang=eng (accessed 6 March 2015).

[139] The broader question of whether ISDS in investment agreements may have a "chilling effect" on regulatory authorities in signatory states was recently considered in a study of the EU (in the context of the ongoing TTIP negotiations between the EU and the US):

notion of "legitimate expectations" (often raised in complaints of violations of FET) of investors to judicial measures.

Apart from substantive issues, other aspects of pharmaceutical-related investor-state arbitrations include the question of transparency of outcomes, especially where a national regulatory system has been challenged by a private investor. For example, in *Les Laboratoires Servier, S.A.A., Biofarma, S.A.S., Arts et Techniques du Progres S.A.S. v. Republic of Poland*, a complaint brought under the France–Poland BIT, challenging Poland's regulatory system for pharmaceutical

see 3.5.2, EU Directorate-General for Internal Policies Study, *Legal Instruments and the Practice of Arbitration in the EU* (2014). In relation to the concern of a possible "chilling effect" of investment agreements and their ISDS arbitrations, the Study concludes as follows:

> On the positive side, this system of selection of arbitrators entails that States can guarantee that at least one member of a tribunal holds views favourable to respecting State policy freedom. Consequently, it is *simply inaccurate* to picture the situation as being one of States drafting agreements that are then handed over to unaccountable arbitrators. Rather, States *retain significant influence on the interpretation of those agreements through the mechanism of arbitrator selection.*
>
> On the negative side, arbitrator selection also means that *there is unlikely to be much certainty in interpretation of investment agreements in the future. This would require uniformity amongst arbitrators across the field, rather than just a decision by a Supreme tribunal that all others must follow. This kind of uniformity is extremely unlikely.*
>
> Ultimately, then, the question of "chill" needs to be seen in the context of the *reality of international investment agreements. An international investment agreement is not like a domestic law that can simply be redrafted at will if it produces unexpected results.* IIAs, like all treaties, are long-term agreements that are rarely renegotiated, as such negotiations can only occur with the agreement of all participating States and require a considerable commitment by the States concerned. *In this light, the flexibility of the arbitrator selection mechanism, while admittedly reducing certainty of interpretation of treaties, and thereby potentially increasing regulatory "chill",* also serves as a means of ensuring ongoing input from States into the interpretation of treaties.
>
> Certainty of interpretation is, after all, only a positive thing if the interpretation adopted is a desirable one. *All judicial supervision, even by domestic courts, involves a State handing over an element of its sovereignty to a judicial body, and potentially being chilled as a result. However, through the ongoing role of States in arbitrator selection, States retain the ability to ensure that their preferred interpretation of any disputed treaty provision is represented in the tribunal's deliberations.*
>
> [Italic emphasis added]

products, the arbitral award was heavily redacted, presumably to protect confidential information.[140]

Generic drug entities have also initiated ISDS arbitration in efforts to protect their rights as investors under investment treaty provisions. For example, in Apotex Holdings Inc and Apotex Inc, US subsidiaries of a Canadian manufacturer of generic drugs, have initiated at least two sets of investor-state arbitration proceedings against the US.[141] One of these was brought under the NAFTA, in which the US prevailed; the other has been brought under ICSID arbitration, and remains pending at the time of writing.

#### 4.7.1.5 Interpretative divergences and ambiguities

The investor-state arbitration system arising from dispute settlement provisions in investment treaties has seen a steady increase in cases, amidst growing criticism. These criticisms are well-documented and debated elsewhere, and the present discussion will not repeat them.[142] One matter worth highlighting here, however, is the concern which has been expressed in relation to interpretative divergences and ambiguities that have been observed in arbitral tribunal decisions. Given that many FTAs that include TRIPS-plus obligations subject such obligations to their ISDS provisions, the concerns over such arbitrations cannot be ignored or confined to the decided cases.

#### 4.7.1.6 Potentially large sums in damages and very public damage to state's reputation

The quantum of damages payable in a BIT dispute can be staggering. For example, in a March 2012 arbitral award, the Libyan government was held liable to pay an investor the sum of US$1.2b inclusive of lost profits and *moral damages* (permitted under Libyan law). The dispute arose in

[140] E.g., the entire texts on pp. 17–19, 24–27, 45–56, 72–75, 77–78, 80–82, 91–99, 113–118, 140–142, 144–151, 169–173, 178–185, and significant portions of several other pages, have been blackened out; see award at: www.italaw.com/sites/default/files/case-documents/italaw3005.pdf (accessed 6 March 2015). The author is grateful to Professor Valentina Vadi for drawing her attention to this award.

[141] *Apotex Inc. v The Government of the United States of America (UNCITRAL Arbitration), pending; and Apotex Inc. v The Government of the United States of America*, ICSID Case No. ARB(AF)/12/1, Award rendered 25 August 2014.

[142] See, e.g., the Hon. the Chief Justice, Sundaresh Menon, Singapore, Transnational Commercial: Realities, Challenges, and a Call for Meaningful Convergence, *SJLS* (2013), pp. 231–252, and Jean E. Kalicki and Anna Joubin-Bret, eds., *Reshaping the Investor-State Dispute System* (Brill/Nijhoff, 2015).

relation to a tourist resort which, due to events described in the award, was never constructed at all. In another case, possibly the largest award granted in ICSID was that in *Occidental v Ecuador*[143]: US$1,769,625,000 (after 25% deduction for the investor's own breach). It is therefore not surprising that host states are wary of investor disputes and the potential damages payable.

In Asia – and in ASEAN, in particular – ISDS claims have been slowly increasing. Table 4.1 shows examples of such claims and the approximate claim sums.

A number of practical problems in respect of treaty negotiations and treaty arbitrations are faced in particular by developing countries/LDCs, such as Cambodia, Lao PDR and Myanmar. These are relatively recent entrants to the international legal environment with respect to trade and investment treaties.[144] The level of legal expertise (both in general and in relation to international trade and investment rules) and legal infrastructure is in great need of improvement. In Cambodia, for example, there has even been a shortage of lawyers.[145]

A meeting of the "treatified" patent bargain and the ISDS treaty "bargain"?
Connecting the dots – a meeting of the "treatified" patent bargain and the ISDS treaty "bargain"?

One idea alluded to earlier[146] is the possibility of considering the inclusion of explicit obligations of patent owners in an FTA (or BIT). This may be done in a variety of contexts, such as a compliance

[143] *Occidental Petroleum Corporation v The Republic of Ecuador*, ICSID Case No. ARB/06/11 Award, 5 October 2012.

[144] Myanmar has been a party to the GATT since its inception in 1948, and became a founding member of the WTO when the organization was established in 1995, but due to its domestic policies it has not been very engaged as such until its recent decisions (starting from the late 1980s and in particular more recently, between 2008-present) to reform its economy; see WTO Secretariat Report, Trade Policy Review, Myanmar, 21 January 2014, WT/TPR/S/293, p. 17, and Report of the Myanmar government, WT/TPR/G/29, pp. 5–6. Cambodia joined the WTO in October 2004, while Lao PDR was the final member of ASEAN to join the WTO, only in February 2013. Cambodia, Lao PDR, Myanmar and Vietnam (or CLMV) as less developed members of ASEAN are the subject of differential treatment and capacity-building schemes under ASEAN's Initiative for Integration (IAI); see www.asean.org/communities/asean-economic-community/category/initiative-for-asean-integration-and-narrowing-the-development-gap (accessed 13 April 2015).

[145] See e.g. Bridgette Dunlap, *The Rule of Law without Lawyers: American Legal Reformers and the Cambodian Lawyer Shortage*, http://papers.ssrn.com/sol3/papers.cfm?abstract_id=2424255 (accessed 13 April 2015).

[146] See Chapter 2, at note 88 and accompanying text.

Table 4.1. *Examples of recent investor-state claims brought against ASEAN states*

| Country | Case | Approximate size of claim | Outcome |
|---|---|---|---|
| Cambodia | *Cambodia Power Co. v Kingdom of Cambodia* | US$300 m | Claim dismissed |
| Lao PDR | *Thai-Lignite v Lao PDR* | US$57 | Claim succeeded; enforcement challenges pending. |
| | *Sanum Investments v Lao PDR* | US$235 m | Jurisdictional challenge succeeded in Singapore High Court decision |
| The Philippines | *Fraport v Philippines* | US$425 | Two arbitrations initiated:<br>The first resulted in an award in favour of the claimant, but which was subsequently annulled.<br>The second could not proceed for want of jurisdiction. |
| Thailand | *Walter Bau (in liquidation) v Thailand* | €30 m | Claim succeeded; various enforcement proceedings brought. |

requirement for a patent holder to be able to avail himself of ISDS under the treaty. Several FTAs/BITs already provide for ISDS to be available subject to certain requirements being fulfilled by an investor, such as the observance of domestic laws when making the investment and the obtaining of prior written approval from a domestic investment authority as an approved investor.[147] This approach would, in effect, be a meeting of the "treatified" patent bargain as discussed in Chapter 2, and the bargain in FTAs/BITs where the host states or investors consent to participate in international arbitration with a qualifying investor.

In a relatively small number of FTAs, such as the Australia-US FTA and the Japan-Australia Economic Partnership Agreement, there are no provisions for investor-State arbitration. In such cases, private investors would need to resort to national courts (or other national dispute mechanisms) or rely on their home State to initiate State-to-State proceedings.

Interestingly, Myanmar's proposed new Investment Law includes an "Investment Grievance Mechanism". This appears to draw inspiration from the Korean Investment Ombudsman system, providing foreign investors with an early-stage forum for airing grievances before the grievances escalate into legal disputes and before the investor resorts to treaty or court proceedings.[148]

## 4.8 "Counterweight" provisions relevant to the regulation of the pharmaceutical industry

Given the complexity of treaty negotiations, their politics, and the need to weigh economic and non-economic considerations, it may sometimes be overlooked that certain "counterweight" provisions can be useful in addressing some of the concerns surrounding the economic-benefits-versus-public health-and-access to medicines debate. One such area is the behaviour of private parties in marketing of pharmaceutical products; inappropriate or unethical behaviour – such as the use of bribery – can distort rational physicians' prescribing decisions and officials' procurement decisions.

[147] See, e.g., the Germany–Philippines BIT and Article 4(a) of the ASEAN Comprehensive Investment Agreement. See also, e.g. *Fraport AG Frankfurt v Republic of the Philippines*, ICSID Case No. ARB/11/12, Award, 10 December 2014, paras. 319–333, 467–468.

[148] See section 20 of the draft law of 2015, note 118 above. Section 21 of the draft law further provides for dispute settlement mechanisms such as investor-State arbitration. For information on the Korean Foreign Investment Ombudsman system, see www.i-ombudsman.or.kr/eng/index.jsp (accessed 5 October 2015).

Such distortions in turn can lead to greatly inflated pharmaceutical prices because the more expensive medicines sold by such marketing parties are chosen over less costly but effective alternatives. While such conduct is often seen as a matter for the national enforcement of anti-corruption laws, treaties can play a part in addressing it too.

It is interesting to note that Chapter 5 of the KORUS FTA contains the following (relatively unusual) provision, to address ethical issues with regard to manufacturers of pharmaceutical products or medical devices:

> **Article 5.5: Ethical Business Practices**
>
> 1. Each Party shall adopt or maintain appropriate measures to prohibit pharmaceutical product or medical device manufacturers and suppliers from providing *improper inducements* to healthcare professionals or institutions for the listing, purchasing, or prescribing of pharmaceutical or medical device products eligible for reimbursement under healthcare programs operated by its central level of government.
> 2. Each Party shall adopt or maintain appropriate penalties and procedures to enforce the measures that it adopts or maintains in conformity with paragraph 1.[149]
>
> [Italic emphasis added]

The above provision appears to have been seen as a necessary counterbalance such reimbursement commitments, providing a safeguard of sorts against forces that may "improperly" affect decisions on eligibility for reimbursement in a healthcare system. In this regard, this provision stands out as protection for the signatories that they may rein in any such "improper inducements". The fact that what is "improper" is undefined and that inducements can take many forms both provide a signatory with flexibility, as well as difficulty in determining the scope of prohibited conduct or inducements. Not many other FTAs contain such a commitment, despite complaints about such conduct or practices in many countries. In the next chapter, further safe harbour and exception provisions are explored to provide further balance to TRIPS-plus provisions.

## 4.9 Conclusion

The WTO members in Asia are exposed to legal risks attendance with the various agreements signed, including the TRIPS Agreement. In addition, FTAs which contain ISDS provisions in conjunction with IP commitments bring separate legal risks, particularly where broad or

[149] Art. 4 of the EU–Korea FTA contains similar provisions.

"trade-off" language is incorporated and will be subject to interpretation by a future arbitral tribunal. ISDS interpretations in general remain a vexed and complex area, with *ad hoc* tribunals interpreting a diversity of investment treaty provisions, sometimes without consistency of reasoning.[150] ISDS may also lead to a successful claimant being awarded a very sizeable compensation sum, if a violation of the relevant treaty is found to have occurred.

The complications and juridical differences that have arisen from the many international arbitral awards have led to some calls to remove ISDS altogether from the treaties under negotiation.[151] Negotiations on the ISDS aspects of the TTIP, which is under negotiation between the US and the EU, have been suspended in order for a public consultation to be carried out by the EU.[152] In an effort to inject greater certainty into some common investment treatment guarantee provisions which have been litigated in several international arbitrations, the EU also released a "fact sheet" on such provisions.[153] Most recently, the EU has put forward a proposal to establish an "investment court" instead of investor-State arbitration, with a view to presenting it to the US in its TTIP negotiations, and using it in "other ongoing and future negotiations".[154]

In an effort to assuage critics of the risks of ISDS in relation to the TTIP, the EU Trade Directorate representative Mr Ignacio Garcia Bercero wrote with confidence and optimism – rightly or wrongly – regarding

[150] See the following related works by this author: *The Rule of Law and Foreign Investment: Advancing Towards an Ideal through Investment Treaties*, chapter in publication by the Singapore Academy of Law (forthcoming); *Investor-State Dispute Settlement Reform – Examining the Formative Aspect of Investment Treaty Commitments: Lessons from Commercial Law and Trade Law*, Transnational Dispute Management (2014); Asian Treaty-Makers and Investment Treaty Arbitration: Negotiating with a Wary Eye, *Contemp. Asia Arb. J.*, Vol. 5, No. 2 (2012), 243–263; and MFN and Dispute Settlement – When the Twain Meet, *Journal of World Investment*, Vol. 7, No. 1 (2006), 25.

[151] See, e.g., Dan Ikenson, *Eight Reasons to Purge Investor-State Dispute Settlement from Free Trade Agreements*, 4 March 2014, Forbes, www.forbes.com/sites/danikenson/2014/03/04/eight-reasons-to-purge-investor-state-dispute-settlement-from-trade-agreements/. See also a related Cato Institute paper: www.cato.org/publications/free-trade-bulletin/compromise-advance-trade-agenda-purge-negotiations-investor-state.

[152] See EU website: http://europa.eu/rapid/press-release_IP-14-56_en.htm (accessed 23 July 2014).

[153] EU Fact Sheet, *Investment Protection and Investor-to-State Dispute Settlement in EU Agreements*, November 2013, http://trade.ec.europa.eu/doclib/docs/2013/november/tradoc_151916.pdf (accessed 23 July 2014).

[154] See "Commission proposes new Investment Court System for TTIP and other EU trade and investment negotiations", 16 September 2015, E website, http://europa.eu/rapid/press-release_IP-15-5651_en.htm (accessed 25 September 2015).

ISDS and the UK, in the context of the UK's right to regulate health services, as follows:

> In order for an investor to bring a successful ISDS case, it would need to prove that its rights under the relevant treaty have been breached, for example by expropriation without compensation, a denial of justice or manifestly arbitrary treatment. In such cases, an ISDS tribunal can award compensation but it cannot overturn national regulation, nor can it order repeal or reversal of a government's decision [related to the organization and the management of health services]. *Regardless of the content of the TTIP, such cases are unlikely to happen in the UK.* Member States are already required to respect applicable domestic and EU law . . .[155]
>
> [Italic emphasis added]

He appears to have overstated the case with the rather simplistic argument above, particularly by using the words "regardless of the content of the TTIP". Treaty content, and its interpretation, make all the difference, particularly when one applies the customary international law interpretative directions in Arts 31 and 32 of the Vienna Convention on the Law of Treaties.

Notably, there is an ongoing debate over whether to include ISDS provisions in the TTIP, with vociferous objections to such inclusion being raised in certain EU member states.[156] Australia, too, has been more circumspect with regard to ISDS recently.[157]

Already, ASEAN countries with relatively limited economic resources, such as Cambodia, Lao PDR (a least developed country), Thailand and Vietnam, have been facing investor challenges under BITs, albeit not in the area of health regulation measures. A number of ASEAN countries

[155] Letter to the Rt Hon John Healey, MP, Chair, All-Parliamentary Group on the TTIP, dated 8 July 2014; http://trade.ec.europa.eu/doclib/docs/2014/july/tradoc_152665.pdf (accessed 23 July 2014).

[156] See EU, http://trade.ec.europa.eu/consultations/index.cfm?consul_id=179 and *Euractiv*, www.euractiv.com/sections/trade-society/isds-decision-delayed-end-ttip-talks-311234 (accessed 13 April 2015).

[157] In 2011, the government of Australia (then under Prime Minister Julia Gillard) announced that it would no longer include ISDS in future trade agreements; see *Gillard Government Trade Policy Statement: Trading Our Way to More Jobs and Prosperity* (April 2011), available at www.acci.asn.au/getattachment/b9d3cfae-fc0c-4c2a-a3df-3f58228daf6d/Gillard-Government-Trade-Policy-Statement.aspx. This has shifted to a case-by-case approach under Prime Minister Tony Abbott; see stance set out by Department of Foreign Affairs and Trade, at www.dfat.gov.au/trade/topics/pages/isds.aspx (accessed 13 April 2015). See also generally, Locknie Hsu, Asian Treaty-Makers and Investment Treaty Arbitration: Negotiating with A Wary Eye, *Contemp. Asia Arb. J.*, Vol. 5, No. 2 (2012), 243, and especially p. 259.

have also recently signed (possibly TRIPS-plus) provisions in the TPP;[158] it remains to be seen what the final text contains. It is therefore important to examine next the role and importance of exceptions and flexibilities in negotiating treaties that bring together provisions on trade, investment, intellectual property protection and medicines.

[158] See, e.g., the following information on the TPP negotiations: *TPP demands on opposition proceedings*, www.pharmatimes.com/article/13-04-03/US_holds_back_on_pharma_IP_in_Asia-Pacific_trade_talks.aspx and *Pharmaceutical firms' battle over generic drugs threatens TPP*, www.japantimes.co.jp/news/2013/03/12/business/pharmaceutical-firms-battle-over-generic-drugs-threatens-tpp/. The TPP negotiating states announced on 5 October 2015 that they had signed the agreement but the text is not publicly available at this point of time.

# 5

## The importance of exceptions and flexibilities

### 5.1 Introduction

Given the legal risks arising from treaty commitments explained in previous chapters, together with requests for TRIPS-plus commitments in FTA negotiations (such as those of the TPP), the value of exceptions provisions cannot be underestimated. Indeed, as explained in a recent joint study published by the WHO, WIPO and WTO, there has been a "ratcheting-up" of intellectual property standards through the extensive global web of FTAs.[1]

A key factor in such "ratcheting-up" of IP requirements through FTAs is that these commitments have to be extended to other WTO members on a most-favoured nation (MFN) basis. Article 4 of the TRIPS Agreement provides:

> With regard to the protection of intellectual property, any advantage, favour, privilege or immunity granted by a Member to the nationals of any other country shall be accorded immediately and unconditionally to the nationals of all other Members.[2]

The effects of such commitments therefore transcend the trade relations of the immediate FTA parties; they often require amendments to domestic laws so that they are of general applicability and not just to apply to trade with the FTA partner.[3] In effect, therefore, a state that makes such commitments can expect system-wide "ratcheting-up" and, concomitantly, a significant loss of regulatory space if suitable exceptions are not carefully negotiated for inclusion.

[1] WIPO, WHO and WTO, *Promoting Access to Medical Technologies and Innovation – Intersections between Public Health, Intellectual Property and Trade* (2012); the survey is available at: .

[2] Article 4 provides a number of very limited exemptions to this requirement which do not apply here.

[3] See WIPO, WHO and WTO study, footnote 1 above, p. 84.

The US' continuing push, in particular, towards TRIPS-plus provisions in bilateral and regional FTAs is well documented.[4] In a recent report by the Congressional Research Service (CRS), the following was explained:

> The United States has sought increased intellectual property rights (IPR) protection in its FTAs. IPR negotiating objectives in the last U.S. trade promotion authority (P.L. 107–210) in effect between 2002 and 2007 included, among others (1) the application of existing IPR protection to digital media; and (2) negotiation of trade agreements in terms of IPR that "reflect a standard of protection similar to that found in U.S. law." This phrase *opened the door to the negotiation of provisions that go beyond the level of protection provided in the WTO Trade Related Aspects of Intellectual Property (TRIPS) Agreement, most recently with the TPP negotiations.*[5]
>
> [Italic emphasis added]

A number of FTAs entered into by the US already contain such TRIPS-plus commitments.[6] Such commitments in FTAs and their effect on access to pharmaceuticals have been debated not only in trade fora

[4] See, for example, Ruth Lupert and Deborah Gleeson, The High Price of "Free Trade": US Trade Agreements and Access to Medicines, *Journal of Law, Medicine and Ethics* (Spring 2013), 199–223; Hiroko Yamane, *Understanding TRIPS – Globalisation of Intellectual Property Rights and Access to Medicines* (Hart Publishing, 2011); Beatrice Lindstrom, Scaling Back Trips-Plus: An Analysis of Intellectual Property Provisions in Trade Agreements and Implications for Asia and the Pacific – especially Part C – *International Law and Politics*, Vol. 42 (2010), 917–980; UNDP, *The Potential Impact of Free Trade Agreements on Public Health*, May 2012; Richard D Smith, Carlos Correa and Cecilia Oh, *Trade, TRIPS and Pharmaceuticals, The Lancet* (January 2009), 684–691; Joseph Stiglitz, Trade Agreements and Health in Developing Countries, *The Lancet*, Vol. 373, No. 9661 (January 2009), 363–365; Deborah Gleeson and Sharon Friel, Emerging Threats to Public Health from Free Trade Agreements, *The Lancet*, Vol. 381, No. 9876 (April 2013), 1507–1509; Jonathan Berger and Achal Prabhala, "Assessing the Impact of TRIPS-Plus Patent Rules in the Proposed US-SACU Free Trade Agreement, www.who.int/hiv/amds/capacity/tza2_oxfamreport_pricing_financing.pdf; and Jakkrit Kuanpoth, *TRIPS-Plus Rules Under Free Trade Agreements: An Asian Perspective, Chapter 2, Intellectual Property and Free Trade Agreements*, ed. Christopher Heath and Anselm Kamperman Sanders (Hart Publishing, 2007).

[5] Ian F. Ferguson, William H. Cooper, Remy Jurenas and Brock R. Williams, *The Trans-Pacific Partnership Negotiations and Issues for Congress*, 15 April 2013, at p. 32; available at: www.fas.org/sgp/crs/row/R42694.pdf.

[6] Examples include the US' FTAs with Australia, Central America, Jordan, Morocco, Peru and Singapore. By contrast, China, a relatively recent entrant to the WTO and, subsequently, to the FTA "race", has signed nine FTAs to date which do not include TRIPS-plus provisions.

such as the WTO, but also in the WHO and UN General Assembly's Human Rights Council.[7]

With respect to the current TPP negotiations, the USTR has referred to the use of a "differential approach" and the aim to "identify ways to tailor potential flexibilities based on countries' existing laws and international obligations".[8] Meanwhile, an example of a TRIPS-plus request that has been raised by a number of US Senate members for inclusion in the negotiations is a 12-year protection period for regulatory data for biologic medicines or biopharmaceuticals.[9] This is currently the period of protection under US domestic law.

The attraction of trade and investment benefits in such agreements may lead negotiating states to accept such provisions that significantly expand patent holders' protection (beyond just patent rights), reduce regulatory space and increase dispute risk for the signatory states; if this is so without commensurate guarantees either of appropriate "flexibilities", or of alternative avenues for increased access to medicines committed by the requesting country, it could reduce, rather than increase, access. This makes it imperative for Asian states negotiating new FTAs and similar agreements to pay particular attention to the scope and kinds of exceptions to be negotiated, and other *quid pro quo* guarantees addressing public health.

## 5.2 Exception provisions – general considerations

Exception provisions exist in most major trade and investment treaties, such as those agreed under the auspices of the WTO. Reading through exception provisions in FTAs, one wonders if these are not often inserted as an afterthought, without as much attention being applied to them as to negotiation of the treaties' substantive-obligation provisions. A number of these exceptions are in the nature of "general exceptions", encompassing a range of reasons for which states are expressly permitted – subject to stated conditions – to derogate from treaty commitments. General exceptions may include measures pertaining to the protection of public morals,

[7] See, for example, the Report of the Special Rapporteur on the right of everyone to the enjoyment of the highest standard of physical and mental health: Anand Grover, A/HRC/11/12, 31 March 2009.

[8] USTR website, https://ustr.gov/about-us/policy-offices/press-office/blog/2013/November/stakeholder-input-sharpens-focuses-us-work-on-pharmaceutical-IP-in-TPP (accessed 6 May 2015).

[9] See, for example, the statement by Senator Orrin Hatch, US Senate Finance Committee hearing, 24 April 2013, text available at: www.finance.senate.gov/imo/media/doc/04%2024%2013%20Hatch%20Opening%20Statement%20on%20TPP%20hearing.pdf.

human health and the environment. A well-known example is Article XX of the GATT. For present purposes, the most pertinent part of Article XX is paragraph (b), which permits measures for the protection of human health. The conditions for applying this paragraph are found in the chapeau, which states:

> Subject to the requirement that such measures are not applied in a manner which would constitute a means of arbitrary or unjustifiable discrimination between countries where the same conditions prevail, or a disguised restriction on international trade, nothing in this Agreement shall be construed to prevent the adoption or enforcement by any contracting party of measures: . . .

The General Agreement on Trade in Services contains a near equivalent of this in its Article XIV. The WTO TBT Agreement does not contain an Article XX-equivalent but the case law shows that similar considerations can arise under the TBT provisions. In *US – Clove Cigarettes* dispute,[10] Indonesia successfully challenged US regulatory treatment of menthol and clove cigarettes under the TBT Agreement. Significantly, the Appellate Body noted in its report that

> we observe that the *TBT Agreement* does not contain among its provisions a general exceptions clause. This may be contrasted with the GATT 1994, which contains a general exceptions clause in Article XX.[11]

The Appellate Body did, however, note the recital language of the TBT Agreement, which it found to reflect the same "balance" as in the general exceptions provision of Article XX of the GATT:

> *Recognizing* that no country should be prevented from taking measures necessary to ensure the quality of its exports, or for the protection of human, animal or plant life or **health**, of the environment, or for the prevention of deceptive practices, at the levels it considers appropriate, subject to the requirement that they are not applied in a manner which would constitute a means of arbitrary or unjustifiable discrimination between countries where the same conditions prevail or a disguised restriction on international trade, and are otherwise in accordance with the provisions of this Agreement.[12]
>
> [bold emphasis added]

[10] *US – Measures Affecting the Production and Sale of Clove Cigarettes*, WTO/DS406/AB/R, decision of the Appellate Body (22 March 2012).

[11] Para. 101, *ibid.*

[12] See paras. 94–96 and 109. The Appellate Body noted at para. 109: "in the GATT 1994 this balance is expressed by the national treatment rule in Article III:4 as qualified by the exceptions in Article XX, while, in the *TBT Agreement*, this balance is to be found in Article 2.1 itself, read in the light of its context and of its object and purpose." For a more detailed examination of WTO interpretations of exceptions such as Article XX and TRIPS flexibilities, see Yamane, footnote 4 above.

This is useful as it shows that preamble language can also support health regulation flexibility. This is a noteworthy point for Asian negotiators in considering the language to include in FTA chapter preambles.

More generally, while decisions such as *US – Clove Cigarettes* provide important guidance in the interpretation of the TBT provisions, the specific language of Asian FTA exceptions, if any, that apply to permit health and tobacco regulation will have to be examined in accordance with treaty interpretation principles to determine the degree of flexibility available to the signatories.

A number of FTAs of Asian countries contain GATT Article XX-like (or GATS Article XIV-like) provisions.[13] As a matter of negotiating "precedent" – something valued by many trade negotiators – it would not be difficult to argue for inclusion in an FTA of a general exceptions provision using the above GATT and GATS provisions as "templates", since the language is already familiar. In addition, the prerequisites of some of the Article XX "limbs", and its chapeau, have been explained and applied in a number of WTO cases, therefore providing some guidance on the scope and applicability of such provisions.[14] Unsurprisingly, therefore, many of the FTAs entered into by Asian states contain similar general exceptions, or expressly incorporate the GATT and GATS articles as part of the FTA. Where such general exceptions are so included, they can provide important safe harbours for health and other measures, insofar as the treaty conditions are satisfied. As will be seen in the next part, many of the FTAs signed by Asian states do contain such provisions, particularly in relation to trade in goods commitments. Indeed, this would be a minimum expectation when discussing exceptions during negotiations.

Similarly, given the multilateral consensus reflected in the TRIPS and Public Health Declaration, a provision giving express recognition to its principles should also be expected to be included. Indeed, several post-Declaration FTAs have done this.[15]

At the level of national laws, a number of Asian patent laws themselves – such as those of India, Philippines and Vietnam – refer specifically to policies of protecting public health interests.

[13] See, for example, Art. 8.15, China-Switzerland FTA, which incorporates elements of Art. XIV of the GATS.

[14] See, for example, *EC – Measures Affecting Asbestos and Asbestos-Containing Products*, WT/DS135/AB/R, report of the Appellate Body, adopted on 5 April 2001.

[15] See, for example, the US-Peru FTA. See generally, Julien Chaisse, Exploring the Confines of International Investment and Domestic Health Protections – Is a General Exceptions Clause a Forced Perspective? *Am. J. L. & Med.*, Vol. 39 (2013), 332–360.

## 5.3 Applicability of GATT 1994 Article XX to WTO Accession Protocol commitments: lessons learnt

Recent accessions to the WTO have seen a number of commitments that transcend those in the original "package" which WTO members accepted via the Marrakesh Agreement at the end of the Uruguay Round. China, which has been involved in a number of WTO disputes since acceding to the WTO in 2001, has found that application of Article XX to its accession commitments was not an obvious point.[16]

In the present context, a useful lesson is that negotiating parties to bilateral or regional agreements who wish to have access to general exceptions must make this very clear in the treaty language being

[16] See, for example, *China – Publications and Audiovisual Products* and *China – Raw Materials.*

The Appellate Body in *China – Rare Earths* put to rest the question of the relationship between China's Protocol of Accession and the WTO multilateral agreements (which include GATT 1994 and its Art. XX), holding that it was *possible* for an Art XX exception to apply, but that whether an exception would apply in a given case would depend on various factors. The Appellate Body stated:

> 5.74 Paragraph 1.2 of China's Accession Protocol serves to build a bridge between the package of Protocol provisions and the package of existing rights and obligations under the WTO legal framework. Nonetheless, neither obligations nor rights may be automatically transposed from one part of this legal framework into another. The fact that Paragraph 1.2 builds such a bridge is only the starting point, and does not in itself answer the questions of whether there is an objective link between *an individual provision* in China's Accession Protocol and existing obligations under the Marrakesh Agreements and the Multilateral Trade Agreements, and whether China may rely on an exception provided for in those agreements to justify a breach of such Protocol provision. Such questions must be answered through a thorough analysis of the relevant provisions on the basis of the customary rules of treaty interpretation and the circumstances of the dispute. The analysis must start with the text of the relevant provision in China's Accession Protocol and take into account its context, including that provided by the Protocol itself and by relevant provisions of the Accession Working Party Report, and by the agreements in the WTO legal framework. The analysis must also take into account the overall architecture of the WTO system as a single package of rights and obligations and any other relevant interpretative elements, and must be applied to the circumstances of each dispute, including the measure at issue and the nature of the alleged violation.

Note that in para. 5.50 the analysis required was explained further in the following terms: "... this bridge does not dispense with the need to analyse, *on a case-by-case basis*, the specific relationship between an individual provision in the Protocol, on the one hand, and provisions of the Marrakesh Agreement and the Multilateral Trade Agreements, on the other hand." [italic emphasis added]

negotiated, particularly if the agreement contains "integral" parts found in different documents.

Further, in some agreements, the "general exceptions" are made to apply only to commitments in certain chapters or areas (with the excepted chapters containing no separate exceptions of their own); in negotiating a new treaty, if such limited provisions are proposed as "precedents", careful consideration needs to be made as to whether the limitations of applicability are acceptable in the new negotiating context.

## 5.4 Safe harbours: Asian treaties' general exceptions

While general exception clauses have been said to be no "panacea",[17] properly drafted, they can serve an important function as a safe harbour in certain circumstances.

Safe harbours in the form of exceptions differ from carve-outs. While exceptions generally excuse in a broad fashion certain categories of measures, carve-outs exclude the areas in question from any commitment at all. For example, a certain sector or product may be singled out for exclusion from a treaty's commitments. Negotiating to exclude certain areas at the outset is the most extreme position to take – if that fails, the next best option is a strong exception provision. A third alternative would be to exclude a particular type of complaint or dispute from the treaty's dispute settlement system.

Various Asian FTAs contain some form of an equivalent to Article XX. Table 5.1 provides a summary of some examples, both regional and bilateral. The provisions below are general exceptions that are expressly spelt out within the FTA; where a provision incorporates Article XX GATT by reference only, the provision is marked with "*".

If one examines the general exceptions in such FTAs, one can discern differences in scope and applicability within them, even though they generally repeat or incorporate by reference, the exceptions in Article XX.

The general exception of broadest applicability – it applies to the entire treaty – is Article 71 of the ANZSCEP, the first FTA signed by Singapore.[18]

While many FTAs include Article XX-like provisions or incorporate Article XX provisions by express reference, these may not always apply

17 Chaisse, *ibid.*, at p. 359, in the context of investment treaties.

18 The text of Art. 71 is available at: www.fta.gov.sg/anzscep/Part_11_dispute%20settlement (duplicate).pdf.

Table 5.1. *Examples of Asian trade treaties incorporating general exceptions*

| Trade agreement | General exception provision |
|---|---|
| ASEAN Trade in Goods Agreement (ATIGA) | Article. 8 |
| Korea-US Free Trade Agreement (KORUS) | Article 23.1 |
| Malaysia-Australia FTA (MAFTA) | Articles. 12.18 and 18.1 |
| South Asia Free Trade Agreement (SAFTA) | Article. 14 |
| Malaysia-India Comprehensive Economic Cooperation Agreement (MICECA) | Article. 12.1 |
| Singapore-India Comprehensive Economic Cooperation Agreement (CECA) | Article 2.13 |
| China-New Zealand FTA (CNZFTA) | Article 200 |
| China-Singapore FTA(CSFTA) | Article 105 |
| Agreement between New Zealand and Singapore on a Closer Economic Partnership (ANZSCEP) | Article 71 |

to *all* commitments within the treaty. An example of a general exception that applies to limited chapters is Article 21.2(1) of the Korea-Singapore FTA, which applies only to the following stated commitments:

> 1. Article XX of GATT is incorporated into and made part of this Agreement, for the purposes of:
>    (a) Chapters 3 (National Treatment and Market Access for Goods), 4 (Rules of Origin), 5 (Customs Procedures), 6 (Trade Remedies), and 14 (Electronic Commerce), except to the extent that a provision of those Chapters applies to services or investment; and
>    (b) Chapter 16 (Government Procurement), except to the extent that any of its provisions applies to services.[19]

Such an exception would significantly not, for instance, apply to other chapters in the treaty, such as those on Investment (Chapter 10) and Intellectual Property (Chapter 17).

Likewise, the USSFTA and KORUS both contain a general exception provision of limited application.[20] The additional reference to the TRIPS

[19] The full text of the Exceptions chapter is available at: www.fta.gov.sg/ksfta/korea-singapore%20free%20trade%20agreement%20(ksfta)_chap21.pdf.

[20] See Art. 21.1.

and Public Health Declaration in Article 18.11 in KORUS offers protection as stated in the Declaration.[21] There is however no incorporation of other language relating to the overarching principles and objectives of TRIPS Agreement vis-à-vis the regulation of pharmaceuticals. For FTAs that have significant commitments impinging on the regulation of pharmaceuticals (found in the intellectual property chapters of these agreements), the limited applicability of general exceptions to exclude intellectual property or pharmaceutical regulation measures raises significantly the risk of legal exposure for the signatory states in regulating in these areas.

A general exception provision (modelled after Article XX) of narrow applicability is found in the Japan-Singapore Economic Partnership Agreement (JSEPA) which is embedded in Chapter 2 on trade in goods, applying only to that chapter.[22] The EFTA-Singapore FTA also contains a goods-specific general exception.[23] It is logical that in these earlier FTAs, the FTA general exception copying Article XX – which clearly was designed to apply to trade in goods issues within the GATT – would be limited to the trade-in-goods chapter, without equivalents in other chapters. More recent FTAs such as the USSFTA and the Korea-Singapore FTA show a change to the approach, recognizing the need to extend general exception protection over commitments beyond those in the goods chapter.

A number of BITs signed by Asian countries contain a GATT Article XX-style exception. An example is the Canada-China bilateral investment treaty.[24]

A noteworthy point is the role played by general exceptions which may be found in domestic investment statutes. Myanmar's proposed new investment law, for example, provides for general exceptions which do not explicitly refer to public health reasons.[25]

## 5.5 TRIPS principles and objectives in FTAs

The EU-Singapore FTA whose negotiations were concluded in 2012 and which awaits EU ratification, for example, contains a provision which

[21] As the USSFTA was signed prior to the Declaration, the FTA does not include any reference to the Declaration, unlike Art. 18.11 KORUS.

[22] Text of the Chapter is available at: www.fta.gov.sg/jsepa/fta_jsepa_agreement3.pdf.

[23] See www.fta.gov.sg/esfta/part_2_esfta.pdf.

[24] Art. 33.2, Agreement between the Government of Canada and the Government of the People's Republic of China for the Promotion and Reciprocal Protection of Investments.

[25] On the proposed new law, see Chapter 4, note 118; the general exceptions are found in Section 22, which addresses the right to adopt or maintain "reasonable non-discriminatory measures for prudential reasons".

incorporates the matters in Arts. 7 and 8 of TRIPS.[26] This permits a treaty interpreter to draw on the subject matter of these two TRIPS provisions when interpreting other provisions in the IP Chapter.

## 5.6 TRIPS exceptions and flexibilities

### *5.6.1 TRIPS exceptions*

The TRIPS Agreement offers members a number of exceptions, including the Principles in Article 8 (with an express reference to public health) and the general exception provision in Article 30.[27] In addition, a number of provisions have built-in flexibilities available to WTO members even though they are not called "exceptions".[28] The exceptions are discussed in the WTO's TRIPS Council as well as regularly in other fora, such as the WIPO's Standing Committee on the Law of Patents.[29]

Article 30 of the TRIPS Agreement, set out below, contains an important exception that applies to patents:

> **Article 30**
>
> **Exceptions to Rights Conferred**
>
> Members may provide limited exceptions to the exclusive rights conferred by a patent, provided that such exceptions do not unreasonably

[26] Art. 11.1 provides as follows:

**Objectives**

1. The objectives of this Chapter are to:
   (a) facilitate the production and commercialisation of innovative and creative products and the provision of services between the Parties; and
   (b) increase the benefits from trade and investment through the adequate and effective level of protection of intellectual property rights and the provision of measures for the effective enforcement of such rights.
2. The objectives and principles set forth in Part I of the TRIPS Agreement, in particular in Article 7 (Objectives) and Article 8 (Principles) shall apply to this Chapter, *mutatis mutandis*. See http://trade.ec.europa.eu/doclib/docs/2013/september/tradoc_151761.pdf (accessed 25 July 2014).

[27] On TRIPS exceptions, see also generally *The TRIPS Agreement and Pharmaceuticals*, 2000, at p. 28; the full report is available at: http://apps.who.int/medicinedocs/en/d/Jh1459e/6.3.html.

[28] For discussions on the TRIPS flexibilities, see, for example, Henning Ruse-Khan, The International Law Relation between TRIPS and Subsequent FTA-Plus Free Trade Agreements: Towards Safeguarding TRIPS Flexibilities? *JIPL*, Vol. 28, No. 2 (2011), 325, and his suggestions of clauses at pp. 358–363.

[29] See WIPO, www.wipo.int/patents/en/topics/exceptions_limitations.htm (accessed 9 April 2015).

conflict with a normal exploitation of the patent and do not unreasonably prejudice the legitimate interests of the patent owner, taking account of the legitimate interests of third parties.

One of the issues is that this provision does not expressly refer to R&D or instances of applicability. Proposals have therefore been made to bring to the fore in more concrete terms the applicability of research exceptions to allow better use of the flexibility under Article 30.[30] This important exception was examined by a WTO panel in *Canada – Pharmaceuticals*, and the decision remains the only discussion in WTO case law available on this provision (there was no appeal to the Appellate Body in this case). The case involved two aspects of Canada's pharmaceutical patent laws: provisions permitting use of a patented product (such as a medicine) before expiry of the patent, for regulatory review purposes, and a provision permitting stockpiling of medicines which were under patent. Canada argued that these provisions were protected as "limited exceptions" under Article 30. This is discussed further below, in relation to what has become known as the *Bolar* exception.

#### 5.6.1.1 The *Bolar* exception

Generic pharmaceutical manufacturers aim to market their products upon expiry of the patent of a branded, originator product. In the period leading up to such expiry, they may already initiate regulatory approval applications for their products. In such applications, they may show that their products are therapeutically equivalent (or bioequivalent) to the originator product. This is sometimes known as "springboarding".[31]

In a US case, *Roche Products Inc. v Bolar Pharmaceutical Co.*,[32] it was found that a generic manufacturer could not rely on a research exemption to show bioequivalence of its drug which was intended to be sold upon expiry of the branded drug's patent. The US government subsequently enacted the Drug Price Competition and Patent Term

[30] See the proposals of Edson Beas Rodrigues Jr. and his criticisms of proposals by others (such as Rochelle Dreyfus), *The General Exceptions Clauses of the TRIPS Agreement*, Chap. 5 (Cambridge University Press, 2012).

[31] See, for example, Australia Productivity Commission Report, *Compulsory Licensing of Patents* (2013), p. 190.

[32] 733 F. 2d 858 (Fed. Circ. 04/23/1984). See generally, Anthony Tridico, Jeffery Jacobstein and Leythem Wall, *Facilitating Generic Drug Manufacturing: Bolar Exemptions Worldwide*, June 2014, www.wipo.int/wipo_magazine/en/2014/03/article_0004.html (accessed 16 April 2015), and Daniel Gervais, *The TRIPS Agreement: Drafting History and Analysis*, 4th ed. (Sweet & Maxwell, 2012) at pp. 472–475.

Restoration Act of 1984 (also known as the Hatch-Waxman Act)[33] to reverse this situation, and the law now permits generic manufacturers to use and test patented materials for the purpose of obtaining regulatory approval. Legislative exceptions based on such use are hence now commonly known as "*Bolar* exceptions".

Canada's legislation permitting such use for regulatory approval came under challenge at the WTO in *Canada – Pharmaceuticals.*[34] Canada defended the provisions by relying on the TRIPS Agreement exception in Article 30. The panel found that Canada's relevant provisions satisfied Article 30 as a limited exception within its terms. In its decision (from which there was no appeal), the panel explained that to satisfy Article 30, three separate and cumulative requirements must be met. The complaint by the EC against Canada focused on two main aspects of the Canadian Patent Act: provisions on a regulatory review provision and a stockpiling exception. The former permitted generic pharmaceutical producers to use patented inventions during the period of effect of the patent solely for purposes of developing and submitting information for product approval. The panel explained that for Article 30 to apply, three separate but cumulative conditions must be fulfilled, namely:

> 1) the exception must be "limited"; (2) the exception must not "unreasonably conflict with normal exploitation of the patent"; (3) the exception must not "unreasonably prejudice the legitimate interests of the patent owner, taking account of the legitimate interests of third parties".[35]

The panel found this exception under Canadian law to be compliant with TRIPS, being covered as a "limited exception" by Article 30. (The stockpiling exception, on the other hand, was found not to have satisfied Article 30 and was therefore not TRIPS-compliant.) For countries that are host to generic pharmaceutical manufacturers, this finding was a watershed confirmation of the right to permit such manufacturers to use a patent – even during its lifetime – for purposes of obtaining regulatory approval. The panel explained the significance of this succinctly:

> Without the regulatory review exception, the patent owner might be able to prevent potential competitors from using the patented product during

[33] 98 Stat. 1585.

[34] *Canada – Patent Protection of Pharmaceutical Products*, WT/DS114/R, Panel report adopted on 7 April 2000. See also Chapter 8, and, generally, Edson Beas Rodrigues Jr., *The General Exception Clauses of the TRIPS Agreement* (Cambridge University Press, 2012), and especially at pp. 90–116.

[35] Para.7.20, Panel Report.

> the term of the patent to comply with testing requirements, so that competitors would have to wait until the patent expires before they could being the process of obtaining marketing approval. This, in turn, would prevent potential competitors from entering the market for the additional time required to complete the regulatory approval process, in effect extending the patent owner's period of market exclusivity beyond the end of the term of the patent.[36]

As a result, Canada's law permitting use for regulatory review passed muster under Article 30. Such provisions have come to be known as *Bolar* exceptions.[37]

In interpreting Article 30, the panel had occasion to explain the interaction between that provision and the patentability provision in Article 27. It also acknowledged the important role of Articles 7 and 8 in interpreting Article 30:

> Both the goals and the limitations stated in Articles 7 and 8.1 must obviously be borne in mind when [examining the words of the conditions in Article 30] as well as those of other provisions of the TRIPS Agreement which indicate its object and purposes.[38]

There has also been no interpretation and application of these provisions by the WTO Appellate Body. While one might have thought that these articles would inform the interpretation of others such as Article 30, the panel referred to the well-known treaty interpretation principles in Articles 31 and 32 of the Vienna Convention on the Law of Treaties (VCLT) but did not relate them explicitly to Articles 7 and 8 of TRIPS.[39] In the WTO dispute, there was disagreement between Canada and the EU on interpreting the very balance that these provisions were meant to preserve.[40]

36 Para. 7.2, Panel report, footnote 33 above. See also para. 7.3 for further elaboration by the panel on this point.

37 The name derives from a US case, *Roche Products Inc. v Bolar Pharmaceutical Co.*, 733 F. 2d 858 (Fed. Circ. 04/23/1984). Ironically, in the *Bolar* case, the US Court of Appeals for the Federal Circuit had held that such regulatory review use prior to expiry of the relevant patent was not permitted; Congress subsequently enacted the so-called Hatch-Waxman legislation to undo the decision and to permit such use; see Margo A. Bagley, *Patent Term Restoration and Non-patent Exclusivity in the US, in Pharmaceutical Innovation, Competition and Patent Law*, ed. Josef Drexel and Nari Lee (Edward Elgar Publishing 2013), at pp. 112–115.

38 Panel report, para. 7.26.

39 It did refer in a brief manner to the provisions; see e.g. Panel report, paras. 7.23-7/26 and 7.89-7.92.

40 See, for example, Panel report, paras. 4.12–4.21, 4.30–4, 4.33–4.41. Third parties such as Brazil and Thailand also made submissions on the "balance".

Giving up the flexibility of a *Bolar* exception in an FTA can mean such a delay in entry to the market of generic pharmaceuticals in a signatory country. In the absence of generic competition, such a country can expect to face higher prices for patented pharmaceuticals during such a delay. More broadly, where an FTA incorporates the full flexibilities permitted by Article 30, it can afford a significant safe harbour for health-related measures – such as the *Bolar* exception – that may be justified under its requirements. An example of a provision reflecting Article 30 flexibilities is Article 16.7(3) of the USSFTA:

> 3. Each Party may provide limited exceptions to the exclusive rights conferred by a patent, provided that such exceptions do not unreasonably conflict with a normal exploitation of the patent and do not unreasonably prejudice the legitimate interests of the patent owner, taking account of the legitimate interests of third parties.

As will be discussed further later, certain countries in Asia have already legislated equivalents of a *Bolar* exception, including Brunei Darussalam,[41] China,[42] Hong Kong,[43] India,[44] Malaysia,[45] the Philippines,[46] Thailand[47] and Singapore.[48] The Singapore legislation provides, too, that experimental use is not patent infringement,[49] and contains further exceptions applicable to individual or specific patient use that are deemed to be non-infringements.[50] Lao PDR laws, on the

[41] *Patents Order 2011*, section 64(2)(b) (experimental purposes) and (2)(g) (for meeting marketing approval requirements).

[42] Third Revision of China's Patent Law (2008), Article 69.5. On China's *Bolar* exception and practice, see Li Feng, Jiancheng Jiang and Yuan Wang, *Experimental Use Exemption of Patent Infringement: A Brief Comparison of China and the United States*, January 2015, American Intellectual Property Association, especially pp. 5–8, www.aipla.org/committees/committee_pages/Biotechnology/international/Shared%20Documents/International_Buzz_201501c.pdf (accessed 9 April 2015). See also Yamane, footnote 4, at 381–2.

[43] See Section 75, *Patents Ordinance* (www.asialaw.com/Article/1972094/Search/Results/Hong-Kong-Pharmaceutical-Patents-Breaking-the-Ranks.html?Keywords=Bolar+provision).

[44] Section 107A, *Patents Act*, 1970. [45] Section 37(1A), Patent Act 1983.

[46] *Vide* Universally Accessible Cheaper and Quality Medicines Act of 2008, Republic Act No. 9502, June 6, 2008, amending Section 72 of Republic Act No. 8293, i.e. the Intellectual Property Code of the Philippines.

[47] Section 36, Patent Act (No. 2) B.E. 2535 (A.D. 1992).

[48] Section 66(2)(h), Patents Act, Cap. 221, 2005 Revised Edition. Section 66(1)(b) contains a separate exception for acts done for experimental purposes.

[49] Section 66(2)(b), Patents Act, *ibid.*

[50] Section 66(2)(c) and (i), Patents Act, footnote 48 above.

other hand, do not include a *Bolar* exception provision. Indonesian law provides for a research exception.[51]

Given the WTO's endorsement of this specific application of Article 30, it is permissible and desirable for those in Asia to ensure its inclusion in the FTAs under negotiation, to ensure that this flexibility is preserved. However, provisions for such an exemption need to be drafted clearly as to scope and application. Even in the EU, for example, some doubts have arisen as to the scope of the *Bolar* exemption there. In 2013, certain questions on the scope of the exemption – governed by European Directive 2001/83/EC on the Community Code relating to medicinal products for human use as amended by Directive 2004/27/EC – have been referred to the EU Court of Justice.[52] In response, the UK Intellectual Property Office has invited public comment on these. The period for comments ends on 8 April 2014.[53]

It should be noted that the *Bolar* exception is not formulated in identical ways throughout the EU.[54] In fact, a case testing the scope of the German version of the *Bolar* exception, referred from the Düsseldorf Court of Appeal to the Court of the Justice of the EU (CJEU), is pending at the time of writing.[55] In the UK, at the time of writing, a legislative amendment has been tabled to ensure that the *Bolar* exception currently in section 60(5) of the UK Patents Act 1977 permits clinical trials for both innovator and generic medicines.[56]

[51] Art. 16(3).

[52] See www.ipo.gov.uk/pro-policy/policy-information/ecj/ecj-2013/ecj-2013-c66113.htm.

[53] The *Bolar* exemption under EU law is governed by European Directive 2001/83/EC on the Community Code relating to medicinal products for human use as amended by Directive 2004/27/EC.

[54] The EU directive implemented by member states is Article 10(6) of Directive 2001/83/EC of the European Parliament and of the Council of 6 November 2001 on the Community Code relating to medicinal products for human use (OJ 2011 L 311, p. 67), as amended by Directive 2004/27/EC of the European Parliament and of the Council of 31 March 2004 amending Directive 2001/83/EC on the Community Code relating to medicinal products for human use (OJ 2004 L 136, p. 34). See generally Liz Cohen and Laura Peirson, The UK Research and Bolar Exemptions: Broadening the Scope for Innovation? *JIPL & Pract.*, Vol. 8, No. 11 (2013), 837–845, and Taylor Wessing, *Not All* Bolar *Exemptions Are the Same*, at: www.taylorwessing.com/synapse/september13.html (accessed 28 May 2014).

[55] *Astellas Pharma Inc. v Polpharma SA Pharmaceutical Works*, Case C-661/13, application to CJEU lodged on 13 December 2013.

[56] See *The Legislative Reform (Patents) Order 2014*, text of which is available at: www.legislation.gov.uk/ukdsi/2014/9780111114537 (accessed 28 May 2014). For the background of this proposed amendment, see the UK Intellectual Property Office informal consultation of 2011, formal consultation of 2012, and the UK government response to the consultation of February 2013, available at: www.ipo.gov.uk.

### 5.6.2 Bolar *and R&D exceptions in Asia*

As mentioned, several Asian countries have legislated general research-related exceptions in their patent laws, with some including, additionally, an explicit exception relating to acts carried out for the purpose of obtaining marketing approval or *Bolar* exceptions. Among the *Bolar* exception provisions, many do not contain any explicit, specific exemption for the conduct of *clinical trials* from patent infringement. This is significant as the clinical trial stage is when data may become the subject of protection of Article 39 of TRIPS as undisclosed information, or as exclusive data under TRIPS-plus FTA provisions. Article. 39 does not define "unfair commercial use", and it has been a matter of debate whether this includes use for obtaining regulatory approval. The lack of exemption for use of patented material for this stage therefore lays open the possibility of arguments of violation by WTO members (in the case of Article 39) and by patent holders (in the case of FTAs, possibly as investors).

Within Asia, *Bolar* exceptions exist with varying degrees of coverage. For example, the laws of China, Vietnam, the Philippines, Taiwan and Thailand include exemptions that specifically exclude from infringement acts done in connection with obtaining regulatory approval, in some cases with specific reference to acts such as trial production and testing that may apply to drugs.

For example, Articles 69(4) and (5) of the Patent Law of the People's Republic of China exclude the following from patent infringement:

> (4) Any person uses the relevant patent specially for the purpose of scientific research and experimentation;
>
> (5) Any person *produces, uses, or imports patented drugs* or patented medical apparatus and instruments, *for the purpose of providing information required for administrative examination and approval*, or produces or any other person imports patented drugs or patented medical apparatus and instruments especially for that person.[57]
>
> [Italic emphasis added]

[57] Para. (5) was added in 2008, *vide* Decision of the Standing Committee of the National People's Congress on Amending the Patent Law of the People's Republic of China (adopted at the 6th Meeting of the Standing Committee of the Eleventh National People's Congress on 27 December 2008).

Chinese Taipei Patent Act provides:

> **Article 60 Limitations of Patent Right: Research and Trials of Pharmaceuticals**
>
> 1. The effects of the patent right shall not extend to *research and trials, including their practical requirements*, necessary for obtaining registration and market approval of drugs under the Pharmaceutical Affairs Act or obtaining market approval of pharmaceuticals from a foreign country.
>
> [Italic emphasis added][58]

Article 125(a) of the Vietnamese Law on Intellectual Property is broad, and provides that industrial property right owners cannot prevent others from performing the following acts:

> *Using inventions*, industrial designs or layout-designs in service of their personal needs or for non-commercial purposes, or for purpose of evaluation, analysis, *research, teaching, testing, trial production* or *information collection for carrying out procedures of application for licenses* for production, importation or circulation of products; . . .
>
> [Italic emphasis added]

Article 36 of Thailand's Patents Act is more general but broad, and provides:

> A patent-holder's exclusive rights do not extend to:
>
> (1) any act for the purpose of *study, research, experimentation or analysis, provided* that it does not unreasonably conflict with a normal exploitation of the patent and do not unreasonably prejudice the legitimate interests of the patent owner;
>
> . . .
>
> (4) *any act concerning an application for drug registration*, the applicant intending to produce, distribute or import the patented pharmaceutical product after the expiration of the patent term; . . .
>
> [Italic emphasis added]

Under Singapore's Patents Act, Section 66(2)(h) exempts an act that might otherwise be an infringement under Section 66(1) from being an infringement, if:

> (h) it consists of the doing of any thing set out in subsection (1) in relation to the subject-matter of the patent to support any application for marketing approval for a pharmaceutical product, provided that any thing produced to support the application is not –

[58] Patent Act, as amended in 2013.

(i) made, used or sold in Singapore; or
(ii) exported outside Singapore,

other than for purposes related to meeting the requirements for marketing approval for that pharmaceutical product; . . .[59]

Further, while some patent legislation exceptions do not refer to drug trials explicitly, they might instead refer to uses that are *reasonably* related to drug development.[60]

[59] Section 66(1) provides: "66.—(1) Subject to the provisions of this Act, a person infringes a patent for an invention if, but only if, while the patent is in force, he does any of the following things in Singapore in relation to the invention without the consent of the proprietor of the patent:

(a) where the invention is a product, he makes, disposes of, offers to dispose of, uses or imports the product or keeps it whether for disposal or otherwise;
(b) where the invention is a process, he uses the process or he offers it for use in Singapore when he knows, or it is obvious to a reasonable person in the circumstances, that its use without the consent of the proprietor would be an infringement of the patent;
(c) where the invention is a process, he disposes of, offers to dispose of, uses or imports any product obtained directly by means of that process or keeps any such product whether for disposal or otherwise."

[60] For example, the Malaysian Patents Act provides as follows:

**Section 37. Limitation of rights.**

(1) The rights under the patent shall extend only to acts done for industrial or commercial purposes and in particular not to acts done only for scientific research.

(1A) The rights under the patent shall not extend to acts done to make, use, offer to sell or sell a patented invention *solely for uses reasonably related to the development and submission of information to the relevant authority which regulates the manufacture, use or sale of drugs.*

[Italic emphasis added]

By comparison, Indian law also includes a general regulatory approval-related exclusion. Art. 107A of the Indian Patents Act, 1970, as amended in 2005, provides an exclusion from infringement in the following terms:

107A. Certain acts not to be considered as infringement. – For the purposes of this Act, – (a) any act of making, constructing, using, selling or importing a patented invention *solely for uses reasonably related to the development and submission of information required under any law for the time being in force, in India, or in a country other than India, that regulates the manufacture, construction, use, sale or import of any product*; . . .

[Italic emphasis added]

Indonesian law does not contain an explicit *Bolar* exception. Indonesia's Law on Patents excludes from a patent holder's exclusive rights such use as is described below (Article 16(1), referred to below, provides for the patent holder's exclusive rights)[61]:

> Article 16
>
> ...
>
> (3) Exempted from the provisions as referred to in paragraph (1) and paragraph (2) if the use of said Patent is for the sake of education, research, experiment, or analysis, as long as it does not harm the normal interest of the Patent holder.

One argument that has been advanced is that the existing provisions may provide sufficient flexibility to permit *Bolar*-type use in the name of public health.[62] It should be further noted that under a 2008

Section 47 also subjects a patent to certain conditions, including the following:

> (3)any machine, apparatus or other article in respect of which the patent is granted or any article made by the use of the process in respect of which the patent is granted, may be made or used, and any process in respect of which the patent is granted *may be used, by any person, for the purpose merely of experiment or research* including the imparting of instructions to pupils; . . .
>
> [Italic emphasis added]

Finally, the Philippines Cheaper and Quality Medicines Act provides as follows:

> 72.3. Where the act consists of making or using exclusively for experimental use of the invention for scientific purposes or educational purposes and such other activities directly related to such scientific or educational experimental use;
>
> 72.4. In the case of drugs and medicines, where the act includes *testing*, using, making or selling the invention including any data related thereto, *solely for purposes reasonably related to the development and submission of information and issuance of approvals by government regulatory agencies required under any law of the Philippines or of another country that regulates the manufacture, construction, use or sale of any product: Provided*, That, in order to protect the data submitted by the original patent holder from unfair commercial use provided in Article 39.3 of the Agreement on Trade-Related Aspects of Intellectual Property Rights (TRIPS Agreement), the Intellectual Property Office, in consultation with the appropriate government agencies, shall issue the appropriate rules and regulations necessary therein not later than one hundred twenty (120) days after the enactment of this law; . . .
>
> [Italic emphasis added]

[61] Indonesian Law No. 14 of 2001 Regarding Patents.

[62] See, for example, Christoph Antons, Michael Blakeney, and Christopher Heath, eds., *Intellectual Property Law Harmonisation within ASEAN and APEC* (Kluwer Law International, 2004), p. 115.

decree,[63] which introduces patent linkage (Article 12), parties that are not the innovator/pharmaceutical patent holder may *apply* for approval of their product 2 years before expiry of a patent (Article 13) but may *distribute* it only upon expiry of the patent. However, this provision does not address what such an applicant may do to gain approval during the 2-year period, which leaves one to conclude it is the 2001 law which governs the issue.

In the case of Cambodia, as a relatively new LDC member of the WTO, transitional flexibility has been provided; under Article 4 read with Article 136 of Law on the Patents, Utility Models Certificates and Industrial Design, pharmaceutical products which are mentioned in Article 4 are excluded from patent protection until 1 January 2016.[64] In addition, presently under Article 44(iv), the rights of patents shall not extend to "acts done only for experimental purposes relating to a patented Invention".

### 5.6.3 *TRIPS flexibilities*[65]

Patentability provisions under TRIPS afford members with a degree of flexibility within which to craft their law and policy. It is, as mentioned, in Chapter 2, an important "gateway" or threshold to be crossed by inventors who wish to gain the rights which a patent brings.[66]

#### 5.6.3.1 Article 27 – patentability

While Article 27 of the TRIPS Agreement requires members to provide for patent protection for "any inventions", it also provides (in paragraph 1) certain minimum requirements for patentability, including the requirements that an invention must be new, involve an inventive step and be capable of industrial application. Beyond the basic requirements provided in paragraph 1, paragraphs 2 and 3 permit areas which members may choose to exclude from patentability. Article 27.2 of the TRIPS Agreement, in particular, provides:

[63] Decree of the Minister of Health, No 1010/MENKES/PER/XI/2008 on Drug Registration.

[64] Decree NS/RKM/0103/005.

[65] See generally Henning Grosse Ruse-Khan, The International Law Relation between TRIPS and Subsequent TRIPS-Plus Free Trade Agreements: Towards Safeguarding TRIPS Flexibilities? *JIPL*, Vol. 18, No. 2 (2011), 1.

[66] There is support for this view, that patentability is one of the flexibilities available under TRIPS; see, for example, Daniel Gervais, *Patentability Criteria as TRIPS Flexibilities: The Examples of China and India, in Patent Law in Global Perspective*, ed. Ruth L. Okediji and Margo A. Bagley (Oxford University Press, 2014).

> Members may exclude from patentability inventions, the prevention within their territory of the commercial exploitation of which is necessary to protect *ordre public* or morality, **including to protect human**, animal or plant life **or health** or to avoid serious prejudice to the environment, provided that such exclusion is not made merely because the exploitation is prohibited by their law.
>
> [Bold emphasis added]

Such an exception is broad in scope, especially since matters such as public morality are difficult to define. In the context of the issue of patentability of new technologies relating to genes, their isolation and sequencing, the issue of public morality may arise in the event of a dispute over patentability of such subject matter.[67]

However, one finds that not every FTA preserves this flexibility on patentability in its entirety. The Malaysia-Australia FTA offers an example of a provision that seeks to preserve the parties' right to retain exceptions to patentability; Article 13.11(4) "freezes" and preserves existing exceptions in the parties' domestic laws. The provision states:

> Nothing in this Article will limit the scope of exceptions to patentability available in each Party's laws and regulations at the time that this Agreement enters into force.[68]

The USSFTA, on the other hand, is an example of an FTA which reduces what the parties may exclude from patentability, contrary to the flexibility in Article 27.3. Article 16.7 provides for the Parties to exclude such inventions from patentability as are provided for in Articles 27.2 and 27.3(a) of the TRIPS Agreement. This therefore disallows them from excluding those inventions that are mentioned in Article 27.3(b) of the TRIPS Agreement, namely:

> (b) plants and animals other than micro-organisms, and essentially biological processes for the production of plants or animals other than non-biological and microbiological processes. However, Members shall provide for the protection of plant varieties either by patents or by an effective *sui generis* system or by any combination thereof. The provisions of this subparagraph shall be reviewed four years after the date of entry into force of the WTO Agreement.

[67] See, for example, the *Patent Examination Guidelines* of the Intellectual Property Office of Singapore (IPOS) which refers to this possibility.

[68] The full text of the FTA is available at: www.miti.gov.my/cms/documentstorage/com.tms.cms.document.Document_7257fc06-c0a81573-2ce72ce7-76d00623/MAFTA.pdf.

As a result of the USSFTA, Singapore enacted the Plant Varieties Protection Act 2004.[69] Other Asian FTAs also require protection for plant varieties. Examples include the KORUS FTA[70] and the EU-Korea FTA.[71] By contrast, the Japan-Switzerland Agreement preserves a version of Article 27.3(b).[72]

Patentability and generic producers' rights of course raise a wider question than what is found in FTAs. Significantly, a few recent US Supreme Court cases show there is a need to balance the rights of generic drug manufacturers with those of drug patent holders under the law.[73]

Another issue in patentability, which has also appeared in some FTAs, concerns the practice of "evergreening". Pharmaceutical companies holding patents have been known, for example, to engage in this practice. Some authors have explained this as follows:

> Evergreening, although not a formal legal concept, is a term referring to the numerous ways in which patent owners of pharmaceutical products use the patent laws to extend their monopoly privileges beyond periods that are normally allowed by law, particularly over high-revenue-earning drugs.[74]

As part of patent holders' "evergreening" efforts to obtain further protection based on their existing patents, new uses or methods of using of existing patented pharmaceutical products are claimed as new inventions, and argued as meriting a new patent. While patents formerly claimed mainly rights over the original product or process, increasingly, pharmaceutical companies are claiming patents for such new uses or methods.[75]

Article 27 of the TRIPS Agreement does not expressly address "evergreening", but provides broad discretion within its terms to members as to whether, for instance, new uses of known products may be excluded from patentability under domestic laws. Article 18.8(1) of the

[69] Cap. 232A, 2006 Rev. Ed. See generally, CRS Congressional Research Report 2004 on the USSFTA and its IPR provisions.
[70] See Art. 18.8(2).
[71] See Art. 10.39 of the EU-Korea FTA. The full text of the FTA is available at: http://ec.europa.eu/trade/creating-opportunities/bilateral-relations/countries/korea/.
[72] See Art. 117.3 of the Agreement of Free Trade and Economic Partnership between Japan and the Swiss Confederation, footnote 42 above and accompanying text.
[73] See, for example, http://newsandinsight.thomsonreuters.com/Legal/News/2012/03_-_March/Supreme_Court_sets_aside_Myriad_gene_patent_ruling/.
[74] Gaurav Dwivedi, Sharanabasava Hallihosur and Latha Rangan, Evergreening: A Deceptive Device in Patent Rights, *Technol. Soc.*, Vol. 32 (2010), 324–330.
[75] *Ibid.*

KORUS FTA, however, expressly provides for inclusion of *new uses* of a known product.

India has enacted legislation to limit the scope of patentability of new uses in section 3(d) of its Patents Act.[76] In 2013, India's Supreme Court ruled in a landmark decision, *Novartis AG v Union of India & Ors*, that "evergreening", where therapeutic efficacy of an applicant drug was absent, was not permitted under its patent legislation.[77] The Supreme Court also hastened to clarify that this *did not mean* India's patent legislation excluded protection for all incremental inventions for pharmaceutical and chemical substances.[78]

**5.6.3.1.1 General saving of national law exceptions** The Malaysia-Australia FTA contains the following express saving clause for existing exceptions to patentability in the signatories' national laws:

[76] See the new section 3(d) introduced by the Indian Patents (Amendment) Act 2005:

> *(d)* the mere discovery of a new form of a known substance which does not result in the enhancement of the known efficacy of that substance or the mere discovery of any new property or new use for a known substance or of the mere use of a known process, machine or apparatus unless such known process results in a new product or employs at least one new reactant.
>
> *Explanation.* – For the purposes of this clause, salts, esters, ethers, polymorphs, metabolites, pure form, particle size, isomers, mixtures of isomers, complexes, combinations and other derivatives of known substance shall be considered to be the same substance, unless they differ significantly in properties with regard to efficacy; . . .

[77] Supreme Court judgment. With respect to the meaning of "efficacy" in section 3(d) (*ibid.*), the Supreme Court opined, at pp. 90–91:

> Efficacy means "the ability to produce a desired or intended result". Hence, the test of efficacy in the context of section 3(d) would be different, depending upon the result the product under consideration is desired or intended to produce. In other words, the test of efficacy would depend upon the function, utility or the purpose of the product under consideration. Therefore, in the case of a medicine that claims to cure a disease, *the test of efficacy can only be "therapeutic efficacy"* . . . With regard to the genesis of section 3(d), and more particularly the circumstances in which section 3(d) was amended to make it even more constrictive than before, we have no doubt that *the "therapeutic efficacy" of a medicine must be judged strictly and narrowly . . . What is evident, therefore, is that not all advantageous or beneficial properties are relevant*, but *only such properties that directly relate to efficacy, which in case of medicine*, as seen above, is its *therapeutic efficacy.*
>
> [Italic emphasis added]

[78] *Ibid.*, p. 95.

> Nothing in this Article will limit the scope of exceptions to patentability available in each Party's laws and regulations at the time that this Agreement enters into force.[79]

This provides a degree of assurance that existing exceptions remain intact, and that the FTA does not dilute patentability with regard to the excepted subject matter.

In the context of FTA negotiations, therefore, there presently appears to be a divergence of views in Asian states as to whether such new uses or methods of using existing patented pharmaceuticals can and should be excluded from patentability, and, indeed, whether such exclusion is TRIPS-consistent. The more restrictive view would exclude them under domestic law, and, accordingly, exclude this possibility under an FTA.

#### 5.6.3.2 Exhaustion of rights

Article 6 of the TRIPS Agreement permits members to provide for exhaustion of rights, thereby opening the way for members to permit parallel imports of pharmaceutical products where the patent holder (or an authorized party) has placed the product on the market already. Exhaustion may refer to international, regional or national exhaustion,[80] depending on where the legislation states the patent holder is to have already placed his product. For pharmaceutical products, this is an alternative source of supply and can lead to lower prices for consumers. Parallel imports are permitted in several Asian countries, including Malaysia[81] and Singapore.[82]

The Declaration on the TRIPS Agreement and Public Health reiterated this flexibility in paragraph 5(d):

> The effect of the provisions in the TRIPS Agreement that are relevant to the exhaustion of intellectual property rights is to leave each member free to establish its own regime for such exhaustion without challenge, subject to the MFN and national treatment provisions of Articles 3 and 4.

Some Asian FTAs have taken advantage of this flexibility and expressly reflect the right to provide for such exhaustion. An example is the Hong Kong-New Zealand FTA.[83]

[79] Art. 13.11.4, Malaysia-Australia Free Trade Agreement.

[80] See Musungu and Oh, *The Use of Flexibilities in TRIPS by Developing Countries: Can They Promote Access to Medicines?* (2005), p. 27; available at: www.who.int/intellectualproperty/studies/TRIPS_flexibilities/en/. See also Chapters 2 and 7.

[81] Section 37(2)(i) and section 58A(1), Patents Act 1983.

[82] Section 6(2)(g) read with section 66(3), Patents Act; footnote 47 above.

[83] See Chapter 11, Art. 3(3)(a). The provision permits the parties to provide for international exhaustion of rights.

#### 5.6.3.3 Compulsory licensing under the Paris Convention

Under the Paris Convention, compulsory licensing had already been recognized as a permitted deviation from patent rights to be used to deal with abuses in the exercise of patent rights.[84]

Article 5(A)(2) of the Paris Convention states:

> Each country of the Union shall have the right to take legislative measures providing for the grant of compulsory licenses to prevent the abuses which might result from the exercise of the exclusive rights conferred by the patent, for example, failure to work.

The context of this provision is important; essentially, while the provision above permits the grant of a compulsory licence for circumstances such as a failure to work, such a grant is proscribed by the surrounding provisions, particularly, sub-paragraph (4). The Article also deals (in sub-paragraph (3)) with the possibility of forfeiture of a patent "in cases where the grant of compulsory licences would have been sufficient to prevent" abuses mentioned in sub-paragraph (2). Sub-paragraph (3) read with sub-paragraph (1) suggest that where such abuses occur, they may be first remedied by the use of compulsory licences, before resorting to forfeiture.

Two other observations are pertinent. First, a failure to work – though not defined in the article – is recognized as a ground for grant of a compulsory licence.[85] Secondly, under sub-paragraph (4), a patent owner may fend off forfeiture if he "justifies his inaction by legitimate reasons". Again, what amounts to such reasons is undefined in the Convention.

#### 5.6.3.4 Compulsory licensing under TRIPS

Article 31 of the TRIPS Agreement expressly permits compulsory licensing by members, albeit subject to much more stringent conditions than those under the Paris Convention.[86]

The scope of Article 31 is wide, as shown by the chapeau:

[84] The full text of the Paris Convention is available at: www.wipo.int/treaties/en/ip/paris/trtdocs_wo020.html#P123_15283 (accessed 29 July 2014).

[85] Certain time requirements apply; see Art. 5(A)(4).

[86] For an explanation of how compulsory licensing came to be included in the TRIPS Agreement despite earlier unsuccessful attempts to exclude in discussions relating to the Paris Convention, as well as an overview of Article 31 of TRIPS and its related waiver, see Jerome H Reichmann, *Compulsory Licensing of Patented Pharmaceutical Inventions: Evaluating the Options*", *J. L., Med. & Eth.* (2009), 247–263.

> Where the law of a Member allows for other use[7] of the subject matter of a patent without the authorization of the right holder, including use by the government or third parties authorized by the government, the following provisions shall be respected . . .
>
> [7] "Other use" refers to use other than that allowed under Article 30.

Interestingly, it has been observed that Belgium has justified its compulsory licensing regime with reliance on Articles 8 and 30 and not Article 31.[87]

Article 31 leaves members to determine the circumstances under domestic law in which compulsory licensing may be permitted.

It may apply, for example, to "a national emergency or other situations of extreme urgency or in cases of public non-commercial use".[88] National legislation differs on the forms and criteria for use of compulsory licensing.[89] It also sets out a number of conditions for exercise of the right under it.

**5.6.3.4.1 Examples of national compulsory licensing provisions** The following common types of national compulsory licensing grounds have been identified[90]:

- Non-working of a patent by the patent holder
- For working a dependent patent
- To correct abusive use of patent
- In public interest
- For government use
- Pursuant to paragraph 6 of the Doha Declaration

**5.6.3.4.2 Non-working of a patent as a ground for the grant of compulsory licence** As can be seen from Article 5A(2) of the Paris Convention (above), "failure to work" a patent is explicitly stated as an

[87] See Jerome H. Reichman, Compulsory Licensing of Patented Pharmaceutical Inventions: Evaluating the Options, *J. L. Med. & Eth.* (2009), 247–263, at p. 250.

[88] Art. 31(b); in such circumstances there is a waiver of the requirement to make prior efforts to obtain a license from the patent holder.

[89] See WIPO, CDIP/5/4, Annex II.

[90] Annex to document CDIP/5/4, WIPO Committee on Development and Intellectual Property (CDIP). See also generally, John R. Thomas, Compulsory Licensing of Patented Inventions, CRS paper, 20 January 2014, available at: ipmall.info/hosted_resources/crs/R43266_140114.pdf (accessed 17 January 2015). See also the comparison of Australia's compulsory licensing regime with those of other countries, in Chapter 10, Productivity Commission report, *Compulsory Licensing of Patents*, (2013).

example of an abuse which could result from exercise of the exclusive rights of a patent. Notably then, this shows the drafters considered this to a clear type of abuse of a patent. The provision does not define what amounts to such a "failure to work". This leaves it open for national systems to determine in their national laws when such an abuse is deemed to arise. The TRIPS Agreement does not make any such explicit mention of a failure to work; instead, Article 8.2 of TRIPS leaves it to WTO members to take "appropriate measures" to "prevent the abuse of intellectual property rights by right holders or the resort to practices which unreasonably restrain trade or adversely affect the international transfer of technology".

Some national systems include the failure of a patent holder to "work" the patent (perhaps within a stated period after its grant or subject to other conditions) among the grounds for the grant of a compulsory licence. In the negotiations of the TRIPS Agreement, a provision on non-working of a patent without justification as a possible ground for grant of a compulsory licence was in fact included in earlier drafts.[91]

A pharmaceutical patentee may choose, after grant of his patent, not to "work" the patent in the country of grant. This could be due to various reasons, including the desire to maintain high prices of his product which is imported into that country. Article 5 permits, within a stipulated minimum period, for grant of a compulsory licence to remedy this. In a number of countries, such as Brazil and India, patent legislation retains provisions equivalent to this. However, it has not been internationally settled whether purely importing (as opposed to manufacturing domestically, or importing together with manufacturing domestically) alone should constitute a "failure to work". In the case of Brazil, this issue formed the subject of a WTO dispute.

**5.6.3.4.3 Brazil** The interpretation of the flexibility with regard to "failure to work" has led to different results in national patent laws, and to different opinions on whether these might be compliant with TRIPS. For example, the non-working ground under Brazilian law formed the basis of a complaint at the WTO brought by the United States against Brazil.[92] In 2000, the US requested consultations with Brazil under the

[91] See Chairman's Report to the Group of Negotiations on Goods, MTN.GNG/NG11/W/76, at p. 35; and Daniel Gervais, *The TRIPS Agreement – Drafting History and Analysis*, 4th ed. (Sweet & Maxwell, 2012), pp. 532–546.

[92] *Brazil – Measures Affecting Patent Protection*, WT/DS199. For examples of national provisions relating to "failure to work" ground, see Appendix in Thomas Cottier,

WTO system, arguing that the "local working" requirements in Brazil's industrial property law were not in compliance with the TRIPS Agreement.[93] The law provided for grant of a compulsory licence for a failure to work, and this meant either "failure to manufacture or incomplete manufacture of the product" or "failure to make full use of the patented process". As the matter was settled, there was no panel ruling on the compliance status of these requirements with TRIPS.[94]

**5.6.3.4.4 India** India's Patents Act permits the grant of a compulsory licence to remedy a failure to work. Under s. 84(1) read with s. 84(7) of India's Patents Act, compulsory licensing may be ordered where the reasonable requirements of the public are not being met, by reason, for example, of the patentee refusing to grant a licence on reasonable terms or by reason of imposing certain restrictive conditions in licences. These provisions were a matter of dispute in *Bayer Corporation v NATCO Pharma Ltd.*[95] In this case, NATCO was granted a compulsory licence (the first such licence in India after the TRIPS Agreement and Doha Declaration) by the Indian Controller of Patents for use of a drug patented by Bayer, a US corporation, namely tosylate (compound of carboxyaryl substituted diphenyl ureas) sold under the brand name Nexavar. The drug was used in the treatment of kidney and liver cancer. Under s. 84 of the Indian Act, a party may apply for a compulsory licence for a failure to work a patent, 3 years after grant of a patent. The grounds for a grant of such a compulsory licence are as follows:

(a) that the reasonable requirements of the public with respect to the patented invention have not been satisfied, or

Shaheeza Lalani and Michelangelo Temmerman, *Use It or Lose It? Assessing the Compatibility of the Paris Convention and TRIPS with Respect to Local Working Requirements*, Working Paper No. 2012/11, June 2013, NCCR Trade Regulation. See also generally Bryan Mercurio and Mital Tyagi, Treaty Interpretation in WTO Dispute Settlement: The Outstanding Question of the Legality of Local Working Requirements, *Minn. J. Int. L.* (2010), 275–326; and Michael Halewood, Regulating Patent Holders: Local Working Requirements and Compulsory Licenses at International Law, *Osg. H. L.J.*, Vol. 35 (1997), 243.

[93] DS 199.

[94] See discussion in ed. Carlos M. Correa, Research Handbook on the Interpretation and Enforcement of Intellectual Property under WTO Rules, (2010) Edward Elgar, at pp. 235–236.

[95] Judgment of the Bombay High Court, pronounced on 15 July 2014, available at: http://bombayhighcourt.nic.in/generatenewauth.php?auth=cGF0aD0uL2RhdGEvanVkZ2VtZW50cy8yMDE0LyZmbmFtZT1PU1dQMTEyODEzLnBkZiZzbWZsYWc9Tg== (accessed 17 January 2015).

(b) that the patented invention is not available to the public at a reasonably affordable price, or
(c) that the patented invention is not worked in the territory of India.

After having approached Bayer for a voluntary licence to sell the drug at a reduced price under what NATCO considered as reasonable terms and conditions and not succeeding, NATCO applied for a compulsory licence. The Controller granted a compulsory licence in 2011. Bayer appealed the decision to the Appeal Tribunal and was unsuccessful. The Indian Intellectual Appeals Board (IPAB) upheld the Controller's decision to grant a compulsory licence for use of Bayer's Nexavar medicine.[96] One of the grounds relied on by the IPAB was the ground of non-working of the Nexavar patent by Bayer. The IPAB was also careful to state that non-working may arise from mere importing but this might not be so in every case as much would depend on the facts and evidence in each case. In so holding, the IPAB related this case-by-case reasoning to the requirement in TRIPS Art 31(a) that "authorization of such use shall be considered on its individual merits".

Bayer sought to quash the IPAB's decision in the Bombay High Court, and, again, failed.[97] The Court declined to interfere with the IPAB's decision that NATCO had made efforts to first obtain a voluntary licence from Bayer and had, upon failure to obtain such a licence, petitioned for a compulsory licence. The Court further found that NATCO had made a case establishing all three grounds in s. 84(1) above.

With regard to ground (c), it agreed with the IPAB that a failure to work was to be determined on a case-by-case basis, and this was to be carried out in accordance with the general principles set out in s. 83. (The Board had found that "working" meant "worked on a commercial scale and not a token working".[98]) While the Court held that "worked in India" did not mean the drug had to be only manufactured in India, it held that Bayer had the burden of showing to the authorities why it had not manufactured in India (particularly since it had manufacturing facilities there), which it did not satisfy. It was for Bayer to show that it was "impossible/prohibitive" to manufacture the patented drug in India.

Apart from the fact that this marked the first compulsory licence granted under s. 84 against a patent holder, the case also provides insight

[96] *Bayer v NATCO*, IPAB Order No. 45 of 2013, www.ipab.tn.nic.in/045-2013.htm (accessed 17 January 2015).
[97] See Judgment of the Bombay High Court, pronounced on 15 July 2014, note 95, above.
[98] p. 21, *ibid.*

into the thinking of the Indian High Court and the IPAB in interpreting s. 84 and the related "principles" in s. 83. It also provides insight into the process of determining whether a drug has been supplied to the public in India at reasonable prices,[99] and what make be taken into consideration in determining terms such as adequate remuneration for the patent holder when a compulsory licence is granted under s. 84.[100] Finally, the High Court judgment provides an explanation of its perception of the patent bargain as it exists under Indian law.[101]

Subsequent to the NATCO decision, the Indian Controller of Patents rejected a petition by BDR Pharmaceuticals International Pvt. Ltd. for a compulsory licence for another patented cancer treatment drug (dasatinib, sold under the brand name Sprycel) in 2013, whose patentee is Bristol Myers Squibb Company, a US company.[102] It has been further reported that the Indian Ministry of Health is (at the time of writing) separately seeking the grant of a compulsory licence from the Department of Industrial Policy and Promotion (DIPP).[103]

Other countries whose laws include local "working" of patents include Indonesia,[104] Malaysia,[105] Thailand[106] and Vietnam.[107] In the case of

[99] On the question of whether the patented drug was being supplied to the public at "reasonably affordable" prices, the Court rejected Bayer's argument that the determination of such prices had to take into account the R&D costs of the drug as well as of failed drugs (pp. 41–42). The Court observed that the drug in question was an "orphan" drug in the US (so that it was entitled to receive tax credits or otherwise of up to 50% of its R&D costs), and that NATCO's evidence pointed to recovery of R&D costs within a year (the Court stated that Bayer did not bring evidence to contradict this).

[100] The Controller of Patents had fixed the royalty at 6% of net sales by NATCO; the IPAB increased this to 7%, which rate the Bombay High Court upheld.

[101] Note 97 above, at pp. 24–25.

[102] See report at: http://articles.economictimes.indiatimes.com/2013-10-31/news/43561264_1_voluntary-licence-compulsory-licence-dasatinib (accessed 17 January 2015).

[103] See, for example, http://articles.economictimes.indiatimes.com/2014-10-19/news/55197364_1_dasatinib-blood-cancer-drug-compulsory-licences and www.wsj.com/articles/india-divided-on-whether-to-break-bristol-myers-drug-patent-1416304089 (accessed 17 January 2015).

[104] See Articles 19 and 75(2) of the Indonesian Patents Law, No. 14 of 2001.

[105] S. 49, Malaysian Patents Act. [106] S. 46, Thailand's Patent Act 1979.

[107] Art. 145.1(b), Vietnam's Law of Intellectual Property, No. 50/2005/QH11, provides for the following situation:

> Where the holder of exclusive right to use such invention fails to fulfill the obligations to use such invention provided for in Clause 1, Article 136 and Clause 5, Article 142 of this Law upon the expiration of a 4-year duration as from the date of filing the invention registration application and the expiration of a 3-year duration as from the date of granting the invention patent.

Singapore, the grant of a compulsory licence is permitted to remedy anti-competitive practices through the failure to supply the local market on reasonable terms.[108]

Chinese law provides for the use of compulsory licensing in relation to local working. This is found in Chapter VI of the PRC Patent Law of 2008. In 2003 and 2005, China issued measures on compulsory licenses to address public health problems.[109] In 2012, the revised Measures for the Compulsory Licensing for Patent Implementation were promulgated.[110] Articles 5 to 8 of the Measures set out the grounds for grant of a compulsory licence.[111] Article 5 supports Article 48 of the Patent Law,

Art. 136.1 provides:

> *Article 136.* - Obligation to use inventions and marks
> 1. Owners of inventions are obliged to manufacture protected products or apply protected processes to satisfy the requirements of national defense, security, disease prevention and treatment and nutrition for the people or to meet other social urgent needs. When there arise the needs mentioned in this Clause but invention owners fail to perform such obligation, the competent state agency may license such inventions to others without permission of invention owners according to the provisions of Articles 145 and 146 of this Law.

For a discussion of compulsory licensing in Vietnam, see Tran Viet Dung, Le Thi Nam Giang and Nguyen Nhu Quynh, Vietnam: A Case Study for Sustainable Technology Transfer, in *Sustainable Technology Transfer, A Guide to Global Aid and Trade Development*, ed. Hans Henrik Lidgard, Jeffery Atik and Tu Thanh Nguyen (Kluwer Law International, 2012).

By way of comparison, Israel's patent legislation lists a failure to exploit an invention as one of the grounds for showing abuse of a patent monopoly, in the absence of a reasonable justification; see ss. 117–119, 120 and 122 of the Patents Law 5727-1967.

108 S. 55(2), Patents Act.

109 *Measures of January 1, 2006, for Compulsory License on Patent Implementation Concerning Public Health Problems* (promulgated by Order No. 37 of the State Intellectual Property Office (SIPO).

110 Order of the Director of the State Intellectual Property Office No. 64, promulgated 15 March 2012 and entered into force on 1 May 2012.

111 Articles 5–8 state:

> Article 5
> If a patentee has without good reason not implemented or fully implemented their patent within three years from the date of grant of the patent right and within four years from the date of filing of the patent, then an entity or individual which is able to implement the patent can petition for a compulsory license based on the provisions of Article 48 paragraph 1 of the Patent Law. If the actions of a patent right holder in exercising their patent right are legally deemed to be monopolistic actions, then in order to eliminate or reduce the negative effects of said actions on competition, an entity or individual which is able to implement the patent can petition for a

and refers to the situations where a patentee fails to "implement" or to "fully implement" its patent "without good reason" within 3 years from date of its grant, or within 4 years from the date of filing the application. This appears analogous to the non-working provision in other national compulsory licensing laws except that it does not explicitly require local manufacture. It further provides that where a patentee's actions are "legally deemed as monopolistic actions", a compulsory licence may also be granted to address negative effects of the actions. Article 6 refers to a situation of emergency or "irregular event of the state" and use for "purposes of the public interest". Article 7 permits a compulsory licence pursuant to Article 50 of the Patent Law. Finally, Article 8 permits grant of a compulsory licence where, "for the benefit of public health", manufacture for export abroad is permitted under international treaties. This would include the flexibility provided by the WTO's Paragraph 6 mechanism, that is, the waiver of Article 31(f) of the TRIPS Agreement.

> compulsory license based on the provisions of Article 48 paragraph 2 of the Patent Law.
>
> Article 6
> In an emergency or irregular event of the state, or for the purposes of public interest, the competent department under the State Council is entitled to recommend the State Intellectual Property Office to grant a compulsory license for implementing the invention patent or utility model patent of the petition based on the provisions of Article 49 of the Patent Law.
>
> Article 7
> For purposes of public health, an entity which is able to implement the patent can petition for a compulsory license based on the provisions of Article 50 of the Patent Law for the manufacture of patented medicines and export of the same to the following countries or regions:
>
> 1. The most underdeveloped countries or regions.
> 2. Developed or developing members of the World Trade Organization that express via said organization that they wish to be importers based on the relevant international treaties.
>
> Article 8
> If a patented invention or utility model is an important technical advancement over a previously patented invention or utility model and has prominent economic significance, and if the implementation thereof depends on the implementation of the earlier invention or utility model, the patent holder can request that the compulsory license be issued for the previous patent based on the provisions of Article 51 of the Patent Law. If the State Intellectual Property Office grants a compulsory license to implement the previous patent, the patent holder of the previous patent may request that the compulsory license be issued for a later patent.

Article 8 includes the possibility of export to "developing countries", whereas the waiver envisages LDCs.[112]

By way of comparison, under the UK Patents Act 1977, separate grounds for the grant of compulsory licences are set out for patent proprietors who are from WTO members and those who are not.[113] The latter proprietors are subject to more available grounds, including local working requirements in the UK.[114]

The position of the EU has been that the non-discrimination requirement in Article 27(1) applies to the grant of compulsory licences under Article 31. It has relied on, *inter alia*, certain remarks of the WTO panel in *Canada – Patent Protection for Pharmaceutical Products*.[115] It should be noted that the panel's remarks on Article 31 were in the nature of *dicta* as the provision in question there was Article 30; also, Canada had in that case conceded that Article 31 would be subject to the requirements in Article 27(1).[116] This panel decision was not appealed to the Appellate Body, which therefore did not have the opportunity to comment on this matter. The EU recognizes that this is a contested point by developing countries.[117]

In June, 2001, for example, a number of developing countries – including several in Asia – submitted the following arguments in support of local working requirements for pharmaceutical products, emphasizing the objectives of TRIPS and the imperative of transfer of technology:

[112] The waiver defines an "eligible importing member" to mean "*any least-developed country Member, and any other Member that has made a notification to the Council for TRIPS of its intention to use the system as an importer*, it being understood that a Member may notify at any time that it will use the system in whole or in a limited way, for example only in the case of a national emergency or other circumstances of extreme urgency or in cases of public non-commercial use . . ." [italic emphasis added]; full text available at WTO website, www.wto.org/English/tratop_e/trips_e/implem_para6_e.htm (accessed 11 March 2015).

[113] See generally, s. 48, and specifically, s. 48A and s. 48B, UK Patents Act 1977. An explanation offered for this is the requirement in Art. 27(1) of the TRIPS Agreement, that patent rights are to be provided by WTO members "without discrimination as to the place of invention, the field of technology and *whether products are imported or locally produced*" [italic emphasis added]; see *Terrell on the Law of Patents*, 7th ed., Sweet & Maxwell, at pp. 585–586.

[114] See especially s. 48B(1)(a) and (b), Patents Act 1977.

[115] WT/DS114/R; see paras. 7.91–7.92; see EU paper, *Legal Issues related to Compulsory Licensing under the TRIPS Agreement and Questions on TRIPS and Data Exclusivity*, 2001, trade.ec.europa.eu/doclib/html/122031.htm, at footnote 8 and accompanying text (accessed 11 March 2015). See also Daniel Gervais, *The TRIPS Agreement – Drafting History and Analysis*, 4th ed. (Sweet & Maxwell, 2012) pp. 432–433.

[116] See para. 7.90, panel report, *ibid.*

[117] See EU paper, note 115 above, at p. 6.

> *The objective of the* ***promotion of technological innovation and the transfer and dissemination of technology*** *places the protection and enforcement of IPRs in the context of the interests of society. Such an objective is essential for the promotion of health policies, as it* ***encourages the development of domestic production of pharmaceutical products.*** *Whenever economically feasible, local production of pharmaceutical products is extremely important to ensure that medications are more readily available in the market, and at more affordable prices. Local manufacturing of pharmaceutical products also encourages sustainable access to medications by insulating the price of patented medicines against currency devaluations, as well as supporting the development of local expertise, which is vital in addressing local needs. As mentioned above, these objectives can be obtained by the normal exercise of patent rights.* ***Where the patent holder fails to meet the objectives of the TRIPS Agreement and of public health policies, however, Members may take measures to ensure transfer and dissemination of technology to provide better access to pharmaceuticals.***
>
> [Emphases are as in the original.][118]

The recent Korea-Australia FTA expressly permits Korea to revoke a patent which has not been continuously worked in Korea for 2 years or more from the date of award of the patent.[119] The IP chapter also does not contain any data exclusivity provision.

### 5.6.3.5 The TRIPS and public health declaration

The unsatisfactory provisions in Article 31 of TRIPS (specifically, Article 31(f), in not permitting export of products manufactured where countries facing scarcity of such products could not manufacture these products themselves) and, more generally, the debate over public access to pharmaceuticals led to adoption on 14 November 2001 by the WTO Ministers of the *Declaration on the TRIPS Agreement and Public Health*, to address "the gravity of public health problems afflicting many developing and least-developed countries, especially those resulting from HIV/AIDS, tuberculosis, malaria and other epidemics".[120] The Declaration affirmed members' right to "take measures to protect public health" and to grant compulsory licences. It also reaffirmed the fact that each member has the right to determine what constitutes a national

[118] Para. 20, TRIPS and Public Health – Submission by the African Group, Barbados, Bolivia, Brazil, Cuba, Dominican Republic, Ecuador, Honduras, India, Indonesia, Jamaica, Pakistan, Paraguay, Philippines, Peru, Sri Lanka, Thailand and Venezuela, document IP/C/W/296.

[119] Korea-Australia Free Trade Agreement (KAFTA), entered into force 12 December 2014, at footnote 74, Chapter 13, Intellectual Property.

[120] WT/MIN(01)/DEC/2.

emergency or other circumstances of extreme urgency (the language found in Article 31 (b) of the TRIPS Agreement). The important language of paragraph 5(a)–(c) of the Declaration is reproduced here:

> 5. Accordingly and in the light of paragraph 4 above, while maintaining our commitments in the TRIPS Agreement, we recognize that these flexibilities include:
>
> a. In applying the customary rules of interpretation of public international law, each provision of the TRIPS Agreement shall be read in the light of the object and purpose of the Agreement as expressed, in particular, in its objectives and principles.
> b. *Each member has the right to grant compulsory licences and the freedom to determine the grounds upon which such licences are granted.*
> c. Each member has the *right to determine what constitutes a national emergency or other circumstances of extreme urgency*, it being understood that public health crises, including those relating to HIV/AIDS, tuberculosis, malaria and other epidemics, can represent a national emergency or other circumstances of extreme urgency . . .[121]
>
> [Italic emphasis added]

Paragraph 6 clearly recognizes the inadequacy of Article 31 in addressing the needs of countries which have no manufacturing capacity in pharmaceuticals.[122] This Declaration resulted in an effort to extend the compulsory licensing system of Article 31. Following debates on how to take this forward, WTO members agreed in August 2003 to establish a waiver of Article 31(1)(f) in order to implement the above.[123] Regulating provisions were drafted for this waiver. It was subsequently decided that this waiver would be incorporated eventually as an amendment of the TRIPS Agreement, through insertion of its provisions as Article 31*bis*.[124]

[121] *Declaration on the TRIPS Agreement and Public Health*, 14 November 2001.

[122] Paragraph 6 of the Declaration states:

> 6. We recognize that WTO members with insufficient or no manufacturing capacities in the pharmaceutical sector could face difficulties in making use of compulsory licensing under the TRIPS Agreement. We instruct the Council for TRIPS to find an expeditious solution to this problem and to report to the General Council before the end of 2002.

[123] WT/L/641, 8 December 2005; Implementation of Paragraph 6 of the Doha Declaration on the TRIPS Agreement and Public Health, Decision of the General Council, 30 August 2003. For a detailed description of the waiver and its implementation, see Ho, note 151.

[124] Decision to Amend the TRIPS Agreement, Decision of the General Council, 6 December 2005.

It is therefore open to WTO members to incorporate compulsory licensing in their legislation, and, indeed, many countries have done so. However, such legislation would have to comply with the conditions under Article 31, subject to the Article 31(f) waiver.

**5.6.3.5.1 Compulsory licensing in Asia** The power to make use of compulsory licensing is not of interest just to developing countries, but is of concern in some developed countries as well. Some European countries have recently reviewed their legislation by updating the compulsory licensing provisions, to better address current public health issues (such as gene patenting and research). For instance, France, Belgium and Switzerland have introduced legislative changes to address these concerns.[125]

Many countries in Asia have compulsory licensing legislation, including China,[126] Indonesia,[127] India,[128] Malaysia,[129] Thailand[130] and Singapore.[131] While the circumstances for invocation may differ, such provisions provide exceptional access to pharmaceutical products through what is essentially a means of "forced" or non-voluntary licensing. The countries which have utilized this means are still relatively few, although in recent years use has been increasing.

In 2010, a petition by Korean civic groups for a compulsory licence for the drug Fuzeon, used to treat HIV/AIDS, was refused on the ground that the price set by the drug manufacturer was not unreasonable.[132]

[125] Esther van Zimmeren and Geertrui Van Overwalle, *A Paper Tiger? Compulsory License Regimes for Public Health in Europe* (2010), available at: http://papers.ssrn.com/sol3/papers.cfm?abstract_id=1717974.

[126] Patent Law of the People's Republic of China (2008). Article 50 permits grant of a compulsory licence "for the benefit of public health".

[127] Law No 14 of 1 August, 2001 regarding Patents – see in particular Articles 75–76. Article 99 also provides for exploitation of a patent by the government itself where "it is very important for the conduct of defence and security of the State and for an urgent need for the sake of public interest".

[128] Section 84, Patent Act 1970, as amended in 2002 by the Patents (Amendment) Act 2002.

[129] Section 48, Patent Act 291 of 1983; text available at: www.myipo.gov.my/paten-akta.

[130] Section 46, Patent Act, footnote 47 above. On Thailand's use of compulsory licenses, see Ho, footnote 123 above, Chapter 6, at pp. 167–179.

[131] Section 55, Patents Act, footnote 48 above; this is for the purpose of remedying an anti-competitive practice. Section 56 provides for government use of a patented product for a public non-commercial use and in national emergencies or other circumstances of extreme urgency; this reflects the flexibility provided by Article 31(f) of the TRIPS Agreement.

[132] See Bong-young Min and Hye-young Kwon, *TRIPS and New Challenges for the Pharmaceutical Sector in South Korea*, in *The New Political Economy of Pharmaceuticals*, ed. Hans Löfgren and Owain David Williams (Palgrave Macmillan,

A number of Asian governments have begun issuing compulsory licences in recent years. Malaysia was the first to do so (2004) with respect to three pharmaceutical formulations.[133] Thereafter, Taiwan (2005),[134] Thailand (2006–2007)[135] and India (2012)[136] issued compulsory licences for pharmaceutical products as well. In 2004 and 2012, Indonesia issued presidential decrees for "government use" orders for the government to exploit certain antiretroviral pharmaceutical products.[137, 138]

Table 5.2 summarizes the compulsory licensing actions taken in Asian countries to date.[139]

It is critical for such countries to preserve the right to make use of these TRIPS-permitted means in any FTA they may be negotiating. During the Thai-US FTA negotiations (which were subsequently suspended), for instance, it has been reported that the US had requested a restriction on compulsory licensing.[140]

2013), at pp. 209–210, and Young Kim and John Kim, Refusal of Compulsory License for AIDS Drug Patents Becomes Final, *WIPR* (February 2010).

133 See Musungu and Oh, *The Use of Flexibilities in TRIPS by Developing Countries: Can They Promote Access to Medicines?* (2005), pp. 23–25; available at: www.who.int/intellectualproperty/studies/TRIPS_flexibilities/en/ (accessed 13 April 2015).

134 See Taiwan Employs Compulsory Licensing for Tamiflu, *Financial Times*, 25 November 2005 and James P. Love, *Recent Examples of the Use of Compulsory Licenses on Patents*, Knowledge Ecology International Research Note 27:2, available at: www.keionline.org/misc-docs/recent_cls_8mar07.pdf

135 See reports at www.reuters.com/article/2012/10/12/us-indonesia-hiv-idUSBRE89B0O620121012 and www.keionline.org/content/view/90/1, http://keionline.org/content/view/41/1 and www.wipo.int/edocs/mdocs/mdocs/en/wipo_ip_econ_ge_4_12/wipo_ip_econ_ge_4_12_ref_saggi.pdf.

136 See http://keionline.org/node/1657.

137 See the 2004 and 2012 President Decrees at, respectively, http://lists.essential.org/pipermail/ip-health/2004-December/007233.html and www.citizen.org/documents/PresidentalDecree20121.pdf. See also www.ihs.com/products/global-insight/industry-economic-report.aspx?id=1065972339.

138 Articles 74–77, 79 and 99–103 of the Indonesian Patent Law No. 14 of 01/08/2001 govern the use of compulsory licensing.

139 See generally, Sakda Thanitcul and Matthew Lim Braslow, note 145, Siraprapha Khim Rungpry, Compulsory Licensing Issues and Trends in Asia, *Pharm. Pat. Analyst*, Vol. 2, No. 6 (2013), 681–683, and Chapter 11,Prabodh Malhotra and Bhajan Grewal, *TRIPS-Plus: Free Trade Agreements Jeopardizing Public Health in Developing Nations, in Regional Trade Agreements in Asia*, ed. Tran Van Hoa and Charles Harvie (Edward Elgar Publishing Limited, 2008), esp. at pp. 223–224.

140 Jakkrit Kuanpoth, *Patent Rights in Pharmaceuticals in Developing Countries, Major Challenges for the Future* (Edward Elgar Publishing, 2010), at p. 184.

Table 5.2. *Compulsory licenses granted in Asia*

| Country | Pharmaceutical name and patent holder | Year of issue of compulsory licences |
|---|---|---|
| Indonesia | Nevirapine | 2004 |
| | Lamivudine | ------- |
| | ------- | 2007 |
| | Nevirapine | |
| | Lamivudine | |
| | Efavirenz (Merck) | |
| India | Sorafenib tosylate (Bayer: Nexavar) | 2012 |
| Malaysia | Didanosine/ddI (Bristol-Myers Squibb) | 2004 |
| | Zidovudine, and lamivudine combined with zidovudine (GlaxoSmithKline) | |
| Taiwan | Tamiflu | 2005 |
| Thailand | Efavirenz (Merck) | 2007 |
| | Lopinavir and ritonavir (Abbott: Kaletra) | 2008 |
| | Clopidogrel (Sanofi-Aventis: Plavix) | |
| | Letrozole (Novartis) | |
| | Docetaxel (Sanofi-Aventis) | |
| | Erlotinib (Roche) | |

**5.6.3.5.2 Singapore** Singapore's Patents Act contains provisions for compulsory licensing. By way of terminology, it should be noted that under the Act, a distinction is made between "compulsory licensing" and "use of patented products for services of government".[141]

The *Doha Declaration Implementation Decision*, which embodies the WTO members' decision on a waiver of Article 31(f) of the TRIPS Agreement, is implemented in Singapore via the "government use" provisions in sections 56–62 of the Patents Act.[142] Section 56 permits the Singapore government to:

141 Governed by Parts XI and XII of the Act, respectively.

142 That these are the relevant provisions is evident from the statements of the then Senior Minister of State for Law in Parliament, 25 August 2008, at the second reading of the Patents (Amendment) Bill 2008, which introduced amendments to these provisions to align them with the requirements of the Waiver; see *Hansard*, available in the subscription database at www.lawnet.com.sg.

... do anything in relation to a patented invention –

(a) for a public non-commercial purpose; or
(b) for or during a national emergency or other circumstances of extreme urgency ...

Section 62(1) of the Patents Act further provides that where the government acts under Section 56, it 'shall pay such remuneration to the patentee as may be agreed, or as may be determined by a method agreed, between the government and the patentee having regard to "the economic value of the patented invention ..."', or, in default of agreement, as may be determined by the court under Section 58.

Certain WTO members, such as Chinese Taipei, Hong Kong, Korea, Macao and Singapore, announced in 2003 that they will not to use the WTO waiver except in cases of emergency or other circumstances of extreme urgency[143]; the provision is nonetheless an inroad for the US in its efforts to control more strictly compulsory licensing use.[144]

**5.6.3.5.3 Thailand** In the case of Thailand, a military coup occurred which deposed the then Prime Minister, Thaksin Shinawatra, in 2006.[145] Between November 2006 and January 2007, the Thai government which took over issued compulsory licences for three patented medicines, namely efavirenz (Stocrin of Merck Sharp and Dohme), lopinavir/ritonavir (Kaletra of Abbott Laboratories) and clopidogrel (Plavix of Sanofi-Aventis). These licences were granted under Article 51 of the Thai Patents Act as "Government Use" licences.[146] The first two are

[143] Other countries were Israel, Kuwait, Mexico, Qatar, Turkey and the United Arab Emirates; see WTO press release, 30 August 2003, www.wto.org/english/news_e/pres03_e/pr350_e.htm (accessed 16 January 2015); and Musungu and Oh, at p. 63; footnote 132 above.

[144] Through its Bilateral Trade Agreement (BTA) with the United States, signed on 13 July 2000, Vietnam has also committed to restricting use of compulsory licensing to emergencies and situations of extreme urgency: see Art. 7.8 of the BTA.

[145] See BBC report, *Thai PM Deposed in Military Coup*, 20 September 2006, at: http://news.bbc.co.uk/2/hi/world/asia-pacific/5361512.stm (accessed 28 May 2014).

[146] See Sakda Thanitcul and Matthew Lim Braslow, Compulsory Licensing of Chronic Disease Pharmaceuticals in Thailand, *Thai J. Pharm. Sci.*, Vol. 37, No. 2 (April–June 2013), 61– 83, at pp. 63 and 66. Article 51 provides as follows: "51. In order to carry out any service for public consumption or which is of vital importance to the defense of the country or for the preservation or realization of natural resources or the environment or to prevent or relieve a severe shortage of food, drugs or other consumption items or for any other public service, any ministry, bureau or department of the government may, by themselves or through others, exercise any right under Section 36 by paying a royalty to the patentee or his exclusive licensee under paragraph 2 of Section 48 and shall notify the

antiretroviral drugs while the third is an anti-platelet drug for preventing coronary obstruction, that is, it differs from the other two in that it addresses a non-communicable disease.

The Thai government set out its position in a White Paper setting out its position on the compulsory licences in February 2007.[147] In it, it stated, regarding the three compulsory licences: "The Thai Ministry of Public Health views these decisions on the government use of patents as a form of social movement that aims at improving access to essential medicines and the health of the people."[148]

The compulsory licences were greeted with both strong international criticism as well as support. In the case of the EU, Thailand met with mixed signals in the form of contrary messages from then Trade Commissioner Peter Mandelson and the European Parliament.[149] On one hand, the Commissioner wrote to the Thai Minister for Public Health discouraging use of compulsory licences, and on the other, the European Parliament passed a resolution in the same period which not only recognized the difficulties of access to medicines in developing countries, but *specifically* "[e]ncouraged the developing countries to *use all means available to them under the TRIPS Agreement, such as*

patentee in writing without delay, notwithstanding the provisions of Sections 46, 46*bis* and 47. The author is grateful for the extensive and interesting face-to-face discussion with Professor Sakda Thanitcul, May 2014, at the Chulalongkorn University on these and other related matters. On the Thai compulsory licenses, see also generally, Mishka Glaser and Ann Marie Murphy, *Patients v Patents: Thailand and the Politics of Access to Pharmaceutical Products*, (2010) J. Third World Studies, Vol. XXVII, No. 1, pp. 215–234.

In the circumstances under the above paragraph, the ministry or bureau or department shall submit its offer setting forth the amount of remuneration and conditions for the exploitation to the Director-General. The royalty rate shall be as agreed upon by the ministry or bureau or department and the patentee or his licensee, and the provisions of Section 50 shall apply *mutatis mutandis*."

147 Ministry of Public Health and National Health Security Office, Thailand, *Facts and Evidences on the 10 Burning Issues Related to the Government Use of Patents on Three Patented Essential Drugs in Thailand – Document to Support Strengthening of Social Wisdom on the Issue of Drug Patent*, February 2007, available at: www.moph.go.th/hot/White%20Paper%20CL-EN.pdf (accessed 28 May 2014).

148 *Ibid.*, in *Preface*, by the then Thai Minister for Public Health, Dr Mongkol Na Songkhla.

149 For a detailed account of the exchanges between then Commissioner Mandelson and Thailand's Minister for Public Health, see Gaëlle Krikorian, The Politics of Patents: Conditions of Implementation of Public Health Policy in Thailand, in *Politics of Intellectual Property: Contestation Over the Ownership, Use, and Control of Knowledge and Information*, ed. Sebastien Haunss and Kenneth C. Shadlen (2009). See also, generally, Richard A. Epstein and F. Scott Kieff, Questioning the Frequency and Wisdom of Compulsory Licensing for Pharmaceutical Patents, *Univ. Chic. L. R.*, Vol. 78, No. 71 (2011).

*compulsory licences and the mechanism provided by Article 30* thereof"[150] [italic emphasis added].

**5.6.3.5.4 Compulsory licensing as a remedy for anti-competitive conduct** Not many FTAs signed by Asian states include express provisions on compulsory licensing. The USSFTA is one agreement which expressly recognizes a right to resort to compulsory licensing, but imposes restrictions as to its use, thereby narrowing the flexibility originally available under Article 31 of the TRIPS Agreement. The following sub-paragraph of Article 16.7 of the FTA (note the limiting words are "except in the following circumstances" in the opening sentence below) provides:

> 6. Neither Party shall permit the use[16–11] of the subject matter of a patent without the authorization of the right holder except in the following circumstances:
>
> (a) to remedy a practice determined after judicial or administrative process to be anti-competitive under the competition laws of the Party;[16–12]
>
> (b) in the case of public non-commercial use or in the case of a national emergency or other circumstances of extreme urgency, provided that:
>
> (i) such use is limited to use by the government or third parties authorized by the government;
>
> (ii) the patent owner is provided with reasonable and entire compensation for such use and manufacture; and
>
> (iii) the Party shall not require the patent owner to transfer undisclosed information or technical "know how" related to a patented invention that has been authorized for use without the consent of the patent owner pursuant to this paragraph . . .
>
> 16–12 The Parties recognize that an intellectual property right does not necessarily confer market power upon its owner.

Under the Singapore Patents Act, "compulsory licensing" under section 55 refers to the situation where an application is made to the court by "any interested person" on the ground that "the grant of the

150 Para. 9, European Parliament Resolution, P6_TA(2007)0353, 12 July 2007. In the Resolution, the European Parliament also asked the European Council (comprising EU heads of state, Council President and President of the European Commission, and which sets the EU policy agenda) to "support the developing countries which use the so-called flexibilities built into the TRIPS Agreement and recognized by the Doha Declaration in order to be able to provide essential medicines at affordable prices under their domestic public health programmes" (para. 8).

licence is necessary to remedy an anti-competitive practice". This reflects the requirement in Article 16.7.6(a) of the USSFTA, above.

### 5.6.3.6 Exceptions relevant to data exclusivity and patent linkage

**5.6.3.6.1 Data exclusivity** The TRIPS Agreement generally provides members with freedom to impose regulatory approval requirements. Such approval differs from patent regulations in that regulatory approval deals with the government approval process that must be satisfied for a pharmaceutical product to be placed on the market. Considerations taken into account during this process often include safety and, perhaps, whether there has been prior approval of the product already given in other markets. In recent years, some FTAs have made incursions into states' regulatory approval processes in two main ways: in imposing data exclusivity requirements on an FTA party, and in patent linkage.

Pharmaceutical companies often seek through FTAs to restrict a state's ability to use their research data for approval of alternative products.[151] Such provisions vary in scope, with some exceeding the requirements under Article 39 of the TRIPS Agreement. An example is Article 3 of the EFTA-Korea FTA, which provides as follows:

> **Article 3**
> **Undisclosed Information**
>
> The Parties shall protect undisclosed information in accordance with Article 39 of the TRIPS Agreement. The Parties shall prevent applicants for marketing approval for pharmaceuticals and agricultural chemical products from relying on undisclosed test or other undisclosed data, the origination of which involves a considerable effort, submitted by the first applicant to the competent authority for marketing approval for pharmaceuticals and agricultural chemical products, utilising new chemical entities, for an adequate number of years[2] from the date of approval, except where approval is sought for original products. Any Party may instead allow in their national legislation applicants to rely on such data if the first applicant is adequately compensated.[3]
>
> [2] The number of years shall be determined by the relevant laws and regulations of the Parties.
> [3] The Parties note that Korea currently does not have such a compensation system.

[151] See Cynthia M. Ho, Chapter 9, *Access to Medicine in the Global Economy: International Agreements on Patents and Related Rights* (Oxford University Press, 2011).

Although the provision leaves it to parties to determine the duration of protection under their national laws, it exceeds the requirements of Article 39 of TRIPS by requiring specific prevention of use of undisclosed test data for pharmaceuticals for marketing approval of pharmaceutical products. It also does not provide for the exceptional situations under which Article 39 permits disclosure. What is notable, however, is that in its last sentence, Article 3 permits disclosure/reliance by applicants other than the first applicant on such data "if the first applicant is adequately compensated". The interpretation of "adequate" is critical in this context; if the compensation required is too high, it may ultimately render it uneconomical for the subsequent applicant to proceed with the use.

The obligation on undisclosed information in the China–Australia Free Trade Agreement reflects much of the language found in Article 39 of the TRIPS Agreement but *omits* the health-related words in Article 39(2). Article 11.18 of the Agreement provides:

**ARTICLE 11.18: PROTECTION OF UNDISCLOSED INFORMATION[2]**

1. In the course of ensuring effective protection against unfair competition, each Party shall protect undisclosed information in accordance with paragraph 2.
2. Natural and legal persons shall have the possibility of preventing information lawfully within their control from being disclosed to, acquired by, or used by others without their consent in a manner contrary to honest commercial practices[3], so long as such information:
   (a) is secret, in that it is not, as a body or in the precise configuration and assembly of its components, generally known among or readily accessible to persons within the circles that normally deal with the kind of information in question;
   (b) has commercial value because it is secret; and
   (c) has been subject to reasonable steps under the circumstances, taken by the person lawfully in control of the information, to keep it secret.[152]

[2] It is understood that, consistent with the TRIPS Agreement, protection of undisclosed information includes protection of confidential information that is otherwise undisclosed and provided in commercial situations, such as architectural drawings or proprietary commercial-in-confidence information.

[3] For the purposes of this provision, "a manner contrary to honest commercial practices" shall mean, at least, practices such as breach

[152] Signed 17 June 2015. The text of the FTA is available at http://dfat.gov.au/trade/agreements/chafta/official-documents/Pages/official-documents.aspx (accessed 13 October 2015). The FTA also contains provisions for investor-State arbitrations.

> of contract, breach of confidence and inducement to breach, and includes the acquisition of undisclosed information by third parties who knew, or were grossly negligent in failing to know, that such practices were involved in the acquisition.

However, Article 11.7 refers to the principles of the Doha Declaration on the TRIPS Agreement and Public Health and the "Paragraph 6" waiver mechanism.

By contrast, some FTAs simply reiterate the obligations in Article 39 of the TRIPS Agreement; an example can be found in Article 15.19 of the China–Korea Free Trade Agreement.[153]

**5.6.3.6.2 Patent linkage** While patent linkage exists in the patent legislation of some countries (such as Singapore and the US), some other countries do not make such linkages. The EU is such an example.[154]

An example of an FTA that includes restrictions in both these areas of regulatory approval is the USSFTA. Article 16.8(1) for instance prohibits the use of data of an originator pharmaceutical company in assessing safety and efficacy for marketing approval of the same product or similar products, for 5 years from the date of approval of the originator's pharmaceutical product. (This is not applicable if the applicant has the originator's consent.) Article 16.8(4) provides for patent linkage. These requirements have since been translated into legislative requirements under Singapore's Medicines Act.[155] However, the Singapore legislation also includes some important exceptions to the data exclusivity provisions. One of this is in section 19B(1) of the Act, which provides as follows:

> **Circumstances where protection under section 19A does not apply**
>
> 19B. – Notwithstanding section 19A, the licensing authority may, during the protected period in relation to confidential supporting information –
>
> (a) disclose that confidential supporting information, or use that confidential supporting information for the purposes of determining

153 Text available at http://fta.mofcom.gov.cn/korea/annex/xdzw_en.pdf (accessed 13 October 2015).

154 See EU Directive 2001/83/EC. See also EU Commission Pharmaceutical Sector Inquiry Report, 2009, at p. 23; text available at: http://ec.europa.eu/competition/sectors/pharmaceuticals/inquiry/communication_en.pdf.

155 Cap. 176, Sections 19A(1) and 19D, and Section 12A, respectively. By comparison, Art. 10.36 of the EU-Korea FTA imposes a period of 5 years for the protection of data with respect to pharmaceutical products, while the Japan-Switzerland Agreement, however, imposes a period of 6 years from time of approval of the original application.

> whether to grant any application other than the application to which it relates or related, as the case may be –
> (i) with the consent of the applicant who made the application to which the confidential supporting information relates or related; or
> (ii) *if that disclosure or use is, in the opinion of the licensing authority, necessary to protect the health or safety of members of the public; . . .*
>
> [Italic emphasis added]

Section 19B(1)(a)(ii) is significant as it confirms the authorities' right to override the data exclusivity requirements of Section 19A where it is "necessary to protect the health or safety of members of the public" – a broad discretion retained without definition of the circumstances in which this may apply.[156]

### 5.6.3.7 Other aspects in Asian FTAs

Express recognition of the flexibilities referred to in the TRIPS Declaration have been made in some FTAs, such as the KORUS (Article 18.9(3)) and US-Peru FTA (Article 16.13).

In the KORUS Agreement, the signatories included a "confirmation letter" regarding disputes over patent linkage as required in the intellectual property chapter. The letter in effect put in place a moratorium over the use of dispute settlement processes with regard to patent linkage, for a period of 18 months from entry into force of the agreement.[157] This is significant as it protects the parties from FTA dispute settlement (investor-state arbitration in particular) for complaints over implementation of patent linkage, even if it is for a limited period.

Finally, the China-Chile FTA does not provide for TRIPS-plus commitments, but rather, contains an explicit statement seeking to affirm the need to balance intellectual property right owners and broader public interests.[158]

[156] Sections 19B(1)(b) and (c) provide for other exceptions, such as disclosure to other government departments, or to bodies such as the WHO.

[157] See the text of the letter at: www.ustr.gov/trade-agreements/free-trade-agreements/korus-fta/final-text.

[158] Article 111.1(c) provides as follows:

> The aim of cooperation on intellectual property rights will be: . . .
> (c) to achieve a balance between rights of right holders and the legitimate interests of users and the community with regard to protected subject matters . . .

#### 5.6.3.8 Examples of public health protection provisions in FTAs

An example of an attempt at providing regulatory discretion for health by way of specific carve-outs for certain measures from dispute settlement, to reduce the risk of legal exposure, is found in the following, in the US-Colombia FTA:

> **Article 16.13: Understandings Regarding Certain Public Health Measures**
>
> 1. The Parties affirm their commitment to the Declaration on the TRIPS Agreement and Public Health (WT/MIN(01)/DEC/2).
> 2. The Parties have reached the following understandings regarding this Chapter.
>    (a) The obligations of this Chapter do not and should not prevent a Party from taking measures to protect public health by promoting access to medicines for all, in particular concerning cases such as HIV/AIDS, tuberculosis, malaria, and other epidemics as well as circumstances of extreme urgency or national emergency. *Accordingly, while reiterating their commitment to this Chapter, the Parties affirm that this Chapter can and should be interpreted and implemented in a manner supportive of each Party's right to protect public health and, in particular, to promote access to medicines for all.*
>    (b) In recognition of the commitment to access to medicines that are supplied in accordance with the Decision of the General Council of 30 August 2003 on the Implementation of Paragraph Six of the Doha Declaration on the TRIPS Agreement and Public Health (WT/L/540) and the WTO General Council Chairman's statement accompanying the Decision (JOB(03)/177, WT/GC/M/82) (collectively, the "TRIPS/health solution"), this Chapter does not and should not prevent the effective utilization of the TRIPS/health solution.
>    (c) With respect to the aforementioned matters, if an amendment of the TRIPS Agreement enters into force with respect to the Parties and a Party's application of a measure in conformity with that amendment violates this Chapter, the Parties shall immediately consult in order to adapt this Chapter as appropriate in the light of the amendment.
>
> [Italic emphasis added]

Article 17.4 of the Korea-Peru FTA also provides some useful and interesting language to safeguard regulatory space with respect to protection of public health:

4. The Parties recognize the principles established in the *Declaration on the TRIPS Agreement and Public Health* (WT/MIN(01)/DEC/2) (hereinafter referred to as the "Declaration"), adopted on November 14, 2001 by the WTO Ministerial Conference, and the *Implementation of Paragraph 6 of the Doha Declaration on the TRIPS Agreement and Public Health* (WT/L/540) (hereinafter referred to as the "Decision"), adopted on August 30, 2003 by the WTO General Council. Likewise, the Parties recognize the importance to promote the implementation and full use of *Global Strategy and Plan of Action on Public Health, Innovation and Intellectual Property* (WHA61.21), adopted on May 24, 2008 by the 61st World Health Assembly.
5. In accordance with Article 8.1 of the TRIPS Agreement, the Parties may use, in formulating or amending their laws and regulations, the exceptions and flexibilities allowed by multilateral agreements related to the protection of intellectual property, and in particular, adopt measures necessary to protect public health and nutrition, taking into consideration the TRIPS Agreement and the principles established in the Declaration and the Decision.
6. Nothing in this Chapter shall hinder a Party from adopting the necessary measures to prevent the abuse of the intellectual property rights by its holders or the resort to practices that unreasonably restrain trade or international transfer of technology.

The US-Peru FTA, on the other hand, contains a provision that reflects Article 30 of the TRIPS Agreement.

**Article 16.9: Patents**

...

3. Each Party may provide limited exceptions to the exclusive rights conferred by a patent, provided that such exceptions do not unreasonably conflict with a normal exploitation of the patent and do not unreasonably prejudice the legitimate interests of the patent owner, taking account of the legitimate interests of third parties.

Article 11.5 of the China-Switzerland FTA provides that the signatory countries "recognize the principles established in the Doha Declaration on the TRIPS Agreement and Public Health adopted on 14 November 2001 by the Ministerial Conference of the WTO" and that the provisions of the chapter are "without prejudice" to the declaration.[159]

[159] The FTA entered into force on 1 July 2014. See also generally, Article 15.5 of the China-Korea Free Trade Agreement; text available at http://fta.mofcom.gov.cn/korea/annex/xdzw_en.pdf (accessed 13 October 2015), and Article 11.7 of the China-Australia Free Trade Agreement; text available at http://dfat.gov.au/trade/agreements/chafta/official-

#### 5.6.3.9 Explicit linkage of MFN treatment to exceptions and limitations

One of the interesting recommendations in the *Principles for Intellectual Property Provisions in Bilateral and Regional Agreements* is to include MFN treatment with regard to exceptions and limitations in an FTA, so that a party may enjoy such exceptions and limitations extended to its treaty partner's other partners, based on similar situations (e.g. level of economic development).[160]

### 5.7 Using a non-violation provision to safeguard the use of exceptions and limitations

The non-violation ground of action is one stemming from Article XXIII:1(b) of the GATT 1947 (now GATT 1994),[161] which, through case law of GATT panels, WTO panels and the WTO Appellate Body, has been interpreted to rarely be applicable.[162] In the context of TRIPS, the non-violation ground of action has been suspended since the inception of TRIPS and has been periodically renewed by members of the

documents/Documents/chafta-chapter-11-intellectual-property.pdf (accessed 13 October 2015).

[160] See Principles 25–26, H. Grosse Rush-Khan, *et al.*, Principles for IP Provisions in Bilateral and Regional Agreements, *IIC*, Vol. 44, No. 8 (2013) 878–883.

[161] Article XXIII:1 of GATT 1994 provides as follows:

**Nullification or Impairment**

1. If any contracting party should consider that any benefit accruing to it directly or indirectly under this Agreement is being nullified or impaired or that the attainment of any objective of the Agreement is being impeded as the result of
   (a) the failure of another contracting party to carry out its obligations under this Agreement, or
   (b) *the application by another contracting party of any measure, whether or not it conflicts with the provisions of this Agreement*, or
   (c) the existence of any other situation,

the contracting party may, with a view to the satisfactory adjustment of the matter, make written representations or proposals to the other contracting party or parties which it considers to be concerned. Any contracting party thus approached shall give sympathetic consideration to the representations or proposals made to it. [Italic emphasis added]

[162] See, for example, *Japan – Measures Affecting Consumer Photographic Film and Paper*, WT/DS44/R, Panel report adopted 22 April 1998, and *European Communities – Measures Affecting Asbestos and Asbestos-Containing Products*, WT/DS135/AB/R, Appellate Body report adopted 5 April 2001.

WTO.[163] Non-violation provisions have been included in some FTAs, but so far they are not tailored to protect the right to rely on exceptions and limitations.[164]

At the WTO, members have so far agreed to exclude use of non-violation complaints in relation to TRIPS. It is suggested here, however, that a "non-violation" provision could be explored for inclusion in the FTA, specifically to ensure that *exceptions and flexibilities* are not derogated from by a treaty partner by using means external to the FTA, such as threats of trade sanctions and of withdrawal of benefits (such as those under a GSP programme) or other unilateral pressure mechanism provided for under its domestic law (such as listing a country as a "delinquent" country in a "watch" list). Such a clause would, for instance, permit a country to bring the other to FTA dispute settlement if the latter carried out such actions in response to the former country's reliance on an exception or limitation. GATT and WTO jurisprudence have limited the scope and applicability of non-violation complaints. However, in the few instances where a panel or the Appellate Body has had to deal with such a complaint, the notion of "legitimate expectations" of benefits accruing from GATT concessions has been considered as one of the relevant factors in such complaints.[165] In the present context, developing countries could argue that they have a *legitimate expectation* that they would be able to exercise the flexibilities provided for by the TRIPS Agreement, and that such expectation should not be permitted to be stymied by non-violative but effective measures of another country – such as the use of GSP withdrawal threats as a means of pressure, or as measures taken to produce a chilling effect.

[163] See WTO background information, www.wto.org/english/tratop_e/trips_e/nonviolation_background_e.htm, and the latest renewal of the suspension of such claims until 31 December 2015 at the Bali Ministerial Conference, December 2013: www.wto.org/english/news_e/news13_e/trip_10oct13_e.htm#nonviolation (accessed 23 July 2014). See also Cottier and Schefer, *Non-Violation Complaints in WTO/GATT Dispute Settlement: Past, Present and Future*, in *International Trade Law and the GATT/WTO Dispute Settlement System*, ed. Ernst-Ulrich Petersmann (Kluwer Law International, 1997).

[164] See, for example, this author, Non-violation Complaints – WTO Issues and Recent Free Trade Agreements, *Journal of World Trade*, Vol. 39, No. 2 (2005), 205. Examples of FTAs that have non-violation complaint provisions are the US-Morocco FTA (Art. 20.2), US-Australia FTA (Art. 21.2), NAFTA (Art. 2004 and Annex 2004), the Korea-Chile FTA (Art. 19.2 and Annex 19.2) and the USSFTA (Art. 20.4).

[165] See, for example, *Japan – Measures Affecting Consumer Photographic Film and Paper*, WT/DS44/R, WTO panel report adopted on 22 April 1998, at paras. 10.50 and 10.61–10.81.

## 5.8 Investment treaty general exceptions and expropriation exceptions relevant to health

General exceptions resembling GATT Article XX are generally not commonly found in bilateral investment treaties (BITs). Notably, the more recent regional investment treaties signed by ASEAN do contain general exceptions.

However, some FTAs do provide a measure of protection to the state against complaints within the investment chapters themselves. Examples are the KORUS investment chapter (Article 11.8(3)(b)) and the Agreement of Free Trade and Economic Partnership between Japan and the Swiss Confederation (Articles 95 and 101).[166] Another example is Article 11.8(3)(c)(ii), in the same chapter, which contains an important exception on *environmental measures* that may be taken to protect human health. The investment chapter of the Malaysia-Australia FTA also provides a number of "General Exceptions" in its Article 12.18. It further provides in Article 12.7.3 that a breach of another provision is not necessarily a breach of FET.

The FTA signed between Singapore and India, or CECA, contains both a provision of measures "in the public interest" as well as a GATT-like General Exceptions provision.[167]

[166] The text of the Japan-Switzerland Agreement is available at: www.mofa.go.jp/region/europe/switzerland/epa0902/agreement.pdf.

[167] India–Singapore Comprehensive Economic Cooperation Agreement, signed on 29 June 2005, entered into force on 1 August 2005. The two relevant provisions provide as follows:

**ARTICLE 6.10: MEASURES IN THE PUBLIC INTEREST**

Nothing in this Chapter shall be construed to prevent:

(a) a Party or its regulatory bodies from adopting, maintaining or enforcing any measure, on a non-discriminatory basis; or
(a) the judicial bodies of a Party from taking any measures; consistent with this Chapter that is in the public interest, including measures to meet health, safety or environmental concerns.

**ARTICLE 6.11: GENERAL EXCEPTIONS**

1. Subject to the requirement that such measures are not applied in a manner which would constitute a means of arbitrary or unjustifiable discrimination against the other Party or its investors where like conditions prevail, or a disguised restriction on investments of investors of a Party in the territory of the other Party, nothing in this Chapter shall be construed to prevent the adoption or enforcement by a Party of measures:
   (a) necessary to protect public morals or to maintain public order;
   (b) necessary to protect human, animal or plant life or health;

In the investment chapter of the EFTA-HK FTA, Article 4.6 provides for measures relating to health concerns as a "Right to Regulate":

> Article 4.6 Right to Regulate
> 1. Nothing in this Chapter shall be construed to prevent a Party from adopting, maintaining or enforcing any measure consistent with this Chapter that is in the public interest, such as measures to meet health, safety or environmental concerns and reasonable measures for prudential purposes . . .[168]

The ASEAN-China and ASEAN-Korea investment agreements each contain an Article XX-like provision, from whose applicability investment measures are not excluded.[169] On the other hand, while Article XX of the GATT 1994 does not apply to the investment chapter of the ASEAN-Australia-New Zealand FTA (i.e. Chapter 11), Article XIV of the GATS – which contains a health exception-like Article XX(b) of the GATT – *is* applicable to the investment chapter.[170]

FTA investment commitments often include those relating to expropriation. As expropriation may take a direct or indirect form, some FTAs contain an explicit exception for health-related measures of the state in the definition of "indirect expropriation".

An example is found in Annex 11-B, paragraph 3(b) exception, of the KORUS FTA.[171] The 2012 US Model BIT, announced in April 2012 and

(c) necessary to secure compliance with laws or regulations which are not inconsistent with the provisions of this Chapter including those relating to:
 (i) the prevention of deceptive and fraudulent practices to deal with the effects of a default on a contract;
 (ii) the protection of the privacy of individuals in relation to the processing and dissemination of personal data and the protection of confidentiality of individual records and accounts;
 (iii) safety;
(d) imposed for the protection of national treasures of artistic, historic or archaeological value;
(e) relating to the conservation of exhaustible natural resources if such measures are made effective in conjunction with restrictions on domestic production or consumption.

168 The text is available at: www.efta.int/free-trade/free-trade-agreements/hong-kong.aspx. See also the EU's stance in the TTIP negotiations on the right to regulate, discussed in Chapter 4.

169 See Arts. 16 and 11, respectively, of the texts at: www.fta.gov.sg/acfta/asean-china_inv_agreement(certified_copy).pdf and www.fta.gov.sg/akfta/article11.pdf.

170 See Chapter 15, Arts. 1.1 and 1.2.

171 Text available at: www.ustr.gov/trade-agreements/free-trade-agreements/korus-fta/final-text.

which replaces the 2004 Model BIT of the US, now contains an exception that reflects the language of this annex.[172] In addition, there is preamble and other language that refer to health-related measures.[173]

[172] 2012 Model BIT: Annex B, on indirect expropriation

> " ... (b) Except in rare circumstances, non-discriminatory regulatory actions by a Party that are designed and applied to protect legitimate public welfare objectives, such as **public health**, safety, and the environment, do not constitute indirect expropriations."
>
> [Bold emphasis added]

[173] The following excerpts of the Model BIT provide for health-related rights:

> **Preamble**
>
> *Desiring* to achieve these objectives in a manner consistent with the protection of **health**, safety, and the environment, and the promotion of internationally recognized labor rights ...
>
> [Bold emphasis added]
>
> **Article 8.3(c)**
>
> (c) Provided that such measures are not applied in an arbitrary or unjustifiable manner, and provided that such measures do not constitute a disguised restriction on international trade or investment, paragraphs 1(b), (c), (f), and (h), and 2(a) and (b), shall not be construed to prevent a Party from adopting or maintaining measures, including environmental measures:
>
> (i) ...;
> (ii) necessary to protect human, animal, or plant life or **health**; or ...
>
> [Bold emphasis added]
>
> **Article 12.4**
>
> 4. For purposes of this Article, "environmental law" means each Party's statutes or regulations, or provisions thereof, the primary purpose of which is the protection of the environment, or the prevention of a danger to human, animal, or plant life or **health**, through the:
>
> (a) prevention, abatement, or control of the release, discharge, or emission of pollutants or environmental contaminants;
> (b) control of environmentally hazardous or toxic chemicals, substances, materials, and wastes, and the dissemination of information related thereto; or
> (c) protection or conservation of wild flora or fauna, including endangered species, their habitat, and specially protected natural areas,
>
> in the Party's territory, but does not include any statute or regulation, or provision thereof, directly related to worker safety or **health**.
>
> [Bold emphasis added; original footnote omitted]
>
> **Art. 13.3**
>
> 3. For purposes of this Article, "labor laws" means each Party's statutes or regulations, or provisions thereof, that are directly related to the following:
>
> ...

The annex language relating excluding measures relating to health, safety and other public welfare grounds from being "indirect expropriation" is being adopted by negotiators in more recent non-US FTAs, such as the Malaysia-Australia FTA (signed in May 2012, containing a modified version of the US Model BIT Annex wording).

There does not therefore appear to be a consistent practice emerging as yet, at least in the agreements of ASEAN, with regard to the treatment of general exceptions as they apply (or not) to investment commitments.

Since BITs and investment chapters of FTA can carry significant commitments of open-ended scope, which are likely to apply to regulatory measures regulating pharmaceuticals, an exception on health is critical to preserve regulatory space for the host state. This can be done by either extending an Article XX provision in the FTA to such a chapter, or (as the examples above do) including a separate, specific exception clause in the investment chapter itself.

An additional consideration is that, given the intrusive nature of compulsory licensing to an investor's patent rights, it may be advisable to include a specific provision exempting such licensing (as permitted under TRIPS Article 31 and the related public health waiver) from forming the subject matter of challenge under investor protection provisions relating to fair and equitable treatment and expropriation.

By way of comparison, the Common Market for Eastern and Southern Africa (COMESA) investment agreement provides an interesting example which seeks to explicitly protect permissible compulsory licensing from the scope of its expropriation provisions. Article 20(6) provides as follows:

> 6 This Article shall not apply to the issuance of compulsory licences granted in relation to intellectual property rights, or to the revocation, limitation or creation of intellectual property rights, to the extent that such issuance, revocation, limitation or creation is consistent with applicable international agreements on intellectual property.[174]

> (f) acceptable conditions of work with respect to minimum wages, hours of work, and occupational safety and **health**.
>
> [Bold emphasis added; original footnote omitted]

[174] Investment Agreement for the COMESA Common Investment Area, 2007, COMESA website, http://programmes.comesa.int/index.php?option=com_content&view=article&id=111&Itemid=149; and TRALAC, www.tralac.org/wp%20...%20/blogs%20...%20/**Investment**_agreement_for_the_CCIA.pdf (accessed 22 August 2014). A variant of Art. 20(6) can be found in the Panama-Singapore FTA, Art. 9.7(4):

Such provisions have begun to appear in FTAs involving Asian signatories. An example can be found in Article 14.11(6) of the Japan-Australia Economic Partnership Agreement.[175]

The COMESA Agreement also includes not only an Article XX GATT-like health provision, but also a very broad exception category (in paragraph 1(d), below) that is not found in Article XX.[176]

> This Article does not apply to the issuance of compulsory licenses granted in relation to intellectual property rights, or to the revocation, limitation or creation of intellectual property rights, to the extent that such issuance, revocation, limitation or creation is consistent with the *Agreement on Trade-Related Aspects of Intellectual Property Rights* in Annex 1C to the TRIPS Agreement.

Art. 20(8) of the COMESA Agreement further excludes from the scope of indirect expropriation "bona fide regulatory measures taken by a Member State that are designed and applied to protect or enhance legitimate public welfare objectives, such as public health, safety and the environment", these being seen as "[c]onsistent with the right of states to regulate and the customary international law principles on police powers".

[175] Signed on 8 July 2014 and entered into force on 15 January 2015. Text available at http://dfat.gov.au/trade/agreements/jaepa/official-documents/Pages/official-documents.aspx (accessed 13 October 2015).

[176] Art. 22 provides as follows:

**ARTICLE 22**

**General Exceptions**

1. Subject to the requirement that such measures are not applied in a manner which would constitute a means of arbitrary or unjustifiable discrimination between investors where like conditions prevail, or a disguised restriction on investment flows, nothing in this Agreement shall be construed to prevent the adoption or enforcement by any Member State of measures:
   (a) designed and applied to protect national security and public morals;
   (b) designed and applied to protect human, animal or plant life or health;
   (c) designed and applied to protect the environment; or
   (d) *any other measures* as may from time to time be determined by a Member State, subject to approval by the CCIA Committee.
2. Nothing in this Agreement shall be construed to prevent a Member State from adopting, maintaining or enforcing *any measure that it considers appropriate to ensure that investment activity in its territory is undertaken in a manner sensitive to the principles outlined in subparagraphs 1(a) to (c) above.*

[Italic emphases added]

The italicized parts provide further safeguards for policy space under this agreement. Art. 22(1)(d) departs from the usual Art. XX GATT formulaic list and includes a wide discretion for action by a member state (subject to the chapeau, much like under

With regard to the IP Chapter, Article 11.1(2) applies the principles and objectives of TRIPS *mutatis mutandis*. Further, under Article 11.30(1), the parties "recognize the importance of the Declaration on the TRIPS Agreement and Public Health", and in interpretation of the FTA's patent and test data provisions, the parties "shall ensure consistency with this Declaration", while under Article 11.30(2), the parties agree to "respect" the WTO decisions on the "paragraph 6" mechanism.

In the negotiated text of the EU-Singapore FTA, the investment chapter contains a national treatment provision which carries a GATT Article XX-like provision, found in Article 9.3(3).[177]

## 5.9 Security exceptions

Under the CECA, India and Singapore have included an interesting exclusion from dispute settlement jurisdiction in the area of national security. The parties have agreed in Annex 5 that with respect to disputes arising under Article 6.21(3)(b) or (c), the investment chapter of that FTA, assertion of a defence of a security exception provided for under Article 6.12 will be non-justiciable, in that

> . . . it shall not be open to any arbitral tribunal to review the merits of any such decision, even where the arbitral proceedings concern an assessment

GATT jurisprudence) to take any other measures so long as they are approved of by the COMESA Common Investment Area Committee, defined and established under Art. 7 (1) of the agreement, comprising the Meeting of Ministers responsible for investment of each member state. Art. 22(2), on the other hand, provides an auxiliary safeguard for measures that a member considers "appropriate to ensure that investment activity . . . is undertaken in a manner sensitive to" the excepted areas in paragraph (1). It is auxiliary in the sense that it may not be a primary measure under paragraph (1) but will be related to and will support the objectives of paragraph (1).

177 In relation to investor-state dispute settlement, there are also limits on who may make a claim. For example, the definition of "juridical person" for the purpose of who is an "investor" requires "substantial business operations" in the EU and in Singapore. This is to exclude entities which, for example, have only a registered address but not substantial business within the two territories. The provision therefore limits what types of entities may be "investors" enjoying the treaty's protection. In addition, the provision on standard of treatment appears to delimit FET to an exhaustive list of acts (Art. 9.4). Further, the expropriation provisions limit both the scope of indirect expropriation and expropriation of intellectual property rights (Annexes 9-A and 9-C, respectively). There is also preamble language that reaffirms parties' right to adopt measures to pursue "legitimate policy objectives such as social, environmental, security, *public health and safety*, promotion and protection of cultural diversity" (italic emphasis added), which would form part of the objectives of the treaty when interpreting its provisions.

> of any claim for damages and/or compensation, or an adjudication of any other issues referred to the tribunal.[178]

This probably arises from arguments arising elsewhere regarding whether an assertion of national or essential security reasons should be self-determining by the asserting state. The Annex 5 provision removes any doubt for the asserting state with regard to the CECA.

In terms of measures relating to public health and to pharmaceuticals in particular, such a provision offers food for thought. For example, a non-justiciability provision could serve to confirm an FTA state's independence in determining what constitutes the grounds for use of compulsory licensing, as well as what constitutes situations of urgency or extreme emergency when issuing a compulsory licence. After all, these are matters confirmed as being within the province of each state to decide, as shown in paragraphs 5a and 5b of the Declaration on the TRIPS Agreement and Public Health, explained above.

Such a provision could also preserve a state's right to determine, through its domestic agencies, what constitutes an anti-competitive licensing practice of a patent holder when enacting remedial measures. This is an area of flexibility clearly recognized even under the TRIPS Agreement, and in which determination is left to the relevant domestic agency. Even if the FTA requires such domestic determination to be made judicially or be subject to judicial review, it removes any doubt as to whether such a determination can still be challenged in investor-state arbitration under the FTA. This would, in turn, significantly reduce the risk of such dispute settlement being raised in respect of such determinations.

### 5.9.1 *Foreign policy measures*

In the Malaysia-India FTA, the security exception provides (*inter alia*) a different type of immunity:

> the measures which a Party adopts or maintains with respect to a non-Party or investors of a non-Party shall not impinge upon the other Party's sovereign rights to conduct its foreign policy nor shall it prohibit

[178] See text of Annex 5 at: www.fta.gov.sg/ceca/annex%205%20-%20non-justiciability%20of%20security%20exceptions%20(relating%20to%20chapter%206).pdf. Note that India's new draft Model BIT also contains such a provision (Annex 1, para. (ii)); see footnote 186 below and accompanying text.

> enterprises of foreign enterprises that are subject to such measures from establishing themselves in the other Party.[179]

## 5.10 Interpretational treaty signals

In addition to explicit provisions dealing with exceptions or flexibilities in policy-making, some treaties contain language that may aid in interpreting the scope and application of these provisions. Preamble language of a treaty often sets out overarching objectives of signatories.[180] Some treaties have drawn on the following language used in paragraph 4 of the Ministerial Declaration on TRIPS and Public Health:

> that the TRIPS Agreement does not and should not prevent Members from taking measures to protect public health. Accordingly, while reiterating our commitment to the TRIPS Agreement, we affirm that the Agreement *can and should be interpreted and implemented in a manner supportive of WTO Members' right to protect public health and, in particular, to promote access to medicines for all.* In this connection, we reaffirm the right of WTO Members to use, to the full, the provisions in the TRIPS Agreement, which provide flexibility for this purpose.
>
> [Italic emphasis added][181]

Specific "objectives" provisions may also provide interpretational cues for a tribunal that has to deal with a dispute that raises such provisions. An example can be seen in TRIPS (Article 7). Another is in the COMESA treaty, in which Article 11 states:

> The objectives of Part Two of this Agreement are to provide COMESA investors with certain rights in the conduct of their business *within an overall balance of rights and obligations between investors and Member States.*
>
> [Italic emphasis added; Part Two deals with rights and obligations under the treaty.]

The FTA between Brunei Darussalam, Chile, New Zealand and Singapore contains one of the most detailed "objectives" provisions

[179] Para. (b), Annex 12–1, Malaysia-India Comprehensive Economic Partnership Agreement (MICECA).

[180] For some examples, see Suzanne A. Spears, *Making Way for Public Interest in International Investment Agreements*, Chap. 13, at pp. 290–4, in *Evolution in Investment Treaty Law and Arbitration*, ed. Chester Brown and Kate Miles (Cambridge University Press 2011).

[181] Para. 4.

with respect to IP obligations.[182] Notably, the ACIA does not contain such "balance" language. Apart from preambles, language that lends itself to aiding interpretation of substantive IP obligations in FTAs may also be found within the IP chapters. Another example is the Colombia-Korea FTA, in which Article 15.5(1) provides as follows:

> The Parties recognize the need to maintain a balance between the rights of the right holders and the public interest.[183]

Even more pertinently, Article 17.4(5) in the IP chapter of the Peru-Korea FTA provides as follows:

> In accordance with Article 8.1 of the TRIPS Agreement, the Parties may use, in formulating or amending their laws and regulations, *the exceptions and flexibilities allowed by multilateral agreements related to the protection of intellectual property, and in particular, adopt measures necessary to protect public health and nutrition,* taking into consideration the TRIPS Agreement and the principles established in the Declaration and the Decision.[184]

More recently, the Korea-Australia FTA, signed in April 2014, provides as follows:

**ARTICLE 13.10: UNDERSTANDINGS REGARDING CERTAIN PUBLIC HEALTH MEASURES**

1. The Parties recognise the importance of the *Declaration on the TRIPS Agreement and Public Health*, adopted on 14 November 2001 (hereinafter referred to as the "Doha Declaration") by the Ministerial Conference of the WTO. In interpreting and implementing the rights and obligations under Article 13.8, the Parties are entitled to rely upon the Doha Declaration.
2. Each Party shall contribute to the implementation of and shall respect the *Decision of the WTO General Council of 30 August 2003*

[182] Article 10.2, *Trans-Pacific Strategic Economic Partnership Agreement*, signed on 18 July 2005, in force in Brunei Darussalam 12 July 2006, in Chile 8 November 2006, and in New Zealand and Singapore 28 May 2006. By comparison, the Model BIT of Norway refers to corporate social responsibility in the preamble and in Article 32. The preamble contains other relevant language on equality and mutual benefit, protection of health, improvement of living standards and promotion of sustainable investments, all of which can be useful examples for developing countries to consider; see www.italaw.com/sites/default/files/archive/ita1031.pdf (accessed 9 April 2015).

[183] Signed on 21 February 2013, not yet in force at the time of writing.

[184] Signed 14 November 2010, entered into force 1 August 2011. For a variant, see also Article 15.5(1) of the Colombia-Korea FTA IP chapter. This provision is critical as the General Exceptions in the FTA do not apply to the IP Chapter (Chapter 15).

on paragraph 6 of the Doha Declaration, as well as the *Protocol amending the TRIPS Agreement*, done at Geneva on 6 December 2005.[185]

Paragraph 1 is significant in that it expressly refers to interpretational *reliance* on the Doha Declaration. This express linkage of the Doha statements with interpretation of provisions of the FTA and explicit direction to treaty interpreters avoid any argument of their relevance under, for example, Articles 31 or 32 of the Vienna Convention on the Law of Treaties, when interpreting the FTA.

In the Korea-Australia FTA, interpretative guidance has been provided on test data exclusivity provisions in the IP chapter, linking such interpretation to the Doha Declaration.[186]

FTAs may also include important language regarding the objectives of the treaty in their preambles. An example is the Korea-India agreement, which includes the following in its preamble:

> FURTHER REAFFIRMING their rights to pursue economic philosophies suited to their development goals and their rights to realise their national policy objectives
>
> ...

Finally, it is noteworthy that India's new draft model BIT also contains provisions with interpretation instructions and "signals" for tribunals.[187]

## 5.11 Negotiating space

It is evident from the above discussion that there remains considerable negotiating space for Asian states in terms of exceptions provision relating to health and, particularly, to pharmaceuticals regulation.

[185] The full text of the IP chapter of the FTA in which this provision is found is available at www.dfat.gov.au/fta/kafta/downloads/KAFTA-chapter-13.pdf (accessed 9 October 2014).

[186] See, for example, Art. 11.30.1, EU-Singapore FTA (signed but not yet in force at the time of writing), which provides:

> The Parties recognise the importance of the *Declaration on the TRIPS Agreement and Public Health*, adopted on 14 November, 2001 by the Ministerial Conference of the WTO, at Doha. In interpreting and implementing the rights and obligations under Sub-Section E (Patents) and Sub-Section F (Protection of Test Data Submitted to Obtain an Administrative Marketing Approval to put a Pharmaceutical Product on the Market), the Parties shall ensure consistency with this Declaration.

[187] See, for example, Arts. 14.9 and 19.3 of the draft text; see footnote 186 below and accompanying text.

Apart from the TRIPS flexibilities and existing GATT and GATS language, tailored provisions are possible, depending on the TRIPs-plus commitments being requested.

In realization that past commitments made in treaties may not be appropriate for their regulatory objectives, some countries are reviewing and renegotiating these treaties. These include South Africa, which is reviewing its BITs. Australia has also recently reviewed its investor-state dispute settlement stance in relation to international investment treaties.

In an announcement made in 2011, the Gillard government even announced that Australia would discontinue the practice of including provisions on investor-state dispute settlement in future negotiations.[188] This followed the initiation of such arbitration action against Australia for its tobacco product "plain packaging" law by Philip Morris under the Australia-Hong Kong BIT.[189] Subsequently, separate actions have been brought against Australia by Honduras, the Dominican Republic, Ukraine and Cuba at the WTO in respect of this legislation, and these are now pending.

### *5.11.1 India's new draft Model BIT*

In early 2015, the Indian government released a draft new *Model Text for the Indian Bilateral Investment Treaty.*[190] The draft is noteworthy first as a result of India's recent exercise to review its existing BIPAs, and second for its new provisions, including exceptions and other provisions that have implications for, *inter alia*, trade and investment in pharmaceutical products. The draft contains a number of limitations on the scope of liability and on dispute settlement. Several of these have the potential, if incorporated in a BIT by India, to affect investor claims relating to public health, pharmaceutical and intellectual property grievances.

On the scope of liability, for example, under Article 3, the standard of treatment offered to investors does not mention the contentious phrase "fair and equitable" treatment, but rather states plainly that each party

188 Gillard Government Trade Policy Statement, April 2011, www.dfat.gov.au/publications/trade/trading-our-way-to-more-jobs-and-prosperity.html. See generally, also, article by this author, Tobacco Regulation and Its Discontents", *Transnational Dispute Management*, Vol. 9, No. 5 (November 2012), 1–11. The present approach is a case-by-case one.

189 *Philip Morris Asia Limited (Hong Kong) v Australia*, proceedings reported at: www.ag.gov.au/Internationalrelations/InternationalLaw/Pages/Tobaccoplainpackaging.aspx.

190 Text available at: https://mygov.in/sites/default/files/master_image/Model%20Text%20for%20the%20Indian%20Bilateral%20Investment%20Treaty.pdf (accessed 9 April 2015).

shall not subject investors of the other party to three categories of measures.[191] Article 16.3 excludes treaty protection for measures taken by local bodies or authorities of India at the "district, block or village level".

A number of provisions also make specific reference to public health as a matter excusing particular measures (e.g. in Article 4.5 in relation to national treatment, Article 5.4 in relation to expropriation and Article 16.1(v) under general exceptions). The draft also directs dispute resolution tribunals on interpretation of its provisions, including the following instruction:

> This treaty shall be interpreted in the context of the high level of deference that international law accords to States with regard to their development and implementation of domestic policies.[192]

Article 2.6(v) also excludes from the treaty

> the issuance of compulsory licences granted in relation to intellectual property rights, or to the revocation, limitation or creation of intellectual property rights, to the extent that such issuance, revocation, limitation or creation is consistent with the Law of the Host State.

The draft text clearly reflects a desire to address a number of concerns and criticisms that have arisen in relation to the international investor-state dispute process in general, and specifically as it might relate to adjudication on claims with national public health implications. Whether this will be adopted by India or whether it will serve to provide ideas for other developing countries remains to be seen.

### 5.11.2 *Exceptions for the future?*

With Asian states being involved in large-scale FTAs negotiated by Asian states such as the TPP and the RCEP, a thoughtful and well-defined negotiating strategy for dealing with exceptions is necessary. To provide a better balance with TRIPS-plus provisions, perhaps it is perhaps time to argue for an inclusion of an "Article XX(b)-plus" provision, as a counter-balance. While a general exceptions provision may be relatively easy to argue, this is no assurance that regulatory measures taken in respect of

[191] These are "denial of justice under customary international law", "un-remedied and egregious violations of due process" and "manifestly abusive treatment involving continuous, unjustified and outrageous coercion or harassment" (Art. 3.1).

[192] Art. 14.9(i).

pharmaceuticals will necessarily fall within it. Is there a case for arguing a more specific exception that expressly addresses these products and their regulation, and if so, what form could it take? While the text of the TPP is not yet available at the time of writing, there are indications that a GATT Article XX-like provision will be included.[193] There also appear to be "right to regulate" provisions.[194]

Apart from formulating and advocating appropriate exceptions and exemptions, including those permitted under the TRIPS Agreement and related decisions, another consideration is the exclusion of appropriate areas from treaty dispute settlement, as alluded to above. In addition to, say, possibly non-justiciable areas which could be expressly stated in a treaty, exclusion of particular types of disputes is another consideration. One example is the non-violation claim, a relatively obscure and seldom-used type of claim originating in Article XXIII of GATT 1994. This type of claim has been very limited in use in the WTO, and case reports indicate that they are likely to arise only very exceptionally.[195] In relation to TRIPS in particular, WTO members have agreed to a moratorium on the application of such claims. In the case of FTAs, such a moratorium does not apply, and depending on the dispute settlement provision of the FTA, this type of claim may yet be brought. An examination of the dispute settlement provisions, and possibly an express exclusion of such claims being brought, would therefore be appropriate during negotiations.

### *5.11.3 Options in negotiating trade and investment exceptions and flexibilities*

Given the various exceptions and flexibilities recognized under the WTO agreements, particularly the GATT and TRIPS agreements, a number of options may be considered in negotiating FTA commitments, particularly those that have an impact on pharmaceutical products. These include the following:

193 See para. 29, Singapore Ministry of Trade and Industry, *Summary of the TPP Agreement*, at https://www.mti.gov.sg/MTIInsights/SiteAssets/Pages/TPP/Summary%20of%20TPP%20Agreement.pdf (accessed 13 October 2015).

194 Ibid., paras. 9 and 25.

195 See Locknie Hsu, *Non-Violation Complaints – WTO Issues and Recent Free Trade Agreements* (2005) JWT. See also Thomas Faunce, Submission for the Review of Export Policies and Programmes to the Australian Department of Foreign Affairs and Trade, 2 May 2008, at p. 42.

- Inclusion of a general exception *à la* Article XX GATT. Such a provision can be strengthened with an explicit application to the intellectual property/pharmaceutical commitments. A further consideration may be inclusion of express mention of pharmaceuticals under an equivalent provision to Article XX(b), in exempting measures for protection of human health.
- Inclusion of a provision equivalent to Article 30 of the TRIPS Agreement to ensure that the full force of its flexibilities apply to the FTA.
- Inclusion of the flexibilities provided for in relation to compulsory licensing by the TRIPS Agreement and the related waiver of Article 31(f).
- Negotiation of separate express provisions providing exemptions for specific measures by states, such as those with respect to patentability, evergreening, deregistration, data exclusivity and patent linkage. On patent linkage, for instance, inclusion of an express exception to data exclusivity requirements where public health or safety requires it (such as that mentioned in section 19B of the Singapore Medicines Act, as discussed above) is a possible consideration.
- Inclusion of a specific investment-related exception, to limit or remove risk of investor-state arbitration in respect of pharmaceutical-related regulations in protection of public health.
- Inclusion of customized provisions such as non-justiciability provisions in areas of permitted flexibility under the TRIPS agreement, such as those mentioned above.

## 5.12 Conclusion

In negotiating FTA exceptions, exemptions and exclusions, Asian states need a clear negotiating strategy for commitments affecting trade and regulation of pharmaceuticals. The pressure to give TRIPS-plus commitments is undeniable. Instead of mere copying of "standard" exception clauses, one should go further to examine in detail what specific exceptions and exemptions could be tailored for every TRIPS-plus commitment that may be given. "Article XX-plus" exceptions may be appropriate in order to provide regulatory flexibility and a balance in regulating trade in pharmaceuticals.

A study of existing tailored FTA clauses can yield useful ideas for how to tailor such clauses. Adaptation and ingenuity are needed to provide negotiating possibilities that are mutually acceptable. The provision found in the USSFTA mentioned above, which translated into Section 12A of the Singapore Patents Act, is a good example of a result of creative thinking and in-depth discussion of how to accommodate competitiveness in the face of a TRIPS-plus commitment.

The examples show that the crafting of suitable exceptions and exclusions often requires a substantive policy discussion and attention of the implications of a TRIPS-plus commitment and how state regulatory priorities on access to pharmaceuticals need to be accommodated despite such a new commitment. Clearly, on the subject of exceptions it should not, for instance, be left to the legal team to merely insert a general exception at the end of the negotiating process without careful thought.

# 6

## Innovation and competition

### 6.1 Innovation, competition: seeking a balance

Discussions on innovation necessarily reach into diverse areas of policy-making, of which trade, investment and legal policies form but a part, and to which globalization has contributed new dimensions. While "innovation" has been explained in various ways, the complexity in crafting policy to promote it is undeniable. The OECD's *Oslo Manual* explains this complexity as follows:

> ... innovation policy has only recently emerged as an amalgam of science and technology policy and industrial policy. Its appearance signals a growing recognition that knowledge in all its forms plays a crucial role in economic progress, that innovation is at the heart of this "knowledge-based economy", and also that innovation is a more complex and systemic phenomenon than was previously thought. Systems approaches to innovation shift the focus of policy towards an emphasis on the interplays between institutions, looking at interactive processes both in the creation of knowledge and in its diffusion and application ...[1]

Pharmaceutical innovation in particular is of great interest as public health, whose advancement depends on such innovation, is a universal concern.[2] In Asia, some countries have made strategic policy choices to position themselves as "hubs" for such innovation. As countries in Asia chase the innovation dream, they need to increase understanding on the usefulness of an area of law that is receiving increasing prominence in the region: competition law. What contributions should and could competition law provide within the innovation-healthcare regulatory matrix?

The roles and interactions of competition, intellectual property and innovation are neither new, nor merely a developing country query.

[1] OECD, *Oslo Manual*, 2005, at para. 2.1.8.

[2] For a good historical account of pharmaceutical innovation, see OECD, Nobuo Kiriyama, *Trade and Innovation: Pharmaceuticals*, OECD Trade Policy Working Papers, No. 113 (2011), pp. 12–18.

OECD countries, for example, regularly conduct discussions on these interfaces in various economic sectors, including the pharmaceutical sector.[3]

Governments use a number of policy tools to encourage innovation, such as rewarding R&D activity with tax incentives.[4] Legal protection for those who invest in research and innovative activity to generate new products and processes, in the form of patents, is one such tool. As has been explained previously, the state makes a "bargain" with such innovators, that in return for the disclosure of the invention for others to learn from it, the state guarantees a legal monopoly for a limited duration, which forms the innovator's incentive. This thinking certainly applies to pharmaceutical inventions. In the *Human Genome Sciences v Eli Lilly Company*, Lord Neuberger explained the legal monopoly and the incentive it carries in the following manner:

> it is worth remembering the purpose of the patent system, namely to provide a temporary monopoly as an incentive to innovation, while at the same time facilitating the early dissemination of any such innovation through an early application for a patent, and its subsequent publication. Although this is true in any sector, it has particular force in the pharmaceutical field, where even many of those who are sceptical about the value of intellectual property rights accept that there is a public interest in, and a commercial need for, patent protection.[5]

However, just as in other industries, pharmaceutical patent holders may seek to impose further entry barriers to competitors. These may take a variety of forms, including various arrangements to significantly delay, or block altogether, competitors' market entry. Examples include highly restrictive licensing conditions and anti-competitive price arrangements, and, in some cases, refusing to grant a licence altogether without

[3] See OECD Policy Roundtable, *Competition, Patents and Innovation*, 2006. See also OECD, *Competition and Regulation Issues in the Pharmaceutical Industry* (2000). www.oecd.org/daf/competition/sectors/1920540.pdf.

[4] See Chapter 7, and, for example, the Singapore Inland Revenue Authority's menu of incentives at: www.iras.gov.sg/irasHome/page04.aspx?id=13838#Registration_of_Patents,_Trademarks,_Designs_and_Plant_Varieties (accessed 21 March 2014), and that of the Malaysian Investment Development Authority, at: www.mida.gov.my/env3/index.php?page=manufacturing-sector-2 (accessed 21 March 2014).

[5] [2011] UKSC 51, at para. 99. Many others support such a view; see for instance Wendy H. Schacht and John R Thomas, Congressional Research Report, "Patent Reform in the 112th Congress: Innovation Issues", 30 June 2011, pp. 3–5. The report goes on to highlight a number of problems arising from the patents system, including barriers to entry and the emergence of "patent speculators" or "patent trolls" (pp. 5, 31).

sufficient reason. A refusal to deal or licence *per se* is not necessarily anti-competitive. When a refusal to deal – such as a refusal to licence use of a patented pharmaceutical – *is* a breach of competition policy is, however, a debatable issue. Although refusals to licence have been found to be anti-competitive in other industries, a consensus has not yet emerged on precisely how to apply competitive principles to such refusals by pharmaceutical enterprises.

Patent law and competition law have largely operated under separate authorities in many jurisdictions, without the latter necessarily acting as a check on the former to safeguard competition.[6] With the advent of aggressive patent-plus requirements under FTAs as well as certain commercial practices of pharmaceutical patent holders precipitated by the expiry of patents on "blockbuster" prescription pharmaceuticals, a review of the relationship between these two areas of law as they apply to pharmaceutical products and their roles in promoting innovation has become increasingly relevant.[7]

Already, on both sides of the Atlantic, competition authorities have begun to examine some of these practices. These include the so-called pay-for-delay agreements and reverse payment settlement agreements.[8]

The interface between existing competition law concepts and pharmaceutical arrangements is not without problems or tensions.[9] Until recently, case law in the US, for example, was not clear as to whether such arrangements fell under antitrust scrutiny. In addition, certain concepts such as the "relevant market", "refusal to deal" and "essential

[6] See, for example, the discussion by Herbert Hovenkamp in Antitrust and Innovation: Where Are We and Where Should We Be Going? *Ant. T. L. J.*, Vol. 77, No. 3 (2011), 749, arguing forcefully against "over-protection" in the patent system without other regulatory checks (such as antitrust law) to ensure the most competitive outcomes.

[7] On the interface between pharmaceutical and competition, see generally Josef Drexl and Nari Lee, *Pharmaceutical Innovation, Competition and Patent Law, A Trilateral Perspective* (Edward Elgar, 2013).

[8] See discussions such as those in Michael J. Clancy, Damien Geradine and Andrew Lazerow, *Reverse-Payment Patent Settlements in the Pharmaceutical Industry: An Analysis of US Antitrust Law and EU Competition Law*, 27 October 2013, available at: http://papers.ssrn.com/sol3/papers.cfm?abstract_id=2345851 (accessed 17 March 2014), Frank & Kerber, *Patent Settlements in the Pharmaceutical Industry: An Antitrust Perspective*, 2013, http://papers.ssrn.com/sol3/papers.cfm?abstract_id=2277508 and Pat Treacy and Sophie Lawrance, Intellectual Property Rights and Out of Court Settlements, in *Intellectual Property and Competition Law – New Frontiers*, ed. Steven Anderman and Ariel Ezrachi(Oxford University Press, 2011). See Federal Trade Commission (FTC) Study (2010), *Pay-for-Delay: How Drug Companies Pay-Offs Cost Consumers Billions.*

[9] See, for example, Rudolph J.R. Perritz, *Competition within Intellectual Property Regimes: The Instance of Patent Rights*, footnote 8 above.

facilities", which have been applied in non-pharmaceutical contexts, require careful analysis as to how they might be applied to pharmaceutical contexts.[10]

The law on the pharmaceutical patent-competition interface is evolving, and principles applicable to such arrangements are not clear. This is because, first, it is a relatively new aspect in many competition law systems (particularly in Asia), and second, because national competition laws (and where they exist, specific guidelines on IPRs) may differ. What is clear from recent developments, however, is that the US and EU authorities have clarified that such practices or agreements are fair game under competition law scrutiny. These developments provide those who regulate trade in pharmaceutical products elsewhere – such as in Asia – with important lessons to consider.

This chapter examines some of these major developments against the background that for many Asian countries competition law is a new regulatory area. It will discuss how Asian competition law systems – many of which are relatively recent and lack interpretative case law – could respond to such developments. The relative nascence of this field in Asia provides regulators with an important opportunity to shape the role of their competition laws to achieve the desired balance between encouraging innovation in pharmaceuticals development and competitive activity.

## 6.2 Patents and competition law – necessary bedfellows[11]

IPRs and competition policy are viewed as sharing a common objective: the promotion of creative endeavour and commercialization of new products and processes.[12] They are therefore not incompatible and may play complementary roles in promoting innovation. In relation to

[10] See, for example, the EU case, *AstraZeneca AB & AstraZeneca plc v European Commission*, C-457/10P, Judgment of the Court of Justice.

[11] For support for the view that competition and IP rights must coexist, see Cecilio Madero Villarejo and Thomas Kramler, in *Intellectual Property Rights and Competition Rules, a Complex But Indispensable Co-Existence*, in *Intellectual Property and Competition Law – New Frontiers*, footnote 8 above. See also, generally, Thomas Cheng, A Developmental Approach to the Patent-Antitrust Interface, *Nw. J. Int'l L. & Bus.*, Vol. 33, No. 1 (2012), 1–79.

[12] See, for example, para. 2.1, Singapore's *CCS Guidelines on Treatment of Intellectual Property Rights*:

> Both intellectual property ("IP") and competition laws share the same basic objective of promoting economic efficiency and innovation. IP law does this

pharmaceutical patents, competition law and policy has been becoming an important companion policy instrument to address abuses of IPRs. The TRIPS Agreement itself recognizes the role of competition policy in this respect.[13] It expressly addresses anti-competitive practices in several articles. Article 8(2) provides as follows:

> 2. Appropriate measures, provided that they are consistent with the provisions of this Agreement, may be needed to prevent the abuse of intellectual property rights by right holders or the resort to practices which unreasonably restrain trade or adversely affect the international transfer of technology.

Article 40 of the TRIPS Agreement separately permits members to deal with licensing practices and conditions that may be considered as an "abuse of intellectual property rights having an adverse effect on competition in the relevant market".[14] Further, under Article 31(k) of the TRIPS Agreement which permits WTO members to utilize compulsory licensing, compliance with the conditions in Articles 31(b) and (f) is waived where the measure is used to address anti-competition.

Given the existence of a number of patent-related commercial practices which may inhibit competition by generic pharmaceutical manufacturers, an important question faced by regulators is, how should competition law and policy be applied in the context of pharmaceutical patentees' practices, so that *both* competition and IP policy can be best harnessed to promote innovation?

> through the provision of incentives for innovation and its dissemination and commercialisation, by establishing enforceable property rights for the creators of new and improved products and processes. Competition law does this by helping to promote competitive markets, thereby spurring firms to be more efficient and innovative.

[13] See, for example, Articles 8(2), 31(k) and 40. See generally Nuno Pires de Carvalho, *The TRIPS Regime of Undisclosed Information* (Kluwer Law International, 2008), at pp. 161–188. For a discussion of the competition law-IP nexus in the modern economy, see Robert Pitofsky, *Antitrust and Intellectual Property – Unresolved Issues in the New Economy*, in *Intellectual Property Rights – Critical Concepts in Law*, Vol. I, ed. David Vaver (Routledge, 2006), 166–185.

[14] In the context of negotiating any TRIPS-plus commitments on pharmaceuticals in an FTA, inclusion of a reference to the rights in Article 40, along with such commitments, would expressly preserve this right to allow a signatory state to control any such abuses which may be adversely affecting competition for the pharmaceutical product in question in the market of that state. An example of an FTA provision that preserves the right to permit use of a patent to remedy an anti-competitive practice is Article 16.7.6(a) in the USSFTA. Arguably, an FTA provision that affirms the TRIPS obligations implicitly incorporates the obligation to observe the flexibility in Article 40.

A preliminary point is that the grant of the legal monopoly of a patent to exclude others from exploiting a pharmaceutical is not equal to creation of a monopoly in the economic sense – the former does not necessarily result in the latter, since there may be competitors supplying competing, legitimate products. What a patent does provide is a *legal* monopoly – for a limited duration – over the rights it covers.

While the legal monopoly does not necessarily create an economic monopoly, it may permit monopoly pricing of medicines in certain circumstances.[15] The issue is what view should be taken with regard to extra-patent efforts to block competitor products, with such efforts riding on patents meant only to provide "temporary monopolies" within the boundaries of their claims. Corporate strategies by originator companies such as paying generic manufacturers to delay marketing their generic products are practices which have often been reported.[16] Such means artificially expand the scope and effects of the monopoly, creating significant delays and barriers to entry of would-be competitors. These strategies do not necessarily demonstrate new, innovative activity deserving of expanded protection. Such "patent-plus" rights may in some cases be in fact contributing to an "innovation-minus" situation, in that they may result in negating opportunities for earlier innovative activity and market entry by others. This makes an argument for greater scrutiny from the point of competition law and policy compelling, if they result in impeding efficiency and innovation.

One learned academic has therefore eloquently argued:

> The critical point is *not whether* competition policy may interfere with IPRs, but *when* such intervention is justified.[17]
>
> [Italic emphasis added]

15 See, for example, Sean Flynn, Aidan Hollis and Mike Palmedo, An Economic Justification for Open Access to Essential Medicine Patents in Developing Countries, *J. L. Med. & Ethics* (2009), 184–208, at pp. 186–187.

16 See EIU report, which also points out that 2012 in particular is a very significant year, as many "blockbuster" drug patents will expire, sending the products "over the patent cliff": see *Big Pharma Struggles to Protect Its Blockbusters as They Lose Patent Protection*, at: www.eiu.com/index.asp?layout=ib3Article&article_id=308640015&pubtypeid=1152462500&country_id=1510000351&fs=true.

17 Carlos M. Correa, (2007) *Intellectual Property Law and Competition Law: Exploration of Some Issues of Relevance to Developing Countries*, ICTSD IPRs and Sustainable Development Programme Issue Paper No. 21, International Centre for Trade and Sustainable Development, Geneva, Switzerland, at p. 6.

Recent judicial decisions on both sides of the Atlantic appear to support this, by confirming that pharmaceutical patent holders' practices are not immune from anti-competition scrutiny. As discussed below, a certain momentum appears to be building in some countries to give competition law a greater role in scrutinizing anti-competitive practices in the pharmaceutical industry.

## 6.3 Emerging case law on competition law and pharmaceutical patentees' practices

### *6.3.1 The United States: the tide turns*

The applicability of US antitrust law to pharmaceutical litigation agreements has not been always been an obvious one. As one author argues, antitrust plays an "underappreciated role in the American innovation regime".[18]

There had been, for some time, a division of opinions on the extent of use of antitrust law in dealing with pharmaceutical entities' conduct. The Federal Trade Commission (FTC), a federal agency whose antitrust responsibilities overlap in some ways with those of Antitrust Division of the Department of Justice (DOJ), has taken a strong view that antitrust law can help maintain a proper balance between patent rights, innovation and competition.[19] A detailed FTC report issued in 2003 stated that[20]

[18] Christopher R. Leslie, Antitrust and Patent Law as Component Parts of Innovation Policy, *J. Corp. L.*, Vol. 34, No. 4 (2009), 1259–1289; Mr Leslie further argues that patent and antitrust are interdependent parts of innovation policy (at Part III). See also Robin Feldman, Patent and Antitrust: Differing Shades of Meaning, *Va. J.L. & Tech.*, Vol. 13 (2008), 5. For a historical account of the US case law on patent misuse and antitrust up to 2003, see also Robin Feldman, The Insufficiency of Antitrust Analysis for Patent Misuse, *Hastings L.J.*, Vol. 55 (2003), 399.

[19] See Maher M. Dabbah, Chap. 5, *US Competition Law Regime*, in *International and Comparative Competition Law* (Cambridge University Press, 2010), especially pp. 233–237; Robert F Leibenluft, Eric J. Stock, Leigh L. Oliver and Lauren E. Battaglia, *United States*, in contributing ed. Marleen Van Kerkhove, *Getting the Deal Through – Pharmaceutical Antitrust in 31 Jurisdictions Worldwide* (Global Competition Review, 2013), pp. 182–188; and generally Herbert Hovenkamp, Patents, Property and Competition Policy, *J. Corp. L.*, Vol. 34, No. 4 (2009),1243–1258.

[20] US Federal Trade Commission report, *To Promote Innovation: The Proper Balance of Competition and Patent Law and Policy*, October 2003, at p. 2. Footnote 7 within the quotation cites to *Atari Games Corp. v Nintendo of Am.*, 897 F.2d 1572, 1576 (Fed. Cir. 1990). See also the important remarks of the US Supreme Court in *Bonito Boats, Inc. v Thunder Craft Boats, Inc.*, 489 U.S. 141, 146 (1989), quoted at p. 3 of the report, such as the following: "[F]ree competition" is 'the baseline' on which "the patent system's incentive to creative effort depends." For FTC media releases on pay-for-delay

> Competition and patents are not inherently in conflict. Patent and antitrust law "are actually complementary, as both are aimed at encouraging innovation, industry, and competition."

An important feature of the US system is that settlement agreements concerning specified subject matter between a generic drug applicant which has made an ANDA application with a brand name drug company, or with another generic drug applicant which has submitted an ANDA application for the same listed drug, must be filed with the FTC.[21] Failure to file can lead to a heavy civil penalty being imposed.[22] This feature provides the FTC the opportunity to scrutinize such agreements – which might otherwise be confidential to the parties – for potential antitrust implications.[23]

The Antitrust Division of the DOJ enforces US antitrust law through both criminal prosecutions and civil actions.[24] Under the Obama administration, the DOJ has taken an approach that was more aligned with that of the FTC towards pay-for-delay agreements.[25]

A serious division among the US Circuit Courts had developed over the years over certain patent settlement agreements (e.g. in which a patentee agrees to pay a generics producer to delay entry into the market and not challenge its patent). US judicial decisions on such settlements

settlements, see www.ftc.gov/news-events/media-resources/mergers-and-competition/pay-delay (accessed 31 March 2014).

21 Sec. 1112 of the Medicare Prescription Drug, Improvement, and Modernization Act of 2003, Pub. Law, pp. 108–173.

22 *Ibid.*, Sec. 1114. The penalty is capped at US$11,000 *per day* of violation. See FTC report of 2012 settlement agreements: www.ftc.gov/news-events/press-releases/2013/01/ftc-study-fy-2012-branded-drug-firms-significantly-increased and www.ftc.gov/reports/agreements-filed-federal-trade-commission-under-medicare-prescription-drug-improvement.

23 In 2009, the FTC supported enactment of *the Protecting Consumer Access to Generic Drugs Act of 2009* which would prohibit pay-for-delay agreements; H.R. 1706; see testimony on behalf of the FTC by the then Bureau of Competition Director, Richard A. Feinstein, 3 June 2009, at: www.ftc.gov/news-events/press-releases/2009/06/ftc-testifies-support-bill-banning-pay-delay-settlements-between and speech by the then FTC Chairman, Jon Leibowitz, 23 June 2009, www.ftc.gov/news-events/press-releases/2009/06/ftc-chairman-leibowitz-eliminating-pay-delay-pharmaceutical (accessed 23 December 2014). See also FTC Study 2002, *Generic Drug Entry Prior to Patent Expiration*, and related statement of Daniel E. Troy, then Chief Counsel, Food and Drug Administration, before the Senate Committee on the Judiciary, 17 June 2003.

24 See role of the DOJ at the DOJ website at: www.justice.gov/atr/about/mission.html (accessed 19 March 2014). For an overview of the US dual-agency system of antitrust enforcement, see William Blumenthal, Models for Merging the US Antitrust Agencies, *J. Anti. Enf.* (2012) pp. 1–28, at pp. 4–6.

25 See www.wsj.com/articles/SB124691728092502381 and www.justice.gov/atr/cases/f247700/247708.htm (accessed 6 February 2015).

ranged from those which found such agreements to be presumptively illegal, to those which upheld them "unless" they exceeded the scope of the patent.[26]

In a landmark decision of 2013, *FTC v Actavis*,[27] the US Supreme Court sent a clear signal ending any assumption that "reverse payment" agreements[28] – a type of settlement agreement – enjoyed a *per se* immunity from antitrust action. The Drug Price Competition and Patent Term Restoration Act of 1984 (commonly known as the "Hatch-Waxman Act") contains a procedure for a generic drug manufacturer seeking approval for its drug to certify that the manufacture and sale of the drug would not infringe an existing patent of an originator (because, for example, the patent is invalid or has expired).[29] The originator then has 45 days to file an action claiming infringement, and if the matter is litigated within a 30-day period and the court does not find infringement, the FDA may proceed to approve the generic product. In *Actavis*, Solvay Pharmaceuticals had obtained a patent for its drug. Actavis and another company applied for marketing approval for a generic drug,[30] leading to Solvay bringing a patent infringement suit against them. The parties eventually came to a settlement on the basis that Solvay would pay the defendants "millions of dollars" for, *inter alia*, Actavis and the other party not to bring the generic to market until 31 August 2015 (65 months before Solvay's patent expired) and for Actavis to promote Solvay's patented drug to urologists.[31] In respect of the settlement agreement, the FTC filed suit under the Federal Trade Commission Act,[32] arguing that the defendants had violated the act "by unlawfully agreeing to

[26] See Clancy, Geradin and Lazerow, footnote 8 above, at pp. 4–9. For a further overview of the state of US case law on such agreements as of 2009, see also Remarks of J. Thomas Rosch, then Commissioner, Federal Trade Commission, *Pay-for-Delay Settlements, Authorized Generics, and Follow-on Biologics: Thoughts on the How Competition Law Can Best Protect Consumer Welfare in the Pharmaceutical Context*, World Generic Medicine Congress, 19 November 2009, Washington, D.C.

[27] *Federal Trade Commission v Actavis Inc*, 133 S. Ct. 2223.

[28] So called because the patent litigation *plaintiff* (the patent holder) may, in such agreements, arrange to pay the defendant not to market the latter's goods for a period, unlike a conventional litigation settlement agreement where a *defendant* (who is, for example, sued for patent infringement) may settle a lawsuit by making payment.

[29] Public Law 98–417.

[30] The application was made under what is known as the "Paragraph IV procedure", which permits a generic drug producer to certify (in an "abbreviated" application process) that a known patent is not valid, unenforceable or will not be infringed by his manufacturing, use or sale of his generic product.

[31] Note 27 above, p. 2225.

[32] 15 U.S.C. §§41–58, as amended.

abandon their patent challenges, to refrain from launching their low-cost generic drugs, and to share in Solvay's monopoly profits".[33] The District Court dismissed the action, and the Court of Appeals of the Eleventh Circuit affirmed the decision. The majority decision of the Supreme Court, which resolved the state of disagreement of the Circuit Courts, held that such agreements were not necessarily immune from antitrust scrutiny. It stated (disagreeing with the view of the Court of Appeals of the Eleventh Circuit):

> Solvay's patent, if valid and infringed, might have permitted it to charge drug prices sufficient to recoup the reverse settlement payments it agreed to make to its potential generic competitors . . . *But we do not agree that that fact, or characterization, can immunize the agreement from antitrust attack.*[34]
>
> [Italic emphasis added]

The majority rejected finding such agreements to be *presumptively legal* based on precedents using the so-called "scope of patent" test.[35] On the other hand, it also declined to take up the FTC's suggestion to find such agreements *presumptively illegal*, or to apply a "quick-look" analysis (which shifts the burden to the defendant to show pro-competitive effects). Instead, it preferred the application of *a rule of reason* approach, given that the agreements in question had the potential for anti-competitive harm but may in some cases be justifiable.[36] In doing so, it made it clear that it would look at "traditional antitrust factors such as likely anticompetitive effects, redeeming virtues, market power, and potentially offsetting legal considerations present in the circumstances, such as . . . those related to patents".[37]

In a spirited dissent led by the Chief Justice, the minority considered the majority decision to be crossing a "Rubicon".[38] It was clearly disturbed, describing the effects of the majority decision in the following terms:

> The majority today departs from the settled approach separating patent and antitrust law, weakens the protections afforded to innovators by patents, frustrates the public policy in favor of settling, and likely undermines the

[33] Note 27 above, at p. 2225. [34] Note 27 above, at p. 2230. [35] *Ibid.*, at p. 2237.

[36] Footnote 34 above, at p. 2237. For examples as to when such agreements might be justifiable, see p. 2236. See also Joshua D. Wright, FTC v Actavis *and the Future of Reverse Payment Cases*, remarks at the Concurrences Journal Annual Dinner, New York, 26 September 2013.

[37] Note 27 above, at p. 2231. [38] Note 27 above, at p. 2241.

> very policy it seeks to promote by forcing generics who step into the litigation ring to do so without the prospect of cash settlements. I would keep things as they were and not subject basic questions of patent law to an unbounded inquiry under antitrust law . . .[39]

Nonetheless, this Rubicon has now been crossed, at least under US law with respect to reverse payment settlement agreements. As to the role of patents *vis-à-vis* antitrust scrutiny, following the *Actavis* judgment, one learned academic has argued as follows:

> The patent system is not a deity to which we must respectfully defer. It is a living, breathing part of the organism that is our legal system. If we continue to treat patents with exceptionalism, we have only ourselves to blame as we walk willingly into the volcano . . . patent exceptionalism is alluring because it makes life so simple. Moving away from patent exceptionalism means that we have to worry about the messy question of what is acceptable and what is not acceptable patent behavior.[40]

Pay-for-delay transactions clearly remain very much on the radar of the FTC.[41] In her November 2013 testimony to US House of Representatives Committee on the Judiciary Subcommittee on Regulatory Reform, Commercial and Antitrust Law, Ms. Edith Ramirez, Chairwoman of the FTC, stated that pay-for-delay agreements have been a "top priority" of the FTC in the last decade.[42] She explained the rationale for such agreements – and their costs to consumers – as follows:

> The agreements can profit both the branded manufacturers, who continue to charge monopoly prices, and the generic manufacturers, who receive

[39] Note 34 above, at p. 2247.

[40] Robin Feldman, *Ending Patent Exceptionalism and Structuring the Rule of Reason: The Supreme Court Opens the Door for Both*, 29 September 2013, at p. 3; available at: http://papers.ssrn.com/sol3/papers.cfm?abstract_id=2333291; http://papers.ssrn.com/sol3/papers.cfm?abstract_id=2333291 (accessed 19 March 2014). See also the practice of "product hopping" addressed by the FTC and, recently, by a US Federal Court (in order of 22 May 2015 in *The People of the State of New York v Actavis plc, Forest Laboratories, LLC*, No. 14-4624, and judgment of Sweet D.J., SDNY, 14 Civ. 7473) in which a pharmaceutical company seeks to withdraw its drug in order to replace it with a higher-cost, newer version, before a generic competitor may enter the market, discussed in OECD, *Generic Pharmaceuticals – Note by the United States*, 10 June 2014, DAF/COMP/WD(2014)51, at Part 3.1.

[41] See generally Farasat A. S. Bokhari, What Is the Price of Pay-for-Delay Deals?, *JCLE*, Vol. 9, No. 3 (2013), 739–753.

[42] *Oversight of the Enforcement of the Antitrust Laws*, 15 November 2013; text of the testimony is available at: www.ftc.gov/sites/default/files/documents/public_statements/prepared-statement-federal-trade-commission-oversight-enforcement-antitrust-laws-presented/131115antitrustlawtestimony.pdf (accessed 8 July 2014).

> substantial compensation for agreeing not to compete – all at substantial cost to consumers, federal and state governments, and other purchasers of prescription drugs, all of which are already struggling to contain increasing healthcare costs.[43]

The *Actavis* decision has been hailed as "an important victory for consumers and a vindication of basic antitrust and free market principles".[44] Other strategies, such as abuses of safety protocols and "product hopping", which impede generic competition are also under scrutiny by the Commission.[45]

At the time *Actavis* was decided, predictions were being made as to whether the EU would take a similar rule of reason approach as in *Actavis* for reverse payment settlement agreements.[46]

## 6.4 The European Union: towards increased competition enforcement in pharmaceuticals sector

In 2008, the EU Commission opened an inquiry into competition in the pharmaceutical sector. The inquiry focused on indications that competition in pharmaceutical markets in Europe might not be working well, with fewer new pharmaceuticals being brought to market, and the entry of generic pharmaceuticals sometimes seeming to be delayed.[47] The inquiry gathered findings and recommendations which resulted in a

[43] *Ibid.*, p. 6. [44] *Ibid.*

[45] *Ibid.*, p. 8. Currently, the FTC is pursuing in two court cases, the separate agreements by Actavis, and by Cephalon, Inc. See *FTC v Cephalon, Inc.*, No. 08-cv-2141 (E.D. Pa. complaint filed 13 February 2008). For other variant agreements being scrutinized by the FTC, see *Generic Drug Entry Prior to Patent Expiration, an FTC Study* (2002), Chapter 3. One writer has gone so far as to argue that actions such as the US use of its Special 301 procedure to apply pressure on countries to take certain actions with regard to IPRs can amount to an "abuse of dominant position" by the US; see Fred M. Abbott, *Trade Diplomacy, the Rule of Law and the Problem of Asymmetric Risks in TRIPS*, "IPRs, Trade and Challenges for Development", Cancun Trade and Development Symposium, 11 September 2003, at p. 6: "Conceptually, threats against governments lawfully preparing to take action under the TRIPS Agreement might be treated as a form of abuse of dominant position, and remedial measures taken under national law against the patent holders that instigate the threats. Thus, for example, a threat instigated under Section 301 in the United States might act to initiate a government action in the targeted country for abuse of dominant position by the patent holder (or its industry association)."

[46] See, for example, Gönenç Gürkaynak, Ayşe Güner and Janelle Filson, The Global Reach of *FTC v. Actavis* – Will Europe Differ from the US Approach to Pay-for-Delay Agreements?, *IIC*, Vol. 45 (2014), 128–160.

[47] See announcement on the opening of the inquiry: http://europa.eu/rapid/press-release_IP-08-49_en.htm?locale=en (accessed 6 February 2015).

Final Report which was adopted in 2009.[48] In the report, it was clearly acknowledged that

> both intellectual property rights and competition are necessary to promote innovation and ensure a competitive exploitation thereof. If the existence and exercise of an industrial property right are not of themselves incompatible with competition law, they are not immune from competition law intervention.[49]

One of the key findings of this report was that patent settlements could present anti-competitive issues.[50] This was because such agreements could delay entry of generic competitors.

From 2010, the Commission began a formal process of monitoring patent settlements, reporting on progress annually.[51] Agreements which limit the generic producer's right to market its own medicines and a value transfer from the patent holder (or originator) to the generic producers would attract scrutiny in particular.

Prior to that, in 2005, the Commission had in fact begun to aggressively prosecute patent holders in respect of practices relating to such settlements, and other practices.[52] The first case of abuse of dominance initiated by the Commission that year was against AstraZeneca. Just 6 months before the Supreme Court ruling in *Actavis* (discussed above), the EU Court of Justice held in *AstraZeneca v Commission* that AstraZeneca, the originator of the patented drug Losec, was liable under EU competition law for its conduct both in acquiring its patent and in blocking generic products from the market, levying a fine of €52.5 m.[53] Notably, the EU

[48] See http://ec.europa.eu/competition/sectors/pharmaceuticals/inquiry/.

[49] Original footnote omitted; p. 18, Final Report, *ibid.*

[50] See paras. 3.2–3.3 of the report, *ibid.* [51] Footnote 48 above.

[52] See cases reported, at EU website: http://ec.europa.eu/competition/sectors/pharmaceuticals/case_law.html (accessed 6 February 2015). See also Nadia de Souza, Competition in Pharmaceuticals: The Challenges Ahead post AstraZeneca, *Competition Policy Newsletter* (Spring 2007), pp. 29–43; and generally David W. Hull, The Application of EU Competition Law to the Pharmaceutical Sector, *J. Eur. Comp. L & Pr.* (2013), 426–436, Oxford University Press. For a good overview, see also Elena Cortés, Adam Dawson and Catriona Hatton, Squaring the Circle: The EU's Quest for Balance between Antitrust and Intellectual Property – Antitrust Limits on Intellectual Property Rights, Eur. Antitrust Review (2014).

[53] Case C-457/10 P and Case T-321/05. Note that the complaint did not related to a pay-for-delay agreement but for misrepresentations made in obtaining the patent, and conduct in deregistering a patent to block competition by generic manufacturers; see judgment, December 2012, available at: http://curia.europa.eu/juris/document/document.jsf?text=&docid=131490&pageIndex=0&doclang=EN&mode=req&dir=&occ=first&part=1&cid=15589. Prior to this decision, in 2007, the EU authorities had issued a *Pharmaceutical Sector Inquiry Report* which identified a number of practices by

Commissioner explained the approach of increased scrutiny of the Commission to the European Parliament as follows:

> Due account is taken of the need for the industry to recover its research and development costs, given the industry's heavy dependence on innovation for its further competitiveness. In that context, . . . the Commission is not circumspect about rigorously applying the anti-monopoly provisions in the pharmaceutical sector, for generic competition is an area which has suffered from under-enforcement in the past. However, the importance of the generic segment for the provision of affordable medicines in the enlarged Union cannot be ignored. That is why the Commission will give greater priority to competition in the generic sector in the immediate future.[54]

Other actions have since been taken by the Commission, including the imposition of fines against Lundbeck in June 2013[55] and in December 2013 against Johnson & Johnson (€10.8 million) and Novartis (€5.5 million) for reverse payments.[56] In July 2014, the Commission fined Les Laboratoires Servier and five other drug makers (including Teva) €427.7 million in relation to "pay-for-delay" agreements.[57]

A number of national competition enforcement actions have also been brought in various EU members, including actions involving pay-for-delay agreements.[58] In 2013, the UK fined GSK in 2013 for its pay-for-delay agreement.[59]

pharmaceutical companies that should be better regulated. Among these was the pay-for-delay type of agreement or settlement; EU report available at: http://ec.europa.eu/competition/sectors/pharmaceuticals/inquiry/. Soon after that report, the European Commission initiated competition law action against some companies, amongst which was AstraZeneca. See generally Stanislas de Margerie, "Pay-for-Delay" Settlements: In Search of the Right Standard, *W. Comp. L. & Econ. Rev.* (2013), 85–97.

[54] Commissioner Neelie Kroes' answer to question 49, European Parliament, 13 June 2006, minutes available at: www.europarl.europa.eu/sides/getDoc.do?type=CRE&reference=20060613&secondRef=ITEM-015&language=EN#2-252 (accessed 6 February 2015).

[55] *Commission v Lundbeck*; see press releases at http://europa.eu/rapid/press-release_IP-13-563_en.htm and http://europa.eu/rapid/press-release_MEMO-12-593_en.htm?locale=en (accessed 6 February 2015).

[56] See http://europa.eu/rapid/press-release_IP-13-1233_en.htm (accessed 6 February 2015).

[57] See http://europa.eu/rapid/press-release_IP-14-799_en.htm (accessed 6 February 2015). It has been reported that Teva is appealing this decision: see www.roschier.com/sites/default/files/newsletters/00bda391bd7ccaab3f3c478a6b509b19.html#article1 (accessed 6 February 2015).

[58] See Carlos M. Correa, *Intellectual Property Law and Competition Law: Exploration of Some Issues of Relevance to Developing Countries*, footnote 17 above, at p. 12.

[59] See UK Office of Fair Trading website: www.oft.gov.uk/news-and-updates/press/2013/36-13#.UyfCyYVLVco (accessed 18 March 2014). Note that from 31 March 2014 this office ceased operations and the UK competition authority is now the UK Competition

## 6.5 Asia's first pay-for-delay case

### *6.5.1* Korea Fair Trade Commission v Dong-A/GlaxoSmithKline

In Korea, refusals to deal may amount to unfair practices under the Monopoly Regulation and Fair Trade Act, Art. 23, read with Part V Part B of the Guidelines for Assessment of Unfair Practices.[60]

In 2012, the Korea Fair Trade Commission (KFTC) published new guidelines relating to possible anti-competitive use of patents, dealing particularly with standard-setting issues.[61] These complement the earlier 2010 Guidelines on Unfair Exercise of Intellectual Property Rights. Under the latter guidelines, refusals to license patent rights are specifically addressed, as follows:

> The patent system grants exclusive right to patentees for the working of relevant inventions in order to provide a fair reward for innovative invention and to promote the development of new technology. Therefore, the ability of the patentee within reasonable bounds to refuse to grant a license to protect its rights, in general, is deemed to be a fair exercise of its patent right. However, the acts, as described below, of unfairly refusing to grant a license, and thereby threatening to restrict fair trade can be determined to be outside the bounds of fair exercise of patent rights.
>
> Refusing to grant a license includes not only the act by patentees of refusing to grant a license, but also the acts of causing a third party to refuse to grant a license or of demanding excessive consideration or conditions which are practically or economically impossible to satisfy, or any act which has same effect as a refusal to grant a license as well as the refusal to supply, refusal to purchase, refusal to commence or continue trade.[62]

In 2013, Korea became the first Asian jurisdiction in which the antitrust authorities challenged a pay-for-delay agreement. The KFTC initiated the case against GSK and Dong-A. The matter, which reached the Korean Supreme Court, led to the Court's decision finding violations – not long after the *Actavis* decision in the US Supreme Court. In *KFTC v GSK/Dong-A*, a pay-for-delay agreement was found to violate Arts. 19(4) and (9)

and Markets Authority. Other types of action include "groundless threats" of litigation in patent cases; see UK 2014 report: http://lawcommission.justice.gov.uk/docs/lc346_patents_groundless_threats.pdf.

60 Available at: http://eng.ftc.go.kr/bbs.do?command=getList&type_cd=62&pageId=0401 (accessed 8 July 2014).

61 See *Guidelines for Fair Patent License Agreements* and generally *Review Guidelines on Unfair Exercise of Intellectual Property Rights*, 31 March 2010.

62 WIPO Secretariat, *Refusals to License IPRs – A Comparative Note on Possible Approaches* (August, 2013), p. 34.

of the act, which prohibits "unjust concerted acts".[63] In anticipation of settlement agreements *à la* that in *FTC v Actavis*, KFTC has proposed to introduce a reporting system for drug patent settlement agreements.[64]

## 6.6 Developments in other jurisdictions

### *6.6.1 South Africa*

In South Africa, while the issue has not revolved around pay-for-delay agreements, competition law has been used to discipline pharmaceutical patent holders' pricing practices.

In *Glaxo Wellcome (Pty) Ltd & Others v National Association of Pharmaceutical Wholesalers & Others*, the South African Court of Competition Appeal held that the denial of access to an essential facility and a refusal to deal were not two sides of the same coin.[65] It added the following caution with respect to Section 8(b) of the South African competition law:

> The legislature intended, from the clear architecture of the Act, that there should be limits to the essential facilities doctrine. To demand that a dominant firm should grant access to its facilities is a substantial intervention on the part of a competition authority. The widening of the application and scope of the essential facilities doctrine can have harmful economic effects such as discouraging investment in infrastructure. An investor might be reluctant to invest for fear of a third party demanding a "free ride" on the fruits of such investment.[66]

[63] English translation of the Korea Supreme Court decision on file with author; the author is grateful to Dr Hi-Taek Shin of Seoul National University and Mr J. W. Kim of GSK Korea for assistance in obtaining this translation. For a discussion of this case up to the decision of the Seoul High Court, see Hwang Lee, Pay-for-Delay: The Korean Experience, *J. Eur. Comp. L. & P.*, Vol. 5, No. 4 (2014), 221–226. See also, OECD, *Note by Korea – Developments in Law Enforcements Related to the Adoption of System Monitoring Anticompetitive Behaviours in the Pharmaceutical Sector*, DAF/COMP/WE(2014)58, pp. 3–5, 4 June 2014, www.oecd.org/officialdocuments/publicdisplaydocumentpdf/?cote=DAF/COMP/WD%282014%2958&docLanguage=En (accessed 17 April 2015).

[64] See www.lexology.com/library/detail.aspx?g=27d53a33-e1df-4bf5-8bc1-c4a8c10fd11e (accessed 3 April 2014). See generally, Hackhyun Kim, *Korea's Antitrust Enforcement Strategy in Medical & Pharmaceutical Markets*, 2008, – http://eng.ftc.go.kr/files/static/International_Relations/Seoul_Forum/2008/Hackhyun%20KIM,%20Korea.pdf (accessed 6 February 2015).

[65] Judgment of the Competition Appeal Court of South Africa, 21 October 2002, CASE NO: 15/CAC/Feb02; see paras. 39–56.

[66] Judgment, at para. 54.

The Court stated categorically that it did not consider pharmaceutical products to be essential facilities or "resources" for antitrust purposes under the legislation.[67]

In the *cause célèbre* of South African pharmaceutical litigation, *Hazel Tau and Others v GlaxoSmithKline and Boehringer Ingelheim*,[68] a complaint was brought in 2004 by individuals before the Competition Commission of South Africa against two pharmaceutical companies for their pricing of antiretroviral therapy (ART) drugs used for treating HIV/AIDS. The plaintiffs utilized the South African Competition Act[69] to challenge the pricing practices, under arguments of abuse of dominance through excessive pricing. The complaint – supported by efforts of individuals and non-profit organizations – was eventually settled by agreement, with the grant of voluntary licenses to produce generic versions of the drugs in question.[70] The *Hazel Tau* case is also significant as the supporting testimonies expressly made a link between the essential facility doctrine and a pharmaceutical patent.[71]

In 2014, a market inquiry into the private healthcare market relating to competition was launched, with a final report expected at the end of 2015.[72] Among the terms of reference of the inquiry is the aim to "identify all factors that prevent, distort or restrict competition, including

67 See judgment, para. 53.

68 See www.section27.org.za/wp-content/uploads/2010/10/TauvGSKevidenceAndLegalSubmissions.pdf. For an overview on the regulatory situation and access to medicines with regard to Africa, see: http://ictsd.org/i/news/bridges-africa-review/134384/. See also Tu Thanh Nguyen, Chapter 4, Technology Transfer and Competition Law: Options for Developing Countries, in *Sustainable Technology Transfer – A Guide to Global Aid & Trade Development*, ed. Hans Henrik Lidgard, Jeffery Atik and Tu Thanh Nguyen (Wolters Kluwer, 2012), pp. 73–76; and Jonathan Berger, *Advancing Public Health by Other Means: Using Competition Policy*, in eds. Pedro Roffe, Geoff Tansey and David Vivas-Eugui, *Negotiating Health – Intellectual Property and Access to Medicines*, at pp. 197–200.

69 Act 89 of 1998. Information on the settlement agreements is available here: www.tac.org.za/newsletter/2003/ns10_12_2003.htm#Contents.

70 See South Africa Competition Commission newsletter, March 2004, pp. 1–2, available at: www.compcom.co.za/wp-content/uploads/2014/09/March-04-Newsletter.pdf, and settlement agreement at www.cptech.org/ip/health/sa/settlement12092003.pdf (accessed 22 April 2015).

71 See, for example, the expert reports of Carlos Correa and the Consumer Project on Technology (CPTech), available at http://keionline.org/node/2074. See also commentary in Mor Bakhoum, Chapter 16, Balancing "Incentive to Innovate" and "Protection of Competition": An African Perspective on Intellectual Property Rights and Competition Law, in *Patent Law in Global Perspective*, ed. Okediji and Bagley (Oxford University Press, 2014) .

72 See www.compcom.co.za/healthcare-inquiry/ (accessed 5 May 2015).

any evidence of market failure, regulatory failure or *competition concerns*" (italic emphasis added).[73]

### 6.6.2 *Australia*

The Australian government is a strong proponent of competition law and policy. It has focused attention in recent years on the relationship between this area of law and pharmaceutical products in recent times, as their costs increase.

In 2012, the government initiated a review on pharmaceutical patents, and in 2014, the government released the *Pharmaceutical Patents Review Report 2013*.[74] The report examined a number of aspects of pharmaceutical patents, costs of pharmaceuticals in Australia, biologics and implications of requirements in trade agreements, including the US-Australia FTA and the TPP negotiations. The report made a number of important observations and recommendations, including delays to generic competition in Australia.[75] The government released a rather terse statement on the report, stating that it had "no plans to respond to the report at this stage but may take information in the report into account when considering future policy." It added that the views in the report were those of the review panel and "did not necessarily reflect government policy".[76]

Separately, Australia's Productivity Commission issued a report on compulsory licensing of patents in 2013, addressing competition law issues in this context.[77]

In February 2014, the Australian competition authority, the Australian Competition and Consumer Commission (ACCC), announced that it had initiated legal action against Pfizer Australia Pty Ltd for alleged misuse of market power and exclusive dealing.[78]

[73] Terms of Reference, at p. 81.

[74] See IP Australia, www.ipaustralia.gov.au/about-us/ip-legislation/ip-legislation-changes/review-pharmaceutical-patents/ (accessed 16 April 2015). See T. Harris, D. Nicol, and N. Gruen, *Pharmaceutical Patents Review Report*, (Canberra, 2003). In 2000, the Ergas Committee on Intellectual Property and Competition Policy had undertaken a review of the nexus between IPRs and competition law.

[75] See the report, at xv–xix, and especially pp. 49, 107.

[76] Report and statement available at IP Australia website, at footnote 74 above.

[77] Productivity Commission report, *Compulsory Licensing of Patents*, (2013).

[78] See ACCC media release, *ACCC Takes Action against Pfizer Australia for Alleged Anti-Competitive Conduct*, 13 February 2014, at: www.accc.gov.au/media-release/accc-takes-action-against-pfizer-australia-for-alleged-anti-competitive-conduct (accessed 6 February 2015). This matter remains pending at the time of writing. Note that Australia's National

Finally, the Australian Law Reform Commission (ALRC) recognizes that in the licensing of genetic materials, anti-competitive practices may arise:

> A patent holder could seek to protect its monopoly by refusing to license its patented genetic invention. Alternatively, it could charge prohibitive licence fees, constituting a constructive refusal to licence. For example, the owner of a broad patent over an upstream genetic research tool, such as a DNA sequence, could refuse to license the tool to other researchers. This refusal to license could affect competition within the relevant market for the research tool, or within downstream markets for goods and services developed using the tool.[79]

In March 2015, an independent review of Australia's competition law was released.[80] Reform in the area of intellectual property rights is the subject of one of its recommendations. The panel has recommended the establishment of negotiating mandates and cost-benefit analyses in relation to intellectual property provisions in free trade agreements, and a separate independent review by the Productivity Commission on IP laws and competition policy issues arising from new technology.[81]

### 6.6.3 *Canada*

In Canada, focus has also recently fallen on IPRs and firm practices in the pharmaceutical sector. In 2014, the Canadian authorities released a set of

Access Regime regulates access to "essential" infrastructure facilities but this does not appear as yet to have been linked to the notion of access to, for example, clinical data as "essential facilities" for competition law purposes in the pharmaceutical sector; see Australian Productivity Commission, report on the National Access Regime 2013, at www.pc.gov.au/inquiries/completed/access-regime/report/access-regime.pdf (accessed 16 April 2015). See also, generally, Valentine Korah, Access to Essential Facilities under the Commerce Act in the Light of Experience in Australia, the European Union and the United States, *VUWLR*, Vol. 31 (2000), 231–253.

79 Para. 24.42, *Genes and Ingenuity: Gene Patenting and Human Health* (citing in the next paragraph, problems with licensing in relation to the *BRCA1* gene; see also Chapter 3 of this book), ALRC Report No. 99, at: www.alrc.gov.au/publications/24-competition-law-and-intellectual-property/application-competition-law (accessed 16 April 2015).

80 The so-called "Harper Review" Final Report (named after the review panel's chair, Professor Ian Harper), released on 31 March 2015; see www.australiancompetitionlaw.org/reports/2014harper.html and http://competitionpolicyreview.gov.au/final-report/ (accessed 20 April 2015). For information on prior competition law and policy reviews, see pp. 167 and 424 of the report.

81 See Chapter 9, Final Report, *ibid.*, and Recommendation 6. Chapter 24 addresses essential infrastructure or "bottleneck" facilities and Australia's National Access Regime, but does not include IPRs in its discussion.

draft Intellectual Property Enforcement Guidelines for public consultations,[82] and the final guidelines were released in September 2014.[83] In 2014, the Competition Bureau released a White Paper addressing pharmaceutical industry litigation settlements.[84] Unlike the US system, there is no notification system in Canada for patent settlement agreements. However, the Canadian Competition Bureau may proactively seek leave to intervene and make representations in court cases to bring a competition perspective to proceedings.[85] Notably, the Canadian Competition Bureau also signed an MOU with the Canadian IPO. It is expected that with these initiatives and the increased attention on the IP-competition interface, the Canadian anti-competition scrutiny over pharmaceutical practices will be heightened.[86]

## 6.7 Connecting refusals to licence patents and the essential facilities doctrine: the next logical step?

Refusals to deal and the related essential facilities doctrine have evolved in US and EU case law.[87] Competition law generally regards that a right holder's refusal to deal (e.g. refusal to supply a party or to license) is not necessarily illegal since there is no absolute obligation to "deal".[88] Several

[82] See www.competitionbureau.gc.ca/eic/site/cb-bc.nsf/eng/03689.html (accessed 7 February 2015).

[83] See www.competitionbureau.gc.ca/eic/site/cb-bc.nsf/eng/03808.html (accessed 7 February 2015).

[84] *Patent Litigation Settlements: A Canadian Perspective*, www.competitionbureau.gc.ca/eic/site/cb-bc.nsf/eng/03816.html (accessed 7 February 2015).

[85] Section 125, Canadian Competition Act, and IP Guidelines, *ibid.*, at p. 20.

[86] See remarks of Competition Bureau Commissioner, www.competitionbureau.gc.ca/eic/site/cb-bc.nsf/eng/03817.html (accessed 7 February 2015). See generally OECD, *Global Forum on Competition, "Competition Issues in the Distribution of Pharmaceuticals", Contribution from Canada* (Competition Bureau), DAF/COMP/GF/WD(2014)16, 16 January 2014. On the matter of a refusal to deal, see Canada's 2014 *Intellectual Property Enforcement Guidelines*, which expressly envisage dealing with refusals to license in the context of IPRs (see pp. 12–13 and Example 8, p. 31).

[87] See the informative accounts in Carlos M. Correa, *Intellectual Property Law and Competition Law: Exploration of Some Issues of Relevance to Developing Countries*, footnote 17 above, and Frederick Abbott, Sean Flynn, Carlos Correa, Jonathan Berger and Natasha Nyak, *Using Competition Law to Promote Access to Health Technologies: A Guidebook for Low- and Middle-Income Countries* (UNDP, 2012).

[88] For an overview of refusals to deal, see OECD, *Refusals to Deal*, DAF/COMP(2007) 46, 3 September 2009; and WIPO Secretariat, *Refusals to License IP Rights – A Comparative Note of Possible Approaches*, August 2013.

See, for example, Part D, EU Commission Guidance on Article 102 of the Treaty on the Functioning of the European Union. (See also especially para. 78 on refusals to license IPRs.)

systems do not appear to specifically define refusals to deal.[89] A refusal to deal springs from a dominant party's refusal to deal (particularly with a competitor) and may in some jurisdictions sometimes amount to an abuse of that dominant position.[90] Refusals to licence are, of course, not the only ones that are of possible interest to competition authorities.[91]

A particular inroad into a right holder's right to refuse to deal has been established in respect of "essential facilities" in the area of essential infrastructure.[92] Such a refusal may take various forms, and may be an outright or a *de facto* refusal.

### 6.7.1 *Incipient linkages of the "essential facilities" doctrine to refusals to license IPRs*

In competition law, where a party refuses to license what is seen to be an "essential facility" – often some form of essential infrastructure for others to operate with, for example – it may sometimes amount to be anti-competitive. The doctrine of essential facilities has already begun to be argued as applicable to IPRs.[93]

See also Canada Competition Bureau, *Refusal to Deal*, at www.competitionbureau.gc.ca/eic/site/cb-bc.nsf/eng/h_00665.html (accessed 8 March 2015).

89 See International Competition Network, *Report on the Refusal to Deal with a Rival under Unilateral Conduct Laws* (Unilateral Conduct Working Group, 2010) . The report was based on a questionnaire sent to respondents from 43 jurisdictions, of which the majority were non-Asian countries; among them were Indonesia, Japan, Korea, Singapore and Taiwan (note 2, p. 3 of the report). The questionnaire responses are available at: www.internationalcompetitionnetwork.org/working-groups/current/unilateral/questionnaires-responses/refusal-deal.aspx (accessed 8 March 2015). While the concept was largely undefined in statutes, there may be competition agency guidelines on factors to be taken into consideration; see the report, p. 19.

90 It appears that in some jurisdictions, such as Indonesia, Japan, Korea and Taiwan, refusals to deal may also apply to non-dominant firms; see *ibid.*, p. 9.

91 For other practices, see, for example, Frederick Abbott, Sean Flynn, Carlos Correa, Jonathan Berger and Natasha Nyak, *Using Competition Law to Promote Access to Health Technologies: A Guidebook for Low- and Middle-Income Countries*, footnote 87 above.

92 See, for example, OECD, *The Essential Facilities Concept* (1996); Korah, footnote 78 above; and Sergio Baches Opi, The Application of the Essential Facilities Doctrine to Intellectual Property Licensing in the European Union and the United States: Are Intellectual Property Rights Still Sacrosanct?, *Fordham J. Intell. P., Med. & Ent.* (2001), pp. 412–507.

93 See, for example, Sergio Baches Opi, The Application of the Essential Facilities Doctrine to Intellectual Property Licensing in the European Union and the United States: Are Intellectual Property Rights Still Sacrosanct? *Ford. I.P. Media & Ent. L. J.*, Vol. 11, No. 2 (2001), 412–507.

There is even some indication of preliminary support for the view that clinical testing data that are not possible to replicate may amount to an "essential facility" in the context of refusals to licence.[94]

In the context of pharmaceutical patents, it is arguable for instance that a patented drug and more importantly the knowledge and tests that led to the patented drug may be an "essential facility" for other researchers wishing to build on and innovate upon.[95] If such a view were entertained, it would open the door to refusals to licence by pharmaceutical patent holders to be subject to even closer scrutiny under competition principles. Already, the link between the essential facility doctrine and such

[94] See WIPO, footnote 62 above, para. 30 at p. 13, which states (italic emphasis added and original footnotes omitted): ". . . refusals to license must either be associated with certain serious circumstances, or have certain dire consequences so as to be deemed anticompetitive. In general, the anticompetitive nature of refusals to license IPRs is associated with the concept of *essential facilities*. 'A facility is "essential" if it is otherwise unavailable and cannot be "reasonably or practically duplicated".' When it comes to refusals to deal in general, monopolization is not confined to essential facilities. But the essential facilities doctrine has a very narrow impact on refusals to license IPRs, because of their inherent capacity of being alter-created or alter-invented. No IP title or right is – or should be – capable of constituting an insurmountable barrier for competitors who wish to enter the same relevant market. In other words, IP should never be an essential facility. When it is, the IPRs in question are likely to be invalid or exceptional circumstances (which have, in general, a regulatory dimension) transform them into essential facilities. The example that comes immediately to mind is technical standards; *i.e.*, if IPRs are 'technical standards' in a particular relevant market these IPRs are 'essential' for competitors to compete in that market. However, the essential facility doctrine for IPRs has been also invoked in other situations, such as consumers' lock-in in the case of visible parts of car bodies, when covered by industrial designs. *The impossibility of repeating clinical data necessary for obtaining the marketing approval of certain regulated products because of the risks they create for humans, animals and the environment might be another example of an essential facility that constitutes an unsurpassable barrier for subsequent registrants.*"

[95] See Koichi Sumikura, *Intellectual Property Rights Policy for Gene-Related Inventions – Toward Optimum Balance between Public and Private Ownership*, Chapter 3 in *The Role of Intellectual Property Rights in Biotechnology Innovation*, ed. David Castle (Edward Elgar Publishing, 2009), pp. 80–83. Sumikura points to the fact that while under the Japanese Antimonopoly Act, Art. 21 provides that patents, among other recognized IP rights, enjoy an exception under the act, it is possible to argue that a patent may yet violate the Antimonopoly Act "if license refusal results in obstruction of access to an essential facility" (p. 82). However, he further opines that there is "no clear consensus" internationally on this, and that, "should this theory be arbitrarily applied to cases where the exercise of intellectual property rights is interpreted as preventing the public ownership of technology useful for human beings, the stability of rights would be undermined and the intellectual property rights would exist only in name". The key word underpinning this view is "arbitrarily"; if a rational, well-articulated application of the essential facility argument can be developed in the pharmaceutical context, there is no reason why this field should – unlike the infrastructure field from which it has sprung – *a priori* be immune from its application and scrutiny.

patents has been made. However, given the possible implications for innovation, regulators are understandably cautious in applying the "essential facility" doctrine.[96]

Interestingly, during the TRIPS negotiations, the following provision had appeared in earlier drafts but was not ultimately incorporated into the agreement:

> PARTIES shall not discourage or impede voluntary licensing of undisclosed information by imposing excessive or discriminatory conditions on such licenses or conditions which dilute the value of such information.[97]

Note, however, that this draft provision would have addressed *state* action – as opposed to private party action (such as refusals to licence or imposition of unreasonable conditions for a licence) – to "discourage or impede" voluntary licensing of undisclosed information.

In a recent development – even as this author was formulating the above arguments – China's anti-competition authorities have already proceeded to issue rules on abuses of dominant position with respect to IPRs, in which the refusal to licence has been *explicitly* linked to the essential facilities notion.[98]

The following explains some recent developments as well as possible contrary views. In a survey of various national competition laws, the WIPO has identified three possible approaches among its members on the competition law interface with refusals to license IPRs.[99] Some systems require that for the doctrine to apply, duplication of the facilities should be impossible or undesirable for public policy reasons.[100]

[96] For instance, Singapore authorities have expressed the following view:

> The *Competition Commission of Singapore* advised that "[c]are must be taken not to undermine the incentives for undertakings to make future investments and innovations, especially where the essential facility is a result of a previous innovation".

International Competition Network, *Report on the Refusal to Deal with a Rival under Unilateral Conduct Laws* (Unilateral Conduct Working Group, 2010), footnote 89 above, p. 8.

[97] Chairman's Report to the Group of Negotiations on Goods, MTN.GNG/NG11/W/76, p. 41.

[98] See footnotes 157–159 below and accompanying text.

[99] WIPO Secretariat, *Refusals to License IPRs – A Comparative Note on Possible Approaches*, footnote 62 above.

[100] See the report, footnote 89 above, p. 20, citing Singapore as an example.

## 6.8 Competition law and the pharmaceutical sector in Asia – an evolving regional landscape

Several Asian countries have only begun to enact competition laws in the last 10–15 years. Examples include China, HK, Malaysia, Singapore and Vietnam. ASEAN members aim to have in place national competition laws by the end of 2015.[101] It is noteworthy that agreements or conduct *outside* of some of these jurisdictions – such as Malaysia and Singapore – which have an anti-competitive impact within the jurisdiction could potentially come under scrutiny too.[102] Given that agreements (including anti-competitive ones covered by such legislation) may be easily made extraterritorially today, the scope for prosecution is widened as a result.

While the members have not agreed to fully harmonize such laws, Regional Guidelines on Competition Policy were issued in 2010 to provide broad guidance to members in designing and implementing their national laws.[103] These guidelines also address common provisions on competition policy in FTAs.[104] In addition, some Asian countries

[101] See www.asean.org/communities/asean-economic-community/category/competition-policy and *Handbook on Competition Policy and Law in ASEAN for Business 2013*, June 2013 (link available at the same web page, as at 18 March 2014). Singapore has been sharing its experiences in implementing and enforcement of competition law with newer ASEAN member countries, Cambodia, Lao PDR, Myanmar and Vietnam; see www.scp.gov.sg/content/scp/news_events/key_events/2010/201008/event_201008_04.html (accessed 13 March 2015). See also, generally, eds. Michael W. Dowdle, John Gillespie and Imelda Maher, *Asian Capitalism and the Regulation of Competition – Towards a Regulatory Geography of Global Competition Law* (Cambridge University Press, 2013) and ed. R. Ian McEwin, *Intellectual Property, Competition Law and Economics in Asia* (Hart Publishing, 2011). On challenges faced by lower-income countries, see Chapter 5, Frederick Abbott, Sean Flynn, Carlos Correa, Jonathan Berger and Natasha Nyak, *Using Competition Law to Promote Access to Health Technologies: A Guidebook for Low- and Middle-Income Countries*, footnote 87 above. Officials of competition authorities in Asia regularly join those of other countries in the Asia-Pacific region to discuss and report national competition law policies, issues and developments; see, for example, OECD/Korea Policy Centre documents at www.oecd.org/competition/koreacentrenewsletter.htm (accessed 17 April 2015).

[102] See Section 3 of the Malaysian Competition Act 2010, Act 712, and Section 33 of the Singapore Competition Act, Cap. 50B.

[103] See ASEAN website at: www.asean.org/communities/asean-economic-community/category/competition-policy (accessed 8 July 2014). See generally, Thomas K. Cheng, A Developmental Approach to the Patent-Antitrust Interface, *NW J L & Int'l B*, Vol. 33, No. 1 (2012), 1–79.

[104] Chapter 10 provides as follows:

> **10.4 Common Competition Related Provisions in Free Trade Agreements**
> 10.4.1 AMSs will endeavour to take into consideration some common competition related provisions in a competition chapter of Free Trade

have recently begun to issue more specific guidelines to address competition policy and arrangements involving intellectual property and even biotechnology.

Given the relative newness of competition laws in these countries, a body of case law has yet to be built within the region. Even newer to these jurisdictions would be enforcement of the competition laws in an intellectual property context, such as in the pharmaceutical sector where patent holders' practices may be viewed as being anti-competitive. Within systems with more mature competition laws, such as in Korea, the system is only beginning to address certain types of conduct in the pharmaceutical sector, as will be seen below.

In the US and the EU, as explained above, the interface between pharmaceutical patentees' practices and competition law has only begun to be explored more by competition authorities in the last several years. Competition authorities in Asia therefore are no doubt learning from these experiences elsewhere, in considering any future conduct involving pharmaceutical entities.

Rather than describe the competition laws of the region in general terms, the following discussion highlights salient developments, competition law and IPRs, and, in particular, the role (if any) of the essential facilities doctrine.

### *6.8.1 ASEAN*

Competition law in ASEAN is in a state of accelerated evolution due to economic integration under the AEC. Brunei and Cambodia are in the process of preparing comprehensive competition laws.[105] In 2015, three ASEAN countries enacted new comprehensive competition legislation:

Agreements (FTA) such as provisions that effectively address anticompetitive behaviours, based on principles of transparency, due process and non-discrimination on the basis of nationality and provisions on cooperation (*e.g.*, technical assistance and capacity building).

10.4.2 AMSs should have the flexibility to make proper judgement on the best approach for developing provisions in a competition chapter of FTA that is in line with the developmental goals of their economy and which is not in contradiction to any provisions/ approaches already agreed for within the regional level. This is to ensure for a strong support towards the achievement of the ASEAN Economic Community 2015.

[105] See ASEAN competition law portal, www.aseancompetition.org/aegc/aegc-members (accessed 16 April 2015).

Lao PDR,[106] Myanmar[107] and the Philippines. In the Philippines, prior to 2015, a number of sectoral laws had governed anti-competition matters. In July 2015, the cross-sectoral Philippine Competition Act was enacted.[108] The new Act provides for prohibitions against certain types of anti-competitive agreements and mergers and acquisitions. Notably, it also includes language which permits agreements and conduct which allow consumers "a fair share of the resulting benefits" (such as in sections 14(c) and 15(i)) and provisions that permit differential pricing in certain circumstances (such as "socialized pricing for the less fortunate sector of the economy" in section 15(d)), Separately, the Universally Cheaper and Quality Medicines Act[109] amended the Intellectual Property Code to permit use of an invention by the government and the use of compulsory licensing to remedy anti-competitive conduct.[110] It remains to be seen what other specific provisions, if any, to address competition issues in relation to IPRs and pharmaceuticals, might be included in regulations or guidelines associated with the Competition Act.

In 2005, Malaysia laid the foundation for a competition law in the form of the Fair Trade Practices Policy.[111] The policy addressed three key areas of conduct, namely abuse of dominant position, "hard-core" cartels and anti-competitive practices. In 2010, Malaysia enacted the

[106] Decree 15/PMO (4/2/2004) on Trade Competition has existed since 2004 but a new Business Competition Law was enacted in 2015; see http://jclao.com/parliament-passes-business-competition-law/ (accessed 9 October 2015) and ASEAN competition law portal, at footnote 105 above.

[107] The Pyidaungsu Hluttaw Law No. 9 /2015; see http://myanmarbusinessnews.com/about-myanmar/market-guides/analysis-myanmar-competition-law-2015/ (accessed 9 October 2015).

[108] Republic Act No. 10667; text available at: www.gov.ph/2015/07/21/republic-act-no-10667/ (accessed 9 October 2015). See also generally Llewellyn L. Llanillo and Gelina Rose E. Recio, Protection and Liberalization of Investments under Philippines Law, at p. 16, Paper presented at the Workshop on Business Law, 12th General Assembly, ASEAN Law Association, Manila, Philippines, 26–28 February 2015 (paper on file with the author); action of the Senate of the Philippines, 15 December 2014, at www.senate.gov.ph/lis/bill_res.aspx?congress=16&q=SBN-2282; and Marvin Sy, *Senate OKs Proposed Fair Competition Act, The Philippine Star*, 16 December 2014, (accessed 8 March 2015).

[109] Republic Act No. 9502; text available at www.ipophil.gov.ph/images/Patents/IRRs/RepublicAct9502.pdf (accessed 13 April 2015).

[110] See Sec. 74 and Sec. 93.3 of the Code, Republic Act No. 8293.

[111] See http://mycc.gov.my/legislation/ (accessed 16 April 2015). See also related *Guidelines* and other documents at: www.mycc.gov.my/news.asp?page=Handbook. For an account of the development of the policy and subsequent legislation, see the *Annual Report 2012* of the Malaysian Competition Commission, esp. at p. 22.

Competition Act.[112] A separate guideline on intellectual property-related practices is expected to be issued.[113] Other than such excluded/exempted areas, other sectors and activities – including those of the pharmaceutical sector – could come under scrutiny of the act. So far, no cases handled by the Malaysian Competition Commission (MyCC) have involved this sector. Under the Commission's *Guidelines on Chapter 2 Prohibition* (on abuse of dominant position), a refusal to license intellectual property may be considered by the Commission, which will "take account of the difficult trade-off involved in forcing supply which leads to a short-term increase in competition but which may harm longer term incentives for innovation and investment".[114] The guidelines raise the matter of essential facilities but do not explicitly apply them to IPRs.[115]

Singapore law reflects a blend of domestically and FTA-driven IP and competition law provisions, and provides an interesting case study in competition law in the pharmaceutical industry for other Asian countries. The key regulatory instrument is the Competition Act (supported by a number of guidelines, including the Competition Commission Guidelines) administered and enforced by the Competition Commission

[112] Act 712. The act permits investigative and prosecutorial action against anti-competitive agreements and abuses of dominant positions. Relief from liability and exclusions are provided for in Section 5 and Section 13 (read with the Second Schedule). Under Section 11, market reviews may be carried out. Under Section 3 (read with the First Schedule), certain legislation is also exempted from the scope of the act. The Malaysian Competition Commission (MyCC) is also authorized to grant individual exemptions under Section 6 and block exemptions under Section 8 of the act.

[113] See *Guidelines on Chapter 1 Prohibition – Anti-Competitive Practices*, May 2012, at p. 17.

[114] See paras. 3.25–3.27. Para. 3.27 provides:

> 3.27. The remedy for a refusal to supply is to force the supplier to supply at a reasonable consideration. However, forcing supply may reduce the incentive to invest in the product, intellectual property rights or essential facility. The MyCC will take account of the difficult trade-off involved in forcing supply which leads to a short-term increase in competition but which may harm longer term incentives for innovation and investment.

Note that Section 10(3) in the Competition Act also provides the following exception to the abuse of dominance provision:

> (3) This section does not prohibit an enterprise in a dominant position from taking any step which has reasonable commercial justification or represents a reasonable commercial response to the market entry or market conduct of a competitor.

[115] Tu Thanh Nguyen, *ibid.*, at p. 200.

of Singapore (CCS). There are certain sector-specific anti-competition rules.[116] The Patent Act also plays a part insofar as it relates to compulsory licensing. Contract law principles relevant to policing anti-competitive "restraints of trade" would also play a part. While anti-competitive agreements challenged under contract law usually arise from a party challenging such an agreement in a civil lawsuit, the Competition Act provides for voluntary notification of potentially anti-competitive agreements or mergers for clearance.[117] The CCS takes the general view that intellectual property and competition laws aim to promote economic efficiency and innovation.[118] While the law generally treats IPRs as for other areas,[119] the CCS has issued specific guidelines on IPRs.[120,121] Under Section 51 of the

116 An example is found in the telecommunications sector; see the *Code of Practice for Competition in the Provision of Telecommunication Services 2012* (Telecom Competition Code).

117 See, for example, Sections 42–46 of the Competition Act.

118 Para. 2.1 of the CCS *Guidelines on Treatment of Intellectual Property Rights* states:

> Both intellectual property ("IP") and competition laws share the same basic objective of promoting economic efficiency and innovation. IP law does this through the provision of incentives for innovation and its dissemination and commercialisation, by establishing enforceable property rights for the creators of new and improved products and processes. Competition law does this by helping to promote competitive markets, thereby spurring firms to be more efficient and innovative.

119 Report, footnote 89 above, p. 22.

120 See guidelines at www.ccs.gov.sg/legislation/~/media/custom/ccs/files/legislation/ccs%20guidelines/iprjul07final.ashx (accessed 20 April 2015).

121 Singapore provided its response to an ICN questionnaire in 2009 in relation to IPR issues; excerpts of the informative response by CCS are reproduced here:

> CCS considers that undertakings generally have the freedom to decide whom they will supply. Therefore a refusal to supply, even by a dominant undertaking, would not normally be an abuse. However, in certain circumstances, a refusal to supply by a dominant undertaking may be considered an abuse if there is evidence of (likely) substantial harm to competition and if the behaviour cannot be objectively justified.
>
> CCS will evaluate the facts and circumstances of each alleged abuse situation on a case-by-case basis, in order to determine whether the firm in question's refusal to supply a rival (or rivals) has an anticompetitive effect on the market. In its evaluation, CCS's primary focus will be on the actual or potential effect, regardless of whether the practice is threatened or actual, and whether the firm excludes one or more rivals from the market.
>
> For competition law purposes, CCS will regard intellectual property rights as being essentially comparable to any other form of property. The right to exclude is the basis of property rights, and CCS recognizes that ownership of intellectual property ("IP") does not normally impose on the IP owner an obligation to license the use of that IP to others, even where the

Patents Act, certain restrictive conditions applied to licensing of patents are also prohibited. Section 34 of the Competition Act specifically addresses anti-competitive agreements.[122] Sections 34(1)(b) and/or (c)

> IP ownership confers market power on the IP owner. Therefore a refusal to supply a license, even by a dominant firm, is not normally an abuse.
>
> In limited circumstances, a dominant firm's refusal to supply a license may constitute an abuse of dominance, for example, if the refusal concerns an IP right that relates to an essential facility, with the effect of (likely) substantial harm to competition. The analysis does not differ in this instance from a refusal to supply in a situation not involving IP, in that CCS may consider if the dominant firm is able to objectively justify its conduct, and whether the dominant undertaking has behaved in a proportionate way in defending its legitimate commercial interest. However, CCS notes that essential facilities are rare in practice and IP rights are generally unlikely to create essential facilities.
>
> In determining whether a refusal to supply a license constitutes an abuse, the impact on the technology and innovation markets will be considered. CCS is aware that care must be taken not to undermine the incentives for firms to make future investments and innovations.
>
> The definition of "essential facilities" for IP is defined in the same way as a situation involving non-IP products and services.

For the full questionnaire, see www.internationalcompetitionnetwork.org/working-groups/current/unilateral/questionnaires-responses/refusal-deal.aspx (accessed 8 March 2015), p. 3.

122 Section 34 provides as follows:

> 34. – (1) Subject to section 35, agreements between undertakings, decisions by associations of undertakings or concerted practices which have as their object or effect the prevention, restriction or distortion of competition within Singapore are prohibited unless they are exempt in accordance with the provisions of this Part.
>
> (2) For the purposes of subsection (1), agreements, decisions or concerted practices may, in particular, have the object or effect of preventing, restricting or distorting competition within Singapore if they –
>
> (a) directly or indirectly fix purchase or selling prices or any other trading conditions;
> (b) limit or control production, markets, technical development or investment;
> (c) share markets or sources of supply;
> (d) apply dissimilar conditions to equivalent transactions with other trading parties, thereby placing them at a competitive disadvantage; or
> (e) make the conclusion of contracts subject to acceptance by the other parties of supplementary obligations which, by their nature or according to commercial usage, have no connection with the subject of such contracts . . .

are the most likely parts to be directly relevant, subject of course to the particular agreement's content and construction.

The Competition Act contains no *a priori* exclusion of compulsory licensing (e.g. as permitted under the Patents Act) agreements from the scrutiny under s. 34 of the act, nor is there any *a priori* exclusion from scrutiny simply because pharmaceutical patents may be involved. However, paragraph 2(1) of the Third Schedule provides that the s. 34 prohibition shall not apply to "an agreement to the extent to which it is made in order to comply with a legal requirement".[123]

A primary consideration as to whether there is anti-competitive conduct is the relevant "market".

Sections 34 and 47 of the Competition Act refer to the relevant "market" when determining whether these provisions have been infringed. Section 55 of the Patents Act also requires interpretation of the relevant "market" in order to determine if circumstances exist which justify grant of a compulsory licence.[124] Section 47 deals with conduct that amounts to an abuse of a dominant position.

On the purpose of market definition, the CCS stated its approach to market definition in the *Qantas and British Airways* decision:

[123] Agreements made pursuant to grant of a compulsory license under s. 55 of the Patents Act are presumably covered under this exemption as there would be a "legal requirement", that is, an order of court made under s. 55(3) to comply. Other licensing (and non-licensing) agreements not covered by paragraph 2(1) would remain under the scrutiny of Section 34, raising the relevant market inquiry in such cases.

[124] Section 55 of the Patents Act provides:

> 5. – (1) Any interested person may apply to the court for the grant of a licence under a patent on the ground that the grant of the licence is necessary to remedy an anti-competitive practice.
>
> (2) Without prejudice to the generality of subsection (1), the court may determine that the grant of a licence is necessary to remedy an anti-competitive practice if –
>
> (a) there is a **market** for the patented invention in Singapore;
> (b) that **market** –
>   (i) is not being supplied; or
>   (ii) is not being supplied on reasonable terms; and
> (c) the court is of the view that the proprietor of the patent has no valid reason for failing to supply that **market** with the patented invention, whether directly or through a licensee, on reasonable terms . . .
>
> [Bold emphasis added]

The relevant "market" is, however, not defined under the Patents Act.

> Market definition is the first step in a full competition analysis. It is also a key step in providing the framework for analysis by identifying the competitive constraints acting on a seller of a given product.[125]

Under the *Guidelines on Market Definition*, the market is defined by use of a construct comprising the "hypothetical monopolist", and its competitors that may supply (or could potentially supply) the same or substitute products with the ability to constrain its exercise of market power. The relevant market is taken as the

> smallest product group (and geographical area) such that a hypothetical monopolist controlling that product group (in that area) could profitably sustain "supra competitive" prices, i.e. prices that are at least a small but significant amount above competitive levels. That product group (and area) is usually the relevant market for competition law purposes.[126]

The definition assumes therefore that there is a "product group" that can form the market, formed by the hypothetical monopolist and competitors supplying the same or substitutable products.[127]

[125] Footnote 140 below.

[126] CCS *Guidelines on Market Definition*, para. 2.4. There are two broad approaches to defining the relevant market in this field. The first is use of a product market while the second is use of the "innovation market", which has not been universally accepted; see Joshua D. Wright, Anti-Trust, Multidimensional Competition, *and Innovation*, in *Competition Policy and Patent Law under Uncertainty – Regulating Innovation*, ed. Geoffrey A. Manne and Joshua D. Wright (Cambridge University Press 2011), at pp. 235–236.

[127] Under the CCS *Guidelines on Market Definition*, the "buyer" is treated as "the end-user consumer, and/or an undertaking that buys products as inputs for production or for resale, as the context demands". In the case of prescription medicines, whether originator or generic, the "buyer" may be public authorities such as the Ministry of Health, hospitals, pharmacies and clinics. This is significant as these users may vary in purchasing preferences, depending, for example, on price, therapeutic efficacy and, as accepted in the EU's *AstraZeneca* case, even practitioner inertia to switch to another product. The *Guidelines on Market Definition* require consideration of the demand side (including effects of price increases) and the supply side. On the demand side, evidence of substitution of the goods by buyers is relevant. The test is one of whether the hypothetical monopolist "could profitably **sustain** prices above competitive levels . . ." (para. 3.6). Clearly price considerations are an important factor. The CCS has stated generally as follows, with regard to the interpretation of Section 34 ( ):

> The section 34 prohibition is modeled after the Chapter I prohibition of the United Kingdom (UK) Competition Act 1998 and Article 101 (previously Article 81) of the Treaty on European Union (the EC Treaty). Cases from the UK and the European Union (EU) may thus be persuasive or useful in assisting CCS in arriving at its conclusions, although the value

Other documents that assist in determining the relevant market in a pharmaceutical patent case under Section 34 include Annex B of the *Guidelines on the Section 34 Prohibition* which provides factors to be taken into account in considering market power and market shares, the CCS' *Guidelines on Market Definition*, and its *Guidelines on Treatment of Intellectual Property Rights*. The last two are to be read together where intellectual property rights are in question.[128] The last-mentioned set of *Guidelines* direct attention – when assessing competitive effects of licensing agreements – first to the relevant product market, but competitive effects for technology and innovation markets may also be assessed.[129]

The *Guidelines on Treatment of Intellectual Property* recognize that the "possession of an IPR does not necessarily create market power in itself" and the legal monopoly it brings does not necessarily confer an "economic monopoly"; the guidelines also recognize that "there may be sufficient actual or potential close substitutes that constrain the exercise of market power by the IP owner".[130] The underlying thinking of the government is reflected in the Competition Bill Consultation Paper Annex C.[131]

> of any foreign competition case law depends very much on the overall facts and context of the case before CCS, as well as the extent to which the facts of these foreign cases are applicable to the local context.

Taking this into account, EU interpretations on the market definition will be instructive. In the EU, a number of factors have been identified as being relevant to the market definition exercise. Further, notably, in the *AstraZeneca* case, the European Commission had included as a factor in defining the relevant market, inertia of medical practitioners to switch products. This would have had helped to narrow the product market somewhat. Drawing on the Commission Notice on the definition of the relevant market for the purposes of Community Competition law, the General Court accepted this factor, and the Court of Justice affirmed its finding.

Should such an agreement arise in Singapore, what could be the relevant "market" when applying section 34(2) of the Competition Act? Applying the CCS' *Guidelines*, one way of construing the market for this purpose would comprise: the patented drug for treating a particular disease or condition (smallest product group) in Singapore (geographical area) such that a hypothetical monopolist controlling the drug in Singapore (geographical area) could profitably sustain "supra competitive" prices, i. e. prices that are at least a small but significant amount above competitive levels.

[128] Para. 2.6, CCS *Guidelines on Treatment of Intellectual Property Rights*.

[129] Para. 2.7, *ibid.* [130] Para. 2.5.

[131] The Singapore Ministry of Trade & Industry Competition Bill Consultation states:

> 7. IP laws and competition law are not necessarily inconsistent; rather they can work together to help develop Singapore into a knowledge economy. Competition law, by helping to promote efficient markets, ensures that undertakings innovate to the extent dictated by consumers and other market pressures. The rewards to innovation

The Competition Act also subjects mergers to its scrutiny, under Section 54. To date, there have been two cases relating to the pharmaceutical industry examined by the CCS under the merger provision of Section 54. Both were cleared by the Commission after such examination. In the first case, *GSK Trading Services Limited and UCB SA*,[132] a proposed acquisition by GSK of UCB covered certain rights from UCB to distribute and market certain pharmaceutical products in certain emerging markets by way of a Sale and Purchase Agreement (constituting a merger under the act). In its Decision, the CCS ruled that in determining the product market in question, the therapeutic use of the product and the EU's ATC-3 classification (though not conclusive) were relevant factors for consideration. It further accepted that the geographical market in this case was Singapore, given the regulatory regime requiring that, as argued by GSK, "physicians and pharmacies are only able to supply products that are registered with the HSA", and that a licence and, for specialized products, special import permits from the HSA are required for such products to be sold.[133]

In the second case, *Re Novartis AG of Alcon Inc.*,[134] for its market definition analysis, it reiterated it was reasonable to use the "ATC3" category classification (referring to the third level in the Anatomical Therapeutic Chemical classification system used by the European Pharmaceutical Market Research Association (EphMRA))as the starting point for defining the relevant product markets of the parties. However, it

provided by IPR should thus be maintained. The specific rights provided by IP laws, and the business advantages these confer, would thus not in any way be circumscribed by competition law.

8. However, IPR is a reward to innovation for a specific invention or creation, and should not become a tool for engaging in anti-competitive activities. Where the exercise of the IPR is anti-competitive, it would be subject to competition law.
9. In considering whether a business activity involving the exercise of IPR would have any competition concerns, the Competition Commission would adopt an "economics-based cost-benefit analysis" or "rule of reason" approach. This means that the Competition Commission would take a holistic view and look at the overall net welfare effects of the activity to decide whether a particular use of an IPR reduces welfare in Singapore.
10. To help provide further clarity, the Competition Commission will develop guidelines after the enactment of the competition law on how it would view IPR-related business activities.

[132] CCS Case No 400/002/09, [2009] SGCCS 1, Decision dated 23 March 2009.

[133] Para. 19 of the Decision.

[134] CCS Case No. 400/003/10, [2010] SGCCS 2, Decision dated 20 May 2010.

noted that there may be situations in which one may depart from the ATC3 classification, to use the ATC4 or molecule level.[135]

In the US, a narrow reading of the relevant market has been argued by the US Federal Trade Commission when prosecuting antitrust complaints in relation to patented and generic drugs. The federal courts have so far resisted the FTC's arguments, finding that two drugs treating the same condition do not necessarily form the relevant market: a recent example is *FTC v Lundbeck Inc.*, a decision of the US Court of Appeal of the Eight Circuit in August 2011.[136] Demand substitution and demand cross-elasticity are relevant factors in proving the relevant market. In that case, the Court found that the evidence showed that there was no cross-elasticity based on pricing of the two drugs in question, which proved fatal to the FTC's argument that they formed the market in question. The argument may therefore be made, in such a situation, that the originator/patent holder is an absolute monopolist for that patented product and have 100% of the market. There has been similar thinking in at least one US case.[137]

Under the CCS' *Guidelines on the Treatment of Intellectual Property Rights*, the essential facility notion is addressed in relation to refusals to supply a licence. The guidelines do not specifically address pharmaceutical patent rights and whether such patents would represent "essential facilities". However, a number of points are clear from them. First, the guidelines take the stance that IPR ownership does not impose an obligation to licence on the part of the IPR owner. Second, while a refusal to licence, even by a dominant undertaking, "is not normally an abuse" (referring to Section 47), in "limited circumstances", such a refusal by a

[135] *Ibid.*, paras. 21–24.

[136] Decision available at: www.ca8.uscourts.gov/opndir/11/08/103458P.pdf. The FTC has since this decision filed a petition for rehearing *en banc* in October 2011, challenging, *inter alia*, the Court's reliance on Lundbeck's economist's view on the cross-elasticity of the two drugs in question; see FTC, at www.ftc.gov/os/caselist/0810156/111003lundbeckpetition.pdf. See generally Anish Vaishnav, Product Market Definition in Pharmaceutical Antitrust Cases: Evaluating Cross-Elasticity of Demand, *Colum. Bus. L. Rev.* (2011), 586, available at: http://cblr.columbia.edu/archives/11925.

[137] *Biovail Corp.*, FTC Docket No. C-4060 (2 October, 2002) (Complaint, paras. 18–20 and 22). See also FTC statements at: http://ftc.gov/opa/2002/06/biovailelan.shtm. See M. Howard Morse, Product Market Definition in the Pharmaceutical Industry, *Antitrust Law Journal*, Vol. 71, No. 2 (2003), at p. 650, citing this. See also CRAI Memo, at: www.crai.com/ecp/assets/Market_Definition_Pharma.pdf. For a summary of US antitrust pharmaceutical cases up to 2005, see: http://apps.americanbar.org/antitrust/committees/intell_property/course-material/ip-spring05-cm.pdf.

dominant undertaking may constitute an infringement under Section 47. In this connection, the guidelines mention as an example a refusal concerning an IPR which "relates to an essential facility, with the effect of (likely) substantial harm to competition".[138] Third, a facility is considered "essential" only if "there are no potential substitutes (through duplication or otherwise)" and if it is "indispensable to the exercise of the activity in question". The guidelines further hint of a reluctance to find an essential facility in most cases, stating that it would be "rare in practice" and that "IPRs by themselves are generally unlikely to create essential facilities".

The essential facility doctrine has not been explicitly linked to pharmaceutical patents or pharmaceutical test data in the legislation or its guidelines, nor has it been explored in case law. However, it is clear from that there is no *a priori* exclusion of refusals to licence pharmaceutical patents, or "pay-for-delay" agreements, from scrutiny under the Competition Act. This leaves open the possibility, at least for now, to argue in favour of application of the essential facilities doctrine to pharmaceutical patents and, arguably, data – such as clinical test data – arising from research behind the patented product or process, as such data could form the basis of others' further research and innovative work. Such an argument would not be without obstacles. First, in relation to the balance between promoting innovation and competition, the guidelines on IPRs underscore that in assessment of refusals to licence under Section 47, technology and innovation markets will be considered, and that "[c]are must be taken not to undermine the incentives for undertakings to make future investments and innovations". While the language of the guidelines leaves open the possibility that the "essential facility" argument may be explored in the present context, it is not clear if the CCS (or a Singapore court) will be prepared to make such a finding. Hence, an argument of essential facility as argued before the South African Competition Commission in *Hazel Tau* may or may not succeed before the CCS. Finally, there is another important factor in the analysis under the Competition Act of Singapore, that of "net economic benefit". Under the Competition Act, the following factors are to be taken into account in evaluating

[138] The CCS may here consider if the dominant undertaking is "able to objectively justify its conduct", and whether it has "behaved in a proportionate way in defending its legitimate commercial interest".

such "net economic benefit".[139] This exclusion would likewise apply to abuse of dominant position complaints under Section 47, by virtue of Section 48.

While agreements or actions otherwise falling within Sections 34 and 47 which tend to reduce availability of pharmaceuticals production or distribution will not satisfy (a), they may well satisfy (b) given the broad language, "promoting technical or economic progress". The context of the pharmaceutical and biomedical industries in Singapore and the government's heavy emphasis on their development may play a part in determining if (b) is met. However, the sub-paragraphs in turn provide important limitations to these broad criteria, and an agreement that fails on either (i) or (ii) could still be subject to Sections 34 and 47. Section 35 read with the Third Schedule, however, provides a large number of exemptions from s. 34. A broad ground under the Third Schedule for exemption from s. 34 is, for example, "where there are exceptional and compelling reasons of public policy" why it should not apply to a particular agreement. Another covers "vertical agreements" which are exempted by the Minister. Finally, agreements "with net economic benefit" – another broad ground – may be exempted.[140] Section 36 also permits the minister to make provision for

[139] Para. 9 of the Third Schedule of the Competition Act provides as follows:

**Agreements with net economic benefit**

9. The section 34 prohibition shall not apply to any agreement which contributes to –
   (a) improving production or distribution; or
   (b) promoting technical or economic progress, but which does not –
      (i) impose on the undertakings concerned restrictions which are not indispensable to the attainment of those objectives; or
      (ii) afford the undertakings concerned the possibility of eliminating competition in respect of a substantial part of the goods or services in question.

Para. 4 of the schedule also permits the minister to exclude an agreement from s. 34 if there are "exceptional and compelling reasons of public policy".

[140] To qualify as such an agreement, it must be an agreement that contributes to:

(a) improving production or distribution; or
(b) promoting technical or economic progress, but which does not –
   (i) impose on the undertakings concerned restrictions which are not indispensable to the attainment of those objectives; or
   (ii) afford the undertakings concerned the possibility of eliminating competition in respect of a substantial part of the goods or services in question.

block exemptions extending to any agreement which contributes to the same factors as those above. An example of an agreement which may in fact promote competition, or at least not inhibit it, would be a patent pool agreement. The objective of such an arrangement is to permit parties to negotiate prices of patented drugs in the "pool" for more affordable distribution. An example of a patent pool is the Medicines Patent Pool, which aims to ensure more affordable prices for medicines for treating HIV.[141]

At the very least, one may observe that the "net economic benefit" provision in para. 9 of the Third Schedule also does *not* exclude *a priori* pay-for-delay or similar agreements from the purview of Section 34. On the contrary, the guidelines point to factors that may lead to both a finding that there are no efficiency gains as required, and a finding of possible elimination by such an agreement of a substantial part of the goods in question (e.g. generic medicines) – which may be the very reason for the agreement. While a number of CCS decisions offer some insight into the interpretation and application of the net economic benefit requirements, these have not been related to the pharmaceutical industry.[142]

Finally, it is important that the strong promotion of use of ADR to settle disputes (to reduce litigation where appropriate) in Singapore[143]

See the Third Schedule of the Competition Act. In interpreting "net economic benefit", the CCS has stated in its *Qantas and British Airways* decision: "any objective benefits resulting from the Agreement to outweigh and compensate for any detriments to competition"; 2009 Decision.

141 See the organization's website at: www.medicinespatentpool.org/; it offers a pithy summary of how the arrangement works:

"The Pool acts as a 'one-stop shop' for patent holders, manufacturers of generic drugs or other organisations interested in engaging in research and development. This means it 'pools' multiple patents related to HIV medicines in one place, which are then licensed out by the same one entity, in order to cut down on transaction costs for all parties involved." The pooling brings prices down as well as allows earlier access to medicines, well before their patents expire. See also Chapter 7.

142 See, for example, *Decision in Qantas and British Airways Restated Joint Services Agreement*, CCS 400/002/06, *Decision in Qantas and Orangestar Cooperation Agreement*, CCS 400/003/06, and *Decision in Japan Airlines International Co. Ltd. and American Airlines, Inc., Alliance Agreement and Joint Business Agreement*, CCS/400/008/10.

143 See Part VIII, Economic Review Sub-Committee on Service Industries, Report of the Working Group (Legal Services), available at: www.mti.gov.sg/ResearchRoom/Documents/app.mti.gov.sg/data/pages/507/doc/ERC_SVS_LEG_MainReport.pdf (accessed 5 May 2015).

Indeed, in *FTC v Actavis*, too, the Supreme Court had to weigh the merits of settlement agreements against broader interests; see pp. 12–15. Specifically, it said:

should not cloud the assessment of whether particular settlement agreements in the pharmaceutical sector are potentially anti-competitive.

### *6.8.2 Thailand*

Thailand's anti-competition legislation takes the form of its Trade Competition Act.[144] A well-known complaint initiated by health activists was made in 2007 to the Thai Competition Commission under this act against Abbott for its withdrawal of several of its drugs from the Thai market – reportedly, the withdrawal was in retaliation for Thailand's issue of a non-voluntary licence for one of Abbott's drugs, known as Kaletra.[145] The complaint alleged this to be an abuse of dominant position but the complaint failed.[146] The Thai Commission rejected this complaint apparently on the ground that Abbott could not have abused a dominant position with respect to those drugs, as its sales of the drugs fell below the legislated threshold under the act.[147] This raises a problem since a drug that is deliberately withdrawn or withheld from a market will naturally have low or no sales levels for purposes of such a threshold, yet the withdrawing or withholding party might be in a dominant position *vis-à-vis* earlier versions of the drug.

> We concede that settlement on terms permitting the patent challenger to enter the market before the patent expires would also bring about competition, again to the consumer's benefit. But settlement on the terms said by the FTC to be at issue here – payment in return for staying out of the market – simply keeps prices at patentee-set levels, potentially producing the full patent-related $500 million monopoly return while dividing that return between the challenged patentee and the patent challenger. *The patentee and the challenger gain; the consumer loses.*
>
> [p. 15, italic emphasis added]

[144] B.E. 2542 (1999). Proposed amendments to the Act to include state-owned enterprises were delayed in October 2015; see *Trade Competition Faces Another Delay*, 6 October 2015, The Nation, http://www.nationmultimedia.com/business/TRADE-COMPETITION-ACT-FACES-ANOTHER-DELAY-30270259.html (accessed 7 October 2015).

[145] See UNCTAD, *Recent Important Competition Cases Involving More Than One Country*, TD /B/COM.2/CLP/71, 29 May 2008, at p. 3. See also Chapter 8.

[146] See Tu Thanh Nguyen, footnote 68 above, at pp. 199–200, and Thanitcul Sakda and Matthew Lim Braslow, Compulsory Licensing of Chronic Disease Pharmaceuticals in Thailand, *The Thai Joumal of Pharmaceutical Sciences*, Vol. 37, No. 2 (April–June 2013), at p. 75. Sakda and Braslow questioned the withdrawal by Abbott of the drugs as they appeared to have been intended for developing countries (at least those in Africa); see their footnote 34 and accompanying text.

[147] *Ibid.*

### 6.8.3 *Vietnam*

In 2005, Vietnam promulgated the Competition Law and implementing regulations.[148] Articles 8–11 deal with agreements that restrain competition. Articles 11–15 deal with abuse of dominance. Articles 16–24 deal with economic concentration. The law provides for an exemption procedure.[149] Importantly, the Vietnam Competition Authority (VCA) has signed memoranda of understanding with other agencies in Vietnam, such as the Inspectorate of the Ministry of Science and Technology on implementing competition law, consumer protection and intellectual property, and with the Inspectorate of the Ministry of Health on implementing competition law in the health sector.[150]

It has been reported that during the avian influenza pandemic which affected Vietnam in 2003–2005, negotiations by the Vietnamese authorities with Roche, the producer of Tamiflu which is used to treat the disease, met initially with a refusal by Roche to license local production of generic versions of the drug.[151] While this matter appears to have been handled, due to its urgency, from the perspective of compulsory licensing, it is not known whether and how the competition law of Vietnam could have been utilized in this case.[152]

[148] No. 27/2004/QH11. A task force was set up to look into deficiencies of the law; see its report at www.aseancompetition.org/files/documents/content/articles_publications/129-review-report-vietnam-competition-law_1379910496.pdf; and see *Asia Pacific Competition Update: OECD/Korea Policy Center Newsletter*, January 2013, Issue 8, at p. 4; www.oecd.org/competition/koreacentrenewsletter.htm (accessed 7 May 2015).

[149] See Section 4 of the law.

[150] See *Asia Pacific Competition Update: OECD/Korea Policy Center Newsletter*, November 2011, Issue 5, at p. 9.

[151] See Tran Viet Dung, Le Thi Nam Giang and Nguyen Nhu Quynh, *Vietnam: A Case Study for Sustainable Technology Transfer*, in ed. Hans Henrik Lidgard, Jeffrey Atik and Tu Thanh Nguyen, *Sustainable Transfer of Technology* (Kluwer Law International, 2012), at pp. 255–260. According to the authors, it was only after threat by the Vietnamese authorities of use of a compulsory license and after further negotiations that a license was subsequently granted by Roche to a limited number of Vietnamese producers; they point out further that this occurred just prior to Vietnam's accession to the WTO, which took effect in January 2007 (p. 259).

[152] The Competition Law prohibits "competition restriction agreements" and abuse of dominant or monopoly positions. Art. 13.6 in particular prohibits firms in dominant positions from preventing new competitors from entering the market. It does not refer to refusals to license or to the essential facility notion; see text of the law at footnote 148 above. See generally Tu Thanh Nguyen, Chapter 4, *Technology Transfer and Competition Law: Options for Developing Countries*, footnote 68 above; and for a discussion of compulsory licensing in Vietnam, see Chapter 11, Tran Viet Dung, Le Thi Nam Giang and Nguyen Nhu Quynh, *Vietnam: A Case Study for Sustainable Technology Transfer*, in

### 6.8.4 Elsewhere in Asia

China's current Antimonopoly Law came into force in 2008.[153] Art. 55 of the Antimonopoly Law addresses the scrutiny over exercise of IPRs.[154] The State Administration of Industry & Commerce (SAIC), which administers the law, issued guidelines applicable to IPRs in 2014.[155] The pharmaceutical sector has been reported as being one of six targeted sectors for competition enforcement action.[156] China has begun to enforce the AML against pharmaceutical-sector entities, including for price fixing of raw materials for pharmaceuticals.

In April 2015, the SAIC issued rules on abuse of IPRs, which will come into force on 1 August 2015.[157] Under Article 7, for IPRs which form

*Sustainable Transfer of Technology*, ed. Hans Henrik Lidgard, Jeffrey Atik and Tu Thanh Nguyen, (Kluwer Law International, 2012).

153 Adopted at the 29th meeting of the Standing Committee of the 10th National People's Congress of the People's Republic of China on 30 August 2007, and came into force as of 1 August 2008; see Chinese government website, http://english.mofcom.gov.cn/aarticle/policyrelease/announcement/200712/20071205277972.html (accessed 17 April 2015).

154 The law provides:

> This Law does not govern the conduct of business operators to exercise their intellectual property rights under laws and relevant administrative regulations on intellectual property rights; however, business operators' conduct to eliminate or restrict market competition by abusing their intellectual property rights shall be governed by this Law.

Ministry of Commerce, People's Republic of China, http://english.mofcom.gov.cn/aarticle/policyrelease/announcement/200712/20071205277972.html (accessed 8 February 2015).

155 2013 Sixth Draft, *Guidelines on Anti-Monopoly Law Enforcement in Intellectual Property* Rights. See generally, Xiaoye Wang and Adrian Emch, *Five Years of Implementation of China's Anti-Monopoly Law – Achievements and Challenges* (October 2013) J. Antitrust Enf. Other guidelines include the *Guidelines on the Definition of Relevant Market.*

156 *Six Industries Targeted, Including Medicines*, November 2013, South China Morning Post: www.scmp.com/business/economy/article/1365213/beijing-target-antimonopoly-probes-six-industries (accessed 17 April 2015). See also Natasha Khan, *China Probes 60 Drugmakers in Effort to Curb Drug Prices*, Bloomberg, 4 July 2013, www.bloomberg.com/news/articles/2013-07-04/china-probes-60-drugmakers-to-keep-health-expenses-in-check-1- (accessed 17 April 2015).

157 关于禁止滥用知识产权排除,限制竞争行为的规定, Order No. 74, 7 April 2015, Chinese text available at www.saic.gov.cn/zwgk/zyfb/zjl/fld/201504/t20150413_155103.html (accessed 17 April 2015). For an unofficial English translation, see *Patent Law Amendments Translation*, Winston & Strawn LLP, 12 April 2015, at *China IP*, http://chinaipr.com/ (accessed 17 April 2015). For a translation to English, see http://www.wipo.int/wipolex/en/text.jsp?file_id=370868 (accessed 13 October 2015). For a useful commentary on the background for the rule and a draft leading to the final rule, see

"essential facilities for business production activity" ("生产经营活动必需设施"),[158] the IP holder who holds a dominant position may not refuse to license on the IPR on reasonable terms.[159] The provision includes three factors to be taken into account in applying this prohibition and it remains to be seen whether and how these will be applied in practice, particularly in the pharmaceutical sector.

Hong Kong is one of the most recent adopters of a comprehensive competition law, in enacting the Competition Ordinance applying to all sectors in 2012, which came into effect in late 2014.[160] Draft Guidelines

Adrian Emch and Liyang Hou, *Antitrust Regulation of IPRs: China's First Proposal*, 14 August 2014, *Competition Policy International*, www.competitionpolicyinternational .com/antitrust-regulation-of-iprs-chinas-first-proposal/ (accessed 17 April 2015).

158 Author's own translation.

159 Article 7 provides:

> 第七条　具有市场支配地位的经营者没有正当理由，不得在其知识产权构成生产经营活动必需设施的情况下，拒绝许可其他经营者以合理条件使用该知识产权，排除、限制竞争。
> 　认定前款行为需要同时考虑下列因素：
>
> （一）该项知识产权在相关市场上不能被合理替代，为其他经营者参与相关市场的竞争所必需；
> （二）拒绝许可该知识产权将会导致相关市场上的竞争或者创新受到不利影响，损害消费者利益或者公共利益；
> （三）许可该知识产权对该经营者不会造成不合理的损害。

The (imperfect) translation of the above available at the WIPO website (*ibid.*) is as follows:

> Operators have a dominant market position without justification shall not constitute a case of its intellectual property and business activities necessary for the production facility, refusal to license on reasonable terms to other operators to use the intellectual property rights to eliminate or restrict competition. Behavior identified in the preceding paragraph need to consider the following factors:
> (a) the intellectual property rights can not be a reasonable alternative in the relevant market, to participate in the relevant market for other operators necessary to compete;
> (b) the refusal to license the intellectual property rights will lead to the relevant competition or innovation in the market to be adversely affected, harm the interests of consumers or the public interest;
> (3) to authorize the intellectual property of the operator will not cause unreasonable harm.

See also Article 4(5), Order No. 54, 2010, available in Chinese at www.wipo.int/edocs/lexdocs/laws/zh/cn/cn364zh.pdf (accessed 17 April 2015).

160 See www.compcomm.hk/en/competition_ordinance.html (accessed 7 February 2015).

have also been proposed, and the final report after the receipt of submissions is expected in the near future.[161]

### 6.8.5 *Japan*

Japan's Antimonopoly Act does not explicitly deal with refusals to deal in the pharmaceutical patent context.[162] Refusals to trade in general are addressed in guidelines, specifically in the Designation of Unfair Trade Practices.[163] In 2003, a study report had recommended a review of the role of essential facility doctrine to provide "prompt and effective action" in respect of conduct which prevents market entry where essential facilities exist.[164]

### 6.8.6 *India*

India's competition law[165] expressly deals with refusals to deal. Section 3(4) includes a refusal to deal as a potentially anti-competitive agreement if it

[161] See www.compcomm.hk/en/draft_guidelines_2014.html (accessed 7 February 2015).

[162] The text of the act is available at: www.jftc.go.jp/en/legislation_gls/amended_ama09/index.html (accessed 5 May 2015). See also *Joint Comments of the American Bar Association's Section of Antitrust Law, Section of Intellectual Property Law and Section of International Law and Practice on the Report of the Study Group on the Antimonopoly Act of Japan*, 30 January 2004.

[163] Fair Trade Commission Public Notice No. 15 of June 18, 1982, as revised in 2009.

[164] Japan Fair Trade Commission, *Report of the Study Group on the Antimonopoly Act*, 28 October 2003.

[165] Section 4 of India's Competition Act 2002 prohibits an abuse of dominance which includes limiting or restricting

(i) production of goods or provision of services or market therefor; or
(ii) technical or scientific development relating to goods or services to the prejudice of consumers; or
(c) indulges in practice or practices resulting in denial of market access [in any manner]; or
(d) makes conclusion of contracts subject to acceptance by other parties of supplementary obligations which, by their nature or according to commercial usage, have no connection with the subject of such contracts; or
(e) uses its dominant position in one relevant market to enter into, or protect, other relevant market.

For an in-depth discussion of India's competition law and its pharmaceutical sector, see *Competition Law & the Pharmaceutical Industry*, Centre for Trade and Development (2010): www.cci.gov.in/images/media/completed/PharmInd230611.pdf (accessed 17 April 2015).

"causes or is likely to cause an appreciable adverse effect on competition in India". The *Explanation* statement provides:

> "refusal to deal" includes any agreement which restricts, or is likely to restrict, by any method the persons or classes of persons to whom goods are sold or from whom goods are bought.[166]

As India only passed patent legislation protecting product patents in 2005, anti-competition case law dealing with such patents for pharmaceuticals is only now emerging. As has been explained in Chapter 3, India's Supreme Court recently dealt with the interface between competition and pharmaceutical patents in the context of a compulsory licence granted for non-working of a patent. There has so far not been any decision by the Court on agreements found elsewhere, such as pay-for-delay agreements, under the competition legislation.[167] In a price-related case, however, the Competition Commission recently fined the Bengal Chemists and Druggists Association for fixing prices in supplying essential medicines.[168]

#### 6.8.6.1 Developments in the IPR-competition interface in ASEAN

As mentioned, the IP-competition interface is relatively new in ASEAN countries, given that some members are just in the process of introducing

[166] Under the guidance booklet on *Abuse of Dominance* issued by the Indian Competition Commission, for the essential facility doctrine to be invoked (including in relation to IPRs):

- The facility must be controlled by a dominant firm in the relevant market
- Competing enterprises/persons should lack a realistic ability to reproduce the facility
- Access to the facility is necessary in order to compete in the relevant market; and
- It must be feasible to provide access to the facility.

See http://competitioncommission.gov.in/advocacy/Booklet_AbuseOfDominance 11032008.pdf (accessed 23 April 2014). There is discussion on whether the essential facilities principle applies in India; see, for example, Jaivir Singh, *Is There a Case for Essential Facilities Doctrine in India?*, circ.in/pdf/Essential_Facilities_ Doctrine_India. pdf; and 2010 Centad Study. In Pakistan, the essential facilities doctrine has been applied; see *Asia Pacific Competition Update: OECD/Korea Policy Center Newsletter*, November 2014, Issue 12, at p. 13.

[167] See generally Abhimanyu Ghosh and Deep Chaim Kabir, Balance of Competition and Intellectual Property Laws in the Indian Pharmaceutical Sector, *JIPR*, Vol. 12 (May 2007), 293–302.

[168] Katy Oglethorpe, CCI Fines Pharma Association, *Glob. Comp. Rev.* (14 March 2014).

basic competition laws. The relationship between these two areas of law in these national systems is still evolving and shows a great deal of non-uniformity at this point.[169] Singapore, which introduced competition legislation in the early 2000s, has also established guidelines on IPRs, but there are presently no specific rules addressing, for example, pharmaceutical industry pay-for-delay or settlement agreements that have arisen in case law elsewhere. In Indonesia, competition law generally does not apply to IPR agreements, due to a legislative exemption.[170] However, it appears that a more specific guideline addresses IPR and competition law.[171] In the Philippines, anti-competitive IP agreements and related compulsory licensing are governed by the IP Code of 1998.[172]

The ASEAN Regional Guidelines on Competition Policy expressly provide that, in dealing with abuse of dominance, member states may deal with refusals to deal, setting out the following as part of a possible illustrative list of such conduct:

> 3.3.2.4 Limiting production, markets or technical development to the prejudice of consumers (*e.g.*, restricting output or illegitimate refusal to supply, restricting access to/use of/ development of a new technology).[173]

### 6.8.7 *Anti-competitive conduct and compulsory licensing*

Apart from permitting voluntary licences, the Singapore Patents Act also provides for the grant of compulsory licensing of patented products under specific circumstances. The scope of compulsory licensing is relatively narrow under Singapore law, and a number of requirements

[169] See generally materials on ASEAN countries in this regard, WIPO Roundtable on Intellectual Property and Competition Law, 14–16 September 2011.

[170] Art. 50(b), Law No. 5 Year 1999 Concerning Prohibition on Competition, exempts "[a]greements related to intellectual property rights, such as licenses (of) patents, trademarks, copy rights, industrial product design, integrated electronic circuits and trade secrets as well as agreements related to franchise"; see A. Junaidi, Interaction between IP and Competition Law in Indonesia, www.wipo.int/edocs/mdocs/mdocs/en/wipo_rt_ip_sin_11/wipo_rt_ip_sin_11_ref_indonesia.pdf (accessed 8 August 2014).

[171] Commission Regulation No. 2 Year 2009 Concerning Implementation of Article 50 (b); *ibid.*

[172] See Antonio A. R. Mendoza, Interaction between IP and Competition Law in Philippines, www.wipo.int/edocs/mdocs/mdocs/en/wipo_rt_ip_sin_11/wipo_rt_ip_sin_11_ref_philippines.pdf (accessed 8 August 2014).

[173] ASEAN Competition Law and Policy portal, www.asean.org/images/2012/publications/ASEAN%20Regional%20Guidelines%20on%20Competition%20Policy.pdf (accessed 8 February 2015).

must be satisfied. Elsewhere, compulsory licensing has taken on policy and practical significance by being used as another regulatory tool to ensure affordability of medicines.[174] To date, Singapore has not issued any compulsory licence under its current law. In addition, although the WTO "paragraph 6" decision with regard to compulsory licensing permits the use of such licences to produce medicines in certain circumstances in aid of fellow WTO members without production capacity, Singapore joined a number of other WTO members to state that she would not make use of this waiver to import using compulsory licences, except in situations of national emergency or other circumstances of extreme urgency.[175] The compulsory licensing options under Sections 55 and 56 have not been exercised so far in Singapore. However, elsewhere, they have been utilized in order to provide more affordable medicines.[176]

[174] For example, in 2012, India exercised the right to grant a compulsory license for the first time, leading to challenges by various pharmaceutical companies. In March 2012, Bayer AG was ordered by the Indian Patent Controller to issue a compulsory licence for its patented drug Nexavar to Natco Pharma Ltd., an Indian generics drug manufacturer. Bayer AG lodged an appeal against this order in May 2012, and the appeal was dismissed in September 2012 by the Indian Intellectual Property Appellate Board (IPAB).[173] In another 2012 decision by the IPAB, the measure was found to comply with India's Patents Act and therefore legal. In early September 2012, a New Delhi court ruled against the plaintiff Roche that while its patent over an anti-cancer drug, Tarceva, was valid, the sale of a generic version, Erlocip, by an Indian generics company, Cipla Ltd., did not infringe Roche's patent. A separate case brought by Novartis in relation to its patented drug Glivec is being heard by the Indian Supreme Court. A key consideration arising under the *Bayer* decision is whether the medicine in question was being made "available to the public at a reasonably affordable price" under s. 84 of the Indian Patents Act. (See also the discussion of this case in Chapter 5.)

[175] See WTO website at: www.wto.org/english/tratop_e/trips_e/public_health_faq_e.htm.

[176] Under s. 55 of the Singapore Patents Act, an application for grant of a compulsory license may be made under the following circumstances:

5. – (1) Any interested person may apply to the court for the grant of a licence under a patent on the ground that the grant of the licence is necessary to remedy an anti-competitive practice.

(2) Without prejudice to the generality of subsection (1), the court may determine that the grant of a licence is necessary to remedy an anti-competitive practice if –

(a) there is a market for the patented invention in Singapore;
(b) that market –
  (i) is not being supplied; or
  (ii) is not being supplied on reasonable terms; and
(c) the court is of the view that the proprietor of the patent has no valid reason for failing to supply that market with the patented invention, whether directly or through a licensee, on reasonable terms . . .

A key inquiry under Section 55(2) of Singapore's act relates to the meaning of "on reasonable terms". This raises the issue of whether supply by the patent holder of the patented medicine – where such supply exists – is "on reasonable terms".

By comparison, under Section 85 of the Indian Patents Act, the relevant criterion is whether drugs are being made available at "reasonably affordable prices". Section 85 of the Indian Patents Act does not appear to relate the prices to a relevant market but instead requires consideration more generally as to whether the patented invention is "available to the public at a reasonably affordable price".[177] Further guidance on this is provided in Section 90 of that act.[178]

[177] See, for example, the case of *NATCO v Bayer*, footnote 174 above; see also discussion of this case in Chapter 5.

[178] Section 90 provides as follows:

"90. When reasonable requirements of the public deemed not satisfied

For the purposes of sections 84, 86 and 89, the reasonable requirements of the public shall be deemed not to have been satisfied –

(a) if, by reason of the default of the patentee to manufacture in India to an adequate extent and supply on reasonable terms the patented article or a part of the patented article which is necessary for its efficient working or if, by reason of the refusal of the patentee to grant a licence or licences on reasonable terms, –
  (i) an existing trade or industry or the development thereof or the establishment of any new trade or industry in India or the trade or industry of any person or classes of persons trading or manufacturing in India is prejudiced; or
  (ii) the demand for the patented article is not being met to an adequate extent or on reasonable terms from manufacture in India; or
  (iii) a market for the export of the patented article manufactured in India is not being supplied or developed; or
  (iv) the establishment or development of commercial activities in India is prejudiced; or

(b) if, by reason of conditions imposed by the patentee (whether before or after the commencement of this Act) upon the grant of licences under the patent, or upon the purchase, hire or use of the patented article or process, the manufacture, use or sale of materials not protected by the patent, or the establishment or development of any trade or industry in India, is prejudiced; or

(c) if the patented invention is not being worked in India on a commercial scale to an adequate extent or is not being so worked to the fullest extent that is reasonably practicable; or

(d) if the demand for the patented article in India is being met to a substantial extent by importation from abroad by –
  (i) the patentee or persons claiming under him; or
  (ii) persons directly or indirectly purchasing from him; or
  (iii) other persons against whom the patentee is not taking or has not taken proceedings for infringement; or

Under the Singapore Patents Act, there are two main types of compulsory licensing available and they are subject to certain limitations.

The first is compulsory licensing provided for under Part XI. The other is provided under Part XII and is for "services of the government".

The former, governed by Section 55 and permitted under Article 31(k) of the TRIPS Agreement, applies to situations where a compulsory licence is granted "on the ground that the grant of the license is necessary to remedy an anti-competitive practice".[179] The act does not define "an anti-competitive practice", nor does it expressly link it to such practices as understood under the Competition Act. The language in Section 55 appears to simply reflect the obligation of Singapore under Article 16.7(6) of the USSFTA (which does not define "anti-competitive practices").[180]

Section 9A of the Interpretation Act would require an interpreter of s. 55 to apply the "purposive approach" and would permit him to take into account material not being part of the written law in question but capable of assisting in interpretation of the statutory language.[181]

> (e) if the working of the patented invention in India on a commercial scale is being prevented or hindered by the importation from abroad of the patented article by the patentee or the other persons referred to in the preceding clause.

179 Section 55(2) does provide an instance where the grant of a compulsory license for an anti-competitive practice may be justified, but it is "without prejudice to the generality of subsection (1)". Section 55(2) addresses the following specific situations:

(a) where there is a market for the patented invention in Singapore;
(b) that market –
  (i) is not being supplied; or
  (ii) is not being supplied on reasonable terms; and
(c) the court is of the view that the proprietor of the patent has no valid reason for failing to supply that market with the patented invention, whether directly or through a licensee, on reasonable terms.

180 6. Neither Party shall permit the use of the subject matter of a patent without the authorization of the right holder except in the following circumstances:

> (a) *to remedy a practice determined after judicial or administrative process to be and anticompetitive under the competition laws of the Party* . . .
>
> [Italic emphasis added, original footnotes omitted]

181 S. 9A(1)–(4), Interpretation Act, Cap. 1. Use of the material referred to in s. 9A(2) and (3) is subject to s. 9A(4), which provides as follows:

> (4) In determining whether consideration should be given to any material in accordance with subsection (2), or in determining the weight to be given to any such material, regard shall be had, in addition to any other relevant matters, to –

Such material arguably includes the above USSFTA provision. If we take "anti-competitive practices" in Section 55 to refer to those under the Competition Act (as the words in italics above in the USSFTA provision must refer to this act as being part of "the competition laws of" Singapore), these would include the acts prohibited under Sections 34 and 47 of the Competition Act.[182] In theory, therefore, the Competition Act may be drawn on to identify such "practices" for the purpose of permitting compulsory licences to correct them, under Section 55 of the Patents Act.

Hence, a patent holder's unilateral practices may fall under the scrutiny of Sections 34 or 47, and they may potentially be remedied by the grant of a compulsory licence under Section 55 of the Patents Act.

A further question is whether Article 31(k) of the TRIPS Agreement is to be interpreted to permit one WTO member to *export* a patented pharmaceutical under a compulsory licence where *another member's* authorities have found that the patentee has acted in an anti-competitive manner.[183]

(a) the desirability of persons being able to rely on the ordinary meaning conveyed by the text of the provision taking into account its context in the written law and the purpose or object underlying the written law; and
(b) the need to avoid prolonging legal or other proceedings without compensating advantage.

182 These are however subject to the exclusions in the Third Schedule of the Competition Act.

183 See, for example, the discussion in Burton Ong, *Compulsory Licences of Pharmaceutical Patents to Remedy Anti-Competitive Practices under Article 31(k) of the TRIPS Agreement: Can Competition Law Facilitate Access to Essential Medicines?*, at para. 4.2.2, in R. Hilty and K. Liu, eds., Compulsory Licensing: Practical Experiences and Ways Forward, *Max Planck Institute Studies on Intellectual Property and Competition Law*, Vol. 22 (Heidelberg: Springer, 2014), 235–265. The present author is of the view at least under s. 55 read with s. 66 of the Singapore Patents Act, Singapore may not so export a pharmaceutical product, but may only grant a compulsory license under s. 55 sale or use of such a product within Singapore. While s. 55(1) is broad and does not refer merely to domestic sale or use within Singapore, the exemptions from infringement under s. 66 do not include the export of a product under such a s. 55(1) compulsory license from Singapore. By contrast, under the provisions implementing the Doha Declaration Implementation Decision (the "Paragraph 6" mechanism explained earlier), the government (or a party authorized by it) may "do anything" with a patented invention (presumably within Singapore):

(a) for a public non-commercial purpose; or
(b) for or during a national emergency or other circumstances of extreme urgency.

Under s. 55(2), use of an invention *outside* Singapore in pursuance of any agreement or arrangement between Singapore and for the purpose of *defence* of that other country is

### *6.8.8 What will be the role for competition law and policy in the pharmaceutical sector in Asia?*

Three key observations may be made at this time. First, there will be no uniform approach for some time on how to regulate the practices and agreements of the pharmaceutical industry. Indeed, a threshold question is whether they should not be treated like any other industry.

Second, in considering such questions, Asia has the opportunity to draw important lessons from more mature competition law regimes such as those in the US and the EU, while noting any relevant, specific market conditions in Asia that may differ.

Third, it is likely that pay-for-delay and other practices that have already been addressed in these jurisdictions will begin to emerge in Asia, and will need to be addressed. Indeed, as mentioned, Korea has already recorded its first such case.

The relationship between competition law and policy and the pharmaceutical sector in Asian countries is at a stage of infancy. Newer laws in ASEAN countries, such as that in Philippines, which contain provisions for "socialized" pricing arrangements, while recognizing the importance of innovative endeavour and the need to protect intellectual property rights, offer interesting options.[184] This nascence offers a great opportunity to assess how that relationship could be shaped to ensure a balance between competition and innovation, and to put in place laws with clarity and certainty. For many of these states, the pressures to have a competition law may turn out to be a blessing, providing a needed tool to deal with conduct which may be emerging in the growing Asian pharmaceutical sector.[185]

For countries in Asia whose laws require patent linkage without any "Hatch-Waxman" companion mechanism and mechanisms to address

deemed to be "for a public non-commercial purpose" and therefore not an infringement under s. 56(1).

(Further, s. 66(5A) excludes the exemptions in s. 66(2)(g) and (i) for import, sale or offers to sell "any relevant health product produced for export to any country, other than Singapore, which is an eligible importing member of the World Trade Organization". This language was inserted in 2008 is aimed at preventing diversion of such pharmaceutical products into Singapore; see parliamentary statements of the then Senior Minister of State for Law, 25 August 2008.)

184 See text accompanying note 109, above, and generally, sections 15, 24 and 26 of the Philippine Competition Act.

185 See generally Jonathan Galloway, *Driving Innovation: A Case for Targeted Competition Policy in Dynamic Markets*, in *World Competition Law and Economic Review*, ed. José Rivas (Kluwer Law International, 2011, Vol. 34, No. 1), pp. 73–96.

patent holders' conduct which amounts to an abuse of litigation process, etc., the role of competition laws will be even more critical in ensuring healthy competition among pharmaceutical patent holders and subsequent manufacturers of competitive products. Possibly, therefore, competition law may take on an increasing role in the sphere of the pharmaceutical sector to serve to protect the interests of consumers.

## 6.9 Conclusion

As a number of Asian jurisdictions have been working to put in place comprehensive competition laws, divergences in implementation and interpretation can be expected, even though certain common concepts can be found in each system. For ASEAN, member states' move towards enacting such competition laws is linked to the strong economic integration drive in the region. This has presented members with an opportunity to harmonize their laws, and initiatives such as its Regional Guidelines help to promote such harmonization. The guidelines however leave a number of more specific areas to the member states. One of these is how they are to deal effectively with the interface with pharmaceutical patent holders' practices/agreements and competition law. Elsewhere, such as in the US and the EU, this interface has already been keenly debated and litigated, while this is a relatively unexplored question in Asian/ASEAN legal systems.

For Asia, the US Supreme Court has taken the position in *Actavis* that a rule of reason approach applies with respect to the application of anti-competition law to pharmaceutical sector settlement agreements. This strikes a middle ground between two extremes of denying application of anti-competition law to such agreements altogether, and outlawing all such agreements. It presents Asian authorities with an attractive example to follow, given its compelling attraction in its flexible yet not completely unfettered intervention *vis-à-vis* such agreements.[186] The emergence of the first pay-for-delay-agreement litigation case in Korea signals that such agreements may be an emerging issue in the region. Given that many jurisdictions do not require the reporting of patent settlement

[186] In the case of Singapore, the 1993 *White Paper on Affordable Healthcare* set out five key pillars including *intervention* to keep healthcare costs down *where the market fails to do so*; the White Paper is available at: www.moh.gov.sg/content/moh_web/home/Publications/Reports/1993/affordable_healthcare.html (accessed 7 May 2015). See also speech by the Minister of Health in 2012, explaining affordability of healthcare under Singapore's HealthCare 2020 plan.

agreements or pay-for-delay agreements, detection is an issue. Even then, agreements that come to light may or may not be anti-competitive, and, if the *Actavis* approach were to be adopted, a rule of reason analysis would have to be undertaken by the authorities.

The doctrine of essential facilities has not, to this author's knowledge, yet been canvassed in relation to pharmaceutical entities' abuse of dominance. Indeed, the concept may require a redefinition in the context of pharmaceutical products and their markets in order for it to apply. A key question is *how* this notion should be adjusted in this context to better define what amount to essential facilities *vis-à-vis* pharmaceutical products and their associated IPRs, such as patent rights, and proprietary clinical test data.

Finally, competition law may act as an important check on anti-competitive conduct in the pharmaceutical industry and to a limited extent act as a counterbalance to various TRIPS-plus provisions that may disfavour generic drug producers, but it will not entirely level the playing field. This leads us to the next chapter, in which ideas for a new construct are discussed.

# 7

## Setting new directions

### The need for innovative legal and policy solutions

### 7.1 The need for new paradigms and strategies

Important shifts have been occurring in the pharmaceutical arena, whether in business or in the medical field. On the business front, patent expiries and global changes in demand and supply of drug raw materials and products, and changes in procurement and consumer attitudes have played a part in causing the private sector to review business models, strategies and marketing.[1] In the regulatory area, on the other hand, increasing medical costs, aging populations with increasing healthcare needs, budgetary constraints and trade and investment pressures form part of the pool of issues faced by governments in dealing with health- and pharmaceutical-related policies. Some have even argued that there is a new "countervailing transnational legal order" to safeguard social and economic rights in the access to medicines arena.[2]

Increasingly stringent TRIPs-plus requirements being imposed in various bilateral and regional treaties may not be sustainable, in terms of balancing developing countries' regulatory space (*vis-à-vis* access to pharmaceuticals) and innovation aspirations. Some already doubt that *TRIPS itself* has spurred innovation.[3] The relevant, further question for present purposes is whether this trend of TRIPS-plus FTAs indeed spurs innovation as well.

Various proposals and initiatives have been put forward to address aspects of what are seen to be the shortcomings of today's trends. Many of

[1] For an insightful overview, see Michael Edwards, *R&D in Emerging Markets: A New Approach for a New Era*, McKinsey Insights and Publications. See also *Pharma 2020* reports, PWC.

[2] Gregory Shaffer and Susan K. Sell, *Transnational Legal Ordering and Access to Medicines*, in *Patent Law in Global Perspective*, ed. Ruth L. Okediji and Margo A. Bagley, (Oxford University Press, 2014).

[3] See, for example, David S. Abrams, Did TRIPS Spur Innovation? An Empirical Analysis of Patent Duration and Incentives to Innovate, *U. Pa. L. Rev.*, Vol. 157 (2009), 1613.

these are important and have begun to bring about some degree of positive change. Despite the many important initiatives that exist, the pressures of treaty negotiations and other forces require further changes to bring about more balanced policies and approaches, to ensure even better access to medicines.

The WHO's Global Strategy and Plan of Action on Public Health, Innovation and Intellectual Property (GSPOA) is a major initiative which draws together a number of important threads in the access debate in addressing the needs of developing countries. The following excerpts from it aptly summarize the tenuous position today, despite many new initiatives:

> Member States, the pharmaceutical industry, charitable foundations and nongovernmental organizations have taken initiatives in recent years to develop new products against diseases affecting developing countries and to increase access to existing health products and medical devices. However, these initiatives are not sufficient to surmount the challenges of meeting the goal of ensuring access and innovation for needed health products and medical devices. More efforts should be made to avoid suffering and reduce preventable mortality and to meet the health related Millennium Development Goals and to implement States' obligations and commitments arising under applicable international human rights instruments with provisions relevant to health ...
>
> Intellectual property rights are an important incentive for the development of new healthcare products. This incentive alone does not meet the need for the development of new products to fight diseases where the potential paying market is small or uncertain ...
>
> The price of medicines is one of the factors that can impede access to treatment.[4]
>
> [Original footnote omitted]

A pertinent observation directed at the need for coherent and well-informed policy-making in the supply of health products and services is as follows:

> ... in a complex policy world of the 21st century when universal health coverage was becoming a passionately pursued goal and people's expectations and demands were rising for high quality accessible healthcare, the supply side has to readjust itself in terms of innovation and delivery of healthcare, including health technologies, and, in order to make it possible, the policy makers had no choice but to be better informed, evidence based and affirmative in action ... in order to accomplish that

[4] WHO, 24 May 2008, 61st World Health Assembly, WHA 61.21, pp. 3 and 4.

> task, they had to deal with many policy domains to strike coherent policy solutions to respond to challenging solutions.[5]

While relevant stakeholders are already working together to address various aspects of the access to medicines problem, and will have to do so increasingly,[6] it is clear that no single stakeholder or policy is in a position to solve the issues on access unilaterally. A concerted, coherent framework of cooperation, forged out of constructive and inclusive discussion, is necessary.

As it has been succinctly put:

> When it comes to increasing access to drugs through lower prices, we cannot rely on one single solution ... [w]e need a mix of mutually supportive strategies. Such a mix of strategies needs to be geared to the circumstances of individual countries.[7]

Apart from the evolving role and strategies required for policy-makers, the diversity of national and global "players" required to increase access has also been emphasized:

> While States have the primary responsibility for enhancing access to medicines, it is a shared responsibility in which numerous national and international actors have a role to play.[8]

An example of new strategies and alliances is the collaboration found in public–private partnerships (PPPs).[9] In recent years, many countries have begun to establish specific governmental units to handle PPP

[5] Statement by the WTO representative, para. 85, Report adopted by the Standing Committee on the Law of Patents, WIPO, Nineteenth Session, Geneva, Switzerland, 25–28 February 2013, SCP/19/8, document dated 27 January 2014.

[6] See, for example, Leisinger and Wagner, footnote 75 below, on the need of joint action of stakeholders.

[7] Closing remarks, Dr Gro Harlem Brundtland, then Director-General, World Health Organization, WHO-WTO Workshop on Differential Pricing and Financing of Essential Drugs, 8–11 April 2001, Høsbjør, Norway, text available at WTO website, www.wto.org/english/tratop_e/trips_e/tn_hosbjor_closing_e.htm (accessed 29 October 2014).

[8] *Report of the Special Rapporteur on the Right of Everyone to the Enjoyment of the Highest Attainable Standard of Physical and Mental Health*, Anand Grover, on access to medicines, para. 6, A/HRC/23/42, text available at: www.ohchr.org/Documents/HRBodies/HRCouncil/RegularSession/Session23/A-HRC-23-42_en.pdf (accessed 23 February 2015).

[9] On PPPs, see generally Benedicte Bull and Desmond McNeill, Development Issues in Global Governance: Public-Private Partnerships and Market Multilateralism (Routledge 2007), and especially Chapter 4 on *Global Partnerships for Health: Health for All, or More "Big Pharma"?*

projects.[10] Some recent initiatives to accelerate drug discovery bring together government institutions, private pharmaceutical companies and non-profit organizations to increase efficiency and lower the costs of discovered medicines.[11]

In addition, the call for sharing of benefits of scientific research in relation to healthcare has been included in international documents such as the Universal Declaration on Bioethics and Human Rights and International Declaration on Human Genetic Data.[12]

It is therefore timely to consider paradigm shifts in terms of policy- and treaty-making. The next section begins with a brief "stock-take" of major initiatives that are contributing to such shifts.

## 7.2 A brief inventory of major initiatives

In the quest for affordable essential medicines, a large number of ideas have been proposed. These have been detailed elsewhere; this chapter aims not to duplicate such excellent efforts, but, rather, to provide the background against which the author's suggestions may be considered. This section therefore provides an overview of some major initiatives and proposals which form part of the global legal and policy efforts to address access to medicines.[13]

[10] See, for example, the units and activities in UNESCAP, *PPP Units and Programmes in the Asia Pacific*, May 2014, www.unescap.org/sites/default/files/PPP%20Units%20and%20Programmes%20-%20May%202014.pdf (accessed 12 November 2014). I am grateful to my J.D. Directed Research student, Mr Teo Yi Lun, for drawing my attention to this list.

[11] See, for example, the Accelerating Medicines Partnership, which brings together the NIH, ten biopharmaceutical companies and various research foundations to address diseases such as Alzheimer's disease: www.nih.gov/science/amp/ (accessed 9 December 2014). See also *The Lancet*, Feb 2014, on this partnership.

[12] Article 15 and Article 19, respectively, UN Educational, Scientific and Cultural Organization, texts available at: www.unesco.org/new/en/social-and-human-sciences/themes/bioethics/bioethics-and-human-rights/ (accessed 13 April 2015).

[13] For an overview of the various major proposals and initiatives, see Chapter 3, *Report of the Commission on Intellectual Property Rights, Innovation and Public Health* (CIPIH), WHO, *Public Health, Innovation and Intellectual Property Rights* (2006) (especially the recommendations at Study 4C); and the trilateral study by WTO, WIPO and WHO, *Promoting Access to Medical Technologies and Innovation – Intersections between Public Health, Intellectual Property and Trade*, (2012) at pp. 115–124. See also, generally, Alan Lyles, Creating Alternative Incentives for Pharmaceutical Innovation, *Clinical Therapeutics*, Vol. 28, No. 1 (2006), pp. 126–128. More recent initiatives include the Accelerating Medicines Partnership; see nih.gov/science/amp/index.htm (accessed 10 April 2015) and Accelerating Drug Discovery Editorial, *The Lancet*, Vol. 383 No. 9917 (February 2014).

### 7.2.1 *Treaty-related aspects*

#### 7.2.1.1 *Revisiting the TRIPS conversation: "flexibilities" is not a negative word, and intellectual property rights are not* a priori *anti-access*

The use of TRIPS flexibilities has by now been acknowledged to be endorsed repeatedly in a number of significant international and multi-region documents, including several outside the context of the WTO, from which TRIPS and the Doha Declaration originate.[14] There has also been heightened awareness of "TRIPS flexibilities" from discussions in various multilateral fora.[15] Such awareness is reflected in certain FTAs entered into by Asian countries which have begun to include provisions which affirm the importance of such flexibilities as well as those provided for under the *Doha Declaration on the TRIPS Agreement and Public Health*. Examples are the KORUS[16] and the Korea-Australia FTA (KAFTA).[17] The latter even expressly states that in interpreting and implementing its key patent obligations provision, Article 13.8, the parties "are entitled to rely upon the Doha Declaration".

The conversations on access to medicines which have sprung from TRIPS and TRIPS-plus agreements have, at times, become highly negative, and a more constructive global conversation is required. As an example of a starting point, documents such as the WHO Global Strategy (above) advocate promotion of technological innovation and access of medicines in developing countries without denigrating the role of IPRs. On the contrary, it clearly recognizes that IPRs are "an important incentive in the development of healthcare products", while recognizing the limits of this incentive where a potential paying market is small or uncertain. It also recognizes that "several factors contribute to the price of health products and medical devices . . .", naming as examples such factors as competition, reduction or elimination of import tariffs on such products and services which "can contribute to the reduction of prices". On the other hand, it states that IPRs "do not and should not prevent [WHO] Member States from taking measures to protect public health".

[14] See Chapters 2 and 4.

[15] See, for example, discussions in Jerome Reichmann, Compulsory Licensing of Patented Pharmaceutical Inventions: Evaluating the Options, *J. Law, Med. & Eth.* (2009), 247–263; Kremer and Williams, *Incentivising Innovation: Adding to the Tool Kit*, NBER Innovation & Econ. Pol. 1 (2010); and Musungu and Oh, *The Use of Flexibilities in TRIPS by Developing Countries: Can They Promote Access to Medicines?*, CIPIH (2005).

[16] Art. 18.9(3). [17] Art. 13.10.

A more positive and constructive approach on the part of *all* involved could promote better conversations and collaborations towards solving the problem of access.

#### 7.2.1.2 IPRs, trade, investment, innovation and public health: aiming for greater coherence

Importantly, in such conversations – or international negotiations – the Global Strategy points the way: "[i]nternational negotiations on issues related to intellectual property rights and health should be *coherent* in their approaches to the promotion of public health" [italic emphasis added].

This has important implications for negotiation of, *inter alia*, bilateral, regional and multilateral economic treaties. As will be elaborated upon later in this chapter and in Chapter 8, part of achieving such coherence in international negotiations should involve the active inclusion of health officials in such negotiations, to assess the public health impact of potential IPR or trade-related commitments, and ensuring internal policy coherence by including public health to be factored into all government policy-making.

### *7.2.2 Non-treaty initiatives*

#### 7.2.2.1 Health funds and prizes

Philanthropic and donor funds help to provide finance for medicines in poorer countries, as well as to support voluntary licensing arrangements such as the Medicines Patent Pool (MPP).[18]

Another approach has been to promote the use of "prizes", whether monetary or not in nature, to incentivize pharmaceutical innovation.[19]

[18] The Global Fund for AIDS, Tuberculosis and Malaria was established as a private-public partnership in 2002 and targets helping countries to fight AIDS, malaria and tuberculosis through funding provided by government, private and non-government donors. It provides financing for affordable medicines through its Affordable Medicines Facility. A recent initiative spearheaded by it is the Equitable Access Initiative; see www.theglobalfund.org/en/equitableaccessinitiative/ (accessed 7 May 2015). Separately, the GAVI is a private-public partnership alliance for vaccines. There has also been a proposal for a "Health Impact Fund"; see Amitava Banerjee, Aidan Holis and Thomas Pogge, The Health Impact Fund: Incentives for Improving Access to Medicines, *The Lancet*, Vol. 375, No. 9709 (2010), pp. 166–9.

[19] See James Packard Love, *Alternatives to the Patent System That Are Used to Support R&D Efforts, Including Push and Pull Mechanisms, with a Special Focus on Innovation-Inducement Prizes and Open Source Development Models*, especially at part 2.4 and

#### 7.2.2.2 Voluntary licensing arrangements

Pharmaceutical originator companies are facing increasing pressure from civil society groups, developing countries and international fora to increase their roles in ensuring access to medicines in developing countries. The voices have not gone unheeded, and two main developments demonstrate this. Companies participate in increasing such access in various ways, first by voluntary licensing agreements negotiated with generic producers, and second through the Medicines Patent Pool.

For the present discussion, included under this rubric are licensing agreements and formal and informal arrangements by companies not to assert their patent rights in certain situations (which are akin to forbearance rather than active licensing). These are to be contrasted with non-negotiated, compulsory arrangements such as the grant by national authorities of compulsory licences under their respective laws.

In the licence agreements and non-assertion policies or arrangements, there are several main determinants of their effectiveness with respect to access to medicines. These are the rate of payment required for the licence (if any royalty is required), the geographical territory covered and whether any other onerous or restrictive terms exist.[20]

It has been noted that pharmaceutical companies "are beginning to accept the reality that patent protectionism and continued confrontation with proponents of pro-access policies will not work. They are therefore

Annexes A and C, Study commissioned by the WIPO Secretariat, Committee on Development and Intellectual Property, CDIP/14/INF/12, 19 September 2014; CIPIH Report, footnote 13 above, at pp. 88–89; and Fred M. Abbott and M. N. Graham Dukes, Global Pharmaceutical Policy: Ensuring Medicines for Tomorrow's World (Edward Elgar Publishing Limited, 2009), at pp. 53–56. See also Nobel laureate Professor Joseph E. Stiglitz, *Prizes, Not Patents*, 6 March 2007, Project Syndicate, at: www.project-syndicate.org/commentary/prizes–not-patents (accessed 10 December 2014).

In the US, the Medical Innovation Prize Fund Act was introduced in the House in January 2005 by Senator Bernard Sanders but was not enacted (it was introduced in the Senate in March 2013, read twice and referred to the Committee on Health, Education, Labor, and Pensions); see http://olpa.od.nih.gov/legislation/109/pendinglegislation/medicalinnovation.asp and www.congress.gov/bill/113th-congress/senate-bill/627. A separate Prize Fund for HIV/AIDS Act was also proposed: see www.opencongress.org/bill/s1138-112/show. See also, generally, www.keionline.org/node/1147.

[20] See generally Peter Beyer, *Developing Socially Responsible Intellectual Property Licensing Policies: Non-Exclusive Licensing Initiatives in the Pharmaceutical Sector*, Chapter 10 in ed. Jacques de Werra, *Research Handbook on Intellectual Property Licensing* (Edward Elgar Publishing, 2013). The author is very grateful to Dr Beyer for the very useful discussion in Geneva (notes are on file with author) and for pointing me to this chapter and other WHO materials.

making some concessions to promote access to medicines."[21] There is room therefore for optimism that such arrangements will continue, if not increase and/or expand.[22]

Patent pools for supply of pharmaceutical products to advance access are a relatively recent development.[23] The advantage of patent pools is that they facilitate negotiation of voluntary licences from originator drug companies so that generic production and supply of patented drugs at lower costs to various countries become possible. A disadvantage is that because it is voluntary, originator companies may choose not to join the pool to permit licensing of their patented drugs.

One of the most prominent schemes to increase access of generic medicine producers to produce patented medicines to date is the MPP, established in Geneva under the auspices of the United Nations in 2010.[24] The MPP is financed 100% by UNITAID, with funds derived from levies on airline tickets.[25]

MPP has seen a degree of success given its relative infancy, in the form of licence agreements with ViiV Healthcare (a joint venture of GlaxoSmithKline (76.5%), Pfizer (13.5%) and Shionogi (10%)), Bristol-Myers Squibb, Roche, Gilead Sciences and the US National Institutes of Health,[26] with negotiations being under way with other companies such

[21] Poku Adusei, Exploiting Patent Regulatory "Flexibilities" to Promote Access to Antiretroviral Medicines in Sub-Saharan Africa, *JWIP*, Vol. 14, No. 1 (2011), pp. 1–20, at p. 6.

[22] See also Beyer, footnote 20 above.

[23] Early-twentieth-century patent pools were formed in relation to aeroplanes, sewing machines and radio; see E. Richard Gold and Jean Frédéric Morin, Promising Trends in Access to Medicines, *Global Policy*, Vol. 3, No. 2 (May 2012), 231–237, at p. 235. See generally WHO, *Report of the Commission on Intellectual Property Rights, Innovation and Public Health; Public Health – Innovation and Intellectual Property Rights* (2006), pp. 50–54; and trilateral report by the WTO, WIPO and WHO, *Promoting Access to Medical Technologies and Innovation* (2012) at p. 118.

[24] For information on the origins of the MPP, see MPP website at: www.ip-watch.org/2014/04/24/one-year-on-at-the-medicines-patent-pool-interview-with-greg-perry/ (accessed 1 May 2014) and the Médecins Sans Frontières website at: www.msfaccess.org/spotlight-on/patent-pool (accessed 1 May 2014).

[25] For an overview and insights into the MPP, see interview with the current Executive Director, Greg Perry, at www.ip-watch.org/2014/04/24/one-year-on-at-the-medicines-patent-pool-interview-with-greg-perry/ (accessed 1 May 2014). UNITAID itself was formed in 2006 by the governments of Brazil, Chile, France, Norway and the United Kingdom as the "International Drug Purchasing Facility": see www.unitaid.eu/en/who/about-unitaid (accessed 1 May 2014).

[26] Information from MPP website: www.medicinespatentpool.org/current-licences/ (accessed 1 May 2014). See also report at: www.reuters.com/article/2014/04/01/aids-drug-viiv-idUSL5N0MT39620140401 (accessed 1 May 2014).

as AbbVie and Boehringer Ingelheim.[27] These agreements have enabled the MPP to sign sub-licence agreements with generic producers for production and supply of generic HIV medicines at a lower cost. Such production and supply are already in progress.

Other pools exist, such as the private-sector initiative Pool for Open Innovation against Neglected Tropical Diseases.[28]

#### 7.2.2.3 Advance market commitments

These consist of agreements between governments as purchasers and pharmaceutical suppliers.[29] They are known as "advance" market commitments as such agreements tend to contain commitments – in advance of production – by the purchaser as to price or quantity. Such an arrangement aims to provide a financial incentive, in the form of assured returns, for manufacturers to produce medicines, based on criteria specified by the purchaser.

Another tool is the "vaccine bond", an instrument through which investors provide capital to stimulate vaccine research and development.[30]

## 7.3 The evolving role of the state as a provider of incentives

In the medicines landscape, the state presently acts as guardian of patentability, granter of marketing approvals, regulator of safety and genuineness of medicines and public funder for medicines, and, in many cases, general guarantor of affordable medicines under domestic constitutional requirements.[31]

The state may also act as the provider of incentives for medical innovation, through a number of measures. One such measure is the provision of various tax incentives. States may wish to encourage R&D for medicines for neglected diseases, or for developing "orphan drugs" (e.g. to treat rare diseases), as innovator enterprises may otherwise not

[27] Information from MPP website: www.medicinespatentpool.org/company-engagement/ (accessed 1 May 2014).

[28] Gold and Morin, footnote 23 above, at p. 236, and http://healthresearchpolicy.org/content/pool-open-innovation-against-neglected-tropical-diseases (accessed 1 May 2014).

[29] See WTO, WIPO and WHO, *Promoting Access to Medical Technologies and Innovation*, footnote 23 above, at p. 118 and p. 113, and Gold and Morin, footnote 23 above, at p. 232. Gold and Morin also refer to patent buyouts as a possible method of removing patent obstacles to access to producing less costly medicines; see p. 233.

[30] See Rockerfeller Foundation report, *Accelerating Impact Investments: Achievements, Challenges, and What's Next in Building the Impact Investing Industry*, July 2012, at p. 23.

[31] See Chapter 2.

find it attractive to invest in developing such medicines. A non-fiscal incentive may take the form of bestowing prizes or rewards for R&D investment in priority areas of medicines, especially those that help address diseases in poor countries.

A number of suggestions have been made to provide incentives for innovation through means such as prizes or rewards. Such suggestions aim to complement, and perhaps shift emphasis from, patents, as incentives.[32] Apart from private contributions to such prizes and rewards, regulators may also provide incentives for innovators to direct their attention on public health goals.[33] Transferable privileges, such as fast-track reviews, for IPRs, have also been suggested.[34]

### *7.3.1 The US and the EU*

In the US, the US Patent and Trademark Office (USPTO) began consultations in 2010 on a proposal to start a programme of rewards for innovators.[35] In 2012, the USPTO launched a pilot programme, Patents for Humanity, an awards competition for patent owners and licensees

[32] See, for example, James Love and Tim Hubbard, *The Big Idea: Prizes to Stimulate R&D for New Medicines*, www.cklawreview.com/wp-content/uploads/vol82no3/Love.pdf; Hubbard and Love, *A New Trade Framework for Global Healthcare R&D*, www.plosbiology.org/article/info:doi/10.1371/journal.pbio.0020052; *Patents and R&D Incentives: Comments on the Hubbard and Love Trade Framework*, www.who.int/intellectualproperty/news/en/Submission3.pdf; the US Medical Innovation Fund Prize Act 2005; and Marlynn Wei, Should Prizes Replace Patents? A Critique of the Medical Innovation Prize Act of 2005, *Boston U. J. Sc. & Tech. Law* (2007), http://papers.ssrn.com/sol3/papers.cfm?abstract_id=955669; Dean Baker and Noriko Chatani, *Promoting Good Ideas on Drugs: Are Patents the Best Way? The Relative Efficiency of Patent and Public Support for Bio-Medical Research*, 2002, www.cepr.net/Promoting_Good_Ideas_on_Drugs.pdf; James Love, *A New Trade Framework for Global Healthcare R&D*, www.cptech.org/ip/health/rndtf/love12042003.doc, and the response, DiMasi and Grabowski, *Patents and R&D Incentives: Comments on the Hubbard and Love Trade Framework for Financing Pharmaceutical R&D*, www.who.int/intellectualproperty/news/en/Submission3.pdf. Other papers on prize funds and other new models (James Love articles): www.cptech.org/ip/health/prizefund/other-articles.html.

[33] See generally Franck Müller-Langer, *Creating R&D Incentives for Medicines for Neglected Diseases: An Economic Analysis of Parallel Imports, Patents and Alternative Mechanisms to Stimulate Pharmaceutical Research* (Gabler, 2009).

[34] WHO, *CIPIH Report*, footnote 13 above, at pp. 87–88. The report discusses, first, transferable IPRs which work by "rewarding" a company that develops a medicine for a particular disease with, for example, a patent term extension for another ("blockbuster") medicine, and, second, by providing a fast-track review process for medicines which meet certain criteria of potential therapeutic benefit.

[35] USPTO Press Release 10-41, 20 September 2010.

which improve global living and health standards for the less fortunate.[36] The programme includes innovations in medicines. Significantly, the winners of 2013 include Gilead Sciences for their contribution in making HIV drugs available to poor countries in Asia and Africa via a network of generic drug manufacturers.[37]

The 2014 programme provides rewards in the form of an award, public recognition and acceleration of any patent application proceeding of the awardee which is pending with the USPTO. The rewards are given for humanitarian use of US patented technology and for humanitarian research.

On the approval side, the US FDA also provides other forms of incentives for prioritizing R&D, such as the use of a "Fast-Track" process, the "Breakthrough Therapy" designation, an Accelerated Approval process and a Priority Review process for particular medicines.[38]

The EU has outlined in a recent blueprint funding incentives to boost medical R&D over a 7-year period from 2014 to 2020.[39] The plan provides nearly €80 billion in funding and extends to innovative activity in various fields, including health.[40] It also supports the emerging use of "demonstration projects" spearheaded by the World Health Assembly, to showcase innovative activity.[41]

## 7.4 The evolving paradigm in R&D: public–private partnerships to promote public health and innovation

Given the prohibitive costs of developing new medicines, and the need to facilitate greater commercial and public access to the fruits of innovation, public–private collaborative partnerships have sprung up to

[36] USPTO Press Release 12-10, 8 February 2012. See generally information at USPTO website at: www.uspto.gov/patents/init_events/patents_for_humanity.jsp (accessed 13 June 2014).

[37] USPTO website: www.uspto.gov/patents/init_events/patents_for_humanity/awards2013.jsp (accessed 13 June 2014).

[38] See FDA website at: www.fda.gov/forconsumers/byaudience/forpatientadvocates/speedingaccesstoimportantnewtherapies/ucm128291.htm (accessed 2 July 2014). See also Karen Young, *Fast-Tracking Drugs to Market – An Opportunity for All Stakeholders?* 26 June 2014, available at www.mondaq.com/x/323366/Life±Sciences±Biotechnology/FastTracking±Drugs±To±Market±An±Opportunity±For±All±Stakeholders&email_access=on (accessed 2 July 2014).

[39] See EU, *Horizon 2020: The EU Framework Programme for Research and Innovation.*

[40] See *Horizon 2020* health sub-portal; http://ec.europa.eu/programmes/horizon2020/en/area/health (accessed 2 July 2014).

[41] See Health Action International, *A Global Opportunity for Global Health*, 4 November 2013.

handle R&D, commercialization and licensing of pharmaceutical breakthroughs.[42]

This new paradigm also sees an important role for academic institutions.[43] According to a 2012 survey examining collaborations between academic medical centres and the biopharmaceutical industry in the US, emerging practices include flexible risk-sharing arrangements and competitive grants.[44]

The WHO's Global Strategy and Plan of Action on Public Health, Innovation and Intellectual Property advocates the sharing of R&D and "health needs driven" R&D.[45]

### *7.4.1 Singapore's "eco-system" promoting innovation and collaboration*

Singapore uses a combination of fiscal and non-fiscal incentives both to attract as well as to encourage R&D in developing new medicines, including neglected tropical diseases.[46] Its approach aims to provide an attractive combination of resources conducive to innovative activity, business and collaboration, including state-of-the-art infrastructure, funding opportunities and access to international experts and networks,

[42] See WIPO charts on PPP at: www.wipo.int/edocs/mdocs/tk/en/wipo_ip_lss3_08/wipo_ip_lss3_08_www_114643.pdf. There have also been suggestions of global, multilateral funds for drug development; see, for example, the O'Neill Commission Report, *Securing New Drugs for Future Generations: the Pipeline of Antibiotics – The Review of Antimicrobial Resistance*, 14 May 2015, and, generally, M. Balasegaram, C. Bréchot, J. Farrar, D. Heymann, N. Ganguly, M. Khor, *et al.*, A Global Biomedical R&D Fund and Mechanism for Innovations of Public Health Importance, *PLoS Med*, Vol. 12, No. 5 (11 May 2015), e1001831.

[43] See Paul G. Wyatt, The Emerging Academic Drug-Discovery Sector, *Fut. Med. Chem.* (2009), 1013–1017.

[44] See Tufts Center for the Study of Drug Development, *Academic-Industry Partnerships for Biopharmaceutical Research and Developing: Advancing Medical Science in the* US, pp. 14–20. The study states that it was funded in part by a grant from the Pharmaceutical Research and Manufacturers of America (PhRMA).

[45] See paras. 29–30, Annex, p. 1, 24 May 2008, 61st World Health Assembly, WHA 61.21, footnote 4 above.

[46] See Inland Revenue Authority website; http://iras.gov.sg/irasHome/page04.aspx?id=614#Research_&_development_expenses (accessed 2 July 2014). The Novartis Institute for Tropical Diseases was established in Singapore in 2002 to discover new medicines for neglected infectious diseases; see www.nibr.com/research/developing_world/NITD/ (accessed 8 August 2014). It is the product of public–private partnership between Novartis and Singapore's Economic Development Board, and is located at the Biopolis, a cutting-edge biomedical sciences R&D complex (see www.jtc.gov.sg/industries/biomedical/biopolis/pages/biopolis.aspx).

including those in the private sector and in academic institutions.[47] On the fiscal front, generous tax incentives are provided for innovative activity (whether medicines related or not). On the non-fiscal front, infrastructural facilities and networking opportunities (linking private, public, conventional and "impact" funding and academic and non-profit entities) are provided within Singapore's "eco-system".[48] These operate against the backdrop of Singapore's broader advantages, such as its excellent business and physical infrastructure, geographical location within Asia/ASEAN, and a highly educated workforce. The success of the approach is underscored by the presence of a number of global pharmaceutical manufacturing and R&D entities in Singapore's dedicated science hubs.[49] Singapore is host to the Novartis Institute for Tropical Diseases, established in 2002 by a partnership between Novartis and the Singapore Economic Development Board (EDB) as a centre for drug discovery to fight infectious tropical diseases and to provide affordable access to such drugs in developing countries.[50] It is also host to the Global Center of Excellence for Business Model Innovation of Medtronic, a global medical technology business, to

[47] See generally Singapore Economic Development Board, www.edb.gov.sg/content/edb/en/industries/industries/pharma-biotech.html. The Agency for Science, Technology and Research (A*STAR) is a dedicated government agency that promotes innovation in the biomedical sciences and oversees various research entities. See www.a-star.edu.sg/About-A-STAR/Overview.aspx (accessed 8 August 2014).

[48] The author is grateful to Ms. Chu Swee Yeok, Mr Kelvin Wong and Ms. Helen V. Chou of the Singapore EDB for helpful information on various aspects of Singapore's development of its innovation "eco-system".

See generally Economic Development Board (EDB) website, www.edb.gov.sg/content/edb/en/industries/industries/pharma-biotech.html (accessed 2 July 2014). The EDB is the key Singapore agency that plans and implements economic strategies to propel Singapore forward as a "global hub for business, investment and talent" – EDB Core Mission, www.edb.gov.sg/content/edb/en/about-edb/our-strategy/mission-vision.html (accessed 14 April 2015). On R&D infrastructure, the government has built state-of-the-art research hubs in Biopolis and Fusionopolis; see www.a-star.edu.sg/Biopolis-Fusionopolis/A-Vision-for-Convergence.aspx (accessed 15 April 2015) and Marissa Lee, *R&D Health: Big Pharma Bets on Biopolis*, The Straits Times, p. C2. Newer, collaborative funding opportunities have also arisen in the "eco-system", including the Asian Venture Philanthropy Network and Impact Investment Exchange Asia: IIX.

[49] See EDB, at: www.edb.gov.sg/content/edb/en/industries/industries/pharma-biotech.html (accessed 14 April 2015). Singapore also offers incentives for businesses which locate their headquarters there. The policy has succeeded in attracting global pharmaceutical companies to relocate their headquarters, such as that of GlaxoSmithKline; see Ong Kai Xuan and Jacqueline Woo, *Global HQ, Dose of Local Talent for GSK*, 12 March 2015, The Straits Times, p. A2.

[50] See below footnotes 88 and 89 and accompanying text. The author is grateful to the NITD, Singapore, for helpful information.

pilot new business models for better medical product delivery in emerging markets in Asia and beyond.[51]

Singapore's National Research Foundation, a unit under the Prime Minister's Office, oversees research in strategic programmes such as the Biomedical Sciences (BMS) Translational and Clinical Research (TCR).[52] The Ministry of Health's National Medical Research Council (NMRC) also provides funding for medical research.[53] The Singapore Clinical Research Institute (SCRI) – a research organization owned by MOH Holdings – was originally set up in 1996 as the Clinical Trials & Epidemiology Research Unit (CTERU) to provide infrastructure for not-for-profit public-sector clinical research.[54] It acts as a coordinating centre for multi-institutional research.[55]

Such collaborations between private and public partners can help speed up discovery of vaccines, therapeutic drugs and diagnostic products.[56]

51 See Medtronic, at: http://newsroom.medtronic.com/phoenix.zhtml?c=251324&p=irol-newsArticle&ID=1850692 and EDB, at: www.edb.gov.sg/content/edb/en/news-and-events/news/singapore-business-news/Industry/medtronic-launches-global-innovation-centre.html (accessed 15 April 2015).

52 See NRF website at: www.nrf.gov.sg/about-nrf/programmes/strategic-research-programmes (accessed 2 April 2014).

53 IPRs belong to the host institution but NMRC has the right to request a "*perpetual, irrevocable, non-exclusive, royalty-free, worldwide license*" (italic emphasis added) to use the IP first generated or first created from the research being funded (called "Foreground IP") for the following purposes only:

"(i) National public health purposes;
(ii) Public non-commercial purposes; and
(iii) For or during a national emergency or other circumstances of extreme urgency."

See para. 4.5, NMRC *Policy Document on Intellectual Property Rights*, available at: www.nmrc.gov.sg/content/dam/nmrc_internet/documents/policiesAndGuidelines/Policy%20Document%20on%20Intellectual%20Property%20Rights.pdf (accessed 14 April 2015).

54 See www.scri.edu.sg/index.php/2012-12-10-08-28-11/about-our-profile.

55 See www.scri.edu.sg/index.php/about-coordinating-centre.

56 For examples of promising results from such collaborations in Singapore, see Laurent Renia and Lisa F. P. Ng, *Singapore's War on Infectious Diseases*, (28 November 2014) The Straits Times, p. A35. For other examples of Singapore's schemes and collaborative work, see, for example, A*Star, Exploit Technologies and its licensing opportunities for pharmaceuticals made available from innovation, www.etpl.sg/innovation-offerings/technologies-for-license&func=summary&type=category&fid=11.

Also: A*Star-AstraZeneca joint research collaboration agreement, 2013, at: www.astrazeneca.com/Research/news/Article/20131104–astar-and-astrazeneca-sign-a-research-collaboration. The Experimental Therapeutics Centre was established in 2006; see www.etc.a-star.edu.sg/index.php?option=com_content&view=article&id=9&Itemid=2.

### 7.4.2 *The US and the EU*

The US FDA's Critical Path Initiative (C-Path Initiative) of 2004 is a programme that aims to identify and prioritize the most pressing development problems in public health and to speed up drug discovery for this purpose.[57]

The European model, the Innovative Medicines Initiative (IMI), is the largest PPP with funds of €2 billion – €1 billion each from the European Union and the pharmaceutical industry through the European Federation of Pharmaceutical Industries and Associations (EFPIA).[58]

## 7.5 Exploring new funding models

### 7.5.1 *Impact investing and public access to medicines: connecting the dots for better access to medicines*

Relatively new networks for investing have emerged in recent years, drawing together hitherto-disconnected investors, government and non-profit service providers to address particular social needs. The Global Impact Investing Network (GIIN) is an example of a worldwide network or "peer community" connecting asset owners/investors, asset managers and service providers to promote investing with measurable social and environmental impact, with financial return.[59] Social impact bonds (SIBs),

Temasek Life Sciences Laboratories (TLSL), funded as a non-profit organization by Temasek Trust, with commercialization through Temasek Life Ventures See www.tll.org.sg/about-us/introduction/, www.temasektrust.org.sg/#beneficiaries and www.tlv.sg/about/ (accessed 2 April 2014).

57 See www.fda.gov/downloads/ScienceResearch/SpecialTopics/CriticalPathInitiative/CriticalPathOpportunitiesReports/ucm113411.pdf and www.c-path.org/pdf/FDAcriticalpathinitiativeinfluenceonnewdrugdevelopmentWoodcockWoosley.pdf (accessed 14 April 2015).

58 See EU website at: www.imi.europa.eu/, EFPIA website at: www.efpia.eu/topics/innovation/innovative-medicines-initiative, and EMA website at: www.ema.europa.eu/docs/en_GB/document_library/Presentation/2013/02/WC500138668.pdf (accessed 2 April 2014). IMI's IP Policy 2007 is available at: www.imi.europa.eu/content/documents#ip_policy. Information on the *Innovative Medicines Initiative 2* is available at: www.europarl.europa.eu/oeil/popups/ficheprocedure.do?reference=2013/0240%28NLE%29&l=en, adopted by EU Parliament in May 2014. See also report at: Elisabetta Vaudano, *The Innovative Medicines Initiative: A Public Private Partnership Model to Foster Drug Discovery*: www.nature.com/news/europe-bets-on-drug-discovery-1.12372 (accessed 2 April 2014).

59 See generally the GIIN website at: www.thegiin.org/cgi-bin/iowa/aboutus/index.html (accessed 26 June 2014). For information on other networks, see Rockerfeller Foundation report, *Accelerating Impact Investments: Achievements, Challenges, and*

or "pay for success" arrangements, were pioneered in the UK, with the first such bond having been issued in 2010. SIBs engage governments, investors and intermediaries who arrange and implement the underlying project; the aim of SIBs is to achieve some aspect of social good (such as ensuring better access to education, etc.) by leveraging on private investor funding. In the UK's pioneering bond, the aim was to address recidivism. The number of countries where such instruments have been announced has since grown to include Australia, Canada, the Netherlands and the US.[60] While SIBs have not had a resounding presence in Asia yet, awareness and debate is growing.[61]

Such bonds work on the principle that the relevant government and the partner non-profit body pay a return to investors if targeted social aims are met. If they are not, the investors would be prepared to forgo any returns, and to treat the investment as a charitable gift. As one of the founding persons of SIBs said, these bonds make use of the "invisible heart" to guide the "invisible hand" of the market.[62]

SIBs may be adapted to fund expensive pharmaceutical R&D research, which currently lies largely in the domain of private pharmaceutical enterprises. Therapeutically successful products of such R&D – with no

*What's Next in Building the Impact Investing Industry*, (July 2012), at p. 39. A related concept, "blended value" in investing, which combines funding with social impact, is attributed to Jed Emerson: see www.blendedvalue.org/ (accessed 26 June 2014). On sustainable investing by SWFs, see, for example, Mirjam Staub-Bisang, Public Pension Funds: Sustainable Investment Pioneers, *The Guardian*, 8 August 2013.

60 See Ashley Pettus, Social Impact Bonds, *Harvard Magazine*, July 2013, Jon Hartley, Social Impact Bonds Are Going Mainstream, *Forbes*, 15 September 2014; and Commerce and Conscience, *The Economist*, 23 February 2013.

61 See, for example, Asian publications on SIBs such as *TODAY* (Singapore), at www.todayonline.com/commentary/time-seriously-invest-social-impact, and *Hong Kong Business* at hongkongbusiness.hk/economy/commentary/financing-hong-kong %E2%80%99s-future-through-social-impact-bonds-part-i (accessed 21 October 2014). It has been reported that India has seen its first SIB project aiming to provide improved access to education for girls; see Social Impact Bonds Make India Debut with Dasra Programme, *AVPN News*, June 2013, www.avpn.asia/resources/newsletters/june-2013-edition/social-impact-bonds-make-india-debut-with-dasra-programme/ (accessed 21 October 2014).

62 Sir Ronald Cohen, Chairman of the UK Social Impact Investment Taskforce, and Director and Co-Founder of Big Society Capital (which launched UK's first SIB), remarks at the World Investment Forum 2014, organized by UNCTAD, Geneva, Switzerland, 15 October 2014. See also www.devex.com/news/want-to-fund-global-development-tap-into-the-heart-of-markets-84330 (accessed 21 October 2014); the remarks are a reference to the task force's report, entitled *The Invisible Heart of Markets – Harnessing the Power of Entrepreneurship, Innovation and Capital for Public Good.*

monopoly on IPRs and pricing – can be made more affordable as investors reap non-market returns.

An early example of a health-related instrument linked to disease control and treatment is the Mozambique Malaria Performance Bond, a development impact bond which aims to increase funds and intervention in the prevention, diagnosis, treatment and monitoring of malaria in Mozambique.[63] Another example is the Global Health Investment Fund.[64]

Interestingly, in April 2014, India's National Institute of Public Finance and Policy invited expressions of interest in the establishment of venture capital "Pharma funds".[65]

Some countries in Asia, such as Singapore, through its Economic Development Board (EDB), have been promoting synergies and networking opportunities among investors, industries, grant-makers and other non-profit organizations for mutually beneficial collaborations and conversations. As a result of the EDB's efforts and Singapore's global position as a wealth management centre, a number of international non-profit organizations have joined the large number of global private enterprises to establish a presence in Singapore.[66]

### *7.5.2 Sovereign wealth funds – can they play a possible role?*

Sovereign wealth funds (SWFs) are defined by the International Monetary Fund as follows:

> SWFs are a heterogeneous group and may serve various purposes. Five types of SWFs can be distinguished based on their main objective: (i) *stabilization funds*, where the primary objective is to insulate the budget and the economy against commodity (usually oil) price swings; (ii) *savings funds* for future generations, which aim to convert nonrenewable assets into

[63] See www.devex.com/impact/partnerships/goodbye-malaria-mozambique-malaria-performance-bond-362 (accessed 28 October 2014). The partners establishing the bond are Dalberg Global Development Advisers, Nando's, Anglo-American plc and the Coca-Cola Foundation. See also generally, Lily Han, Malaria in Mozamique: Trialling Payment by Results, *The Guardian*, 31 March 2014.

[64] See the fund's website at http://ghif.com/ and, generally, www.rockefellerfoundation.org/uploads/files/78e66e85-ac2d-4124-8dcb-61e4149ef49d-impact.pdf at p. 21.

[65] *Detailed Project Report for Investment in Innovative Pharma: Establishing Pharma Funds* (1 April 2014).

[66] For information on the list of international non-profit organizations in Singapore (numbering about 140), see www.edb.gov.sg/content/edb/en/industries/industries/international-non-profit-organisation.html (accessed 21 April 2015). The author is grateful to Mr Kelvin Wong and Ms. Helen Chou for the information. See also EDB "advertorial", Singapore's Social Sector Initiatives Help Change the World, *The Economist*, 8 March 2014, p. 8.

> a more diversified portfolio of assets and mitigate the effects of Dutch disease; (iii) *reserve investment corporations*, whose assets are often still counted as reserve assets, and are established to increase the return on reserves; (iv) *development funds*, which typically help fund socio-economic projects or promote industrial policies that might raise a country's potential output growth; and (v) *contingent pension reserve funds*, which provide (from sources other than individual pension contributions) for contingent unspecified pension liabilities on the government's balance sheet.[67]

Such funds therefore usually invest public savings in long-term investments for return and are not generally philanthropic in objectives. Some thinkers have, however, recently suggested that *public venture funds* be established by governments to invest in new technologies for social benefit.[68]

In the case of SWFs in particular, the incipient link with impact investing (in a broad sense) has in fact already been made, with *philanthropic arms* of some SWFs (such as Singapore's Temasek Holdings) providing funding for health-related projects.[69] In addition, Norway's SWF, reportedly the largest in the world, has invested in renewable energy.[70] The Norges Bank, which manages Norway's Pension Fund Global, engages in responsible and ethical investing, focusing on social issues and the environment.[71] On the broad objectives of the Norwegian fund, it has been said: "As long-term owners . . . we need to

[67] *Sovereign Wealth Funds, a Work Agenda*, International Monetary Fund 5 (2009), available at www.imf.org/external/np/pp/eng/2008/022908.pdf (accessed 28 April 2015).

[68] See, for example, Dani Rodrik, *From Welfare State to Innovation State*, Innovation & Technology, Project Syndicate, 14 January 2015.

[69] See Chapter 7. The notion of SWF investment to promote public health and innovation is to be distinguished from "sovereign IP funds"; see, for example, France Brevets, the French sovereign fund that invests in patents and their commercialization; and OECD, *OECD Reviews of Innovation Policy: France 2014*, p. 210.

[70] See World Wildlife Fund (WWF) report, *Move by Norway Sovereign Wealth Fund to Invest in Renewables Could Have "Global Impact"*, 13 March 2014, at: http://www.wwf.org.uk/about_wwf/press_centre/?unewsid=7062 (accessed 26 June 2014). See also Dimitrij Euler, *Impact Investment: Sovereign Wealth Funds, Corporate Governance and Stock Markets*, 15 April 2013, at: http://papers.ssrn.com/sol3/papers.cfm?abstract_id=2265165 and Natsuko Waki, *Making an Impact May Be New Good*, at: http://blogs.reuters.com/globalinvesting/2013/03/06/making-an-impact-may-be-new-good/ (accessed 26 June 2014). See also report for the Rockefeller Foundation, *Accelerating Impact Investments: Achievements, Challenges, and What's Next in Building the Impact Investing Industry*, *ibid.*

[71] See Norges Bank website at: www.nbim.no/en/responsibility/ (accessed 26 June 2014). Norway's fund has been characterized as being of the "moralist type": see Clark, Dixon and Monk, *Sovereign Wealth Funds: Legitimacy, Governance and Global Power* (Princeton University Press, 2013).

make sure that the companies are long-term profitable, and not only good for the investors, the shareholders, but for society at large."[72] Qatar's SWF, the Qatar Investment Authority, has also reportedly begun considering impact investing.[73]

One possible role would be for SWFs themselves, perhaps through philanthropic arms or focus units, to *collaborate* in creating cross-border cooperative research entities to complement drug discovery and delivery efforts of private pharmaceutical firms.[74] Such collaborative efforts would need to address (and agree on) the handling of critical issues such as those relating to IP, competition, pricing and eligibility of beneficiaries.

## 7.6 The evolving role of business

> The legitimacy of pharmaceutical companies will increasingly depend on them being recognized as fulfilling a socially constructive and ethically responsible role in addressing challenges arising from poverty-related illnesses and premature mortality.[75]

In Chapter 2, the example of GlaxoSmithKline was discussed in the context of pharmaceutical company adjustments to the new global environment. As companies consider corporate culture and organizational changes, it could present a unique opportunity for governments to provide incentives for changes which will produce better results for access to medicines.

### *7.6.1 Industry actions*

As mentioned in Chapter 2, the pharmaceutical industry has had to respond to a number of evolving factors, including globalization,

[72] See Yngve Slyngstad, Chief Executive Officer, Norges Bank Investment Management, which runs Norway's pension fund, in Matthew Price, *Norway: Is World's Largest Sovereign Wealth Fund Too Big?*, BBC, 12 September 2013.

[73] See Vanessa Kortekaas and Simeon Kerr, Qatar Fund Sets Its Sights on Impact Investment Schemes, *Financial Times*, 28 October 2013.

[74] Temasek Holdings, one of Singapore's SWFs, has philanthropic "arms", the Temasek Trust and Temasek Foundation (TF), which already invests in regional healthcare initiatives; see Temasek Foundation website: www.temasekfoundation.org.sg/programmes-partnerships/fa/building-people-healthcare (accessed 26 June 2014). TF already invests in regional healthcare initiatives. See also footnote 127 below and accompanying text.

[75] Klaus M. Leisinger and Anita K. Wagner, *Access to Medicines and Corporate Social Responsibilities of the Pharmaceutical Industry*, Asia-Pacific Biotech, www.asiabiotech.com/articles/readmore/1706-article02.html (accessed 2 October 2014).

emergence of new diseases, emerging markets (including those in Asia) with increasing healthcare needs, government healthcare budget constraints and expiry of "blockbuster" medicine patents. Some enterprises have adapted their internal business structures and philosophies in the wake of such developments. After all, business model innovation is sought and occurring in other fields of business.[76]

## 7.7 Rethinking clinical data protection

The debate on access to medicines (and biopharmaceuticals) has raised questions of what may – or should – form part of the "commons", or property and resources that are available for general use (and therefore often overused),[77] with a concomitant argument that a mirroring concept, an "anticommons" (privatized general property), has been emerging in biomedical research.[78]

In this context, some have called for clinical test data to be viewed as a public good.[79]

Of particular relevance in this debate is the access to and use of valuable data in R&D for pharmaceutical products, such as clinical test data. Clinical trial data, a linchpin to new drug development, has not traditionally been shared information. As has been explained:

> Traditionally, the dissemination of knowledge derived from clinical research has been limited in scope: Investigators who have designed and conducted clinical trials make the decisions about which statistical analyses to conduct and then publish peer-reviewed articles to disseminate their findings. Clinical trial data are considered the property of the

[76] See, for example, Procter & Gamble's "Connect + Develop" model, which aims to attract innovation ideas from a variety of sources to meet existing business needs under an "Open Innovation" plan; see www.pgconnectdevelop.com/. In 2013, Medtronics, a medical parts company, established a Global Center for Excellence for Business Model Innovation in Singapore. More generally, "business model innovation" appears to be a new driver for business growth: see commentary by Rahim Kanani, Business Model Innovation Is the Fastest Path to Greatness, *Forbes*, 4 October 2012.

[77] Garrett Hardin, The Tragedy of the Commons, *Science*, Vol. 162 (1968), 1243–1248.

[78] See Michael A. Heller and Rebecca S. Eisenberg, Can Patents Deter Innovation? The Anticommons in Biomedical Research, *Science*, Vol. 280 (1998), 698–701.

[79] See, for example, Rodwin and Abramson, *ibid.* Others advocate more sharing of such data: see, for example, Harlan M. Krumholz, Open Science and Data Sharing in Clinical Research: Basing Informed Decisions on the Totality of the Evidence, *Cir. Cardiovasc. Qual. Outcomes*, Vol. 5 (2012), 141–142; and Joseph S. Ross, Richard Lehmann and Cary P. Gross, The Importance of Clinical Trial Data Sharing: Toward More Open Science, *Cir. Cardiovasc. Qual. Outcomes*, Vol. 5 (2012), 238–240.

> investigators and the entities that sponsored the research, with little or no opportunity for investigators external to the original study team to access the data. This traditional model is based on dissemination via print publication, the origins of which date back to the 17th century.[80]

A critique of this status quo has been made by the same authors:

> By continued adherence to this model in the age of electronic knowledge exchange, our understanding of clinical interventions is limited by our lack of access to comprehensive data from all clinical trials in several ways.[81]

The increase in public–private collaborations in drug discovery and the presence of public funding (through grants and subsidies, for instance) in drug R&D work have contributed to calls for greater sharing of information and data generated through such R&D.[82]

The handling and sharing of data has been under a significant shift in recent times in some jurisdictions. Private and public bodies have taken steps to greatly increase the amount of data sets generated to better analyse and address user needs and strategic plans, with the establishment of databases which provide free access to certain clinical test data in some cases, as explained below.[83]

### *7.7.1 Towards a sustainable model of clinical data sharing*

Clinical data sharing is at a new point of inflexion in the EU as the EMA attempts to balance information sharing for better R&D and innovation, with private enterprises' interests in protecting their business data.[84] In the US, efforts to promote such data sharing have been described as "modest".[85]

The pharmaceutical industry itself has taken some actions to address the data-sharing issue. Industry-driven "principles" have emerged recently to address criticisms regarding availability of clinical test data.

[80] Rodwin, Marc A. and John Abramson, Clinical Trial Data as a Public Good *Journal of the American Medical Association*, Vol. 308, No. 9 (2012), 871–872.

[81] *Ibid.* [82] See, for example, Rodwin and Abramson, footnote 80 above.

[83] See, for example, the explanation in *Big Data Update*, in *Future Tense*, Singapore Ministry of Trade & Industry, May 2014, available at: www.mti.gov.sg/ResearchRoom/SiteAssets/Pages/Future-Tense-July-2014/Future%20Tense%202014_final.pdf, pp. 71–77 (accessed 10 December 2014).

[84] See Chapter 7.

[85] See Marc A. Rodwin, Five Uneasy Pieces – Institutional Corruption and the Pharmaceutical Industry, *J. Law, Med. & Ethics* (Fall 2013), 581–589, at p. 585, available at http://ssrn.com/abstract=2295002.

The *Principles for Responsible Clinical Trial Data Sharing* for the biopharmaceutical industry, adopted by the Pharmaceutical Research and Manufacturers of America (PhRMA) and the European Federation of Pharmaceutical Industries and Associations (EFPIA) in 2013, are an example.[86] Implementation of these principles began in January 2014. The three key principles are safeguarding patient privacy, respecting the integrity of national regulatory systems and maintaining incentives for biomedical research.

Under the principles, researchers may submit request proposals in which they may request clinical trial data, for evaluation. In such proposals, any conflicts of interest and potential competitive use of the data, among other information, have to be identified. Researchers "must agree not to transfer the shared data or information to parties not identified in the research proposal, use the data for purposes not contained in the research proposal, or seek to re-identify research participants". Researchers who are provided with company data are "encouraged and expected to publish the results of their analysis". Biopharmaceutical companies, on the other hand, are required to establish scientific review boards that will include scientists and/or healthcare professionals who are not company employees. The principles are voluntary in nature, and any data sharing is dependent on the evaluation of qualifications and "legitimacy" of the research proposed. A useful feature in the principles is that the companies are expected to publicly post their data request review process and the identities of the external scientists and healthcare professionals, and relationships with existing external board members. While this voluntary process may not appear to have "teeth", it is a positive first step for the biopharmaceutical industry. Its credibility and actual usefulness will ultimately depend on the implementation of the review process by industry members.

Individual pharmaceutical companies may also have drug discovery information-sharing schemes.[87]

Pharmaceutical entities have themselves undertaken, through foundations and special R&D centres, research which aims to speed up drug

[86] See PhRMA at: www.phrma.org/phrmapedia/responsible-clinical-trial-data-sharing (accessed 26 April 2014). The new principles supplement the PhRMA's *Principles on Conduct of Clinical Trials and Communication of Clinical Trial Results.*

[87] An example is Eli Lilly's Open Innovation Drug Discovery system; see Eli Lilly website: https://openinnovation.lilly.com/dd/docs/oidd_executive_summary.pdf (accessed 20 July 2014).

discovery, especially in neglected areas.[88] An example is the Novartis Institute for Tropical Diseases (NITD) located in Singapore, an arm of Novartis, which undertakes early exploration up to proof-of-concept research in malaria and other tropical diseases. Clinical trials up to Phase 2 are handled by NITD; beyond that, the work is handled by other parties, which will take the research and, if successful, eventual marketing forward. At the time of writing, the NITD has developed two drug candidates which are awaiting marketing approval.[89]

Separately, Novartis has also made available a groundbreaking antimalarial drug without enforcing its patent, in effect (before it expired in October 2014). The WHO and Novartis entered into a 10-year Memorandum of Understanding to supply the drug on a no-profit basis.[90] The 10-year period ended in 2011 but Novartis continues to supply it on the same terms.[91] The drug, Coartem®, based on an ancient Chinese herbal remedy, was developed and commercialized by Novartis in collaboration with Chinese researchers.[92]

[88] Examples of private-firm initiatives include Sanofi-PATH collaboration for non-profit semi-synthetic version, and Pfizer, which established a Center for Therapeutic Innovation.

[89] See Novartis, at www.novartis.com/newsroom/media-releases/en/2014/1843976.shtml and www.novartis.com/newsroom/feature-stories/2011/04/malaria-initiative-1.shtml (accessed 12 April 2015).

[90] See WHO report, www.who.int/malaria/areas/treatment/MoU_termination_ report_ may2011.pdf (accessed 13 April 2015).

[91] See www.malaria.novartis.com/malaria-initiative/access/improving-access-to-essential-malaria-medicines/index.shtm (accessed 6 November 2014).

[92] See www.malaria.novartis.com/downloads/case-studies/hbs-the-malariainitiative-challenge-1.pdf and www.who.int/bulletin/volumes/87/10/09-051009/en/. Professor Zhou Yiqing, whose team in China developed the first artemisinin-based combination therapy (ACT) for malaria, known as Coartem®, shared his approach with WHO as follows:

> Today, our discovery, Coartem, has proven to be highly effective and well tolerated, with high cure rates of over 95%, even in areas of multi-drug resistance. But in an emerging economy like China, nobody has the money to support development of a medicine that has no domestic market and is mostly consumed by poor people outside China. Yet I knew it could cure patients and bring hope and health back to those having malaria. How could I watch such a good medicine die silently in the laboratory and do nothing about it? . . .
>
> In 1990, my team and I applied for the original patent in China . . . The patent now belongs to my institute, IME, and the nation.

See WHO, www.who.int/bulletin/volumes/87/10/09-051009/en/. For a timeline on the Novartis collaboration to produce the medicine, see www.malaria.novartis.com/malaria-initiative/the-malaria-initiative/index.shtml (accessed 6 November 2014); according to the information provided there, 600 million artemisinin-based combination therapy (ACT) treatments have been delivered by Novartis on a no-profit basis. In October

Arguments have been made to analogize the open source approach to innovation in software development to developing biomedicines. There are a number of challenges in making this analogy bear fruit but, despite this, the SNP Consortium has been cited as an example of a successful open-access public resource which promotes innovative activity.[93]

For countries which have to respond to FTA demands for TRIPS-plus data exclusivity provisions and extended data protection periods for new products (such as biologics), it is critical to carefully craft the *quid pro quo* if declining such demands is not an option, as discussed further below.

In the field of copyright, a separate but interesting development has been taking place, with the design of guidelines on assessing the broader societal impact of copyright. The WIPO Draft Guidelines on Assessing the Economic, Social and Cultural Impact of Copyright on the Creative Economy (ESCIA) provide interesting language.[94] The following summarizes the rationale for the guidelines.[95] Two particular features of the

2015, a Chinese scientist, Tu Youyou, was awarded the Nobel Prize in Physiology or Medicine for her work in discovering artemisinin (isolated from a Chinese herb, sweet wormwood, or *qinghaosu*, 青蒿素) through research on ancient Chinese traditional medicines in the 1960s, becoming the first Chinese woman to win a Nobel prize; see http://www.bbc.com/news/blogs-china-blog-34451386 (accessed 7 October 2015). Ms Tu, who had in 2011 been awarded the LaskerDeBakey Clinical Medical Research Award, recounted that she was the first human to test artemisinin; see Kor Kian Beng, *Long, Difficult Journey to Recognition, The Straits Times*, 12 October 2015, p. A14 and http://www.laskerfoundation.org/awards/2011_c_description.htm (accessed 12 October 2015).

93 SNP provides data on single nucleotide polymorphisms in the human genome. See the interesting discussion in *Open Development: Is the "Open Source" Analogy Relevant to Biotechnology?*, Alan G. Isaac and Walter G. Park, in *The Role of Intellectual Property Rights in Biotechnology Innovation*, ed. David Castle (Edward Elgar Publishing Ltd, 2009), at pp. 238–239, and generally, Gudmundur A. Thorisson and Lincoln D. Stein, The SNP Consortium Website Past, Present and Future, *Nucleic Acids Research*, Vol. 31, No. 1 (2003), pp. 124–127.

94 September 2013, www.wipo.int/export/sites/www/copyright/en/performance/.../escia.pdf (accessed 29 July 2014). The author is grateful to Ms. Kimani Goddard, IP Consultant, formerly at WIPO, for valuable information in this area.

95 The preface states:

> ... [C]opyright has a much more important role to play in the constitution of society than is often recognized. Therefore efficient copyright policy, like all socio-economic regulation must focus on balancing economic, social and cultural goals. The Guidelines offer a mechanism to facilitate the gathering and analysis of data to aid this balancing process ... It proposes a system of qualitative and quantitative approaches and techniques that can be specifically tailored to fit the national context of Member States. It is designed to assist governments and stakeholders in the creative economy to consider further and identify economic, social and cultural indicators that

proposed guidelines are relevant in the present context: first, they utilize an "integrated indicator system, which puts forward economic, social and cultural pillars" and, second, they include a "risks/opportunities system designed specifically to allocate value to copyright policies that carry social, cultural and economic risks or opportunities. It is designed to help Member States make decisions that foster beneficial policy outcomes and reduce the likelihood of non-beneficial ones."[96]

The issue of copyright in clinical data documents and collections which are submitted to national authorities for pharmaceutical marketing approval has not entered the mainstream debate on data exclusivity under Article 39 TRIPS and related TRIPS-plus provisions yet and it is timely to have tools – such as these guidelines – to help policy-makers determine the role that copyright might play both in promoting innovation (the creative economy aspect of the guidelines) and in protecting public health interests and facilitating balanced policy-making (the social, cultural and economic risks aspects of the guidelines). It therefore remains to be seen how these interesting guidelines and other future tools may help guide states in achieving a better balance of innovation, economic growth and health goals.

### 7.7.2 *Open source*

In the area of computer programming, a paradigm shift has been taking place due to new technologies and attitudes. One example of such a shift lies in "open source" operating systems, in which access to a programme's source code is made freely available. The transfer of open source thinking to the pharmaceutical sphere has been raised as a

> facilitate the identification of the effects of copyright legislation and policy. The ESCIA also provides a basic blueprint for interpreting research results . . .
>
> A novel approach to understanding the way in which copyright accrues benefits to society is necessary, as utilitarian concepts of economic contribution do not show the full picture of the effects of policies, markets and legislation. The purpose here is to move away from the solely commercial logic of economic studies. What is proposed instead is an approach that opens up analysis to the social and cultural costs and benefits associated with copyright in the creative economy. Furthermore, the intent of the Guidelines is to underline the interlinkage of economic, social and cultural outcomes, and the way in which copyright interacts with these in the creative economy.

Preface, pp. 6–7, *ibid.*, above.

96 Preface, p. 7, footnote 94 above.

possible new drug discovery model but has limitations.[97] The issue is the extent to which clinical trial data should be subject to the notion of open source, that is, the extent of availability for use by researchers for free.

*Open source* arguments with respect to clinical trial data should be distinguished from *open innovation* business models, which rely on a wide range of sources of innovative ideas and solutions to power the businesses, as compared with traditional "in-house" innovation.[98] An example of an international consumer products company utilizing such a model is Procter & Gamble.[99] In the medical field, an example is Stentor Inc.[100] The question being asked is whether there is scope for pharmaceutical innovator companies to adopt such a model.[101]

### 7.7.3 *Recent data-sharing arrangements*

In genomic research, the outstanding success in open collaboration of mapping the DNA sequences in the Human Genome Project within

[97] See, for example, Jagadeesh Napa, *Open Source Drug Discovery – A Feasible Business Model?, Pharma Focus Asia*, at www.pharmafocusasia.com/strategy/open_source_drug_discovery.htm; Kai Bindseil, *Open Innovative Partnerships – A Working Model for the Future of Drug Discovery and Development, Pharma Focus Asia*, at www.pharmafocusasia.com/strategy/open-innovative-partnerships.htm (accessed 6 December 2014); and John A. Spertus, The Double-Edged Sword of Open Access to Research Data, *Cir. Cardiovasc. Qual. Outcomes*, Vol. 5 (2012), 143–144. See also WHO, Report of the Commission on Intellectual Property Rights, Innovation and Public Health, *Public Health – Innovation and Intellectual Property Rights* (2006), footnote 23 above, at pp. 91 and 178; and Pharma Embraces Open Source Models, *Nature Biotechnology*, Vol. 28, No. 7 (2010), 631–634.

[98] Professor Dr Henry W. Chesborough is attributed as having coined the term "open innovation"; H. W. Chesborough, *Open Innovation: The New Imperative for Creating and Benefiting from Technology* (Harvard Business School Press, 2003).

[99] See Procter & Gamble, www.pgconnectdevelop.com/ (accessed 15 December 2014). For other examples, see www.innovationexcellence.com/blog/2014/08/13/15-examples-of-open-innovation-between-big-companies-startups/.

[100] See Arthur A. Boni, A Business Perspective on IP: Open Innovation vs. Open Source in Commercializing Biomedical Opportunities, *J. Comm. Biotech.*, Vol. 19, No. 2 (2013), 55–59.

[101] See, for example, Jackie Hunter and Susie Stephens, Is Open Innovation the Way Forward for Big Pharma?, *Nature Reviews Drug Discovery*, Vol. 9 (2010), 87–88. See also Davide Chiaroni, Vittorio Chiesa and Federico Frattini, Investigating the Adoption of Open Innovation in the Bio-Pharmaceutical Industry – A Framework and an Empirical Analysis, *Eur. J. M. Innov.*, Vol. 12 (2009), 3285–3309; and Ram Subramaniam, Jeffrey H. Toney and C. Jayachandran, The Evolution of Research & Development in the Pharmaceutical Industry: Toward the Open Innovation Model – Can Pharma Reivent Itself?, *Int. J. B. Innov. & Res.*, Vol. 5, No. 1 (2011), 63–74.

13 years – 2 years ahead of schedule – was inspiring.[102] In the medical field, data-sharing systems in clinical research, though still in a relatively nascent stage, already exist in developed countries.[103]

In the pharmaceutical industry, new publicly available databases – an example of a possible paradigm shift – have been emerging. The "unprecedented deposition of hundreds of thousands of potential leads for new malaria drugs into an open source database" by two pharmaceutical companies, GlaxoSmithKline and Novartis, occurred in 2010.[104]

These companies, together with others (such as Merck and Pfizer), are also members of a consortium supporting a WIPO initiative established in 2011, in the form of a publicly searchable database of compounds, data and technologies – royalty-free – known as "Re:Search".[105] The database, which aims to promote R&D in neglected diseases, admits three categories of members, namely "providers", "users" and "supporters".[106] Providers enter into collaboration R&D agreements with users on a voluntary basis. To date, 81 collaboration agreements have been signed by providers and users.[107] Its Guiding Principles aim to ensure royalty-free access to drugs developed through its partnerships to least-developed countries.[108] Other related developments include establishment of "Open Lab" initiatives.[109]

In 2013, ClinVar was launched as a database with free public access, receiving data on important variations on the human genome.[110]

[102] National research institutes, universities and researchers in various parts of the world, such as United States, the United Kingdom, France, Germany, Japan and China, contributed towards a complete mapping of the human genome, with research results posted publicly immediately, and with new data posted every 24 hours; see www.genome.gov/11006943 (accessed 12 December 2014).

[103] See J. P. Ross, Richard Lehmann and Cary P. Gross, The Importance of Clinical Trial Data Sharing: Toward More Open Science, *Circ. Cardiovasc. Qual. Outcomes*, Vol. 5 (2012), 238–240.

[104] See Pharma Embraces Open Source Models, *Nature Biotechnology*, Vol. 28, No. 7 (July 2010), 631–633; and *Learning to Share, Nature* (January 2010), 463, 401. See also the voluntary portal Sharing Clinical Reports at www.iccg.org/about-the-iccg/collaborations/sharing-clinical-reports-project/, which invites sharing of clinical information on variants of *BRCA* breast cancer testing (mentioned in relation to the *Myriad* litigation in Chapter 2), and, generally, Myriad Decision Aftershocks Ripple Through Biotech, *Nature Biotechnology*, Vol. 31, No. 8 (2013), 663–665, at p. 664.

[105] See WIPO, www.wipo.int/research/en/. [106] *Ibid.*

[107] See WIPO, www.wipo.int/research/en/collaborations/collaborationagreements.html (accessed 6 December 2014).

[108] Note 105 above.

[109] See GlaxoSmithKline website, www.gsk.com/en-gb/research/research-funding/tres-cantos-open-lab/ (accessed 6 December 2014).

[110] See www.ncbi.nlm.nih.gov/books/NBK174587/ (accessed 13 April 2015). See also Gina Kolata, DNA Project Aims to Make Public a Company's Data on Cancer

In 2008, an open source-model programme was launched in Asia. India's Open Source Drug Discovery Programme (OSDD) was launched under its Council of Scientific and Industrial Research to explore drug discovery using an open source model, for neglected diseases such as malaria and multi-drug-resistant tuberculosis.[111]

A number of advocates argue for an "open source" approach to pharmaceutical R&D to enable greater access to medicines.[112] However, in the pharmaceutical context, the approach is still in a nascent state and based largely on voluntary participation by those which are "sources" of data and other intellectual property which are the stepping stones to pharmaceutical R&D. Nonetheless "soft" principles have begun to emerge to deal with the sharing and use of such data.[113]

Genes, *New York Times*, 12 April 2013. Another researcher-driven genomic data-sharing initiative is ClinGen, http://clinicalgenome.org/. See also the open-access approach of collaborative work of the Structural Genomics Consortium, University of Toronto.

111 See generally OSDD website at: www.osdd.net/about-us (accessed 20 July 2014); see also Rema Nagarajan, Funds Dry Up for Drug Discovery Project, *Times of India*, 1 April 2014.

112 On discussions of open source approaches, see Harlan M Krumholtz, Give the Data to the People, *New York Times*, 2 February 2014; Cory Doctorow, Why All Pharmaceutical Research Should Be Made Open Access, *The Guardian*, 20 November 2012; the Yale University Open Data Access Project (YODA) at http://medicine.yale.edu/core/projects/yodap/index.aspx; Krumholtz, Ross, Gross, Emanuel, Hodshon, Ritchie, Low and Lehman, A Historic Moment for Open Science: The Yale University Open Access Data Project and Medtronic, *Annals of Internal Medicine*, 18 June 2013; Krumholtz, Open Science and Data Sharing in Clinical Research Basing Informed Decisions on the Totality of the Evidence, *Circ. Cardiovasc. Qual. Outcomes*, Vol. 5 (2012), 141–142; Joseph S. Ross, Richard Lehman and Cary P. Gross, The Importance of Clinical Trial Data Sharing: Toward More Open Science, *Circ. Cardiovasc. Qual. Outcomes*, Vol. 5 (2012), 238–240; John A. Spertus, The Double-Edged Sword of Open Access to Research Data, *Circ. Cardiovasc. Qual. Outcomes*, Vol. 5 (2012), 143–144, http://circoutcomes.ahajournals.org/content/5/2/143.full.pdf+html?sid=04dec1d5-0674-4ee5-8304-613014e28fe2; Kathie Bailey Mathae and Paul F. Uhlir, *Committee on the Case of International Sharing of Scientific Data: A Focus on Developing Countries*, (2012) Board on International Scientific Organizations, Board on Research Data and Information, Policy and Global Affairs, National Research Council; *The EU Framework for Research and Innovation* – Horizon 2020, 11 December 2013, Guidelines on Open Access to Scientific Publications and Research Data in Horizon 2020; and EU Commission background paper, www.recolecta.net/.../background-paper-**open-access**-october-2012_en.pdf.

113 See, for example, OECD, *Principles on Open Access to Research Data from Public Funding*, 2007; and OECD, *2004 Declaration on Access to Research Data from Public Funding*. The UK parliament has discussed such a model; see UK Parliament – Open Access: Fifth Report of Session 2013–14, Vol. 1: Report.

### 7.7.4 *Clinical data sharing: regulatory developments in the EU*

There have been developments in the EU to enhance access to clinical trial data. These have been taking place on different fronts.

First, since 2012, the European Medicines Agency (EMA) has been taking steps towards greater access to such data.[114] On 24 June 2013, it released a draft policy on publication and access to clinical trial data for a 3-month public consultation. A final policy and implementation plan was accepted by its Management Board on 12 June 2014.[115] The revised policy, to be finalized and formally adopted in July 2014, "will not only allow the Agency to proactively publish clinical trial data that are submitted as part of marketing authorisation applications, but also give the possibility to download, save and print the trial data for academic and non-commercial research purposes".[116] The far-reaching implications of this policy are succinctly summarized as follows by the EMA:

> It takes the Agency beyond its legal obligations and provides an unprecedented level of access to clinical trial data that are used as part of decision-making for new medicines.[117]

[114] See EMA website at: www.ema.europa.eu/ema/index.jsp?curl=pages/special_topics/general/general_content_000556.jsp and EMA Press Release, 12 June 2014. See *EU Directive 2001/20/EC of the European Parliament and of the Council of 4 April 2001 on the approximation of the laws, regulations and administrative provisions of the Member States relating to the implementation of good clinical practice in the conduct of clinical trials on medicinal products for human use* and the implementing guideline, *Commission Guideline – Guidance on posting and publication of result-related information on clinical trials in relation to the implementation of Article 57(2) of Regulation (EC) No 726/2004 and Article 41(2) of Regulation (EC) No 1901/2006 (2012/C 302/03).*

[115] *Ibid.*; the EMA Management Board agreed on a set of key principles during its meeting in December 2013. Several concerns were raised about the new policy; see the EMA's response: www.ema.europa.eu/ema/index.jsp?curl=pages/news_and_events/news/2014/06/news_detail_002120.jsp&mid=WC0b01ac058004d5c1 (accessed 27 June 2014).

[116] EMA Press Release, 12 June 2014.

[117] *Ibid.* EMA officers had argued earlier in an academic piece that

> Contrary to industry fears, we argue that access to full – though appropriately deidentified – data sets from clinical trials will benefit the research-based biopharmaceutical industry. We predict that it will help to increase the efficiency of drug development, improve cost-effectiveness, improve comparative-effectiveness analysis, and reduce duplication of effort among trial sponsors . . .
>
> Clearly, however, legitimate interests in intellectual property and the protection of private investments must be weighed against other legitimate interests, such as transparency regarding the outcomes of clinical trials and the protection of public health. Striking the right balance among all such

At the legislative level, the European Parliament issued the new Clinical Trials Regulation on public disclosure of clinical test data on 16 April 2014, with the establishment of an EU web portal and database for such data.[118] The preamble includes the following (italic emphases added):

> (67) In order to ensure a sufficient level of transparency in the clinical trials, *the EU database should contain all relevant information as regards the clinical trial* submitted through the EU portal. *The EU database should be publicly accessible* and *data should be presented in an easily searchable format*, with related data and documents linked together by the EU trial number and *with hyperlinks*, for example linking together the summary, the layperson's summary, the protocol and the clinical study report of one clinical trial, as well as linking to data from other clinical trials which used the same investigational medicinal product. All clinical trials should be registered in the EU database prior to being started. As a rule, the start and end dates of the recruitment of subjects should also be published in the EU database. No personal data of data subjects participating in a clinical trial should be recorded in the EU database. *The information in the EU database should be public, unless specific reasons require that a piece of information should not be published*, in order to protect the right of the individual to private life and the right to the protection of personal data, recognised by Articles 7 and 8 of the Charter. *Publicly available information contained in the EU database should contribute to protecting public health and fostering the*

> interests is a duty for all responsible stakeholders involved, not just for regulators.
>
> A managed-release environment that allows sharing of patient-level data while ensuring patient privacy would create a level playing field for all stakeholders. What is sometimes labeled as "free riding" may ultimately pay dividends for innovative companies and for public health. It is ironic that the organizations that most resist wider access to data are the ones that stand to benefit so much from greater transparency.

See Hans-Georg Eichler, Frank Pétavy, Francesco Pignatti and Guido Rasi, Access to Patient-Level Trial Data – A Boon to Drug Developers, *N. Engl. J. Med.* Vol. 369 (2013), 1577–1579, available at: www.nejm.org/doi/full/10.1056/NEJMp1310771 (accessed 27 June 2014). See also Report, Ben Hirschler, *Ending Drug Trial Secrecy Is a Boon for R&D, Says EU Agency*, 21 October 2013, Reuters, www.reuters.com/article/2013/10/21/us-clinicaltrials-europe-idUSBRE99K11120131021.

118 *Regulation (EU) No 536/2014 of the European Parliament and of the Council of 16 April 2014 on clinical trials on medicinal products for human use, and repealing Directive 2001/20/EC*. For the legislative history of this regulation, see www.europarl.europa.eu/oeil/popups/ficheprocedure.do?reference=2012/0192%28COD%29&l=EN#foreCast. See generally comments at: www.ip-watch.org/2014/03/14/eu-parliament-to-vote-on-compulsory-publishing-of-clinical-trial-data/ and (accessed 27 June 2014).

*innovation capacity of European medical research, while recognising the legitimate economic interests of sponsors.*[119]

The new regulation also promotes clinical trials for "orphan" medicinal products.[120]

Undoubtedly, the EMA decision to have greater transparency in clinical data sharing and the above regulation are significant steps, even though business confidential information is still protected, as an *exception* permitting disclosure exists.[121] The question of clinical data sharing is presently under study in the United States as well.[122]

The EMA policy came into effect on 1 January 2015.[123] Unsurprisingly, litigation has already been initiated by two pharmaceutical companies to

119 Notably, para. (25) in the preamble further provides: "In order to *increase transparency in the area of clinical trials*, data from a clinical trial *should only be submitted* in support of a clinical trial application *if that clinical trial has been recorded in a publicly accessible and free of charge database which is a primary or partner registry of, or a data provider to, the international clinical trials registry platform of the World Health Organization* (WHO ICTRP) . . ." [Italic emphasis added].

120 *Ibid.*, preamble, para. (6). See generally EMA, www.ema.europa.eu/ema/index.jsp?curl=pages/special_topics/general/general_content_000556.jsp (accessed 29 December 2014).

121 Art. 81(4) of the regulation provides as follows:

> The EU database *shall be publicly accessible unless*, for all or part of the data and information contained therein, confidentiality is justified on any of the following grounds:
>
> (a) protecting personal data in accordance with Regulation (EC) No 45/2001;
> (b) *protecting commercially confidential information*, in particular through taking into account the status of the marketing authorisation for the medicinal product, *unless there is an overriding public interest in disclosure*;
> (c) protecting confidential communication between Member States in relation to the preparation of the assessment report;
> (d) ensuring effective supervision of the conduct of a clinical trial by Member States.
>
> [Italic emphasis added]

122 See FDA Notice, *Availability of Masked and De-Identified Non-Summary Safety and Efficacy Data; Request for Comments*, 6 April 2013, FDA (accessed 27 June 2014). The comment period was subsequently extended to 1 October 2013. See also Michelle M. Mello, Jeffrey K. Francer, Marc Wilenzick, Patricia Teden, Barbara E. Bierer and Mark Barnes, Preparing for Responsible Sharing of Clinical Trial Data, *N. Eng. J. Med.*, Vol. 369 (24 October 2013), 17.

123 For the updated version of the EMA policy as of October 2014 and related questions and answers, see, respectively, www.ema.europa.eu/docs/en_GB/document_library/Other/2014/10/WC500174796.pdf and www.ema.europa.eu/docs/en_GB/document_library/Report/2014/10/WC500174378.pdf (accessed 29 December 2014). At the time of writing, the implementation of the Clinical Trials Regulation was underway with

prevent disclosure of certain information.[124] One of them, initiated by AbbVie, was in relation to its drug that is reportedly the "highest earning drug in the world" at present.[125] The two lawsuits have since been withdrawn upon redaction of certain information by the EMA, but the redaction process itself has been criticized.[126]

### *7.7.5 Public funds and a new data-sharing bargain*

In the case of publicly funded research yielding profitable, safe and efficacious medicines, governments should recognize and institute a "data bargain", in parallel with the more well known "patent bargain". The shaping of the bargain could yield different social dividends in public health, including the following:

- Sharing of profits with the funding government, with the government's share being reinvested in further drug discovery for drugs of priority interest to the country, and/or in funding the poorest in the country to allow them access to medicines which are most needed;
- Requiring as part of the terms of funding the licensing of the resultant data to government-funded entities to conduct further research, to produce a "virtuous cycle" in innovation on the original research. The subsequent research entities would likewise be required to further license their data to other similarly funded entities.

There are several advantages to such a model. First, the access would be assured at the point that funding is provided. Second, the sharing of data is strategically limited to other researchers who will be required to submit to the same terms of data sharing when their work produces further data. This promotes the "virtuous cycle" of innovation and data sharing. Third, each research entity that receives funding under this model would receive

completion of a Functional Specifications document; see EMA, www.ema.europa.eu/ema/index.jsp?curl=pages/regulation/general/general_content_000629.jsp&mid=WC0b01ac05808768df (accessed 12 April 2015).

[124] See, for example, Stephanie Bodoni, *AbbVie, Intermune Sue to Block Clinical-Trial Data Release*, Bloomberg, 11 March 2015, www.bloomberg.com/news/articles/2013-03-11/abbvie-sues-eu-regulator-to-block-clinical-trial-data-release (accessed 12 April 2015).

[125] Ben Adams, *Intermune Drops EMA Lawsuit over Data Release*, Pharmafile, 6 October 2014, www.pharmafile.com/news/187640/intermune-drops-ema-lawsuit-over-data-release (accessed 12 April 2015).

[126] See European Ombudsman, *Letter to the EMA*, dated 16 April 2014, www.ombudsman.europa.eu/en/cases/correspondence.faces/en/58319/html.bookmark; and EMA, www.ema.europa.eu/ema/index.jsp?curl=pages/news_and_events/news/2014/11/news_detail_002210.jsp&mid=WC0b01ac058004d5c1 (accessed 12 April 2015).

licence fees for its data as well as have the data safeguarded in their targeted release and the further access to others' new data generated. Fourth, such a model not only does not stifle innovation (since it provides an entity a choice whether to take up such funding and its terms or not) but would rather promote it through a regenerative research strategy. Finally, the government providing the funding can not only be assured of data access for further research and innovation in drug discovery, but also that its funds bring tangible returns from each funded entity in the form of such assured access.

An example of a publicly funded research network that follows this idea – albeit not with respect just to data sharing – already exists in Asia. Temasek Life Sciences Laboratory (TLSL) is a research institute in biomolecular science research and applications, a beneficiary of Temasek Trust, the philanthropic arm of Singapore's sovereign wealth fund, Temasek Holdings.[127] Through agreements for transfer of research materials in its repository, TLSL permits researchers to make use of research findings to promote further R&D. In return, if such use results in any invention, modification or improvement material, TLSL requires a non-exclusive, worldwide, royalty-free licence from the researcher to make, use, sell and offer for sale the improvement, with a right to sub-licence it as well.[128]

## 7.8 Treaty-based approaches

### *7.8.1 Proposals for a new multilateral treaty*

There have been proposals for a new multilateral treaty in the form of either an R&D treaty or a global public health treaty.[129] In 2010, the

[127] TLSL is funded as a non-profit organization with commercialization of applications through Temasek Life Ventures. See www.tll.org.sg/about-us/introduction/, www.temasektrust.org.sg/#beneficiaries and www.tlv.sg/about/ (accessed 2 April 2014).

[128] TLSL's Material Transfer Agreement is available at: www.tll.org.sg/ip-office/material-transfer-agreement-mta/ (accessed 14 April 2015). In the US and EU, government-funded drug research programmes also exist. In the US, the Bayh-Dole Act, P.L. 96–517, Patent and Trademark Act Amendments of 1980, provides for the government to vest IP rights in funded entities with conditions permitting the government to, for example, practise, or allow to be practised on its behalf, use of the funded patent product, worldwide, on the basis of a "paid-up" license (i.e. no further royalty required): §202(c). On the National Institutes of Health research-sharing policies, see http://grants.nih.gov/grants/sharing.htm. In the EU, the Innovative Medicines Initiative (IMI, and more recently IMI2) also provides government funding for pharmaceutical research with conditions; see the IMI2 IP Policy documents at www.imi.europa.eu/content/documents#ip_policy (accessed 29 April 2015).

[129] See, for example, www.who.int/phi/Bangladesh_Barbados_Bolivia_Suriname_R_DTreaty.pdf (accessed 15 December 2014); and WHO CIPIH report, footnote 13

WHO formed a Consultative Expert Working Group (CEWG) to examine R&D financing and coordination.[130] In 2012, the CEWG issued its report.[131] One of its recommendations was that "formal intergovernmental negotiations should begin for a binding global instrument for R&D and innovation for health".[132] Initial responses to a global R&D treaty were not enthusiastic, and, although in 2012 there appeared to be a somewhat better reception of the idea,[133] it is still not clear if such a treaty will materialize, as further discussions have been put on hold until 2016.[134] In April 2014, a "platform" for a Framework Convention on Global Health was launched.[135] One of the key principles envisaged by the proposal, which seeks to protect the right to health in the context of international trade treaties, is as follows:

> h) Ensure that international law in non-health sectors is interpreted consistent with the right to health and protect against measures in treaties in other sectors, including trade and investment, that undermine health.[136]

It remains to be seen whether such a multilateral instrument will gain sufficient support, given the current fractious "conversations" in this area.

above, at p. 90. See also Nicoletta Dentico, *The Courage to Change the Rules: A Proposal for an Essential Health R&D Treaty* (2005). See also Gostin, LO, Friedman, EA, Buse, K., *et al.*, Towards a Framework Convention on Global Health, *Bull. World Health Organ.*, Vol. 91 (2013), 790–793.

130 See WHO, www.who.int/phi/cewg/en/ (accessed 15 December 2014).

131 CEWG, *Research and Development to Meet Health Needs in Developing Countries: Strengthening Global Financing and Coordination* (April 2012).

132 *Ibid.*, p. 120.

133 See Rebecca Herscher, Pharma Backs Latest Attempt at a Global Health R&D Treaty, *Nature Medicine* (2012), 838. For a discussion of possible elements in a treaty, see www.cptech.org/workingdrafts/rndtreaty.html (accessed 15 December 2014).

134 See WHO, *Report of the Open-Ended Meeting of Member States on the Follow-Up of the Report of the Consultative Expert Working Group on Research and Development: Financing and Coordination*, Geneva, 26–28 November 2012, A/CEWG/4, *Annex*, at p. 5; and Rachel Kiddell-Monroe, Johanne Helene Iversen and Unni Gopinathan, Medical R&D Convention Derailed: Implications for the Global Health System, *J. Health Dip.* (June 2013).

135 Eric A. Friedman, K. Srinath Reddy, Juliana Nantaba, Geetanjali Misra and Armando De Negri Filho (on behalf of the Interim Steering Committee of the Platform for a Framework Convention on Global Health), A Platform for a Framework Convention on Global Health, *The Lancet*, Vol. 383, No. 9936 (June 2014), 2207–2208; and www.globalhealthtreaty.org/.

136 Para. (h), *Social Determinants of Health and Global Governance for Health*, www.globalhealthtreaty.org/ (accessed 27 June 2014).

### 7.8.2 *Re-imagining the trade and investment treaty bargain and its relationship to public health*

While the result of treaty negotiations often depends on the bargaining power of the parties, because, for example, one party may be a developing country with less economic clout and resources than the other(s), it is possible to consider negotiating requests which take into account, precisely, the less advantageous position of the former. For example, one could "re-imagine" the patent bargain as it is being shaped at the negotiating table: the more stringent a TRIPS-plus requirement is being requested by the other party, the more of a trade-off in other critical areas and suitable exceptions could be exacted.[137] The objective would be to establish a more balanced "treatified" patent bargain, such as requiring a greater level of *quid pro quo* if more stringent TRIPS-plus provisions are demanded. Of course, negotiating parties will have to examine trade-offs in other parts of the FTA as a whole as well, such as in trade in goods, services and investment.

Another under-explored dimension of re-imagining the treaty *quid pro quo* is for a country to consider – in exchange for data exclusivity protection commitments demanded by a treaty negotiation partner – requirements of some or all of the following to be incorporated into the treaty:

- Access to voluntary licenses at reduced royalty rates of such data for patented pharmaceuticals of interest produced by businesses of the

[137] See, for example, Chapter 5 for examples of exceptions that have been observed in recent FTAs. The WHO's Global Strategy and Plan of Action on Public Health, Innovation and Intellectual Property (GSPOA) has recognized the need for specific action within the implementation time frame of 2008–2015 to

> ... (b) *take into account, where appropriate, the impact on public health when considering adopting or implementing more extensive intellectual property protection than is required by the Agreement on Trade-Related Aspects of Intellectual Property Rights,* without prejudice to the sovereign rights of Member States;
>
> (c) *take into account in trade agreements the flexibilities* contained in the Agreement on Trade-Related Aspects of Intellectual Property Rights, including those recognized by the Doha Declaration on the TRIPS Agreement and Public Health and the WTO decision of 30 August 2003 ...
>
> [Italic emphases added]

Sub-elements 5(2)(b) and (c), p. 31, WHO, (2011). Element 5 relates to the strategy of "Application and management of intellectual property to contribute to innovation and promote public health".

treaty partner (e.g. by reference to a "positive list" of medicines and their improvements, or of medicines for priority diseases, as identified by the country seeking such licenses);

- Access at agreed reduced price rates applicable to a "positive list" of medicines and their improvements, or of medicines for priority diseases, as identified by the country seeking such licenses;
- Where a pharmaceutical product of interest to the country has been invented through public funding or co-funding from the treaty partner, access to such product at highly discounted prices to users in the country.

How the treaty negotiating partner arranges for the above to be implemented on its side may be left to it to negotiate with its own business entities which own patents, so as to be in compliance with its other international obligations.

In terms of investment protection commitments, treaty negotiators in Asia will face decisions on the level of legal exposure that is acceptable in terms of the standard of treatment, what amounts (or does not amount) to expropriation which calls for compensation, the nature and limits of such compensation, and the extent that investor-state dispute settlement is to be offered in an FTA. Presently, at one end of the spectrum, there are treaties in which there are no TRIPS-plus provisions – provisions which exclude patent provisions from ISDS – and at the other, those with TRIPS-plus provisions with commitments which may give rise to ISDS claims by patent holders.[138] Other possible provisions to counterbalance such requests could be those setting out a better articulated role for national competition law, such as its role in checking anti-competitive conduct and arrangements such as unreasonable refusals to licence, pay-for-delay and anti-competitive patent litigation settlement agreements.

Where possible, transparency in negotiations should be promoted, especially where there are provisions that impact public health.

[138] The Korea-Canada FTA, for example, defines "investments" subject to treaty protection broadly; it includes intellectual property rights, which in turn are defined broadly, as follows: "**intellectual property rights** means copyright and related rights, trademark rights, rights in geographical indications, rights in the industrial designs, patent rights, rights in layout designs of integrated circuits, and rights in relation to protection of undisclosed information"; see Art. 8.1 and 8.45.

### *7.8.3 The need for health representatives in the treaty negotiation process and to be in close cooperation with other national agencies*

The GSPOA recognizes that health officials need to be integrated into negotiations of IP-related treaties.[139] A whole-of-government approach to public health policy buttresses such integration, permitting more coherent policy-making.[140]

National agencies in some countries, such as in South America, have formed new cooperative relationships to better address the interface of IPRs and public health. An example is the patent examination guidelines jointly issued by the country's Ministry of Industry, Ministry of Health and National Industrial Property Institute.[141] In Brazil, ANVISA, the health surveillance authority, is given power to analyse pharmaceutical patent applications with the National Industrial Property Institute (INPI, the Brazilian patent office), and it appears that ANVISA's prior consent may be required for such applications.[142]

Another axis of collaboration is between the health, IP and competition agencies. In 2014, for example, the Canadian Intellectual Property Office (CIPO) signed a memorandum of understanding with the Competition Bureau of Canada to work with each other as they fulfil their respective mandates.[143] Examples of such collaboration can also be seen in competition guidelines which specifically address IPRs and related anti-competitive practices.[144] More generally, the Canadian

[139] GSPOA, sub-element 5.1(g), p. 30, footnote 137 above:

> (g) promote active and effective participation of health representatives in intellectual property related negotiations, where appropriate, in order that such negotiations also reflect public health needs

[140] The "whole-of-government" approach has been applied in the areas of e-government (see http://unpan3.un.org/egovkb/Portals/egovkb/Documents/un/2012-Survey/Chapter-3-Taking-a-whole-of-government-approach.pdf) and trade policy (see, for example, India, *Foreign Trade Policy Statement, 2015–2020*, at p. 1, http://dgft.gov.in/exim/2000/ftp2015-20E.pdf and http://dgft.gov.in/exim/2000/FTPstatement2015.pdf (accessed 21 April 2015)).

[141] See http://lists.keionline.org/pipermail/ip-health_lists.keionline.org/2012-May/002145.html (accessed 22 August 2014), and Brazil's Patent Reform (2013), note 4 on p. 15.

[142] Agência Nacional de Vigilância Sanitária, ANVISA website (accessed 22 August 2014); see also Brazil's Patent Reform (2013), Chapter 8. INPI stands for Instituto Nacional da Propriedade Industrial; see www.inpi.gov.br/portal/.

[143] CIPO website, www.cipo.ic.gc.ca/eic/site/cipointernet-internetopic.nsf/eng/wr03772.html (accessed 22 August 2014).

[144] See, for example, *Draft IP Enforcement Guidelines* (2014), Canadian Competition Bureau.

government has also been considering how to balance innovation, patent rights, competition and affordable medicines.[145] This is consistent with a whole-of-government approach recognizing health as a cross-cutting priority policy area, which is discussed below.[146]

#### 7.8.3.1 Balancing with the non-trade attractions of regional trade agreements

FTA participation can lead to greater liberalization and therefore greater movement of goods, services and investments. Extending commitments in such agreements to a select and strategically chosen group or sole trading partner can speed up negotiations, in contrast to the much more (and increasingly) complex trade liberalization negotiations at the WTO. As a result, the successful conclusion of negotiations for a bilateral (or regional) FTA can bring earlier results, in terms of both implementation and economic benefits. However, FTAs usually require parties to extend themselves beyond what they might have already agreed to at the multilateral level. This requires intensive negotiations, both externally and internally, and can give rise to negotiating and policy tension as such a push may be made by placing sensitive areas on the table. Examples of difficult issues that have arisen in some Asian FTA negotiations may include government procurement, intellectual property rights protection and regulation over pharmaceutical and tobacco products.

Apart from real economic benefits that may arise from FTA participation, given the rapid rise in FTAs in Asia, a fear of being left out of preferential trade and investment opportunities can constitute a real incentive for some states to start their own FTA negotiations, as such preferences may seriously erode competitiveness. This is because FTA preferences are not required by states to be extended to other trade partners on a most-favoured nation (MFN) basis (unlike the general case at the multilateral level of the WTO). FTAs are a recognized avenue of non-MFN preferential treatment under WTO rules.[147] In particular, the allure of large preferential trading blocs such as those that might arise potentially under the TPP, TTIP and RCEP encourages states to join the "bandwagon".

[145] See Government's Response to the Third Report of the House of Commons Standing Committee on Industry, Science and Technology, *Intellectual Property Regime in Canada*, June 2013.

[146] For more on a whole-of-government approach, see Chapter 8.

[147] See Art. XXIV, *GATT 1994*.

For some Asian states, FTAs are not simply about economic considerations, but may encompass broader, geopolitical concerns.[148] Often trade and investment concessions are accepted together with non-economic factors as a package.[149]

Public authorities need to consciously strike a balance between trade and non-trade benefits and any "dis-benefits" of entering an FTA, especially where the counter-party's negotiating demands include a significant change to the existing system. An example would be public health regulations such as those affecting how medicines are approved, marketed and sold, and who may do so. Another would be balancing trade liberalization by removal of tariff and non-tariff barriers with

[148] As the Chief Negotiator of Singapore for the USSFTA, Ambassador Tommy Koh, explained:

> Singapore wants an FTA with the U.S. for a combination of economic and strategic reasons. The U.S. is Singapore's largest foreign investor and second largest trading partner. The U.S. is also Singapore's most important source of technology and management know-how. *Singapore's interest in the U.S., however, transcends business and economics. Singapore wishes to entrench the presence of the U.S. in the region because it underpins the security of the whole Asia-Pacific region. Singapore regards the U.S.-Singapore FTA as a symbol of continued U.S. commitment to the region. Therefore, for Singapore, the USSFTA is not just about securing tariff-free entry for Singapore's exports to the U.S. market. It is not just about attracting more Foreign Direct Investment (FDI) to Singapore. It is also about enhancing the prospects of peace and stability in the region.*
>
> [Italic emphasis added]

Tommy T. B. Koh, Chapter 1, *The USSFTA: A Personal Perspective*, in *The United States Singapore Free Trade Agreement – Highlights and Insights*, (2004), Institute of Policy Studies & World Scientific Publishing Co. Pte. Ltd.

[149] As another member of the Singapore negotiating team explained:

> For example, in negotiating the USSFTA, Singapore was mindful that there were some difficult areas, including the area of intellectual property rights where we had to consider committing to TRIPs-plus standards. While this was seen as an opportunity to speed up the review of Singapore's then IP system, clearly there were trade-offs that needed to be taken into consideration when accepting the overall "package" of obligations.
>
> . . . With respect to pharmaceutical products, for example, the negotiations over parallel imports apparently remained a sticking point and this was resolved only at the last day of negotiations.

Liew Woon Yin, then Director-General, Intellectual Property Office of Singapore (IPOS) and USSFTA negotiator, in Chapter 12, *Intellectual Property Rights, The United States Singapore Free Trade Agreement, – Highlights and Insights*, ed. Tommy Koh and Chang Li Lin, (Institute of Policy Studies, Singapore and World Scientific Publishing Co. Pte. Ltd, 2004), at pp. 126 and 130.

respect to tobacco products against existing tobacco control laws, policies and international obligations. The demands of negotiation may require considering changes that bring a risk of higher pharmaceutical prices. Finally, the potential application of investor-state dispute settlement (such as arbitration) to such provisions raise the potential of costly and high-profile disputes brought by investors.

Authorities through internal consultation often make a final decision which consists of a mix of economic and political judgment. As a result, one question that Asian treaty negotiators often have to consider is how the product of the final decision will affect the country in the long term, especially where a treaty carries investor-state dispute mechanisms and associated risks.[150] In coming to a decision on final trade-offs, a negotiator may have to deal with the tension of the potential trade-investment/public health interface, such as where TRIPS-plus provisions – not hitherto part of the national system – are being demanded. An Asian state may therefore aim to agree to necessary trade-offs but find it necessary to ask for safe harbour provisions, to strengthen the *quid pro quo*.

One way to ensure inclusion of non-trade concerns that apply over the longer term to a negotiating state – such as public health issues – is to adopt an approach that consciously and meaningfully enquires as to and properly debates the implications of trade arrangements for public health, such as a Health in All Policies (HiAP), health technology assessment or health impact assessment approach (see later, and Chapter 8), before signing a trade agreement.[151]

#### 7.8.3.2 Assessing trade, investment, innovation and medicines using a whole-of-government approach

The WHO CIPIH Report of 2005 explicitly recommends as follows:

> Developing countries need to decide in the light of their own circumstances, what provisions, consistent with the TRIPS Agreement, would

[150] See Chapter 4.

[151] In the case of Australia, South Australia has adopted the "health lens analysis" approach; see www.sahealth.sa.gov.au/wps/wcm/connect/public±content/sa±health±internet/health±reform/health±in±all±policies/health±lens±analysis±projects and www.health.gov.au/internet/nhhrc/publishing.nsf/Content/458/$FILE/458%20-%20O%20-%20SA%20Health%20-%20SASP%20 . . . through%20a%20health%20lens.pdf.

See generally Deoraj Caussy, Priti Kumar and U. Than Sein, Health Impact Assessment Needs in South-east Asian Countries, *Bulletin of the World Health Organization*, Vol. 81 (2003), 439–443, in which the authors conclude that implementation of health impact assessment in the WHO's Southeast Asia Region group of countries was, in 2004, "at a rudimentary stage" (p. 442).

> benefit public health, *weighing the positive effects against the negative effects.* A public health justification should be required for data protection rules going beyond what is required by the TRIPS Agreement. There is unlikely to be such a justification in markets with a limited ability to pay and little innovative capacity. Thus, developing countries should not impose restrictions for the use of or reliance on such data in ways that would exclude fair competition or impede the use of flexibilities built into TRIPS.
>
> In bilateral trade negotiations, it is important that governments ensure that ministries of health by properly represented in the negotiation, and that the provisions in the texts respect the principles of the Doha Declaration. Partners should *consider carefully any trade-offs they may make in negotiation.*[152]
>
> [Italic emphasis added]

It has also been suggested that public health impact assessments be carried out on FTAs being negotiated, prior to their being signed, to allow governments to better address and negotiate provisions which are expected to affect public health systems in their countries.[153]

This kind of approach lays a strong foundation for health and pharmaceutical policies to be allowed to permeate all relevant areas of policy-making. Increasingly, such areas include international trade law and trade and investment treaty-making. Suggestions have been made, for example, for public health assessments of FTAs.[154]

In the context of trade and investment treaty negotiations, health issues often form but part of the "package", without necessarily getting as full an assessment in impact as the economic benefits of the "package" get. After all, it is often the trade or commerce ministry officials who lead such negotiations; the health ministry colleagues are likely to be consulted, but they may not form part of a core team that provides input in every aspect of the deal.

In 2000, the European Council emphasized the "development of healthy policies in all other areas of Community activity" and "insist[ed] on the need to develop expertise in evaluating the impact of the other

[152] WHO, CIPIH Report, footnote 13 above, recommendations 4.20–4.21 at p. 181. See also Rhona MacDonald and Richard Horton, Trade and Health: Time to Get the Health Sector Involved, *The Lancet*, Vol. 373, No. 9660 (2009), pp. 273–274.

[153] See, for example, Frederick M. Abbott, Toward a New Era of Objective Assessment in the Field of TRIPS and Variable Geometry for the Preservation of Multilateralism, *JIEL*, Vol. 8, No. 1 (2005), 77–100, at pp. 94–97.

[154] See, for example, UNITAID, *The Trans-Pacific Partnership Agreement: Implications for Access to Medicines and Public Health*, March 2014, pp. 101–102.

policies on health and its determinants".[155] In 2006, the Health in All Policies (HiAP) approach was first promoted during Finland's presidency of the EU.[156]

HiAP promotes the assessment of health impact of policies in all areas of government policy by linkage to five main "determinants": The rationale behind this is that health is often seen to be a mere handmaiden in the process of globalization:

> Globalization, economic integration and the strengthening of the commercial legal framework at global and regional levels have also implied that the interests of the private sector, markets and competitiveness are considered to be of greater importance than health. In this policy environment, implementation of such regulatory, public health, social determinant- or equity-oriented policy measures, which restrict the free mobility of goods, services and people or limit commercial and investment opportunities, has become more easily contested.[157]

The objective is therefore to move health higher up in the EU agenda.[158] HiAP also advocates the use of Health Impact Assessment.[159]

The whole-of-government approach to policy-making is not a new phenomenon in Asia. It is employed, for example, in Singapore.[160]

[155] European Council Resolution of 29 June 2000 on Action on Health Determinants (2000/C 218/03).

[156] *Health in All Policies: Prospects and Potentials*, www.euro.who.int/__data/assets/pdf_file/0003/109146/E89260.pdf; and WHO, *Cross-Country Analysis of the Institutionalization of Health Impact Assessment*, Social Determinants of Health Discussion Paper 8 (2013), available at: http://apps.who.int/iris/bitstream/10665/83299/1/9789241505437_eng.pdf (accessed 27 July 2014). As explained by a study by the Finnish Ministry of Social Affairs and Health and the European Observatory on Health Systems and Policies:

> Health in All Policies is an encompassing approach which goes beyond the boundaries of the health sector. It addresses all policies such as transport, housing, the environment, education, fiscal policies, tax policies and economic policies. It is based on values and principles similar to those in the WHO's call for multisectoral action for health, and the concept of building healthy public policies, or the whole government approach.
>
> [Original footnotes omitted; italic emphases added. P. xxvii, *ibid.*]

[157] P. 10, *ibid.* [158] Chapter 2, *ibid.* [159] Chapter 6, *ibid.*

[160] See Peter Ho, former head of the Singapore Civil Service, *Coping with Complexity*, www.mckinsey.com/insights/public_sector/coping_with_complexity, McKinsey *Insights and Publications*, October 2012; and, generally, on the government's *Centre for Strategic Futures*, at www.psd.gov.sg/content/psd/en/csf/csf_aboutus/csf_vision_mission_key.html. See also Kharina Zainal, *Reviewing Whole-of-Government Collaboration in the Singapore Public Service, Ethos* (Singapore Civil Service College publication), Issue 9, June 2011.

Singapore also has a health technology assessment unit in the Ministry of Health.[161]

In 2014, ASEAN health ministers took an important step towards greater use of HiAP, by committing to "advocate" HiAP to the APSC and AEC.[162]

#### 7.8.3.3 Emergence of health technology assessment (HTA)

According to the WHO, HTA "studies the medical, social, ethical, and economic implications of development, diffusion, and use of health technology, as defined by the International Network of Agencies for Health Technology Assessment (INAHTA)".[163]

In 2011, the EU set out its Third Health Programme, in which it advocated health technology assessment (HTA).[164]

161 See Keng Ho Pwee, Health Technology Assessment in Singapore, *Int. J. of Tech. Assmt. H.C.*, Vol. 25, Supp. (2009), 234–240.

162 Para. 12, Joint Statement, 12th ASEAN Health Ministers Meeting, 18 September 2014; see Chapter 1, footnote 136 and accompanying text.

163 WHO, Patricia Whyte and Cameron Hall, *Working Paper 6: The Role of Health Technology Assessment in Medicine Pricing and Reimbursement*, June 2013, *Review Series on Pharmaceutical Pricing Policies and Interventions*, WHO/HAI Project on Medicine Prices and Availability, at p. 1. Other definitions can be found at: www.who.int/hia/about/defin/en/ (accessed 2 July 2014).

164 See para. 1.1, Annex 1, Proposal for a Regulation of the European Parliament and of the Council on Establishing a Health for Growth Programme, the third multi-annual programme of EU action in the field of health for the period 2014–2020; see http://ec.europa.eu/health/programme/docs/prop_prog2014_en.pdf (accessed 27 July 2014). This built on EU Directive 2011/24/EU of 9 March 2011 on the application of patients' rights in cross-border healthcare L88/45, which established support for cooperation in HTA. The EU explains HTA as follows:

> Health Technology Assessment (HTA) is way of assessing the ways science & technology are used in healthcare and disease prevention. It covers *medical, social, economic, and ethical issues.*
>
> It provides policy-makers with objective information, so they can formulate health policies that are *safe, effective, patient-focused and cost-effective.*
>
> HTA should be transparent, unbiased, robust and systematic – firmly rooted in research and the scientific method.
>
> Diagnostic and treatment methods, medical equipment, pharmaceuticals, rehabilitation & prevention methods, but also organisational and support systems used to deliver healthcare are examples of Health Technologies.
>
> [Italic emphases added]

See EU Commission website, http://ec.europa.eu/health/technology_assessment/policy/index_en.htm (accessed 30 June 2014). See also, generally, WHO, www.who.int/medical_devices/assessment/en/(accessed 30 June 2014); and Cliff S. Goodman, *Introduction to Health Technology Assessment*, www.nlm.nih.gov/nichsr/hta101/ta101_c1.html (accessed 30 June 2014).

HTA has been gaining greater acceptance in emerging economies including those of Asia, through actions such as inter-sectoral government committees examining policies. HTA in a formal, institutionalized policy, by this name, is still relatively new in Asia; Korea is the first country in Asia to formally adopt a formal HTA approach in policy-making.[165] Thailand, too, has adopted HTA.[166]

HiAP and HTA may prove to be useful approaches and tools for policy-makers confronted with competing policy area demands, particularly in the context of the trade/investment (economic) and social dimensions nexus.

## 7.9 The use of differential or tiered pricing: affordable prices and facilitative aspects in trade and investment treaties

Differential pricing, a useful policy-maker's tool, employs different price levels for pharmaceutical products according to factors such as the level of income or the ability of a country to pay for the products.[167] Usually,

[165] See Tessa Kennedy-Martin, Beth D. Mitchell, Kristina S. Boye, Wen Chen, Bradley H. Curtis, Jennifer A. Flynn, Shunya Ikeda, Li Liu, Yen Huei Tarn, Bong-Min Yang and Emmanuel Papadimitropoulos, The Health Technology Assessment Environment in Mainland China, Japan, South Korea and Taiwan – Implications for the Evaluation of Diabetes Mellitus Therapies, *Value in Health Regional Issues* (2013), 108–116; Emerging Hub for Pharmacoeconomics and Outcomes Research: Asia-Pacific Region, *Value in Health* (2013), 1–2. See also Bong-Min Yang and Kenneth Lee, Growing Application of Pharmacoeconomics and Outcomes Research in Healthcare Decision-Making in the Asia-Pacific Region, *Value in Health*, Vol. 12, Supp. 3 (2009); Sadasivan Sivalal, *Health Technology Assessment in the Asia Pacific Region, Int. J. Tech. Assessment. Healthcare*, Vol. 25, Supp. 1 (2009), 196–201; and Yang Bong-Min, The Future of Health Technology Assessment in Healthcare Decision Making in Asia, *Pharmacoeconomics*, Vol. 27 (2009), pp. 1–11. For China, see Y. Chen, D. Banta and Z. Tang, *Health Technology Assessment Development in China*, Int. J. Tech. Assess. Health Care, Vol. 25, Suppl. 1 (2009), pp. 202–209, www.ncbi.nlm.nih.gov/pubmed/19575825. For Singapore, see Keng Ho Pwee, *Health Technology Assessment in Singapore*, Int. J. of Tech. Assess. Health Care, Vol. 25, Supp. 1 (2009), pp. 234–240.

[166] See Y. Teerawattananon, S. Tantivess, J. Yothasamut, P. Kingkaew and K. Chaisiri, Historical Development of Health Technology Assessment in Thailand, *Int. J. Technol. Assess. Health Care*, Vol. 25, Suppl. 1 (2009), pp. 241–252.

[167] See generally Prashant Yadav, *Differential Pricing for Pharmaceuticals – Review of Current Knowledge, New Findings and Ideas for Action*, (2010); and Patricia M. Danzon and Adrian Towse, Differential Pricing for Pharmaceuticals: Reconciling Access, R&D and Patents, *Int'l J. H. C. Fin. & Econ.*, Vol. 3, Iss. 3 (2003), 183–205. See also Jayashree Watal, *Background Note* for Workshop on Differential Pricing and Financing of Essential Drugs to the WTO Secretariat, WHO-WTO Workshop on Differential Pricing and Financing of Essential Drugs, 8–11 April 2001, Høsbjør, Norway.

this works by pricing pharmaceuticals for certain countries – such as least developed countries – significantly lower than for developed countries. Negotiations are conducted by procuring governments to secure affordable prices. Where a pharmaceutical product is not subject to a patent in a country (either because there has been no application or an application has been rejected by the state IP authority) and the country has the capacity to produce generic versions, the differential pricing strategy may be less urgent. This is however not true if the domestically produced medicines are of poor quality, so that many still rely on the innovator's product, albeit at a higher price than a generic alternative. In the case where a patent does exist for the innovator's product, the need to negotiate a differentiated, lower price becomes more urgent if the state does not wish to resort to using other options – such as compulsory licensing – to increase affordability and access.

Differential pricing is advocated at the international level. The WHO, for example, advocates that countries use "a combination of different pharmaceutical pricing policies that should be selected based on the objective, context and health system".[168]

The EU supports tiered pricing of pharmaceuticals for better access to developing countries, having articulated the following stance:

> ... tiered pricing for developing countries should no longer be the exception, but the rule. What is lacking, however, is a clear view of how to proceed in a way which is both *global* and *systemic*.[169]

In 2003, the EU established a framework for differential pricing, which allows pharmaceutical companies to utilize such pricing while taking advantage of regulatory commitments which address trade diversion of cheaper medicines back into the union. More is said of this system below.

[168] WHO, *Guideline on Country Pharmaceutical Pricing Policies* (2013), at p. 25. See also UNAIDS' advocated stance in *Getting to Zero: 2011–2015 Strategy*, Joint United Nations Programme on HIV/AIDS (UNAIDS), 2010, available at: www.unaids.org/en/aboutunaids/unaidsstrategygoalsby2015/:

> Partner with the pharmaceutical industry to implement tiered pricing for antiretroviral drugs and other HIV commodities in low- and middle-income countries to increase access to affordable medicines and speed up access to the next generation of treatment.
>
> (at p. 41)

[169] Paul Vandoren, then Head of Unit, Directorate-General for Trade, European Commission, *Financing and Differential Pricing: A Developed Country Government Perspective*, p. 6, WHO-WTO Workshop on Differential Pricing and Financing of Essential Drugs, 8–11 April 2001, Høsbjør, Norway.

Global pharmaceutical companies such as GSK (for vaccines procurement), Novartis, Roche and Sanofi have indicated their support for tiered pricing models.[170] Most recently, Gilead Sciences, a relatively young but important company in the industry, arranged for differential pricing for its new medicine, Sovaldi (see discussion below).[171]

### *7.9.1 Negotiating lower prices – an example: Egypt's negotiated price for Sovaldi*

A recent example is the negotiation of differentiated pricing between Egypt and Gilead Sciences in respect of Sovaldi (sofosbuvir being the non-proprietary name), the first all-oral medicine used for treating hepatitis C, a viral disease which causes inflammation of the liver, leading to liver disease or failure. Egypt has a very high level of incidence of hepatitis C.[172] This medicine was approved for treatment of chronic hepatitis C by the US FDA in December 2013[173] and, soon after, also received EU Commission approval in January 2014.[174] The WHO's April 2014 guidelines recommend the use of sofosbuvir, and in May 2015 it was added to the WHO's Essential Medicines List.[175]

According to the WHO, hepatitis C is a major global health problem and there is currently no vaccine for it.[176] The virus is classified into 11 genotypes: genotypes 1–3 have a worldwide distribution with genotype 3 being endemic specifically in Southeast Asia, while genotypes 6–11 are

[170] See generally Prashant Yadav, *Differential Pricing for Pharmaceuticals – Review of Current Knowledge, New Findings and Ideas for Action*, Study conducted for the UK Department for International Development, August 2010, and, for example, the following statements: www.gsk.com/media/280905/tiered-pricing-for-vaccines-policy.pdf, www.thisisafricaonline.com/Business/Novartis-CEO-Tiered-pricing-key-to-unlocking-African-pharma-markets?ct=true, http://timesofindia.indiatimes.com/business/india-business/Tiered-pricing-to-make-drugs-affordable-Sanofi-CEO/articleshow/23327798.cms and www.sanofipasteur.us/commitment/access (accessed 1 October 2014).

[171] UN agencies such as UNAIDS support the use of tiered pricing; see note 168 above.

[172] WHO website, www.who.int/csr/disease/hepatitis/whocdscsrlyo2003/en/index4.html (accessed 20 August 2014).

[173] See FDA announcement, 6 December 2013.

[174] See Gilead website, http://investors.gilead.com/phoenix.zhtml?c=69964&p=irol-newsArticle&ID=1891746 (accessed 22 July 2014).

[175] See WHO, *Guidelines for the Screening, Care and Treatment of Persons with Hepatitis C Infection*, April 2014, pp. 13–15, 19, 74–77 (referring to sofosbuvir). The author is grateful to Ms. Yuanqiong Hu, MSF Geneva, for the discussion of this issue with the author. On the addition to the EML, see footnote 108, Chapter 1, and accompanying text.

[176] WHO website, www.who.int/csr/disease/hepatitis/whocdscsrlyo2003/en/index3.html (accessed 20 August 2014).

found in various parts of Asia.[177] The stated benefits of Sovaldi include its oral form of use and the fact that it could be effective for persons of genotypes 1–4.[178]

The per-course-of-treatment price of the medicine is US$84,000 in the United States, US$67,000 in Germany, £35,983 (€45,041 or US$57,000) in the UK, and US$55,000 in Canada, and these prices have raised serious concerns on both sides of the Atlantic.[179]

Given that poorer countries would not be able to afford such prices, countries such as Egypt have negotiated for differential prices to make the medicine accessible. As a result of such negotiations, Gilead has agreed to a price of US$900 per course of treatment in Egypt; a patent application by Gilead for Sovaldi had earlier been rejected by the Egyptian patent authorities.[180] Within Asia, India and Myanmar will also benefit from this lowered price through procurement by Doctors without Borders (Médecins San Frontières) from Gilead.[181] These reduced pricing arrangements will exclude middle-income countries such as China, Indonesia and Thailand. Gilead has filed a patent application for sofosbuvir in India, which faces opposition.[182] In August 2014, it was announced that Gilead will also charge US$900 per course of treatment in India.[183] According to Gilead, such generic licensing is expected to give access to the medicine to 91 developing countries and account for 54% of individuals worldwide

[177] WHO website, www.who.int/csr/disease/hepatitis/whocdscsrlyo2003/en/index2.html (accessed 20 August 2014).

[178] See Sovaldi website, www.sovaldi.com/what-is-sovaldi/how-does-sovaldi-work (accessed 20 August 2014).

[179] See *Sovaldi Stumbles as NICE Assesses Hep C Drug's Price*, PMLive, 17 June 2014, www.pmlive.com/pharma_news/sovaldi_stumbles_as_nice_assesses_hep_c_drugs_price_578912; and Steve Miller, *The Sovaldi Tax: Gilead Can't Justify the Price It's Asking for Its Hepatitis C Therapy*, Forbes, 17 June 2014. The high cost of Sovaldi has led policy-makers and regulators in the US and the UK to question the basis of Gilead's pricing strategy; see, for example, Ransdell Pierson, *US Senators Ask Gilead to Explain Cost of Solvadi Hepatitis Drug*, 11 July 2014, Reuters; and Simeon Bennett, *UK Cost-Regulator May Reject Gilead's Sovaldi Treatment*, 16 June 2014, Bloomberg Businessweek.

[180] See Maggie Fick and Ben Hirschler, *Gilead Offers Egypt New Hepatitis C Drug at 99 Per Cent Discount*, 21 March 2014, Reuters. The author is grateful to Dr Peter Beyer, WHO, for the discussion with the author on this example of negotiated pricing.

[181] See Makiko Kitamura, *Developing Countries Seek Lower Price for Gilead Sovaldi*, 11 April 2014, Bloomberg report.

[182] See Zeba Siddiqui, *NATCO Seeks to Block Gilead's Hepatitis C Drug Patent in India: Source*, 10 April 2014, Reuters.

[183] See Rupalee Mukherjee, *New Hepatitis-C Drug 99% Cheaper in India*, 6 August 2014, The Times of India; Ed Silverman, *How Much? Gilead Will Charge $900 for Sovaldi in India*, (7 August 2014), Wall Street Journal; and John LaMattina, *Even at $900 (vs. $84,000 in US) Hep C Cure Sovaldi's Cost Could Be Unacceptable in India*, (8 August 2014), Forbes.

having hepatitis C.[184] In January 2015, Gilead announced that it would extend its licence for generic manufacture – subject to regulatory approval – of an investigational compound as part of a single tablet for treatment of all six genotypes of hepatitis C.[185] In the same month, the Indian authorities rejected Gilead's patent application for Sovaldi,[186] but in an appeal by Gilead heard by the Delhi High Court in February 2015, the matter was remanded for the Controller General to reconsider the application.[187] The patent application is still pending at the time of writing.

Despite the increased access through Gilead's licences, this has been criticized in that middle-income countries – such as Malaysia and Thailand – are excluded from such licensing arrangements. By March 2015, it was reported that a $10-per-pill version of Sovaldi had been produced and distributed by Incepta Pharmaceuticals Ltd. in Bangladesh.[188] As an LDC, Bangladesh does not have to comply with the requirements of patent protection under the TRIPS Agreement until 1 January 2016.

#### 7.9.1.1 Creating a sustainable framework for tiered pricing

Supply arrangements for HIV/AIDS drugs at reduced prices brokered by UN agencies have arisen from the Accelerating Access Initiative (AAI).[189] Differential pricing under the initiative has met with challenges.[190] While such pricing can offer improved access to medicines, it is not without its challenges.

184 Gilead Sciences, Inc., press release, 15 September 2014.

185 *Gilead Expands Hepatitis C Generic Licensing Agreements to Include Investigational Pan-Genotypic Agent*, Gilead Sciences, Inc., website, press release dated 26 January 2015. The regimen would combine the compound and sofosbuvir as a single tablet. At the time of writing, this compound is, according to Gilead, undergoing Phase 3 clinical studies evaluation.

186 See Ketaki Gokhale, *Gilead's India Patent Snag May Spur Low-Cost Sovaldi Copies*, 15 January 2015, Bloomberg.

187 *Gilead Pharmasset, LLC v Union of India & ANR*, WP(C) 687/2015 and CM No. 1222/2015. See also Report, Rupali Mukherjee, *HC Asks Patent Office to Review Gilead's Case*, (3 February 2015), Times of India.

188 See, for example, Ketaki Gokhale and Makiko Kitamura, *$10 Copy of Gilead Blockbuster Sovaldi Appears in Bangladesh* (9 March 2015) Bloomberg.

189 See www.who.int/hiv/pub/progressreports/accelerating/en/ and www.globalhealthprogress.org/programs/aai-accelerating-access-initiative. The AAI comprises a "partnership of five United Nations organizations (UNAIDS Secretariat, WHO, UNICEF, UN Population Fund, and the World Bank) and seven research-based pharmaceutical companies (Abbott, Boehringer Ingelheim, Bristol-Myers Squibb, Gilead Sciences, GlaxoSmithKline, Roche, and Merck & Co., Inc.)" subscribing to a joint statement of intent (AAI Factsheet 2006).

190 See Yadav, *Differential Pricing for Pharmaceuticals*, at pp. 22–23.

### *7.9.2 Disincentives for tiered pricing*

Presently, manufacturers of patented medicines in developed countries are concerned with possible diversion of tiered-price medicines flowing back into their own markets, eroding their profits in such markets.

Nonetheless, to the extent that it can contribute to increasing access to essential medicines in developing countries, it is important to examine whether the legal and regulatory landscape can be improved to allow it to function more effectively. This entails examining policies affecting both the supply and demand sides.

### *7.9.3 Supply-side concerns: trade diversion back into developed country markets*

Price arbitrage leading to flow-back of medicines from low-price markets back into developed country markets (for which the low prices were not meant) is a common concern expressed as being a challenge to successful implementation of tiered pricing systems by pharmaceutical companies, as such flow-backs erode their profits in the latter markets.[191] National and regional laws on exhaustion of patent and trademark rights also have a bearing on this concern. Another issue faced by suppliers is price arbitrage within countries with large income disparities.

Some countries in ASEAN, such as Vietnam, use external reference pricing, which is a source of concern for suppliers of pharmaceuticals.[192] It has been argued, for example, that reference pricing is a blunt instrument and that HTA, with its examination of factors beyond price, is a more appropriate tool for purposes of procurement price negotiations

[191] See, for example, Prashant Yadav, *Differential Pricing for Pharmaceuticals – Review of Current Knowledge, New Findings and Ideas for Action*, Study conducted for the UK Department for International Development, August 2010; Kevin Outterson, *Pharmaceutical Arbitrage: Balancing Access and Innovation in International Prescription Drug Markets*, Yale J. Health Policy, Law, and Ethics, Vol. 5, No. 1, Article 4 (available at: http://digitalcommons.law.yale.edu/yjhple/vol5/iss1/4); and Peter J. Hammer, Differential Pricing of Essential HIV Drugs: Markets, Politics and Public Health, *JIEL*, Vol. 5, No. 4 (2002), 883–912. See also Richard C. Morais, Psst . . . Wanna Buy Some Augmentin? *Forbes*, 12 April 2004. See also The New Drugs War, *The Economist*, January 2014.

[192] See T. A. Nguyen and A. A. Roughead, *Pharmaceutical Pricing Policies in Vietnam*, in *Pharmaceutical Prices in the 21st Century*, ed. Zaheer-Ud-Din Babar (Springer International Publishing, 2015), at pp. 332–333. See also, generally, Vietnam Ministry of Health and Health Partnership Group, *Joint Annual Health Review 2013, Towards Universal Health Care*; and Anh Tuan Nguyen, Rosemary Knight, Andrea Mant, Quang Minh Cao and Martin Auton, *Medicine Prices, Availability, and Affordability in Vietnam*, Southern Med Rev., Vol. 2 (2009), 2–9 (accessed 7 May 2015).

and for reimbursement price levels.[193] In the case of India, in 2007, a committee to examine price negotiations for patented drugs was established to look into pricing of patented medicines. The committee in its report recognized that price negotiations should not be carried out in an arbitrary manner[194] and made the following recommendations for three classes of medicines:

- For medicines with no therapeutic equivalence in India;
- For medicines having a therapeutic equivalent in India; and
- For medicines introduced for the first time in India itself.[195]

For the first two classes, the report recommended reference to procurement prices in certain developed economies, namely the UK, Canada, France, Australia and New Zealand.[196] The committee also recommended against linking marketing approval of pharmaceuticals to pricing.[197] This report was not implemented. In 2014, a new inter-ministerial committee was established, and the government is reportedly considering linking marketing approval to prices of pharmaceuticals.[198]

Under the TRIPS Agreement, WTO members have flexibility in their approach to exhaustion of such rights, and to what extent they wish to permit parallel trade of products. Countries may adopt a policy of non-

[193] See, for example, Lisa Brandt, *Price Tagging the Priceless: International Reference Pricing for Medicines in Theory and Practice*, ECIPE Policy Brief No. 04/2013, at p. 3.

[194] *Report of the Committee on Price Negotiation for Patented Drugs*, (2013); see Chapter 8, para. 8.3.

[195] Chapter 9 of the Report, *ibid.*, para. 9.5.

[196] The method of calculating the price for India would be as follows (see paras. 9.5(a) and (b)):

> The ratio of the per capita income of a particular country [in the list] to the per capita income of India would be calculated. The prices of the medicine would be worked out for India by dividing the price of the medicine in a particular country by this ratio and the lowest of these prices would be taken for negotiation for further reduction.

The reason for selecting these countries was as follows (at para. 9.4):

> The countries like UK, Canada, France, Australia and New Zealand have a wide coverage of health insurance by their governments and therefore have high bargaining power in deciding the price of such patented medicines through negotiations.

[197] See the report, para. 8.7.

[198] See Sushmi Dey, *Govt May Negotiate Price of Drugs before Market Entry*, (23 January 2015), The Times of India. On India's present pricing policy, see generally, India's *National Pharmaceuticals Policy 2012*, and Order S.O. 1221(E) of 15 May 2013, available at http://pharmaceuticals.gov.in/ (accessed 2 May 2015).

exhaustion, or exhaustion on a national, regional or international basis.[199] The EU system, discussed further below, utilizes the principle of regional exhaustion. In Asia, Japan and Singapore employ the international exhaustion principle.[200]

To deal with the concern of flow-back, some systems require manufacturers to use distinctive packaging or other indicia to distinguish products intended for low-price markets.

Arbitrage may also occur on a non-commercial scale in the form of buyers who are users. Examples include American citizens travelling to Canada to purchase medicines which cost less there than at home. Within Asia, it has also been known to happen.[201]

### *7.9.4 Demand-side concerns*

Among developing countries, on the other hand, India has expressed the view that while differential pricing is helpful, it does not see it as a lasting or entire solution, and would not obviate for instance, the need for the Paragraph 6 system.[202] In addition, for developing countries with large discrepancies in income within their territories, in particular, differential pricing may be difficult to implement.[203] Another concern is that small populations may not be targeted.[204] Finally, developing countries would

[199] See WIPO website at www.wipo.int/sme/en/ip_business/export/international_exhaustion.htm (accessed 30 October 2014); and WIPO Secretariat, *Interface between Exhaustion of Intellectual Property Rights and Competition Law*, CDIP/4/4 REV./STUDY/INF/2, 1 June 2011.

[200] For Japan, see WIPO Secretariat document, *ibid.*, paras. 58–61. For Singapore, see Section 66(2)(g) read with Sections 66(3) and (5A) of the Patents Act, and section 29 of the Trade Marks Act, 2008 Rev. Ed.

[201] See, for example, Peggy Teo, Kalyani Mehta, Leng Leng Thang and Angelique Chan, *Aging in Singapore: Service Needs and the State* (Routledge, 2006), p. 81; and Malaysia's Government to Review JB-Singapore Toll Hike, Says DPM, *The Straits Times*, 30 July 2014.

[202] Para. 252, Council for Trade-Related Aspects of Intellectual Property Right, Minutes of meeting held on 26–27 October 2010, IP/C/M/64.

[203] See report on an interview with the CEO of Novartis, (11 December 2013) www.thisisafricaonline.com/Business/Novartis-CEO-Tiered-pricing-key-to-unlocking-African-pharma-markets?ct=true.

[204] See generally Lynne Taylor, *Pharma "Crossing New Frontiers with Global Tiered Pricing"*, Pharma Times, www.pharmatimes.com/article/14-08-25/Pharma_crossing_new_frontiers_with_global_tiered_pricing.aspx(accessed 1 October 2014); and IHS Study report, http://press.ihs.com/press-release/country-industry-forecasting-media/pharmaceutical-industry-crosses-new-frontiers-globa (accessed 1 October 2014).

need to examine how differential pricing arrangements might be affected by their existing competition laws.[205]

### *7.9.5 The EU tiered pricing system for developing countries – a regional model addressing the supply-side concerns?*

While pharmaceutical manufacturers may handle the concern of flow-back on their own, some regulatory authorities have also attempted to provide a legal environment that deals with such risks to give some peace of mind to manufacturers. The EU provides one example where such an attempt has been made, through its tiered pricing regime introduced in 2003.

The EU scheme permits tiered pricing for pharmaceutical products to be exported to developing countries, with legal safeguards against such products flowing back into the EU.[206] Asian countries targeted as beneficiaries of the scheme include Bangladesh, Bhutan, China, India, Indonesia, East Timor, Lao PDR, Myanmar, Pakistan and Vietnam. The model works on the basis that pharmaceutical companies may apply to join the scheme by offering to export their products to a list of developing and least developed countries. In exchange for regulatory safeguards against trade diversion back into the EU, manufacturers or exporters must price their offered products at prices no higher than the permitted formula prices set out by the regulator. These are prices which shall be:

- no higher than 25% of the weighted average ex factory price charged by a manufacturer in OECD markets for the same product at the time of application, or
- a manufacturer's direct production costs plus an addition of a maximum of 15%.[207]

To provide such manufacturers and exporters with protection against trade diversion, the regulation expressly *prohibits* the import of such

[205] See e.g. discussion in Chapter 6, in the text accompanying note 110 on the Philippines' new Competition Act provisions on 'socialized pricing' agreements.

[206] Council Regulation No. 953/2003. See also, generally, EU, http://europa.eu/rapid/press-release_IP-03-748_en.htm?locale=en (accessed 28 October 2014); Stephen Kingah, *Access to Medicines and Vaccines in the South: Coherence of Rules and Policies Adopted by the European Commission* (, VUBPress Brussels University Press, 2011), part 3.4.1; and Sébastien Guigner, *The EU as a Global Health Actor: Myth or Reality?*, in *European Union Health Policy: Regional and Global Trends*, ed. Scott L. Greer and Pauletter Kurzer (Routledge, 2013), at p. 102.

[207] Article 3 and Annex III of the EU Council Regulation. The list of eligible countries, or "countries of destination", is found in Annex II, *ibid.*

products for free circulation in the EU, re-export (unless to the countries of destination in the list) and placing under suspensive procedures or in a free zone or free warehouse.[208] Suspected diversion can attract EU customs enforcement action such as suspension or detention.[209] The scheme also obliges such manufacturers and exporters to affix a designated permanent logo on the subject products and related documents. Such regulatory safeguards are useful in that they are additional to any private infringement actions such manufacturers or exporters may have under general intellectual property laws. The scheme has so far attracted only one manufacturer, which has registered nine products for treating HIV/AIDS under the scheme in 2004.[210] According to the EU's 2008 report, such products were sold on tiered prices to countries in East Asia and the Pacific.[211] The EU, which employs the principle of regional exhaustion, offers an interesting regulatory case study in that it establishes a framework that aims to support tiered pricing based on voluntary participation of pharmaceutical companies.

In the case of regional groups such as ASEAN, a key difference is that ASEAN does not apply regional exhaustion, which is a key feature of the EU regime. In fact, countries in ASEAN appear to prefer international exhaustion. This aspect would require reconsideration if an incentive for a stronger regional system of differential pricing were to be contemplated. At present, ASEAN members largely operate their own policies in this regard. In addition, coherence with the emerging competition laws in the region would have to be considered.

A question which requires further study is whether the pricing options in the EU model offer sufficient incentive or flexibility to manufacturers, particularly where products differ in newness on the market and on efficacy.[212] Tiered pricing is an approach employed by commercial

[208] Art. 2 of the EU Council Regulation, footnote 205 above.

[209] Art. 8 of the EU Council Regulation, footnote 205 above.

[210] The manufacturer is GlaxoSmithKline; for the list of its tiered-priced products, see EU Commission, *Annual Report (2010–2011) on the Application of Council Regulation (EC) No 953/2003 of 26 May 2003 to Avoid Trade Diversion into the European Union of Certain Key Medicines*, COM(2012) 775 final. The EU scheme is currently under review.

[211] EU, COM(2009) 574 final, Annex 2. According to the most recent report (2010–2011), the manufacturer did not provide a regional breakdown of sales but it had declared that "tiered priced products had been supplied to all LDCs, the World Bank's low-income countries, and all of Sub-Saharan Africa"; COM(2012) 775 final, at para. 4.

[212] The EU itself is presently conducting an evaluation of the regime under Council Regulation 953/2003, to review, *inter alia*, its effectiveness and relevance – see http://trade.ec.europa.eu/doclib/docs/2014/august/tradoc_152716.pdf (accessed 2 May

entities (states which are WTO members are required to trade on an MFN basis), but regional treaty arrangements may facilitate tiered pricing in the following ways:

- Ensuring there is no cross-market leakage of significance;
- Ensuring there is no leakage by exports through, for example, requiring use of different markings and packaging (an example can be found in the Doha Declaration and the para. 6 system for exports under compulsory licensing);
- Addressing the status of parallel imports to ensure that the treaty parties' laws and policies do not promote cross-market leakage.

## 7.10 Dispute resolution provisions for public health

A number of examples and suggestions have been put forward in Chapters 4 and 5 regarding the use of FTA provisions, exceptions and ISDS. In addition, what may be considered here is a more concerted effort to move public health-related investor complaints towards alternative dispute resolution mechanisms which are non-adversarial in nature. There are some indications that more thought is being put into this. For example, the EU-Singapore FTA (which is awaiting ratification) contains a mediation mechanism which could be useful in resolving health-related disputes, instead of or before resorting to contentious investor-state arbitration with all accompanying concerns.[213]

Korea has established an ombudsman with respect to investment complaints, which may merit further study and thoughtful modification for other countries in Asia. Vietnam, on the other hand, has established an inter-ministry team to oversee and handle investor disputes.[214]

2015); the author is grateful to Mr Anders Jessen, Director-General, EU Commission, for the information.

213 For the possible usefulness of mediation in such disputes, see Locknie Hsu, Examining the Formative Aspect of Investment Treaty Commitments: Lessons from Commercial Law and Trade Law, Chapter 8, in *Reshaping the Investor-State Dispute Settlement System – Journeys for the 21st Century*, ed. Jean E. Kalicki and Anna Joubin-Bret (Brill/Nijhoff, 2014).

214 T. T. Nguyen and T. C. Q. Vu, Investor-State Dispute Settlement from the Perspective of Vietnam: Looking for a "Post-Honeymoon" Reform, *Trans. Disp. Mgmt.*, Vol. 11, No. 1 (January 2015) (online journal publication), p. 5.

## 7.11 Conclusion

A number of initiatives already exist to promote access to medicines and R&D in medicines, particularly those for treating "neglected" diseases. However, the efforts have been diffuse and largely operating in various different spheres. Against this backdrop, a number of countries are entering cross-border agreements which affect the policy-making in respect of regulation and patenting of medicines. While a global R&D or health treaty may be desirable if multilateral consensus can be achieved through such an instrument to address access issues in a coherent and coordinated manner, there are, in the meantime, a number of areas of immediate interest to developing countries – particularly those in Asia, and ASEAN – which require careful thought and debate. These include the types of national and regional incentives that should be advocated and sought both at the regional and global level, and the treaty and non-treaty actions which can be planned and taken to promote greater access to medicines. ASEAN, with its integration efforts and economic needs, is at an opportune juncture to consider these matters, as it aims to craft a "global voice" on them. Possibilities include the thoughtful design of regional incentives and negotiating strategies for improved and cost-effective drug discovery and delivery to its peoples, while keeping in view the existing multilateral system of obligations and its accompanying flexibilities. Common approaches (such as in patentability criteria as they relate to broader public health and access concerns) of other regions can offer useful ideas for Asian countries.[215]

[215] See, for example, *MERCOSUR Agreement on Patentability Criteria, Access to Medicines, and Public Health*, Meeting of MERCOSUR Health Ministers, Montevideo, Uruguay, in 2009; Argentina has since adopted guidelines on patentability in 2012, pursuant to this initiative. MERCOSUR comprises Argentina, Brazil, Paraguay and Uruguay.

# 8

# What next?

## 8.1 Advancing access: important questions and trends for Asia to consider

As the international business and scientific and regulatory situation evolve, new approaches, alliances and strategies will be required to address access to medicines.[1] Apart from existing and proposed initiatives discussed in the previous chapter, a number of important trends and evolving issues merit the attention of policy-makers in the debate on access to medicines. A number of these are discussed here, with a view to setting the scene for the future.

## 8.2 Clarifying critical matters in the multilateral trade system

### *8.2.1 Some matters for consideration*

A number of important matters remain unclear in the global trading rules affecting public health and pharmaceutical products. One of these is the role that Articles 7 and 8 of the TRIPS Agreement can and should play in interpreting TRIPS obligations and exceptions, and their relationship with subsequent documents such as the Doha Declaration on the TRIPS Agreement and Public Health.

#### 8.2.1.1 Interpreting the TRIPS balance

As we have seen in Chapter 5, in *Canada – Pharmaceuticals*, a WTO panel had the opportunity to address the interpretation and application of an important exception in the TRIPS Agreement, namely Article 30. However, the scope and utility of Articles 7 and 8, which address broader objectives and principles of the Agreement, received relatively little panel analysis, and there has been no interpretation and application of these

[1] See the short but highly insightful editorial by William A. Wulf, Changes in Innovation Ecology, *Science*, Vol. 316 (2007), p. 1253.

provisions by the WTO Appellate Body. The disagreement between Canada and the EU over the interpretation of the "balance" that these provisions were meant to preserve illustrates part of the division of views that exists. The Doha Declaration on the TRIPS Agreement and Public Health provides additional guidance on the interpretation of TRIPS and its objectives and principles, and needs to be taken into consideration in the interpretation of TRIPS and the "balance" it aims at.[2]

#### 8.2.1.2 Implementing Article 39 of the TRIPS Agreement

Article 39 has so far not been the subject of any definitive interpretation by any WTO panel or the Appellate Body. The ambiguities surrounding Article 39, whose language was negotiated within the broader package produced by the Uruguay Round, continue to be a source of debate. Given that this provision represents a fundamental aspect of the access to medicines and research debate, there is, unfortunately, no clear WTO jurisprudential pronouncement on its scope. There has been, on the other hand, a growth of FTA provisions that clearly exceed what Article 39 requires.[3]

In the context of the European Medicines Agency (EMA) and its disclosure regime, the European Ombudsman had occasion to consider – albeit not in a definitive manner as well – Article 39(3).[4] In June 2007, researchers at the Nordic Cochrane Centre, a healthcare research and information centre, applied through the Danish Medicines Agency to the EMA for access to clinical study reports and their trial protocols which had been sent to EMA for marketing approval for certain anti-obesity drugs. Notably, the aim of the request was to obtain and analyse the requested information to assess the safety of the drugs; it was *not* for commercial marketing of any competitor drugs. The EMA rejected the

[2] Para. 5(a) in particular states as follows:

> In applying the customary rules of interpretation of public international law, each provision of the TRIPS Agreement shall be read in the light of the object and purpose of the Agreement as expressed, in particular, in its objectives and principles.

See also the views of the African Group, Barbados, Bolivia, Brazil, Cuba, Dominican Republic, Ecuador, Honduras, India, Indonesia, Jamaica, Pakistan, Paraguay, Philippines, Peru, Sir Lanka, Thailand and Venezuela in a submission to the TRIPS Council, IP/C/W296, at pp. 1–6.

[3] See Chapter 2, footnote 67 and accompanying text, and Chapter 4.

[4] See Draft Recommendation of the Ombudsman, 19 May 2010, at http://www.ombudsman.europa.eu/cases/draftrecommendation.faces/en/4883/html.bookmark (accessed 1 May 2015).

application on the basis that the information fell within exceptions to disclosure to access found in certain EU Rules.[5] The relevant exception permits the EMA to refuse access for protection of "commercial interests of a natural or legal person, including intellectual property".[6] Upon repeated refusal by the EMA, the matter was eventually referred by the researchers in a complaint to the EU Ombudsman. While it did not appear that in its original refusals to the applicants the EMA had referred to Article 39 of TRIPS, it certainly did so in its arguments before the Ombudsman. In May 2010, the Ombudsman reached a conclusion, stating that the EMA had exhibited "maladministration" in its handling of this application. It is not the aim here to analyse the Ombudsman's reasoning and recommendations in detail, nor to opine on the correctness of the maladministration conclusion. Instead, three pertinent – and hopefully useful – observations may be made regarding this case.

First, the complainants' application and the Ombudsman's conclusion on maladministration show the importance of *availability* of an independent, trustworthy review mechanism over the authority which is applying Article 39 TRIPS and related national rules. While an ombudsman may not be the only option, a review mechanism would permit a degree of accountability over decisions of non-disclosure and the possibility of reviewing the correctness of the decision against the situations of permitted disclosure in Article 39(3) (where disclosure is "necessary to protect the public", for instance).

Second, such a mechanism would need to be TRIPS-compliant, independent as well as trustworthy in handling the information being requested. Even if such an agency exercises its duty to interpret its non-disclosure obligation in good faith, it may choose to err on the side of

[5] *Rules for the Implementation of Regulation (EC) No 1049/2001 on access to EMA documents.*

[6] The relevant provision in the EU Rules relied upon by the EMA is found in Article 3(2)(a), which states as follows:

> The Agency shall refuse access to a document where disclosure would undermine the protection of:
>
> (a) Commercial interests of a natural or legal person, including intellectual property.

Note other important aspects of this particular complaint considered by the Ombudsman in coming to his recommendations, such as the nature of the documents being requested, whether they revealed development strategies of those submitting the information and whether redacting was possible or administratively too burdensome.

caution, for administrative convenience and to avoid legal action from those which submitted the information being requested. This may apply in particular to a developing country in Asia which can ill afford to deal with lawsuits, or threats of lawsuits, from deep-pocketed pharmaceutical companies, or incur the displeasure of a major trade partner.[7] Where the agency is the sole arbiter of what may or may not be disclosed, a legal and administrative bottleneck may develop in respect of information which may be legitimately disclosed under Article 39 and national laws and regulations.

Finally, the recommended solution of the Ombudsman that the EMA provide *private access* of the information to the complainants was significant.[8] The complainant's request had not been made for the purpose of obtaining marketing approval for a competitor product, but for examining public safety of the products in question. The private access recommendation for the requested purposes is a breakthrough whose significance has, in this author's view, been underestimated. The decision, though not made by a WTO panel or a judicial authority, may well be seen as showing a degree of official support and confirmation for the view that such access – if transposed to the interpretation of Article 39(3) – would *not* be seen as "*unfair commercial use*".[9] Although limited by the context and nature of the request, the case could contribute towards illuminating what is *not* such use.

7 See, for example, the lawsuits brought against the EMA (but later withdrawn) to challenge disclosure by AbbVie and Intermune; Stephanie Bodoni, *AbbVie, Intermune Sue to Block Clinical-Trial Data Release*, 11 March 2013, Bloomberg, and *AbbVie Drops Legal Action in EU Drug Secrecy Case*, 3 April 2014, Reuters.

8 In this regard, the Ombudsman interestingly drew inspiration from a decision of the European Investment Bank in relation to a complainant seeking access to sections of an audit report which could not be publicly disclosed (see paras. 38–40, Ombudsman's recommendations); there, the EIB granted private access to the complainant to resolve the matter.

9 Although the Ombudsman was primarily concerned with the EU Rules, it was stated as follows, at para. 40:

> . . . the fact that the Rules do not foresee the possibility of granting private access could not exclude *the possibility of granting private access on the basis of Article 39(3) of the TRIPS Agreement.*

For another perspective on this recommendation relating to the public health purpose, see Carlos Correa, *The Impact of Economic Partnership Agreements on WTO Law*, in eds. Josel Drexel, Henning Grosse-Ruse Khan and Souheir Nadde-Flix, *EU Bilateral Trade Agreements and Intellectual Property: For Better or For Worse?*, at p. 105.

#### 8.2.1.3 Reviewing the WTO agreement on pharmaceutical products

The Uruguay Round Trade in Pharmaceutical Products Agreement (known as the "zero for zero" agreement) has not enjoyed any increase in WTO member participation in its commitments.[10] These include the elimination of tariffs on pharmaceutical products and agreement not to employ non-tariff barriers to trade in these goods.

The agreement undergoes periodic reviews.[11] There have been three reviews of the covered products to date, with the most recent having taken place in 2007.[12]

During the Doha Development Round non-agricultural market access (NAMA) negotiations, a proposal for progressive elimination of tariffs on healthcare products including pharmaceuticals was proposed by Singapore, Switzerland, the Separate Customs Territory of Taiwan, Penghu, Kinmen and Matsu and the US.[13] The proposal included special

[10] L/7430.

[11] In 2004, the then EU Trade Commissioner Mr Pascal Lamy's response to a parliamentary question explained the process as well as the EU view on possible extension of the agreement to other WTO members, as follows:

> The "Pharmaceutical Tariff Elimination Agreement" is a non-binding arrangement between the most important pharmaceutical producing countries to reduce their World Trade Organisation (WTO) bindings of duties on certain pharmaceutical products, including active ingredients and intermediate products, to zero. Because new pharmaceutical products are constantly being developed, the arrangement envisages periodic reviews of these products. Any of the participants could have launched the process for the third update (Update III) which was due to be completed in 2002, even if it was indeed the EU that proposed and took forward the previous two updates. Given the interest of the EU Pharmaceutical industry, the Commission is actually considering launching Update III.
>
> As regards the simplification of the updating procedures as well as the extension of the Agreement to other WTO Members, the Commission shall certainly consider such options if they were to be in the wider interests of the EU within the Doha Development Agenda.

See EU, at: http://www.europarl.europa.eu/sides/getDoc.do?pubRef=-//EP//TEXT±WQ±E-2004-0213±0±DOC±XML±V0//EN&language=bg. See also the USTR's remarks on the agreement: https://ustr.gov/issue-areas/industry-manufacturing/industry-initiatives/pharmaceuticals.

[12] USTR, *ibid.*

[13] *Draft Modalities for Open Access to Enhanced Healthcare*, in Fourth Revision of Draft Modalities for Non-Agricultural Market Access, TN/MA/W/103/Rev. 3, 6 December 2008, pp. 107–109; text available at: https://www.wto.org/english/tratop_e/markacc_e/namachairtxt_dec08_e.pdf (accessed 21 April 2015).

and differential treatment considerations for developing countries. While attention in the Doha Round negotiations has more recently focused on trade facilitation,[14] this might be one possible area to revive discussions in, to address tariff barriers to pharmaceutical products and to expand the participation in (and, possibly, even the scope of) this agreement.

#### 8.2.1.4 Royalty rates – the elephant in the room?

One of the unanswered questions in Article 31 of the TRIPS Agreement is what amounts to "adequate remuneration".[15] Some may argue this is a positive sort of ambiguity – "constructive ambiguity", perhaps[16] – in that it allows WTO members flexibility in determining its meaning and implementation. Royalty rates are an important factor in the use of compulsory licensing, as such rates, if set too high, can defeat the ability to effectively use such a licence.

Table 8.1 summarizes the types of compulsory licences and royalty-setting yardsticks in selected Asian countries' patent legislation.

The ambiguity under Article 31(h) has resulted in a transfer of the ambiguity into national laws, because legislatures either have not adequately provided for clear methods of remuneration, or have been reluctant to set any limits in concrete terms.[17] A small number of countries have set a ceiling for the royalty rates payable for compulsory licences.

In the case of Vietnam, it has been observed that until clarifications were issued by the government in 2006, it was not clear what such remuneration entailed.[18] The royalty rate, known as the "remuneration

[14] See WTO, https://www.wto.org/english/tratop_e/tradfa_e/tradfa_e.htm (accessed 21 April 2015).

[15] The UN has published guidelines on possible remuneration rates, including a discussion of possible bases for setting royalty rates. See James Love, Remuneration Guidelines for Non-Voluntary Use of a Patent on Medical Technologies (UNDP, WHO, 2005), available at: www.who.int/hiv/amds/WHOTCM2005.1_OMS.pdf (accessed 11 April 2015). Note that the language on remuneration in the Decision on the "paragraph 6" mechanism and the proposed Article 31 *bis* to amend TRIPS differs; see discussion in Chapter 8.

[16] See, for example, Thu-Lang Tran-Wasescha and Xavier Groussot, *TRIPS Art. 66.2: Between Hard Law and Soft Law*, at p. 22, in Sustainable Technology Transfer – A Guide to Global Aid and Trade Development, ed. Han Henrik Lidgard, Jeffrey Atik and Tu Thanh Nguyen (Wolters Kluwer, 2012). See also Chapter 8.

[17] See for example some of the legislative formulations in Asian patent laws summarized in Table __, Chapter 5.

[18] Tran Viet Dung, Le Thi Nam Giang and Nguyen Nhu Quynh, *Vietnam: A Case Study for Sustainable Technology Transfer*, at pp. 259 and 261, citing Art. 24 of Decree 103/2006/ND-CP, in *Sustainable Technology Transfer – A Guide to Global Aid and Trade Development, ibid.* The Decree (Decree of September 22, 2006, Detailing and Guiding the Implementation of a Number of Articles of the Law on Intellectual Property Regarding Industrial Property) is available at http://www.vcci-ip.com/index.php/en/legal-documents/decree (accessed 29 August 2014); Circular No. 01/2007/TT-BKHCN of February 14, 2007, implements the Decree.

Table 8.1. *Examples of types of public-use/compulsory licensing grounds in Asia*

| Country | National emergency | Government use | Non-working or priced beyond access and minimum period | Other grounds |
|---|---|---|---|---|
| China | ✓<br>Article 49<br>Article 57 – "reasonable" royalty to be paid, to be determined by consultations between the parties, failing which the patent administrative department of the State Council will make a ruling | Article 49<br>(where the public interest so requires)<br>Article 57 – "reasonable" royalty to be paid, to be determined by consultations between the parties, failing which the patent administrative department of the State Council will make a ruling | ✓<br>Article 41<br>(Non-exploitation 3 years from grant or 4 years from filing application or to address anti-competition)<br>Article 57 – "reasonable" royalty to be paid, to be determined by consultations between the parties, failing which the patent administrative department of the State Council will make a ruling | Article 51 – dependent patent<br>Article 57 – "reasonable" royalty to be paid, to be determined by consultations between the parties, failing which the patent administrative department of the State Council will make a ruling |
| India | ✓<br>"Patentees to derive a "reasonable advantage from their patent rights" (S. 92)" | ✓<br>Adequate remuneration in the circumstances of each case, taking into account the economic value of the use of the patent (s. 100)[a] | ✓<br>3 years from grant | Dependent patent<br>Doha "para. 6" mechanism |
| Indonesia | – | ✓<br>"Reasonable compensation" (Article 101) | ✓<br>3 Years from grant<br>Royalty to be determined by the Directorate-General, taking into account "common practices in Patent licensing or | Dependent patent<br>or<br>patent has been implemented "in a form and manner that |

Table 8.1. (*cont.*)

| Country | National emergency | Government use | Non-working or priced beyond access and minimum period | Other grounds |
|---|---|---|---|---|
| | | | similar agreements" (Article 78(2)) | contravenes the public interest" |
| The Philippines | ✓ | ✓ | ✓ | ✓ |
| | "Adequate remuneration" to be paid, taking into account economic value of the authorization<br>For all compulsory licences: may be applied for 3 years from grant; 4 years from application, whichever expires last | "Adequate remuneration" to be paid, taking into account economic value of the authorization | "Adequate remuneration" to be paid, taking into account economic value of the authorization | Public interest so requires (s. 74(a))<br>"Adequate remuneration" to be paid, taking into account economic value of the authorization<br>TRIPS Article 31 ground and Doha "para. 6" mechanism |
| Malaysia | ✓ | ✓ | ✓ | Dependent patent |
| | | (s. 84) | 3 years from grant; 4 years from application | Anti-competitive practice (s. 84) |
| Singapore | ✓ | ✓ | ✓ | ✓ |
| | Remuneration to be agreed, or according to agreed method, or by the Court (s. 62) | Remuneration to be agreed, or according to agreed method, or by the Court (s. 62) | Non-working or inadequate supply (s. 55(2))<br>Remuneration to be agreed, or according to agreed method, or by the Court | Anti-competitive practice (s. 55(1))<br>Remuneration to be agreed, or according to agreed method, or by the Court |

| | | | | |
|---|---|---|---|---|
| Thailand | ✓<br>S. 52 – “fair remuneration” | ✓<br>S. 51<br>Royalty rate to be “as agreed upon by the ministry<br>or bureau or department and the patentee or his licensee”<br>(with Section 50 applying *mutatis mutandis*) | ✓<br>S. 46 and 49 – applicant to propose a “reasonably sufficient” royalty<br>S. 50 – Director-General to fix the royalty – to be “adequate for the circumstances of the case” | Dependent patent<br>Royalty set out by applicant, determined by Director-General (S. 49 and 50) |
| Vietnam | ✓ | ✓<br>Article 135<br>“satisfactory compensation depending on the economic value of such use right in each specific case, and compliant with the compensation bracket set by the Government”<br>(Article 146(1)(d)) | ✓<br>3 Years from grant; 4 years from application (Article 145) | Dependent patent |

[a] See remarks in *NATCO Pharma Ltd. v Bayer Corporation*, Judgment of the Bombay High Court, pronounced on 15 July 2014, at pp. 39–45.

price", is presently capped at 5%.[19] The factors used in determining the rate to be applied are as follows:

a. Invention licensing price under the contract;
b. Funds invested in the creation of the invention, with the support fund from the state budget (if any) taken into account;
c. Profits earned from the use of the invention;
d. Remaining valid term of the protection title;
e. Necessity of the invention licensing;
f. Other elements directly decisive to the economic value of the licensed use right.[20]

In the Philippines, the Intellectual Property Code provides as follows:

> 100.6. The patentee shall be paid adequate remuneration taking into account the economic value of the grant or authorization, except that in cases where the license was granted to remedy a practice which was determined after judicial or administrative process, to be anti-competitive,

[19] Art. 24 of the Circular, *ibid.*, provides:

> The remuneration price shall not exceed 5% of the net sales price of the products manufactured under that invention provided that the principles set out in clause 1 of this Article are conformed with.

See also Vietnam's explanation of its compulsory licensing remuneration system during its accession to the WTO in the WTO Working Party Report, WT/ACC/VNM/48, at paras. 428–431. The following statement by the Vietnamese representative in particular merits mention:

> According to Decree No. 103/2006/ND-CP, the remuneration should take into account the economic value of the right transferred, including the contractual licensing price of the invention, the funds invested for the creation of the invention, the profits gained by using the invention, the remaining duration of validity of the patent, and the need for licensing the invention.

[20] Art. 24 of the Circular, footnote 19 above. See also Vietnam's explanation of its compulsory licensing remuneration system during its accession to the WTO in the WTO Working Party Report, WT/ACC/VNM/48, at paras. 428–431. The following statement by the Vietnamese representative in particular, setting out used factors in determining such remuneration, merits mention:

> According to Decree No. 103/2006/ND-CP, the remuneration should take into account the *economic value of the right transferred, including the contractual licensing price of the invention, the funds invested for the creation of the invention, the profits gained by using the invention, the remaining duration of validity of the patent, and the need for licensing the invention.*
>
> [Italic emphasis added]

> the need to correct the anti-competitive practice may be taken into account in fixing the amount of remuneration.
>
> [Sec. 35-B, R.A. No. 165a]

Rule 12 section 8 of the Implementing Rules and Regulations of the Universally Accessible Cheaper and Quality Medicines Republic Act 9502 (the "Cheaper Medicines Act") also reflects the above in respect of pharmaceutical products. Under a 1977 presidential decree, the general royalty rate for compulsory licensing is not to exceed 5%.[21]

There appear to be no general guidelines within Asia or ASEAN on how such remuneration rates would be dealt with should Article 31 be activated.

In the case of Indonesia and Thailand, compulsory licences have been issued in recent years, with royalty payments set by the respective authorities. In the case of Indonesia, a royalty rate of 0.5% was applied.[22] In Thailand, the rate has also been 0.5% in relation to a compulsory licence

[21] Under a now-repealed patent law, the remuneration rate was addressed in Presidential Decree No. 1263, Amending Portions of Republic Act No. 165 otherwise known as the Patent Law, 1977, http://www.gov.ph/section/executive-issuances/presidential-decrees-executive-issuances/page/74/?sa=U&ei=2AwWU9yqOJCFogTAoICAAg&ved=0CEMQFjAG&usg=AFQjCNFhiW2j5JeV6dBhlOjTJW1TFmgpHA (accessed 11 April 2015), which had provided as follows:

> S 35-B(3) A compulsory license shall only be granted subject to the payment of adequate royalties commensurate with the extent to which the invention is worked. However, royalty payments shall not exceed five per cent (5%) of the net wholesale price (as defined in section 33-A) of the product, substance, or process subject of the compulsory license is involved in an industrial project approved by the Board of Investments, the royalty payable to the patentee or patentees shall not exceed three per cent (3%) of the net wholesale price (as defined in Sec. 34-A) of the patented commodity manufactured under the patented process; the same rate of royalty shall be paid whenever two or more patents are involved, which royalty shall be distributed to the patentees in rates in rates proportional to the extent of commercial use by the licensee giving preferential values to the holder of the oldest subsisting product patent.

For examples of rates set in the Philippines by the Director of Patents, see, for example, James Love, footnote 15 above, pp. 38–39.

Section 239 of the Intellectual Property Code, which repealed Act No. 165, did not however repeal the above decree, and Article XVIII of the Philippines Constitution of 1987 recognizes existing decrees: see http://www.gov.ph/constitutions/the-1987-constitution-of-the-republic-of-the-philippines/the-1987-constitution-of-the-republic-of-the-philippines-article-xviii/.

[22] See James Love, footnote 15 above, at p. 41; and http://www.scielo.org.co/scielo.php?pid=S0121-40042012000300002&script=sci_arttext&tlng=en (accessed 11 April 2015).

for efavirenz.[23] In explaining the factors taken into account in fixing the remuneration rate of this drug (and two others which were the subject of non-voluntary licences under government use), the Thai government explained as follows:

> The level of royalty fees payable to the patent holders have been set at between 0.5 to 2 per cent of the sale value. This is the common range used in most developing countries in the case of public non-commercial use. For those drugs with high retail value, the royalty will be set at the lowest level of 0.5 per cent. For those with low retail value, the royalty will be set at the top level of 2 per cent. For the three drugs that Government Use has been announced, they are all in high demand and the expected retail value is high. So the royalty fees have been set at 0.5 per cent. However, these royalty fees can be negotiated if drug companies are not satisfied with the proposal from the Ministry of Public Health. If the negotiation fails, then the Director General of the Department of Intellectual Properties will determine the fees according to several criteria as established in section 51 of the Thai Patent Act (Document No. 3).[24]

India's first grant of a compulsory licence was accompanied with an applicable royalty rate of, which was subsequently revised by the court to.[25] The Controller of Patents had fixed the royalty at 6% of net sales by NATCO; the IPAB increased this to 7%, which rate the Bombay High Court upheld.

In the only instance of utilization of the "paragraph 6" mechanism to date, Canada's TRIPS Council representative reported on Canada's royalty rates as follows:

[23] See Sakda Thanitcul and M. L. Braslow, *op cit.*, Annex A. Brazil, too, issued a compulsory license for the same drug shortly thereafter with a royalty rate of 1.5%; see http://www.ictsd.org/brazil-issues-compulsory-licence-for-aids-drug; and Eric Bond and Kamal Saggi, *Compulsory Licensing, Price Controls, and Access to Patented Foreign Products*, April 2012, www.wipo.int/edocs/%20...%20ip%20...%20/wipo_ip_econ_ge_4_12_ref_saggi.pdf (accessed 11 April 2015). See also Scherer, F. M., and Jayashree Watal, Post-TRIPS Options for Access to Patented Medicines in Developing Nations, *JIEL*, Vol. 5, No. 4 (2002), 913–939.

[24] Ministry of Public Health and the National Health Security Office, Thailand, *Facts and Evidences on the 10 Burning Issues Related to the Government Use of Three Patented Essential Drugs in Thailand – Document to Support Strengthening Social Wisdom on the Issue of Drug Patent* [*sic*], February 2007, at pp. 11–12, available at: http://apps.who.int/medicinedocs/documents/s18718en/s18718en.pdf (accessed 4 May 2015). The three drugs referred to in the excerpt are efavirenz (Stocrin®) of Merck Sharp and Dohme, lopinavir + ritonavir (Kaletra®) of Abbott Laboratories) and clopidogrel (Plavix®) of Sanofi-Aventis.

[25] *Bayer Corporation v NATCO Pharma Ltd.*, IPAB Order No. 45 of 2013 and Judgment of the Bombay High Court, pronounced on 15 July 2014; see also discussion in Chapter 5 on other aspects of this case.

> With respect to the question asked by Malaysia about royalties, he said that the royalty rates ranged between 0.02 and 3.5 per cent. Under CAMR, the remuneration to be paid by the licensee to the patentee was *calculated by multiplying the monetary value of the supply contract by an amount that fluctuated on the basis of the importing country's standing on the UN Human Development Index.* According to this formula, the lowest country on the index would pay a royalty of approximately 0.02 per cent, and the highest country of 3.5 per cent. Where patentees disagreed with the royalty fixed on the basis of this formula, it was possible to apply to the federal court for an order setting a higher amount. This had not been the case as regards Rwanda, where the patent holders had entirely waived the royalty fee.[26]
>
> [Italic emphasis added. CAMR refers to Canada's Access to Medicines Regime.[27]]

In this case, Canada authorized Apotex Ltd. to manufacture and export the drugs under a compulsory licence to Rwanda, and it appeared to do so at a loss.[28]

While the rate of remuneration under Article 31 (and the proposed Article 31) provides flexibility – the positive side of the wording of the provision – there are no clear definitions or guidelines agreed among WTO members on the setting of these rates. An early FTA which has provided examples of language for negotiators of subsequent trade agreements is the North American Free Trade Agreement (NAFTA).[29] In relation to compulsory licensing, Article 1709.10(h) of the NAFTA provides for "adequate remuneration in the circumstances of each case, taking into account the economic value of the authorization". There has been, thus far, no NAFTA tribunal interpretation of these words. In TRIPS, the relevant words in Article 31(h) are identical: "adequate remuneration in the circumstances of each case, taking into account the economic value of the authorization".

The WTO provisions, however, also leave room for challenging the rates set by member countries in utilizing the compulsory licensing option. Obvious questions include what the meaning of "adequate" is, what circumstances should be taken into account, what is the "economic value of the authorization" and how that should be taken into account in

26 Para. 51, *Annual Review of the Decision on the Implementation of Paragraph 6 of the Doha Declaration on the TRIPS Agreement and Public Health*, 10 December 2010, IP/C/57.

27 See http://www.camr-rcam.gc.ca/index-eng.php (accessed 15 April 2015).

28 See footnote 27 above, at paras. 49 and 60.

29 NAFTA, text available at: https://www.nafta-sec-alena.org/Home/Legal-Texts/North-American-Free-Trade-Agreement (accessed 22 April 2015).

assessing if the rate *is* "adequate".[30] One question that arises, for example, is whether the royalty rate should be applied to the patented product price, or to the generic product price.[31] A related question is whether the market value – such as a commercial royalty rate that the right holder would have charged in granting a voluntary licence for the same drug, perhaps – should be taken as the "economic values of the authorization" in the first place, or whether it should be the value of the authorization to the member country granting the licence, since "economic value" relates to the *authorization* (as opposed to the patent or the underlying drug).[32] One view is that what should be taken into account under Article 31(h) is "what such a licence would normally cost if available from the patent owner".[33]

The proposed provision amending the TRIPS Agreement, Article 31 *bis*, on the other hand, provides as follows:

> 2. Where a compulsory licence is granted by an exporting Member under the system set out in this Article and the Annex to this Agreement, *adequate remuneration* pursuant to Article 31(h) shall be paid in that Member *taking into account the economic value to the importing Member of the use that has been authorized in the exporting Member*. Where a compulsory licence is granted for the same products in the eligible importing Member, the obligation of that Member under Article 31(h) shall not apply in respect of those products for which remuneration in accordance with the first sentence of this paragraph is paid in the exporting Member.
>
> [Italic emphasis added][34]

[30] See Chapters 3, 6 and 8 of WHO, *Remuneration Guidelines for Non-Voluntary Use of a Patent on Medical Technologies* (2005), illustrating the variability of royalty rates in countries.

[31] See *ibid.*, p. 60, on this point, and Chapter 6 generally, on other pertinent questions.

[32] Note that Singapore's Patents Act refers to "reasonable remuneration" rather than "adequate remuneration", and the phrase used in relation to compulsory licensing under Section 55(7) is "taking into account the economic value of the licence" while the phrase used in relation to government use under Sections 58(2) and 62(1) is "taking into account the economic value of the patented invention".

[33] Daniel Gervais, *The TRIPS Agreement – Drafting History and Analysis*, 4th Ed. (2012), Sweet & Maxwell, at p. 496. See also Nuno Pires de Carvalho, *The TRIPS Regime of Patent Rights*, (2010) Kluwer Law International, at p. 467. Note that the WTO acknowledges the lack of a definition of "adequate" and "economic value"; see WTO, https://www.wto.org/english/tratop_e/trips_e/public_health_faq_e.htm (accessed 22 April 2015). During the Uruguay Round negotiations, the US had proposed draft language providing for remuneration "adequate to compensate the right-holder fully for the license, except for compulsory licenses to remedy adjudicated violations of competition law"; MTN.GNG/NG11/W/70, 11 May 1990, draft Article 27.

[34] In the General Council Decision of 8 August 2003, the following was agreed as part of the waiver decision regarding Article 31(f) of the TRIPS Agreement (the "paragraph 6" decision):

While the above provision elaborates upon the remuneration in the context of a "paragraph 6" situation, it explicitly connects the economic value *to the importing state*, that is, the beneficiary of the compulsory licence. In the case of Article 31(h) *simpliciter*, one might argue that the economic value referred to there is the value *to the WTO member granting the compulsory license*, not the economic value to the right holder.

According to a text proposed by a group of developing countries during the TRIPS negotiations, the rights of a patent holder would have included "the right to a reasonable remuneration[1] when the competent authorities of a Party to the present agreement use a patent for government purpose or provide for the granting of a licence of right or a compulsory licence", with footnote 1 reading as follows:

> 1. Such reasonable remuneration will be determined having regard to the economic situation of the Party, the nature of the invention, the cost involved in developing the patent and other relevant factors.[35]

As the TRIPS Agreement does not set out explicit markers for determining "adequate remuneration", states negotiating FTA provisions on such remuneration in relation to compulsory licensing or government use have flexibility to stipulate such markers, including those proposed above, pending any multilateral guidance at the WTO. After all, some national laws (such as those mentioned here) already set limits or terms for such remuneration. It is not necessary to determine *a priori* that "adequate remuneration" equates to market prices; on the contrary, a

> 3. Where a compulsory licence is granted by an exporting Member under the system set out in this Decision, *adequate remuneration* pursuant to Article 31(h) of the TRIPS Agreement shall be paid in that Member *taking into account the economic value to the importing Member of the use that has been authorized in the exporting Member.* Where a compulsory licence is granted for the same products in the eligible importing Member, the obligation of that Member under Article 31(h) shall be waived in respect of those products for which remuneration in accordance with the first sentence of this paragraph is paid in the exporting Member.
>
> [Italic emphasis added]

[35] Article 5 of proposed text in *Communication from Argentina, Brazil, Chile, China, Colombia, Cuba, Egypt, India, Nigeria, Peru, Tanzania and Uruguay*, MTN.GNG/NG11/W/71, 14 May 1990. See also the interesting proposal of a "tiered" royalty system by Elizabeth Siew-Kuan Ng, *Global Health and Development: Patents and Public Interest*, in *Incentives for Global Public Health – Patent Law and Access to Medicines*, ed. Thomas Pogge, Matthew Rimmer and Kim Rubenstein (Cambridge University Press, 2013), pp. 122–123.

contextual interpretation of Article 31 that takes into account the objectives of the TRIPS Agreement, as discernible from the preamble, Article 7 and 8, the "balance" captured in the Agreement (discussed in the WTO panel report of *Canada – Pharmaceutical*) and the entirety of Article 31 itself, would leave room to argue that such remuneration should be "adequate" not from a market perspective but from a "balance" perspective. For FTAs proposing to incorporate provisions similar to Article 31 or to the waiver under the proposed Article 31 *bis*, negotiating states may likewise consider including markers for the level of remuneration.[36]

## 8.3 New incentives – new ways of rewarding conduct that contributes to greater access

According to the Access to Medicines Index 2014 report, licensing activity has increased.[37] One question is how private enterprises may be provided incentives to further encourage such activity. Other positive behaviours of pharmaceutical companies can be further encouraged with different types of incentives. Examples include encouraging them

- not to enforce patents in poor countries;
- to introduce, where there is none, access programs and/or dedicated corporate social responsibility teams or units to address access issues on an ongoing basis[38]; such units or teams should be led by senior

[36] Article 31 *bis* provides:

> 2. Where a compulsory licence is granted by an exporting Member under the system set out in this Article and the Annex to this Agreement, adequate remuneration pursuant to Article 31(h) shall be paid in that Member taking into account the economic value to the importing Member of the use that has been authorized in the exporting Member. Where a compulsory licence is granted for the same products in the eligible importing Member, the obligation of that Member under Article 31(h) shall not apply in respect of those products for which remuneration in accordance with the first sentence of this paragraph is paid in the exporting Member.

[37] See *Access to Medicines Index 2014*, http://www.accesstomedicineindex.org/sites/2015.atmindex.org/files/2014_accesstomedicineindex_patents_0.pdf (accessed 9 March 2015), pp. 104–105.

[38] For instance, Novartis recently announced its initiative to provide greater access to medicines for poor oncology patients. It has a policy of non-enforcement of patents in LDCs and also has in place certain not-for-profit arrangements for anti-malarial drugs and vaccines and cost sharing for certain cancer medications. See http://www.novartis.com/corporate-responsibility/access-to-healthcare/our-key-initiatives/oncology-patient-assistance-programs.shtml and http://www.themaxfoundation.org/GIPAP/Default.aspx (accessed 15 April 2015).

management and also engage regularly with divisions that handle shareholder and investor relations and corporate communications; and
- to enter more voluntary licensing arrangements on terms that improve access for poorer countries and poor patients within middle-income countries.

As such companies are often public companies which are accountable to shareholders for matters such as share value, dividends and profitability, the profit maximization objective and instinct need to be addressed in creative ways. For example, tax incentives may be provided by a government to a publicly listed company if access arrangements that are shown to be valuable and meaningful are made to benefit the country or its poorer citizens. Tax incentives could also be offered to shareholders of such companies, to encourage them to support access programmes of the companies in which they hold shares. On a broader and cross-border scale, a country (A) which wishes to promote access of medicines in another which is an LDC (B) may even agree to provide companies in A that promote access of their medicines in B with tax and other incentives in A. This would be an indirect form of aid which encourages private enterprises to deliver medicines in an affordable way, instead of taking the form of direct financial or material aid. Another way may be to strengthen pooling arrangements such as the Medicines Patent Pool by having such "donor" countries (like the hypothetical A) agree to provide companies which make licenses available to the pool with certain incentives.

## 8.4 Timely evaluations and coherent policy-making in growing areas

A number of emerging areas in pharmaceutical research and regulation merit mention here, as they are likely to give rise to new regulatory and trade negotiating issues in the near future.

### *8.4.1 Biologic and biosimilar products*

Biologic pharmaceutical products make use of biological materials and are already on the market. Biosimilar pharmaceutical products are products which are deemed to be "similar" to a reference biologic product according to regulatory requirements. While they appear to play the role that generics play in relation to "originator" or branded pharmaceuticals, that is, as less costly substitutes, the comparison process in each and the processes of manufacture differ. In the case of generics, they comprise chemicals which need to be shown to be bioequivalent to the originator product. In the case of biosimilar products, bioequivalence is not an appropriate comparison given that they are much more complex molecules and would not be identical to the originator biologic product. As a result, comparison is made on the basis of "similarity" rather than bioequivalence (and hence the use of the moniker "biosimilar").

Due to their powerful potential to deal with diseases in new ways and the potential revenue from these products, research and commercialization of biologics and biosimilar products are increasing.[39] As the law plays catch-up, a number of regulatory concerns have already been identified.[40] This is a relatively new area in medicines. In the US, the Biologics Price Competition and Innovation Act (BPCI Act) of 2009 came into on 23 March 2010,[41] and the first US approval of a biosimilar product was granted in early 2015.[42]

In order to provide a measure of guidance in approvals for biosimilar products, the WHO has issued *Guidelines on Evaluation of Similar Biotherapeutic Products* (SBPs).[43] Regulators in Asia have begun introducing rules governing registration of biosimilar products. For example, Malaysia and Singapore issued guidance documents in 2008 and 2011, respectively.[44] Notably, Singapore's guidance document makes reference

[39] See *The Global Use of Medicines: Outlook through 2017*, IMS Institute for Healthcare Informatics, 2013.

[40] See, for example, Ranjani Nellore, Regulator Considerations for Biosimilars, *Perspect. Clin. Res.*, Vol. 1, No. 1 (January–March 2010), 11–14, and McKinsey & Company – *From Science to Operations – Questions, Choices and Strategies for Success in Biopharma* (2014).

[41] The US government held consultations with industry to produce the resulting federal rules on biosimilars substitution; see *Compromise Reached on US Legislation on Biosimilars Substitution*, Generics and Biosimilars Initiative, 16 January 2015, http://www.gabionline.net/Policies-Legislation/Compromise-reached-on-US-legislation-on-biosimilars-substitution (accessed 27 April 2015).

[42] *FDA Approves First Biosimilar ZarxioTM (filgrastim-SNDZ) from Sandoz*, 6 March 2015, Reuters.

[43] WHO, http://www.who.int/biologicals/areas/biological_therapeutics/BIOTHERAPEUTICS_FOR_WEB_22APRIL2010.pdf.

[44] For Malaysia, see portal.bpfk.gov.my/index.cfm?menuid=46&parentid=15, and for Singapore, see http://www.hsa.gov.sg/content/dam/HSA/HPRG/Western_Medicine/Overview_Framework_Policies/Guidelines_on_Drug_Registration/Appendix%2017_Guidance%20on%20Registration%20of%20Similar%20Biological%20Products%20in%20Singapore%202011.pdf.

to data exclusivity for patented reference products.[45] India issued its guidance document in 2012,[46] while China issued its regulation on biosimilar products on 3 March 2015.[47]

In the TPP negotiations, biologic products were a point of contention as the US had requested an extended duration for data exclusivity protection, as compared with data for non-biologic products. The USTR proposal was for a 12-year period of protection.[48] This proposal has met with objections and concerns from negotiating partners as well as external stakeholders.[49] The panel which authored the *Australia Pharmaceutical Patent Review Report* of 2013, for example, was not convinced that an extension to 12 years was necessary, given current evidence and the fact that this was a new area.[50] In the US, court proceedings already exist in respect of biosimilar products under the US Biologics Price Competition and Innovation Act (BPCIA). An example is the ongoing challenge of Amgen of Sandoz' forthcoming launch of its biosimilar product for treating neutropenia (a disorder where there are abnormally low levels of a type of white blood cells) which recently received FDA

[45] *Ibid.*, at p. 4, referring to portions of the general *Guidance on Medicinal Product Registration in Singapore* on data protection, data exclusivity and patent linkage. Biologic plants have been established in Singapore; see, for example, Amgen Biologics Plant in Singapore, *The Straits Times*, 21 Nov 2011, and New Biologics Plant in Singapore, *The Straits Times*, 8 August 2014. See generally Stanley Lai, The Singapore Patents Act 1994: Whither Biotechnology Patent Law?, *Sing. Ac. L. J.*, Vol. 7 (1995), 397–398.

[46] See http://dbtbiosafety.nic.in/Files%5CCDSCO-DBTSimilarBiologicsfinal.pdf (accessed 29 April 2015).

[47] For China, see CFDA notice: http://www.sfda.gov.cn/WS01/CL0051/115102.html.

[48] See USTR website, https://ustr.gov/about-us/policy-offices/press-office/blog/2013/November/stakeholder-input-sharpens-focuses-us-work-on-pharmaceutical-IP-in-TPP. For examples of US Senate members' views on the duration, see http://keionline.org/node/2069 (accessed 28 April 2015). At the time of writing, the TPP negotiations had been concluded but the final text - which is subject to ratification - is not yet available.

[49] See, for example, http://infojustice.org/archives/31196 (accessed 28 April 2015).

[50] Pp. 178–181; Report available at: http://www.ipaustralia.gov.au/pdfs/2013-05-27_PPR_Final_Report.pdf (accessed 28 April 2015). The relevant finding at p. 181 is as follows:

> There is insufficient evidence to support an increase in data protection beyond the current five-year period for biologics at the present time. However, the Panel acknowledges that the regulatory environment and market for biologic and biosimilar medicines is still developing and that the situation should be revisited when further market experience gives us a better understanding of the relevant issues.

approval.[51] It appears that the minimum period of data protection for biologics under the TPP was ultimately agreed to be below 12 years.[52]

### 8.4.2 *Personalized medicines*

An emerging area of medicine is that of personalized medicines, linked to individual's genes or biomarkers. Such medicines would, unlike existing "mass market" medicines, be much more targeted in approach. Singapore has already identified this as an emerging area in which it is making significant investments.[53]

A number of legal and ethical challenges exist in relation to such medicines.[54] Even the US does not have specific legislation regulating this area as yet. A bill regulating personalized medicines was introduced in the US in 2010. Although this has not been enacted, there has been recent revival of interest on the part of the Obama administration in such legislation, with the 2015 launch of the Precision Medicine Initiative.[55] This initiative is in the process of being fleshed out, and can be expected to raise interesting legal issues in time to come, for both the US and its trade partners.[56]

### 8.4.3 *Traditional medicines*

As countries in Asia continue to develop and assess their positions in being manufacturers and exporters of "conventional" medicines,

[51] *Amgen Inc. v Sandoz Inc.*; see summary of proceedings at http://www.lexology.com/library/detail.aspx?g=ef556243-07d4-454c-aa49-9b11d1bbb6f2 (accessed 29 April 2015). On the FDA approval of Sandoz' product, see Novartis (the parent company of Sandoz Inc.) website, press release, 6 March 2015, http://www.novartis.com/newsroom/media-releases/en/2015/1900097.shtml.

[52] See e.g. http://www.biopharma-reporter.com/Markets-Regulations/TPP-Five-years-data-protection-for-biologics-in-US-Asia-trade-deal (accessed 9 October 2015). The final text of the TPP agreement was not yet publicly available at the time of writing.

[53] See Lynn Kan, Singapore Places Bets on Personalized Medicines, *The Business Times*, 10 November 2012.

[54] See, for example, *Use of "-Omics" Technologies in the Development of Personalised Medicine*, EU Staff Working paper, 25 October 2013, http://ec.europa.eu/health/files/latest_news/2013-10_personalised_medicine_en.pdf (accessed 28 April 2015).

[55] See http://thomas.loc.gov/cgi-bin/query/z?c111:H.R.5440:, https://medium.com/@WhiteHouse/president-obamas-state-of-the-union-address-remarks-as-prepared-for-delivery-55f9825449b2, http://www.nih.gov/precisionmedicine/ and http://www.personalizedmedicinebulletin.com/2015/01/25/president-obama-revisits-interest-in-personalized-medicine/ (accessed 28 April 2015).

[56] See http://www.nih.gov/news/health/mar2015/od-30.htm (accessed 28 April 2015).

traditional medicines are likely to gain stronger interest in future. There are two main reasons for this observation. First, a number of Asian countries – including China, Japan, Korea and India, and within ASEAN, Indonesia and Thailand – are steeped in traditions of herbal and natural medicines.[57] Second, discussions at multilateral fora already take on board the growing significance of traditional medicines.[58] One question, for instance, is the extent to which traditional medicines could or should be integrated with "mainstream", Western medicines. In many jurisdictions, approvals of traditional medicines and accreditation of practitioners are treated separately from those of "conventional" medicine.[59] The WHO *Traditional Medicine Strategy 2014–2023*, for example, aims to integrate traditional medicines in the promotion of universal health coverage.[60] The WHO's GSPOA also encourages R&D in traditional medicines in developing countries.[61]

Traditional medicine provisions have also begun to appear in a small number of FTAs. In the China-Switzerland FTA, for instance, "herbal medicine" is included in the patentability commitment provision.[62] In the China-Peru and China-Costa Rica FTAs, the parties provide for the protection for genetic resources, traditional knowledge and folklore.[63]

Within ASEAN, member countries are in the process of considering adoption of the ASEAN Agreement on Traditional Medicines, together with a number of related guidelines; Singapore completed its public consultation process in January 2015.[64]

57 See information at WHO, SEARO, http://www.searo.who.int/entity/medicines/topics/traditional_medicines_in_sear/en/ (accessed 28 April 2015).

58 See, for example, WHO, http://www.who.int/medicines/areas/traditional/en/. On the link of traditional medicines to a broader debate relating to biodiversity and biopiracy, see, for example, WHO, *TRIPS, CBD and Traditional Medicines: Concepts and Questions*, 2001.

59 See Terry S. H. Kaan, *Traditional, Complementary and Alternative Medicine*, in *Routledge Handbook of Medical Law and Ethics*, ed. Yann Joly and Bartha Maria Knoppers (Routledge, 2015).

60 WHO, http://apps.who.int/iris/handle/10665/92455 (accessed 28 April 2015); see especially parts 3.4 and 4.3, and see also WHA Resolution EB134.R6, January 2014; at p. 8.

61 At Annex, para. 8; see also paras. 31–32.

62 Art. 11.8(1), Free Trade Agreement between the Swiss Confederation and the People's Republic of China.

63 See Art. 145, China-Peru FTA, http://fta.mofcom.gov.cn/bilu/annex/bilu_xdwb_11_en.pdfand; Art. 111, China-Costa Rica FTA.

64 See HSA, http://www.hsa.gov.sg/content/dam/HSA/HPRG/Complementary_Health_Products/Annexes/ASEAN%20Agreement%20on%20Traditional%20Medicines.pdfm; and ASEAN, at: http://www.asean.org/news/asean-secretariat-news/item/harmonising-regulation-of-traditional-medicines-and-health-supplements-ahead-of-asean-integration and http://www.asean.org/communities/asean-economic-community/item/harmonization-of-standards-and-technical-requirements-in-asean (accessed 28 April 2015). For an

### 8.4.4 *Halal pharmaceuticals*

Another emerging type of pharmaceutical that is likely to gain greater interest in the near future is that of halal medicines. Twenty-four per cent of Muslims live in the Asia-Pacific region, and demand for such medicines has been reportedly growing.[65]

As the certification process for such medicines must ensure that certain rules of Sharia law are satisfied with respect to ingredients, manufacture and dispensing, not all medicines may qualify for such certification.[66] In Australia, the first halal medicine was made available in 2013.[67]

A number of countries in ASEAN already have halal certification programmes and production facilities. Brunei Darussalam has set up a halal-certified pharmaceutical production business through a joint venture.[68] Singapore has a halal certification system in place for food and medicines, and has even certified halal traditional medicines.[69] Malaysia has also established a system of halal certification for medicines as well as "halal parks" as infrastructural hubs for manufacture of such products.[70]

overview on traditional medicines in ASEAN, see also, generally, ASEAN, *Harmonization of Traditional Medicines and Health Supplements*, http://www.hsa.gov.sg/content/hsa/en/Health_Products_Regulation/Complementary_Health_Products/Overview/ASEAN_Harmonization_of_Traditional_Medicines_and_Health_Supplements.html.

65 Siddharth Dutta, *Growing Demand for Halal Pharmaceuticals*, 1 September 2014, PharmaAsia, http://www.pharmaasia.com/article/growing-demand-for-halal-pharmaceuticals/10803 (accessed 28 April 2015). See also, generally, Azmi Sarriff and Hadeer Akram Abdul Razzaq, Exploring the Halal Status of Cardiovascular, Endocrine, and Respiratory Group of Medications, *Malays. J. Med. Sci.*, Vol. 20, No. 1 (January–March 2013), and Part III, *The Halal Industry*, Global Islamic Finance Review, 2013.

66 See Kenny Peng, *Halal Pharmaceuticals*, November 2012, *Regulatory Focus*, Regulatory Affairs Professionals Society.

67 See http://www.unwhd.com/component/k2/item/18-first-ever-halal-medication-launches-in-australia/18-first-ever-halal-medication-launches-in-australia.html (accessed 28 April 2015).

68 See EIU, *Brunei: Production of Halal Pharmaceuticals Begins Soon*, The Economic Intelligence Unit, 13 November 2013 (accessed 7 May 2015); and Brunei Guideline for Halal Certification (2007).

69 See Singapore Halal Certification, http://www.muis.gov.sg/cms/services/hal.aspx?id=458; and Ministry of Trade & Industry website, Science Arts Finds Cure in Halal Nod, *The Business Times*, 25 March 2008, http://www.mti.gov.sg/ProEnterprisePanel/Documents/app.mti.gov.sg/data/pages/3753/doc/BT%2025%20Mar%2008%20-%20Science%20Arts%20finds%20cure%20in%20Halal%20nod.pdf (accessed 7 May 2015).

70 See Malaysia Halal Industry Development Corporation, http://www.hdcglobal.com/publisher/alias/?dt.driverAction=RENDER&pc.portletMode=view&pc.windowState=normal&pc.portletId=Newslatest.newsPortlet (accessed 28 April 2015). On Malaysia's halal guidelines, see MS 2424.

Thailand, too, has a national halal certification standard and is a producer of halal products.[71] Such products may also provide an impetus for medical tourism within the region.[72]

In 2014, Indonesia enacted legislation requiring halal certification of pharmaceutical products, which will take effect five years from enactment, that is, in 2019.[73] This has been noted as a potential issue for pharmaceutical exporters.[74] Global pharmaceutical enterprises such as Merck have begun to look into whether their products meet halal requirements.[75]

## 8.5 Can industry do more?

In Chapter 7, a new "soft" law code for the pharmaceutical industry was proposed to advance transparency in the industry's activities in various areas where they may affect access to medicines. A number of instruments and information sources already exist for various aspects of access to medicines in developing countries. For example, information about the Medicines Patent Pool is transparent and available on the internet, including information about licensors, sub-licensees of the pool, and terms of agreements.

Already, thanks to the internet and globalized debate on access, several other tools exist to shed light on gaps and practices which may impede access to medicines. There is even an Access to Medicines Index which seeks to analyse activities of the top 20 research-based pharmaceutical companies which help to improve access in developing countries, and ranks them.[76] The index, started in 2008 by a Dutch entrepreneur, Wim Leereveld, aims to highlight good practices as encouragement to improve

71 National Halal Standard, THS 24000: 2552, The Central Islamic Committee of Thailand General Guidelines on Halal Products. See also Board of Investments, Thailand, at: http://www.boi.go.th/tir/issue/201103_21_3/42.htm (accessed 7 May 2015).

72 See, for example, the discussion at: http://www.reuters.com/article/2014/07/29/us-malaysia-medical-idUSKBN0FY2AT20140729 (accessed 28 April 2015).

73 Halal Product Certification Law, Law No. 33 of 2014, enacted 17 October 2014; see also, generally, Kennial Caroline Laia, Rahajeng K. H. and Harso Kurniawan, Halal Certification a Bitter Pill for Pharma, *The Jakarta Globe*, 6 February 2015.

74 See for example, USTR, 2013 and 2014 National Trade Estimate Reports, Indonesia.

75 See http://magazine.merckgroup.com/en/culture/certificates/Halal_seal1.html; and Takeshi Kumon, Drugmakers Seek New Opportunities in Islamic World, *Nikkei Asian Review*, 21 November 2014, .

76 See information at: http://www.accesstomedicineindex.org/ (accessed 28 July 2014). The index is funded by the Bill and Melinda Gates Foundation and UK and Dutch governments.

access further, instead of adopting an approach of "naming and shaming" companies. The index utilizes four "strategic pillars" – Commitments, Transparency, Performance and Innovation – and under these, it uses seven technical areas: General Access to Medicine Management, Public Policy and Market Influence, R&D, Price, Manufacturing and Distribution, Patents and Licensing, Capability Advancement and Product Donation. The index rankings are published biennially, with the most recent being that of 2012, at the time of writing.

An area that is under discussion is the extent of availability of information on worldwide R&D activities of pharmaceutical companies. To this end, the WHO has been leading discussions on a Global Observatory for Health R&D.[77]

With a slowing of innovative output in the pipeline and the progressive expiry of existing patents on prescription medicines, the pharmaceutical industry will need to continue to consider how to re-engineer their innovation and IPR strategies, leveraging on public, academic and other new resources (such as, as suggested in Chapter 7, new investment fund sources such as philanthropic arms of SWFs, and on the new, dynamic BRICS countries' capabilities).

Unless voluntary licensing activities and efforts in patent pooling reach a level of being able to collectively satisfy public demand for pharmaceuticals in poor countries, one may expect the drumbeat of compulsory licensing to carry on in various parts of the world. The success of voluntary licensing will of course continue to depend on the willingness to enter into licence agreements in the first place, the terms of licensing and the royalty payments requested.

### *8.5.1 New "soft rules" for the pharmaceutical industry?*

Existing codes of conduct for the pharmaceutical industry – national and international – generally deal with promotional and informational aspects of pharmaceutical industry conduct, safety and ethical

[77] See WHO, http://www.who.int/phi/documents/dwp1_global_health_rd_observatory_16May13.pdf; and commentary, *WHO Sets Path for Model R&D Projects for Developing Countries*, IP Watch, at http://www.ip-watch.org/2014/01/25/who-sets-path-for-model-rd-projects-but-doesnt-rock-the-boat/, 25 January 2014, IP Watch (accessed 28 July 2014). See also, generally, WHO, *The WHO Strategy on Research for Health*, http://www.who.int/phi/WHO_Strategy_ on_research_for_health.pdf?ua=1, (accessed 28 July 2014).

research and treatment of patient data.[78] The Association of British Pharmaceutical Industry Code of Practice for the Pharmaceutical Industry 2014,[79] for example, which applies to the promotion of pharmaceuticals to health professionals in the UK, incorporates a number of codes.[80]

The Prescription Medicines Code of Practice Authority (PMCPA), a not-for-profit body, administers and enforces the ABPI Code at arm's length from the ABPI itself. It handles complaints under the code and may impose sanctions for violations.[81] The process comprises a panel and an appeal board.[82]

The APEC Mexico City Principles similarly address primarily ethical business practices in these, although they address ethics in dealing with the government sector specifically in the procurement context.[83]

[78] Examples of codes include the IFPMA Code (2012) and the ABPI Code of Practice 2014. For examples of national codes of conduct relating to medicines, see Medicines Australia's Code of Conduct, China's R&D-based Pharmaceutical Association Committee Industry Code of Practice on the Promotion of Pharmaceutical Products 2003 (in addition to China's Advertising Law of the PRC 1995, Detailed Implementing Rules of the Advertising Law of the PRC 2005, Measures for the Examination of Pharmaceutical Advertisements 2007, Standards for the Examination and Publication of Pharmaceutical Advertisements 2007 and general anti-bribery laws), India's Code of Marketing Practice for Indian Pharmaceutical Industry, Indonesia's IPMG Code of Pharmaceutical Marketing Practices (2013), Japan Pharmaceutical Manufacturers Association Promotion Code for Prescription Drugs, Korea's KRPIA Code of Conduct, Malaysia's Code of Pharmaceutical Marketing Practices for Prescription (Ethical) Products, 2012, and, for Singapore, the Singapore Medical Council Code, Advertising Code and Singapore Association of Pharmaceutical Industries (SAPI) Code of Marketing Practices 2014. See also, generally, Charles De Wet, *Information and Promotion*, in *The Textbook of Pharmaceutical Medicines*, ed. John P. Griffin, John Posner and Geoffrey R. Barker (Wiley-Blackwell, BMJ Books, 2013).

[79] Available at the Prescription Medicines Code of Practice Authority website: http://www.pmcpa.org.uk/thecode/InteractiveCode2014/Pages/default.aspx (accessed 1 May 2014).

[80] *Ibid.*, p. 4. These are the International Federation of Pharmaceutical Manufacturers and Associations' (IFPMA) Code of Practice; The European Federation of Pharmaceutical Industries and Associations' (EFPIA) Code on the Promotion of Prescription-Only Medicines to, and Interactions with, Healthcare Professionals; The EFPIA Code of Practice on Relationships between the Pharmaceutical Industry and Patient Organisations; the World Health Organization's Ethical Criteria for Medicinal Drug Promotion; Directive 2001/83/EC on the Community Code relating to medicinal products for human use, as amended by Directive 2004/27/EC; and The Human Medicines Regulations 2012 (UK).

[81] See cases handled, at: http://www.pmcpa.org.uk/media/Pages/Lundbeck-Limited,-A-Menarini-Pharma,-Bayer-plc,-HRA-Pharma-UK-and-GlaxoSmithKline-named-in-advertisements-for-breaches-of-t.aspx (accessed 1 May 2014).

[82] See http://www.pmcpa.org.uk/aboutus/Pages/default.aspx (accessed 1 May 2014).

[83] The *Mexico City Principles for Voluntary Codes of Business Ethics in the Biopharmaceutical Sector* (2011). Paragraph 13 states:

Such codes are important in addressing the industry-physician/industry-medical facility relationship and the industry-patient relationship. They generally do not, however – probably for good reason as far as the industry is concerned – deal more broadly with other important aspects of the industry-government relationship to address concerns surrounding industry conduct and their impact on access to medicines. It is this latter relationship that encompasses actions of lobbying for better protection at home and overseas (included TRIPS-plus protection in bilateral or regional treaties), actions to influence law and policy reform affecting the industry and pricing actions in the face of public needs and expenditures related to pharmaceutical products. These can all be legitimate actions and certainly have their place in a democratic market-based economy. However, there are presently no global standards for checking or reviewing such actions – whether emanating from the pharmaceutical industry itself or elsewhere – to ensure that such actions conform to certain societal expectations as represented by important international declarations and policies. Examples of these are the MDGs, GSPOA and other regional (e.g. within ASEAN) or national policies.[84] International goals, guidelines and strategies in particular are often linked primarily with government actions and do not focus sufficiently on dealing with the industry directly.

There is scope therefore to develop a new international code of conduct for the pharmaceutical industry to address various matters which have not received requisite attention to date.

**Public Sector Relationships and Procurement**

A. The decision-making process by companies and government during and including the government procurement process, through bidding or any other procedure of government procurement, must be professional and ethical. There should be no attempt to exert inappropriate influence.
B. Companies must provide accurate and balanced information to the government.
C. Companies and government officials should ensure that their relationships and fee-for-service arrangements comply with government ethics rules or procedures.

On APEC, see also http://www.pharmaasia.com/article/apec-endorses-principles-for-industry-codes-of-ethics/9275 (accessed 27 July 2014).

84 See Chapter 2. In September 2015, UN member states adopted the 2030 Sustainable Development Agenda and Sustainable Development Goals; the Goals will take effect on 1 January 2016 (see https://sustainabledevelopment.un.org/post2015/transformingourworld/, accessed 11 October 2015).

The OECD *Guidelines for Multinational Enterprises* (2011) refer to social objectives and public interest: in its "General Policies", where the following is stated:

> Enterprises should:
>
> 1. Contribute to economic, environmental and social progress with a view to achieving sustainable development . . .

In the "Science & Technology" chapter, the guidelines provide as follows:

> 4. When granting licenses for the use of intellectual property rights or when otherwise transferring technology, do *so on reasonable terms and conditions and in a manner that contributes to the long term sustainable development prospects of the host country.*
>
> [Italic emphasis added]

While not forming "soft law", the WHO Good Governance for Medicines programme, on the other hand, aims to address matters such as corruption and transparency of regulation in the pharmaceutical supply chain and supplements the role of national anti-corruption laws and codes of ethics.[85]

One can think of the Santiago Principles on Sovereign Wealth Funds as an example of "soft law" which specifically aims to address SWF entities. While it may be argued that these principles were seen to be necessary to some due to the unique position of SWFs given the sovereign nation involvement and the quantum and type of foreign investments they make, and pharmaceutical entities are generally purely private entities, there are two arguments to introduce such a code for the latter. First, such entities often benefit from the grant of an important patent monopoly that allows them a period in which to recover their R&D investments. This "patent bargain" is traditionally based on the notion that the patentee would disclose enough of the invention to allow others to carry out further innovative activity for the benefit of society at large. In this context, conduct which aims to unfairly change the bargain or to extend that monopoly should be carefully scrutinized by a broader swath of society, and certainly not by industry self-regulation alone. Second, while the law of profit[86] can and does apply to

[85] See WHO, http://www.who.int/medicines/areas/policy/goodgovernance/en/ (accessed 27 July 2014).

[86] This phrase was used by the Delegation of the Holy See during a discussion in the WIPO Standing Committee on the Law of Patents: para. 116, SCP/17/13: "The Delegation stated that the social mortgage on propriety affected also IP and knowledge, and that the law of profit alone could not be applied to what was essential for the fight against hunger, disease and poverty."

pharmaceutical concerns, many of the products produced by this industry have a significant public health dimension in that they have the potential of saving, and in many cases do in fact save lives, and therefore arguably differ from many other commercial goods.[87] Third, such entities often make use of public funds for their R&D activities which may eventually lead to a "blockbuster" or patented medicine that reaps tremendous profits.

This proposal differs from existing calls for a new R&D treaty in two respects: first, it is not an intergovernmental treaty. Second, it addresses broader types of conduct beyond R&D activities, costs and cost sharing. It also does not aim to require patent information or general pricing information, which other organizations or initiatives may currently demand. It also differs from recent initiatives which are IP-centric in focus.[88]

The following provides a scoping outline of possible areas of coverage, for such a code to be explored:

- Explicit linkage of pharmaceutical industry members to the MDGs and GSPOA, and to the Doha Declaration on the TRIPS Agreement and Public Health, for a start;[89]
- Relating this linkage to requiring transparency of pharmaceutical company policies specifically on tiered pricing (if any applies to the particular company), voluntary licensing, non-voluntary licensing, royalty-setting policies in both these contexts and stance on patent pools for medicines;
- Intellectual property policy in collaborative work with the public sector or with academic institutions: licensing policy, IP ownership/ co-ownership, licensing rights and fees and policy with regard to "evergreening" and patent thickets;
- Non-assertion of patent right policies, if any, and to what situations these apply;

[87] See earlier discussion of arguments that information relating to such inventions are even "public goods".

[88] See, for example, *Declaration on Patent Protection, Regulatory Sovereignty under TRIPS*, Max Planck Institute for Innovation and Competition, http://www.ip.mpg.de/en/pub/news/patentdeclaration.cfm; and Principles for IP Provisions in Bilateral and Regional Agreements, H. Grosse Ruse-Khan, *et al.*, *IIC*, Vol. 44, No. 8 (2013), 873–877.

[89] The MDGs will be succeeded by the SDGs in January 2016; see note 84 above. While there are plans for a Framework for Engagement of Non-State Actors at the World Health Assembly (WHA), these have not yet borne fruit at the most recent WHA Assembly, 18–26 May 2015. See WHO, *Framework of Engagement for Non-State Actors, Report by the Director-General*, A68/5, 1 May 2015. Non-state actors may include private enterprises, business associations, academic institutions, NGOs and philanthropic organizations.

- Minimum standard of ethical conduct for industry in interacting with foreign governments and regulatory authorities in law policy reform activities;
- Reporting of innovative individual company policies that uniquely promote access to medicines as an information-sharing exercise;
- Activities that promote TRIPS-plus protection for the industry; and
- Reports of patent litigation between originator and generic producers and broad terms of any conciliatory negotiations and settlements of such litigation.

Already, several pharmaceutical companies have been stepping up to both explain and set policies in many of the above areas, and a code merely ensures that such policies and actions meet certain minimum requirements. Given that it is a "soft law" instrument, no dispute settlement structure is envisaged. In order therefore to give the code more impact, it is suggested that the parties convened to draft such a code could include government representatives from, for example, the largest pharmaceutical importing countries and LDCs, and international pharmaceutical organization representatives such as the IFPMA.[90] To further boost the impact, the code could include an annual review mechanism – much like the Trade Policy Review Mechanism that exists in the WTO – in which the convening governments may compile a joint annual report on the pharmaceutical industry's initiatives and conduct *vis-à-vis* the code's principles in a given year. The funding of the activities for this convening group may be raised from various sources such as philanthropic arms of SWFs, and international organizations such as the WHO, which has a highly relevant mandate to promote public health and access to medicines.

The above provides a scoping list for negotiators to consider and to work with. The "strength" or usefulness of the code will, of course,

[90] In the case of the Santiago Principles on SWFs, developed country states which were recipients of large SWF investments, such as the US, agitated for a process via the IMF to formulate those processes. For the genesis of those principles, see, for example, this author, Multi-Sourced Norms Affecting Sovereign Wealth Funds: A Comparative View of National Laws, Cross-Border Treaties and "Codes", *Journal of World Investment & Trade*, Vol. 10, No. 6 (December 2009), 793–828 and (forthcoming), *Sovereign Wealth Funds – Investors in Search of an Identity in the 21st Century*, article accepted for publication in a special 2014 issue on SWFs of International Review of Law, Qatar University. With the rise of the "BRICS" countries, these may be candidates for the convening group. The BRICS countries recently established a development bank; see http://www.economist.com/news/finance-and-economics/21607851-setting-up-rivals-imf-and-world-bank-easier-running-them-acronym.

depend ultimately on the negotiated content and associated mechanisms for review.

Correspondingly, treaty negotiators may need to revisit a "one-size-fits-all" approach for treaty demands on TRIPS-plus provisions such as data exclusivity. Already, the USTR appeared to have taken a step back during the TPP negotiations in announcing a "flexible" and "differentiated" approach that takes into account the differences between negotiating partners.[91]

## 8.6 Can governments do more?

### *8.6.1 Strengthening national patent systems: establishing clear and appropriate patentability standards*

A few general points will only be mentioned briefly here as they are follow-up points to matters discussed in Chapters 3–5.

The patentability threshold is, as previously discussed, an important one as it gives access to the legal monopoly associated with a patent. The threshold should be applied in a rigorous manner to ensure that only high-quality inventions succeed.[92] Policy-makers in Asia should also provide clarity in their stances (national/regional/global) on the patentability of the output of newer technologies such as genes and gene sequences, and the relevance or applicability of public morals exceptions.

In the area of funding, subsidies and health technology assessment, the innovative impact and therapeutic improvement of a new medicine as compared with existing medicines are taken into account. In the area of patentability of medicines, this is however not always a criterion. As has been seen in Chapter 3, the notion of improved therapeutic efficacy is a criterion for patenting a new form of an existing medicine under Section 3(d) of the Indian Patents Act. The Indian Supreme Court underscored this in the decision *Novartis v The Government of India*. Similarly, the Philippines legislation has introduced improvement in efficacy as a criterion. In China, the patent legislation requires a significant enhancement.

[91] USTR website.

[92] See issues raised in Chapters 3 and 4; and, also, Sanil Khatri, Reformulating the Inventive Step in Patent Law to Achieve an Appropriate Balance between Patent Holders and the Public, *Int. J. Tech. Pol. & L.*, Vol. 1, No. 1 (2012) 69; and Amir Attaran, How Do Patents and Economic Policies Affect Access to Essential Medicines in Developing Countries?, *Health Affairs*, Vol. 23, No. 3 (2004), pp. 155–166.

More broadly, patentability is presently determined at the first instance by national intellectual property offices which do not necessarily apply an HiAP/HTA approach. The result can be a legalistic application of case principles to determine patentability of a new medicine, or a new form or new use of an existing medicine. Whereas in health policy, HTA in particular would require looking at whether the medicine or new form performs in a superior therapeutic manner, patent law criteria may not. The result may be that for such a criterion to clearly apply, one would need to legislate so, as India and the Philippines have chosen to do so.[93]

Under the TRIPS Agreement, a WTO member state would be free to make this choice, or not. Whichever decision is taken, this point merits careful thought by policy- and lawmakers in Asia, particularly before a demand to definitively exclude such an option by treaty is made at the negotiating table by a trade partner. The issue of patentability requirements is being discussed by WIPO members in the Standing Committee on the Law of Patents (SCP).[94]

Under the TRIPS Agreement, pharmaceutical inventions may not be treated differently under Article 27, due to the requirement of technology-field neutrality. This idea has in fact been raised in relation to Article 1709(7) of the NAFTA in the ongoing arbitration case *Eli Lilly v Canada*. It may be time to consider whether this aspect of TRIPS should be reviewed, given the emphasis and special attention already given to pharmaceutical products in the Doha Declaration on the TRIPS Agreement and Public Health.

As seen in Chapter 2, tariff barriers in respect of medicines continue to exist in Asia and need to be reviewed, especially in light of the broader public interest in providing medicines as affordably as possible. The USTR, for example, stated in its 2014 Special 301 Report that "according to the October 2012 WTO report titled *More Trade for Better Health? International Trade and Tariffs on Health Products*, India maintains the highest tariffs on medicines, inputs to medicines, and medical devices among the WTO members identified in the report".[95]

93 See *MERCOSUR Agreement on Patentability Criteria, Access to Medicines, and Public Health* (referred to in Brazil Patent Reform report, 2013, p. 115).

94 The author is grateful to Ms. Aida Dolotbaeva of WIPO for the informative discussion and materials on the work of the SCP in Geneva, Switzerland; notes are on file with the author.

95 See USTR Special 301 2014 Report, p. 26; and pp. 20–21, 24, WTO report, available at: http://www.wto.org/english/res_e/reser_e/ersd201217_e.pdf (accessed 23 July 2014).

Another aspect would be to address whether to regulate mark-ups that add to costs of essential medicines.[96] While not every country will wish to regulate this by imposing a ceiling or a "zero" mark-up policy (since some may prefer prices to be set by the market), there is room for investigating whether mark-ups may be addressed constructively, for example, whether through the lens of trade facilitation in improving the medicines supply chain, or through policy guidelines for public institutions which have a hand in setting final prices.

Governments utilize various means to influence and to control drug prices within their borders. Often, states may use a combination of such means, such as reimbursement caps with competitive procurement tenders. While the objective is generally to keep drug prices affordable to the population, and, intuitively, they appear to be a good way to ensure affordability, their effectiveness can vary due to a variety of a reasons. In some cases, for example, controls may in fact lead to a shortage of certain drugs, due to a lack of incentive on the part of providers to sell at the price levels set.[97]

### *8.6.2 Addressing corruption in the pharmaceutical supply chain*

As mentioned in Chapter 1, corruption can contribute to access-to-medicines problems. According to the WHO, 10–25% of public procurement spending (inclusive of pharmaceuticals) is lost to corrupt

[96] There is some support for this argument; see Chapter 2, UN MDG Working Group Report 2005, at p. 80. See also, generally, WHO, *Increasing Access to HIV Treatment in Middle-Income Countries – Key Data on Prices, Regulatory Status, Tariffs and the Intellectual Property Situation* (2014).

[97] In China, for example, price caps were introduced in 2009 to keep basic medicines at low retail prices. However, this has reportedly led to a shortage due to producers withdrawing their products, so much so that the government recently changed the policy to drop these price caps for a certain number of medicines; see Report, Li Hui and Adam Jordan, *China Backs Away from Price Controls on Basic Medicines*, 17 April 2014, Reuters: and Shan Juan, Government to Give Public Better Access to Low-Cost Drugs, 16 April 2014, *China Daily*. China has plans to move towards trial reimbursement caps and clinical pathways: see *China to Trial Reimbursement Caps and Clinical Pathways in Pilot Schemes from 2014*, IHS, 29 November 2013, http://www.ihs.com/products/global-insight/industry-economic-report.aspx?id=1065984854 (accessed 30 June 2014); and Cheng Tsung-Mei, A Pilot Project Using Evidence-Based Clinical Pathways and Payment Reform in China's Rural Hospitals Shows Early Success, *Health Affairs*, Vol. 32, No 5 (May 2013), 963–973. On clinical pathways, see generally Leigh Kinsman, Thomas Rotter, Erica James, Pamela Snow and Jon Willis, What Is a Clinical Pathway? Development of a Definition to Inform the Debate, *BMC Medicine*, Vol. 8 (2010), 31.

practices.[98] Corruption can take different forms and may occur at different points of the chain of supply, such as in government procurement and in physician prescription.[99] Underpaid doctors may, for example, succumb to corruption and prescribe overpriced medicines.[100] Corruption can take many forms, and in 2013, the EU Commission published a specific study on corruption in the healthcare sector, in which various forms of corruption were identified, including "revolving door corruption".[101]

Given the variables that affect and make up procurement decision-making, the following (rather sensible and useful) view has been expressed that

> The procurement of medicines could be guided by a richer base of information, considering existing and emerging disease burdens, pricing of medicines, IP coverage and other practical data coming from different fields, which might be used in a more integrated way.[102]

[98] WHO Fact Sheet No. 335, *Medicines: Corruption and Pharmaceuticals*, (December 2009). See also generally WHO, *Good Governance for Medicines – Curbing Corruption in Medicines Regulation and Supply*, 2007; www.who.int/healthsystems/%20. . .%20/25GGM.pdf; and ADB/OECD, *Curbing Corruption in Public Procurement and Asia and the Pacific* (2006).

[99] In Thailand, for example, a former Minister for Public Health and his adviser were convicted for corruption in relation to purchase of medical supplies which had resulted in the purchase of highly overpriced medicines by hospitals. The former minister, Rakkiat Sukthana, was convicted and sentenced to 15 years in prison by the Thai Supreme Court for having received 5 m baht to arrange for hospitals to purchase overpriced medicines; see, for example, David Fulbrook, *Thailand Chalks One Up against Corruption*, Asia Times Online, http://atimes.com/atimes/Southeast_Asia/FK09Ae01.html (accessed 26 June 2014). His former adviser, Jirayu Charasathien, was convicted by the Thai Supreme Court for corruption and sentenced to a 6-year prison term; see Report, http://www.abc.net.au/news/2003-10-28/thai-ex-minister-found-guilty-of-corruption/1500788 (accessed 26 June 2014). See also Vinay Bhargava and Emil Bolongaita, *Challenging Corruption in Asia – Case Studies and a Framework for Action* (The World Bank, 2004),pp. 187–192.

[100] See, for example, the discussion about the situation in China, in Physician, Heal Thyself, *The Economist*, 1 February 2014. See also Zhiping Yang and Daiming Fan, How to Solve the Crisis behind Bribegate for Chinese Doctors, *The Lancet*, Vol. 379, No. 9812 (21 January 2012), pp. e13 – e15, and http://www.ncbi.nlm.nih.gov/pubmed?cmd=Link&dbFrom=PubMed&from_uid=18607039. In relation to India, see Dinsa Sachan, Tackling Corruption in Indian Medicine, *The Lancet*, Vol. 382, No. 9905(2013), pp. e23–e24.

[101] European Healthcare Fraud and Corruption Network (EHFCN) study, 2013.

[102] Remarks by the WTO representative at the WIPO SCP, para. 86, SCP/19/8, Nineteenth Session, 25–28 February 2013, document dated 27 January 2014. The statement was made in the broader context of the need for coherence policy-making. See also, generally,

Corruption in the procurement and prescription of medicines remains a problem in many parts of Asia, including large Asian countries such as China, Indonesia and India; dealing with it will be an ongoing challenge. In an effort to tackle the problem, the Chinese government recently issued various regulations and guidelines that specifically address corruption in the procurement of medicines.[103]

At the multilateral level, the WHO's Good Governance for Medicines Approach, launched in 2004, is an initiative which addresses governance in relation to procurement of medicines. The WTO Government Procurement Agreement (GPA) contains a number of disciplines that aim to improve transparency in procurement systems. It is, however, a plurilateral agreement which not all WTO members have chosen to accede to. Within Asia, only Hong Kong, Japan, Korea, Singapore and Chinese Taipei are parties to the GPA, while Indonesia, Malaysia and Vietnam are observers.[104] China is also presently an observer but is negotiating accession to the GPA.

## 8.7 Regional integration and pharmaceutical initiatives

Enhancing recognition and standards of safety through the use of GMP certificates and inspection processes is provided for in the ASEAN Sectoral Mutual Recognition Arrangement for Good Manufacturing Practice (GMP) Inspection of Manufacturers of Pharmaceutical Products (2009) (MRA).[105] The MRA covers "medicinal products" as defined in Article 1, which excludes "biopharmaceuticals, radiopharmaceuticals, traditional medicines and investigational medicinal products". Under Article 19, the written agreement of all ASEAN members is

WHO, *Promoting Access to Medical Technologies and Innovation*, at pp. 80–83, discussing government procurement and medicines.

[103] See, for example, Circular No. 49 of the National Health and Family Planning Commission (26 December 2013) and Guiding Opinions of the General Office of the State Council on Improving the Work of Centralized Procurement of Medicines by Public Hospitals in State Council Gazette Issue No. 7 Serial No. 1510 (10 March 2015). See also *State Council Perfects Medicine Procurement by Public Hospitals*, 28 February 2015, http://english.gov.cn/policies/latest_releases/2015/02/28/content_281475063180344.htm (accessed 3 May 2015). See also, generally, with regard to India, Dinsa Sachan, Tackling Corruption in Indian Medicine, The Lancet, Vol. 382, No. 9905 (2013), e23–e24, and Anita Jain, Samiran Nundy and Kamran Abbasi, Corruption: Medicines' Dirty Open Secret, *BMJ* (2014); 25 June 2015, at http://www.bmj.com/content/348/bmj.g4184 (accessed 3 May 2015).

[104] For the full list of parties, see WTO, https://www.wto.org/english/tratop_e/gproc_e/memobs_e.htm (accessed 3 May 2015).

[105] Done at Pattaya, Thailand, 10 April 2009.

required to review or amend the arrangement – including this provision on scope of the Arrangement. The MRA contains a dispute settlement provision in the event of any dispute or complaint of non-compliance, and these excluded areas would, as a result, not form subject matter for such dispute settlement.[106]

While ASEAN integration has made significant progress to improve opportunities for trade in pharmaceutical products in the region, investment admission and operation rules in some member states may be disincentives to foreign investment (and, possibly, technology transfer opportunities) being made in the sector.[107] This is a difficult area, as the investment rules often represent deep national policy sentiments. Such rules may carry requirements for maximum foreign equity ownership or for a stipulated number of locals to be appointed to senior management or board of directors positions.[108]

### *8.7.1 Reviewing ASEAN trade and investment barriers that apply to the pharmaceutical sector*

A number of trade and investment barriers affecting the pharmaceutical sector exist in ASEAN.[109] Transaction costs of pharmaceuticals increase with the complexity of regulatory requirements processes that may differ

[106] Article 17 refers to use of the ASEAN Enhanced Protocol on Dispute Settlement for disputes under the Arrangement.

[107] See, for example, EU-ASEAN Business Council, *Promoting Trade and Investment between ASEAN and Europe*, August 2014, at pp. 11, 49 and 51.

[108] On ASEAN investment restrictions, see ASEAN, reservation lists in the Schedule to the ACIA, http://www.asean.org/communities/asean-economic-community/item/asean-compehensive-investment-agreement-reservation-list and http://investasean.asean.org/index.php/page/view/acia-reservation-list (accessed 25 March 2015). In the case of Indonesia, for example, there is a maximum foreign equity ownership cap in pharmaceutical industry entities of 75%; see Indonesia, Schedule to the ACIA, p. 19. In the case of Myanmar, only state-owned enterprises under the Ministry of Health are permitted to manufacture pharmaceutical products and provide services related to such manufacture; see Myanmar, Schedule to the ACIA, p. 2. In the case of Malaysia, there is a policy in favour of "bumiputera" or indigenous people of Malaysia, that is, Malays, natives of Sabah and Sarawak and aborigines; see Malaysia, Schedule to the ACIA, p. 7, and note 4 on that page.

[109] See the investment reservations in the ACIA Reservation Lists of ASEAN members, such as Indonesia and the Philippines, at ASEAN, http://www.asean.org/communities/asean-economic-community/item/asean-compehensive-investment-agreement-reservation-list (accessed 29 April 2015). See also, for example, Indonesia's regulation, Ministry of Health Decree No. 1010/MENKES/PER/XI/2008, discussed in USTR, *2014 National Trade Estimates Report on Foreign Trade Barriers*, p. 159.

from country to country. Trade facilitation and harmonization initiatives can serve to eliminate technical barriers to trade, leading to reduced costs of products, such as pharmaceutical products.[110] Pharmaceutical products fall within one of 12 identified economic integration priority sectors of ASEAN. Given their economic and social importance, they also straddle the goals identified for both the economic and social integration "pillars" of ASEAN. Member countries should therefore specifically address plans to further fast-track trade facilitation for these products, as well as improve the regional supply chain for these products.[111]

### 8.7.2 *Reducing marketing approval costs while maintaining safety*

Efforts have been made in various fora to address such administrative costs while maintaining the safety of products to be marketed. Examples include the use of the Common Technical Document (CTD) modular format for applications, developed by the International Conference on Harmonisation (ICH), and the use of electronic documents.[112] ASEAN members have adopted the use of the ASEAN Common Technical Dossier for the Registration of Pharmaceuticals for Human Use (ACTD) and ASEAN Common Technical Requirements (ACTR).[113] In 2004, ASEAN issued Guidelines for the Conduct of Bioavailability and Bioequivalence of Medicines, and in 2013, the group endorsed the adoption of selected ICH guidelines.[114]

110 See ASEAN, http://www.ich.org/about/organisation-of-ich/coopgroup/asean.html (accessed 22 April 2014).

111 On trade facilitation in ASEAN for pharmaceuticals, see generally eds. Basu Das Sanchita, Menon Jayant, Severino Rodolfo C and Shrestha Omkar L, *The ASEAN Community: A Work in Progress* (ISEAS, 2013).

112 See Jacobsen & Wertheimer, *Modern Pharmaceutical Industry, A Primer*, pp. 145–146; and ICH website at: http://www.ich.org/products/ctd.html (accessed 22 April 2014). The CTD has been the mandatory format used in the EU and Japan since 2003.

113 For the texts of the ACTD and ACTR, see *Harmonization of Standards and Technical Requirements in ASEAN*, at: http://www.asean.org/communities/asean-economic-community/item/harmonization-of-standards-and-technical-requirements-in-asean (accessed 22 April 2014); the web page also contains links to other ASEAN guidelines and documents pertaining to harmonization for pharmaceuticals. See also, generally, the ASEAN website at: www.asean.org/archive/actr/9.doc and http://www.asean.org/communities/asean-economic-community/item/accsq-pharmaceutical-product-working-group; and the Health Sciences Authority (HSA) of Singapore website at: http://www.hsa.gov.sg/publish/hsaportal/en/health_products_regulation/western_medicines/guidelines.html (accessed 22 April 2014).

114 See http://www.fda.gov.ph/issuances/99525-fda-circular-no-2013-018 (accessed 22 April 2014).

### *8.7.3 ASEAN integration and health: the* Bali Concord III *and health*

A word should be said here about ASEAN, as it continues its ambitious integration efforts. As explained in Chapter 2, the *Bali Concord III*, the ASCC and ASEAN Declaration of Human Rights are among the major instruments in which ASEAN member states have explicitly signalled affordable healthcare and medicines as an important goal. The ASCC, for instance, states as one of its "Strategic Objectives": "*Ensure access to adequate and affordable healthcare, medical services and medicine*, and promote healthy lifestyles for the peoples of ASEAN" (italic emphasis added).[115] While implementation of the specific steps towards realizing this goal are ongoing, and are likely to be so for a period of time, the direction to be taken in coming years is clear.

In the context of tackling HIV/AIDS infections, access goals which have found explicit expression can be found in the *ASEAN Declaration of Commitment: Getting to Zero New HIV Infections, Zero Discrimination, Zero AIDS-Related Deaths*, which provides a number of access targets, as shown in the excerpts below[116]:

> 7. Recalling that accelerated liberalisation of trade will enhance the region's competitiveness and realise welfare gains for our peoples in the long run, and that *efforts are also needed to ensure that access to affordable healthcare is not undermined* and health policies will be equitable and pro-poor, as noted in the Declaration of the 7th ASEAN Health Ministers Meeting adopted on 22 April 2004;
>
> 8. Concerned that the HIV epidemic continues to threaten the realisation of an ASEAN Community, with socio-economic consequences that pose a formidable challenge in our community-building and *our efforts to ensure access to affordable healthcare*
>
> . . .
>
> 11. Welcoming the findings of recent studies that demonstrate that access to HIV treatment significantly reduces the risk of HIV transmission to a partner; and, that *access to affordable medicines in the context of epidemics such as HIV is fundamental to the full realization of the right of everyone to enjoy the highest attainable standard of physical, social and mental health*;
>
> 12. *Concerned that intellectual property, trade policy barriers and social aspects* such as stigma and discrimination, are hindering prevention activities on HIV and AIDS, *access* to HIV treatments and treatments for co-infections and opportunistic infections, as well as pose as serious threats to the quality of life and livelihood of people living with and affected by HIV; . . .

[115] Para. B4, ASCC. [116] 17 November 2011, Bali, Indonesia.

> 18. Commit to work towards zero AIDS related deaths through the following:
>
> ...
>
> b. By 2015 improve treatment coverage, equity, effectiveness and efficiency by:
>
> ...
>
> ii. Addressing key obstacles such as drug stockouts, financial barriers, . . .;
>
> iii. *Securing and expanding access to affordable and effective HIV diagnostics, ARV and OI drugs, through the full use of existing flexibilities under the Trade-Related Aspects of Intellectual Property Rights Agreement, which are specifically geared to promoting access to and trade of medicines, including in particular the use of compulsory licensing to enable manufacturing or parallel importation of generic drugs;*
>
> iv. Addressing *barriers, regulations, policies and practices that prevent access to affordable HIV treatment by promoting generic competition* in order to help reduce costs associated with life-long chronic care;
>
> [Italic emphasis added]

#### 8.7.3.1 ASEAN's juncture of opportunity

Augmenting access-related goals mentioned above, ASEAN is at a critical juncture of its history, as it seeks to define its voice on a number of issues beyond the region. The *Bali Concord III* takes integration further by posing a challenge to ASEAN countries to speak in one voice in the global community. The following important language was expressed by ASEAN heads of state in the document:

> Consistent with the purposes and principles of ASEAN basic instruments to promote health, science and technology, education, human resources, cultural heritage, and the high quality of life, ASEAN *resolves at the global level* to:
>
> a. *Ensure access* to adequate and *affordable healthcare*, medical services, as well as accessibility to *safe, non-counterfeit, affordable, and effective medication* . . .[117]
>
> [Italic emphasis added]

ASEAN will need to consider carefully how it is to arrive at this voice in the global debate, given the diversity and differing priorities among its members. One thing is clear, however – ASEAN must join its voice with those of others in the larger world community to ensure access to good, safe and affordable medicines for its peoples. It must do so in the context

[117] Para C.3(a), *Bali Concord III.*

of domestic and international legal obligations, as well as help shape even better laws, regulations and policies, to ensure that highly valued economic growth is made sustainable through clear and well-considered healthcare policies.

Clearer national and regional policies in ASEAN regarding generic drug substitution by public health institutions would also be helpful in promoting the use of generic medicines, to help bring down costs and therefore increase access to medicines. At the moment, not every ASEAN country has a clearly articulated or publicized national generic drug substitution policy.[118] Some countries like Japan have policies encouraging the use of generic medicines.[119]

#### 8.7.3.2 What will be ASEAN's emerging global voice on public health and access to medicines?

Despite diverse economic and political situations within ASEAN it is possible – and desirable – to begin working out a template for negotiations and in global discussions on specific issues of common interest within the sphere of public health. The TPP negotiations may already be shaping or reshaping the stances of Brunei, Malaysia, Singapore and Vietnam to some extent, but ASEAN will also need to consider the basis of its "global voice" in accordance with its aspirations in the *Bali Concord III*.

ASEAN's economic integration provides a prime opportunity to forge clearer joint stands on issues that are highly relevant to access to medicines within its member states, including the boundaries of patentability (especially of biotechnological subject matter), evergreening, data exclusivity, treatment of biologics, parallel importation of medicines, generic drug substitution policies, tiered pricing for medicines, ISDS on public

[118] Countries which have clear policies are Indonesia and the Philippines. See generally Andi Hermansyah, Umi Athiyah, Catur Dian Setiawan and Mufarrihah, Are Patients Willing to Ask for Generic Drug Substitutions?, *Int. J. Ph. Teaching & Practices*, Vol. 4, No. 4 (2013), 832–837; Tuan A. Nguyen, Mohamed A. A. Hassalil and Andrew McLachlan, Generic Medicines Policies in the Asia Pacific Region: Ways Forward, *WHO S.E. Asia J. Pub. Health*, Vol. 2, No. 1 (January–March 2013), 72–74; Mohamed Azmi Hassali, Jayabalan Thambyappa, Fahad Saleem, Noman ul Haq and Hisham Aljadhey, Generic Substitution in Malaysia: Recommendations from a Systematic Review, *J. App. Ph. Sci.*, Vol. 2, No. 8 (2012), 159–164; and Amit Banerji and Maulana Azad, Review of Asia-Pacific's Systems with Emphasis on the Role of Generic Pharmaceuticals, *Ac. H. M. J.*, Vol. 9, No. 2 (2013), 53–74.

[119] See Japanese Ministry of Health, Labor and Welfare website, http://www.mhlw.go.jp/english/policy_report/2012/09/120921.html (accessed 29 December 2014).

health measures relating to medicines, how public health might have an overarching place in ASEAN FTAs and in member states' policy-making and the role of competition law in relation to pharmaceutical patent litigation settlement agreements, refusals to licence and pay-for-delay agreements. ASEAN has the opportunity to shape a joint stance on the health-competition-trade-investment-IP law nexus, to achieve its vision of the balance required between economic interests and public health priorities.[120]

### *8.7.4 ASEAN: research sharing*

The push towards regional integration in ASEAN – most visibly in the economic sphere – necessarily raises the question of whether and how its members should deal with access to medicines and R&D issues for pharmaceutical products of relevance to the region. One academic had suggested some level of cooperation between developing countries in dealing with public health issues such as pharmaceutical needs.[121]

In fact, in 2009, the ASEAN Network for Drugs, Diagnostics, Vaccines and Traditional Medicines (ASEAN-NDI) was established with the aim of developing a research network for promotion of R&D in the region, research collaboration and development of, *inter alia*, vaccines and medicines for diseases of common concern. These include medicines for treatment of malaria, tuberculosis and dengue fever.[122] The forerunner to the ASEAN-NDI was the African Network on Drugs and Diagnostics Innovation (ANDI, launched in 2008), and, subsequently, another such regional network, the Pan-American Health Organization's Platform on Innovation and Access, was launched (in 2012).[123]

[120] The intensive regional integration activities will also raise questions of migrant workers working within ASEAN and their ability to access health systems and medicines. In this regard, see generally, the ASEAN Declaration on the Protection and Promotion of the Rights of Migrant Workers, Cebu, the Philippines, 13 January 2007, and Statement on the Establishment of the Committee on the Implementation of the ASEAN Declaration on the Protection and Promotion of the Rights of Migrant Workers, Manila, the Philippines, 3 July 2007.

[121] Jakkrit Kuanpoth, *Patent Rights in Pharmaceuticals in Developing Countries – Major Challenges for the Future* (Edward Elgar, 2010), Chapter 8.

[122] See J. Montoya, *et al.*, *A Look at the ASEAN-NDI: Building A Regional Health R&D Innovation Network* (Infectious Diseases of Poverty, 2014),available at: http://www.idpjournal.com/content/pdf/2049-9957-3-15.pdf (accessed 26 November 2014).

[123] See http://www.andi-africa.org/index.php/aboutus and http://prais.paho.org/rscpaho/ (accessed 26 November 2014). See also Dr Zafar Mirza, WHO, *Message*, ASEAN-NDI Strategic Business Plan.

In 2013, the ASEAN-NDI issued a Strategic Business Model, outlining its own structure, and key funding and deliverables targets. As the network is still new, what remains unclear are the ownership of innovative products (such as patentable medicines) and licensing policies within ASEAN and beyond ASEAN.

### *8.7.5 Dealing with FTA demands with informed policy-making and creativity*

In an era of cost pressures faced by governments in healthcare and external trade negotiation pressure, it is imperative that policy-making be well informed and creative. The information that states must ensure they have at hand should, at a minimum, include the most likely implications (such as interpretative issues and potential for legal disputes) of demanded provisions, repercussions of a finding of non-compliance of such provisions by an international tribunal and the legal options available in other international agreements or, sometimes, what is available in the *demandeur*'s own legal system. Creativity is also required in order to design clever *quid pro quo*s, effective exceptions and counterbalancing mechanisms. A comparative exercise of such agreements, options and mechanisms will provide decision-making on a more sound basis in the face of such demands, and for long-term policy-making.

Indeed, the WHO has encouraged states to carefully consider patentability and its relationship to competition:

> **4.27** Governments should take action to avoid barriers to legitimate competition by considering developing guidelines for patent examiners on how properly to implement patentability criteria and, if appropriate, consider changes to national patent legislation.[124]

For example, in agreeing to patent linkage for the first time under the USSFTA, Singapore's strategy was to include a version of the US' "Hatch-Waxman" process to safeguard the rights of those who might wish to challenge a pharmaceutical patent.[125] Singapore also revamped its patent examination system to introduce substantive examination, to guard against weak or unmeritorious patents.

As mentioned, India is poised to introduce a new generation of BITs with its release of a draft Model BIT in 2015.[126] Many of its provision point to existing criticisms of the ISDS system as it has

[124] See WHO, *Public Health, Innovation and Intellectual Property Rights*, at p. 134.

[125] See Chapter 5. [126] See Chapter 4.

existed for decades, and it remains to be seen whether this draft will be adopted *in toto*.

## 8.8 Watch this space: BRICS and health

Interest in the group comprising Brazil, Russia, India, China and South Africa (BRICS). BRICS actions and initiatives on public health has been growing.[127] The first BRICS Health Ministers' Meeting was held in Beijing, China, in 2011, with the issue of the *Beijing Declaration*.[128] This, and other subsequent documents, refer to affordability of healthcare.

UNAIDS Executive Director Michel Sidibé described at the time the unique contribution which BRICS could bring to the public health discourse, as follows:

> The five BRICS countries are bringing a new voice, a new perspective and new solutions to today's global challenges . . . It is a voice with incredible economic, technological and innovative strength behind it and, at the same time, a voice intimately connected to the needs and interests of the developing world.[129]

Notably, at the first BRICS Science, Technology and Innovation Ministerial Meeting in 2014, the *Cape Town Declaration* was adopted.[130] Notably, paragraph 3 provides as follows:

> We stress the paramount importance of science, technology and innovation for human development. Indeed, while recognizing the role and significance of competitiveness in the rapid technologically changing global environment, we agree *people-centred and public-good driven science, technology and innovation, supporting equitable growth and*

[127] The acronym, BRICS, is said to have been coined by a Goldman Sachs economist; see BRICS website at http://brics6.itamaraty.gov.br/about-brics/information-about-brics (accessed 9 May 2015). As for interest in the BRICS' health initiatives, see, for example, WHO Bulletin, *BRICS and Global Health*, Vol. 92, No. 6 (June 2014), 385–464. See also *WHO Bulletin Focusses on BRICS and Global Health*, 2 June 2014, IP Watch, http://www.ip-watch.org/2014/06/02/who-bulletin-focusses-on-brics-and-global-health/ (accessed 28 July 2014). See also, generally, Peter K. Yu, Access to Medicines, BRICS Alliances and Collective Action, *Am. J. L. & Med.*, Vol. 34 (2008), 345–394.

[128] See *Beijing Declaration*, 11 July 2011, the *Delhi Communiqué*, 11 January 2013, and the *Cape Town Communiqué*, 7 November 2013. The four original countries, Brazil, Russia, India and China, decided in 2010 to invite South Africa to join them; see http://www.brics5.co.za/about-brics/ (accessed 28 July 2014).

[129] See UNAIDS website, http://www.unaids.org/en/resources/presscentre/featurestories/2011/july/20110711bchinabrics/ (accessed 28 July 2014).

[130] *Cape Town Declaration*, done at Meeting of 10 February 2014, Cape Town, South Africa.

> *sustainable development*, shall form the basis of our cooperation within the framework of BRICS.
>
> [Italic emphasis added]

A hint of the collaboration in health by BRICS countries was provided:

> There is already excellent cooperation between BRICS members in the area of health innovation. As part of the global effort to combat diseases such as HIV/AIDS, Brazil, India and South Africa have made significant progress in the production of generic antiretroviral drugs. Rapid drug discovery, drug development, the validation of traditional therapies and advances in diagnostics, genomics, proteomics and stem-cell technology are some of the areas in which the BRICS STI partnership has the potential to make a real difference in the global endeavour to achieve the Millennium Development Goals in health.[131]

The countries also agreed to enter into a memorandum of understanding to, *inter alia*, "co-generate new knowledge and innovative products, services and processes utilizing appropriate funding and investment instruments".[132]

At the 67th World Health Assembly in Geneva in May 2014, the BRICS countries met on the sidelines of the assembly and agreed to continue cooperation in promotion of health among the members of the group under the BRICS Framework for Collaboration on Strategic Projects in Health.[133] In particular, under preparation is a BRICS Monitoring and Evaluation Tool for Universal Health Coverage.[134]

Notably, on TRIPS flexibilities, the following statements form part of a *Joint Communiqué* by BRICS countries, reflecting their stances and concerns:

> 5. Emphasized the importance and need of technology transfer as a means to empower developing countries. They underlined the importance of ensuring access to affordable, quality, efficacious and safe medical products, including generic medicines, biological products, and diagnostics,

[131] Address by the Minister of Science and Technology, Derek Hanekom, at the First BRICS Science, Technology and Innovation Ministerial Meeting, at the Arabella Hotel and Spa in Kleinmond, 10 February 2014.

[132] Para. 4, *ibid.*

[133] *Joint Communiqué of BRICS Countries.* On recent BRICS trade ministers' discussions, see http://brics6.itamaraty.gov.br/media2/press-releases/211-the-4th-meeting-of-the-brics-trade-ministers-joint-communique-fortaleza-14-july-2014 (accessed 27 July 2014).

[134] *Ibid.* For information on the BRICS Health Ministers' Meeting in 2013, see http://www.brics5.co.za/3rd-health-ministers-meeting-south-africa-7th-november-2013-cape-town-communique/ (accessed 28 July 2014).

> through the use of TRIPS flexibilities, for the realization of the right to health. They also renewed commitment to strengthening international cooperation in health, and South-South cooperation in particular, with a view to supporting efforts in developing countries to promote health for all . . .
>
> 8. Reiterated their commitment to use TRIPS flexibilities to promote access to medicines, foster innovation and share these experiences with other developing countries.[135]

Despite the following cautious remark,

> If the group endures, it is likely to promote a global health development discourse informed by its members' experience of cooperation based on partnership and equity.[136]

the signs are, for now, encouraging.[137]

## 8.9 Encouraging new alliances and collaborations

The 2013 Joint Study of the WHO, WIPO and WTO was a watershed publication and collaborative effort, providing a broad base of policy issues, domains, initiatives and trends across the fields and attendant needs of health, trade and IP. The utility of the study arose from the pooling of expertise and analyses from all three areas to address the complexities around access to medicines. This utility and the success of the collaboration signal the continuing collaborative work in all three policy areas. As one representative suggested at the Standing Committee on the Law of Patents in WIPO, given that the WIPO, WHO and WTO each has its own niche of expertise and data, but not with all the information on its own, such collaboration and scope for an "international monitoring mechanism for patent trends, trade rules and the affordability of medicines" would be not only useful but also necessary to help the debate move forward constructively.[138]

[135] *Joint Communiqué of BRICS on the State of Health*, sidelines of the 67th World Health Assembly, 20 May 2014.

[136] Andrew Harmer, *The BRICS Countries: A New Force in Global Health?*, WHO Bulletin, *ibid.*, p. 395.

[137] See *BRICS and Global Health* issue, Bulletin of the World Health Organization, Vol. 92, No. 6 (June 2014), 385–464; see especially in the issue, *The BRIC Countries: A New Force in Global Health?*, at pp. 394–395, and Ilona Kickbusch, *BRICS' Contribution to the Global Health Agenda*, at pp. 463–464.

[138] Statement by the representative of Egypt, para. 137, SCP/17/13, document dated 21 May 2012.

Indeed, perhaps this statement was a hint of a more enduring collaboration to come.[139] After all, as has been observed by the US, there is "no specific single forum in Geneva, able to look univocally at a big framework such as patents and medicine . . ."[140]

In a WIPO discussion in 2012, the Egyptian delegation suggested that a trilateral briefing by the WHO, WIPO and WTO be made a regular agenda item in the Standing Committee on the Law of Patents, and this received the support of the US and South Africa.[141]

In terms of research, greater collaboration across institutions, regions and disciplines to draw on the diversity of expertise and problem-solving techniques should be encouraged.

It is evident that the patent-innovation-incentive, profit-driven, model *alone* is not designed to ensure the affordability of medicines; it has in fact been much criticised as being the reason for the inaccessibility of certain medicines to poor patients.[142] The model alone, however, is not solely responsible for obstacles to access, as national systems also play a part in ensuring access through their healthcare and general infrastructure, financing options, combating corruption which can impede access and national economic laws and policies.

While voluntary licensing is a measure that helps to spread production and distribution of medicines, it remains, first, voluntary (and subject largely to the commercial choice of patent-holders whether to license, and on what terms) and, second, accompanied by remuneration which may render the products of such licensed production still inaccessible.

The landscape has been changing with the emergence of initiatives that address financing of R&D,[143] pricing and their negotiations, licensing, patent legislation, elevation of attention on effects of international treaties on access, greater focus on health and economic policy *vis-à-vis* trade

[139] It was also suggested that past symposia held by the three organizations collaboratively "could be a forum for continuing the dialogue"; statement by the representative of WTO, WIPO SCP/19/8, para. 101.

[140] Para. 158 WIPO, Report of 5–9 December 2011 meeting adopted by the SCP, SCP/17/13, *ibid.* The statement was made in the context of addressing a proposal by the African Group and the Development Agenda Group in WIPO.

[141] WIPO Standing Committee on the Law of Patents, SCP/17/13; see paras. 137, 139 and 142.

[142] See generally Harlan M. Krumholtz, A Note to My Younger Colleagues . . . Be Brave, *Circ. Cardiovas. Qual. and Outcomes*, Vol. 5 (2012), 245–246, at p. 246.

[143] See, for example, WHO, *The Role of Innovative Financing Mechanisms for Health*, World Health Report 2010 http://www.who.int/healthsystems/topics/financing/healthreport/InnovativeBP12FINAL.pdf?ua=1.

and investment treaties, the potential of disputes arising from these treaties, collaboration in the creation of innovative pharmaceutical products and capacity-building for poorer countries in the use of relevant flexibilities under international treaties. Yet, gaps persist, and business models require change.[144] A multi-pronged legal and policy construct is needed to tackle these gaps.

### *8.9.1 A new research agenda: the three Cs*

To tackle the challenge of such a multi-pronged construct, it is timely to establish a new multidisciplinary, multi-institutional and multi-regional research platform to draw together a host of relevant stakeholders, researchers and policy-makers. Such a cross-discipline, cross-border and cross-institution (what this author calls the "three Cs")[145] platform for research is required, to permit Latin American, Asian, European, North American and Australasian collaborative research and maximization of knowledge, experiences and comparative issues to allow for better problem-solving. Emerging approaches such as health in all policies, health technology assessment exercises and pharmaco-economics would be analysed alongside more traditional studies of treaty issues, dispute settlement and negotiating language.

### *8.9.2 Non-adversarial problem-solving*

In Chapter 7, it was mentioned that alternative dispute settlement avenues – possibly non-adversarial ones – for investor complaints should be promoted.

In the broader context, the current debate on access to medicines is often a highly charged one, infused with an adversarial flavour. This is not constructive and can lead to positional stances and, in some cases, intransigence. Non-adversarial problem-solving should be the objective, with mechanisms and institutions working towards this end. Highly

[144] See WTO, *Experts Say Business Models for Pharmaceuticals Must Adapt to New Circumstances*, WHO-WIPO-WTO Symposium, 5 July 2013. See also Modulating the Profit Motive to Meet Needs of the Less-Developed World,*The Lancet*, Vol. 358 (2001), 1638; and Carlos Correa, Towards a New Model for Pharmaceutical Research, *WHO Bulletin*, Vol. 90 (2012), 795–795A.

[145] The author first proposed the "three Cs" idea for the World Investment Forum's Multi-Disciplinary Academic Conference, UNCTAD, Geneva, Switzerland, October 2014.

skilled negotiations making use of interest-based, rather than position-based, strategies should be promoted and valued.[146]

## 8.10 Preparing for the debate ahead: legal education and capacity-building

Given the analysis in this book that the access to medicines issue is one that cuts across disciplines, agreements, organizations and interest groups, it is timely to redesign legal curricula as well as capacity-building programmes for policy-makers.

Legal curricula of many law schools in this field have traditionally focused on distinct subjects of trade law, investment law, health law and IP law. More recently, some curricula have included international economic law to cover trade and investment topics and issues. The gradual blending of lines dividing trade (GATT/WTO) law and investment (BIT, ISDS) law has contributed to this change. The recent raising of TRIPS arguments in ISDS has led to IP-related issues to be pertinent for a trade/investment lawyer. *Vice versa*, traditional pure IP lawyers are now required to pay attention to international trade and ISDS aspects of IP disputes. In the case of access to medicines, the public health dimension is thrown into this growing mix of fields and disciplines.

Beyond this, one needs an appreciation of the political, social and economic realities of the contexts in which these laws and policies need to operate. The law student aspiring to international practice no longer has the luxury of understanding the operations of only one of these areas if he/she is to do service to his/her clients, which may be multinational pharmaceutical companies (whether prescription or generic producers), governments, NGOs and other organizations. It is therefore important to ensure that legal curricula keep up not only with legal trends and developments in international law, but also with those that cut across such diverse fields, to ensure a better appreciation of the law and its place relative to other policy areas. Such a broadened understanding will also allow a young lawyer to better understand and advise on the available legal and policy tools that may be required to support the legal advice or advocacy work undertaken in this area. Cross-disciplinary education for

[146] For the origins and development of this form of negotiation, see Daniel L. Shapiro, Creative Problem Solving: Not Just about the Problem, in *Psychological Components of Sustainable Peace*, ed. Peter T. Coleman and Morton Deutsch (Springer, 2012), at pp. 89–90. See also, Roger Fisher and William Ury, *Getting to Yes – Negotiating Agreement Without Giving In* (Penguin Books, 1991).

lawyers, judges, lawmakers and policy-makers will be increasingly useful for the complex issues that a globalized economy brings. So, too, will capacity-building work based on linked policy areas be increasingly valuable, as technical experts and bureaucrats learn to handle diverse but interconnected policy issues in the access-to-medicines debate.

Under Element 5 of the WHO's GSPOA, education and training in IPRs and public health is a goal.[147] It is therefore hoped that this book will contribute to this target, and add to other efforts at promoting understanding on IPRs through a public health lens, particularly in Asia.

## 8.11 Conclusion

Asia's economic face is changing rapidly, not just from the point of strategic trade and investment agreements. Beyond BRICS, broader developments that are relevant to the economic and legal alliances which will emerge in the next two decades may affect access to economic policy, healthcare and medicines in ASEAN and beyond. These include the evolving development of China's 21st Century Silk Road and Maritime Silk Road, India's Project Mausam, and the US' responses and initiatives beyond the TPP. Apart from the TPP, which, if concluded, will no doubt produce rules calling for close analysis, other significant regional trade and investment arrangements affecting Asia include the RCEP. This arrangement, if concluded, will, unlike the TPP, include both China and India. Other ongoing treaty negotiations in the region include those of China, Japan and Korea for a trilateral FTA.[148] In 2014, the Chinese government proposed within the APEC framework to study the feasibility of launching negotiations for a Free Trade Area of the Asia-Pacific (FTAAP) in future.[149] This would include China and Russia

[147] Para. (e) of Element 5 provides the following targeted action (italic emphasis added):

> (e) *strengthen education and training in the application and management of intellectual property from a public health perspective*, taking into account the provisions contained in the Agreement on Trade-Related Aspects of Intellectual Property Rights, including the flexibilities recognized by the Doha Declaration on the TRIPS Agreement and Public Health and other WTO instruments related to the Agreement on Trade-Related Aspects of Intellectual Property Rights.

[148] See http://fta.mofcom.gov.cn/topic/chinarh.shtml (accessed 7 October 2015). In June 2015, China and Korea concluded negotiations and signed a bilateral FTA; see http://fta.mofcom.gov.cn/topic/enkorea.shtml (accessed 7 October 2015).

[149] See APEC, http://www.apec.org/Meeting-Papers/Leaders-Declarations/2014/2014_aelm/2014_aelm_annexa.aspx (accessed 28 April 2015).

but not India. While there have been indications of the possible issues relating to pharmaceuticals arising in the TPP negotiations, it is not presently clear to what extent these negotiations might be setting new rules on such issues.[150] In the course of FTA and economic integration discussions, ASEAN member countries will also need to reassess any existing national trade and investment barriers affecting pharmaceutical enterprises, as these may have implications on availability and costs of their products.

Institutionally, international organizations such as the UNCTAD, WHO, WIPO and WTO will have to continue working together closely and with creative imagination to produce mechanisms and platforms that contribute constructively to help bridge gaps in the ongoing debate and work on access, particularly in light of the SDGs adopted in 2015. The private and non-profit sectors will need to be engaged in new ways, preferably in non-adversarial problem-solving. In the case of the WTO, as it strives to maintain its status as the multilateral trade negotiations forum, members will have to reassess whether this forum is capable of yielding any instrument and/or process that will contribute further for more access to medicines. As TRIPS and its impending amendment operates alongside a growing number of FTAs containing provisions which affect the TRIPS flexibilities, there will be an increasing need to better define both the delicate "balance" in TRIPS – with its many facets – and the scope of obligations in such FTAs. As the WHO contemplates the extended time frames for its GSPOA plan of action from 2015 to 2022, it will have an opportunity to take into account relevant multilateral and regional developments, including those mentioned above, for

[150] The TPP reportedly will commit signatory governments "to implementing a *robust and balanced* IP regime in the Asia Pacific region" (italics emphasis added); see p. 4, Singapore Ministry of Trade and Industry, *TPP: Seven Things You Should Know*, at https://www.mti.gov.sg/MTIInsights/SiteAssets/Pages/TPP/Seven%20things%20you%20should%20know%20about%20the%20TPP%20%28Final%29.pdf . In addition, the IP Chapter reportedly "contains pharmaceutical-related provisions that facilitate both the development of innovative, life-saving medicines and the availability of generic medicines, taking into account the time that various Parties may need to meet these standards. The chapter includes commitments relating to the protection of undisclosed test and other data submitted to obtain marketing approval of a new pharmaceutical or agricultural chemicals product. It also reaffirms Parties' commitment to the WTO's 2001 Declaration on the TRIPS Agreement and Public Health, and in particular confirms that Parties are not prevented from taking measures to protect public health, including in the case of epidemics such as HIV/AIDS."; para. 18, Singapore Ministry of Trade and Industry, *Summary of the TPP Agreement*, https://www.mti.gov.sg/MTIInsights/SiteAssets/Pages/TPP/Summary%20of%20TPP%20Agreement.pdf (accessed 13 October 2015).

meaningful implementation of the plan.[151] In particular, GSPOA Element 5, which addresses the application and management of intellectual property to contribute to innovation and promote public health, presents an important interface with the work of organizations such as the WTO, WIPO and UNCTAD.

The access-to-medicines debate is far from over, nor will it be in the near future. However, there are distinct steps which can be undertaken, both at the national and multilateral levels, to further improve access, as this book has sought to explain. As organizations such as the United Nations, and those in Asia such as ASEAN, work out their post-2015 agendas,[152] existing law-policy-health and innovation-health-IPR-competition linkages will need to be reviewed, but, more importantly, new linkages need to be considered. IPRs certainly form part of the equation, but addressing them will not be the only solution. Patent systems and activities also need not be the only measure (or claimants) of innovation.[153] Patent-related practices have, in major developed economies, also now been laid open to competition law scrutiny. More effective access to medicines requires health policy to permeate trade, investment and innovation policy, not just the other way around. Finally, as has been pointed out, IPRs should be viewed from a broader, more holistic perspective, with policy-makers appreciating not just their legal and economic roles, but also their political and social roles, particularly in the present context.[154]

[151] *Global Strategy and Plan of Action on Public Health, Innovation and Intellectual Property – Draft Resolution Proposed by Bolivia (Plurinational State of), Brazil, Ecuador, India and South Africa*, A68/B/CONF./1Rev.1, 25 May 2015, presented at the 68th WHA, 18–26 May 2015. The original proposal was by Argentina, Brazil, Ecuador, India and South Africa; see EB136/CONF./7Rev.2.

[152] See, respectively, UN programme of work for post-2015 intergovernmental negotiations, and ASEAN, *Nay Pyi Taw Declaration on the ASEAN Community's Post-2015 Vision*, 12 November 2014.

[153] See the discussion in WIPO Secretariat Document, *Study on the Role of Patent Systems in Promoting Innovative Medicines, and in Fostering the Technology Transfer Necessary to Make Generic and Patented Medicines Available in Developing Countries and Least Developed Countries*, 10 October 2014, SCP/21/8, especially at paras. 5–24.

[154] See generally Sebastian Haunss and Kenneth C. Shadlen, eds., *Politics of Intellectual Property: Contestation, Use, and Control of Knowledge and Information* (Edward Elgar Publishing Limited, 2009); see, in particular, Chapter 7 by Ingrid Schneider, *Can Patent Legislation Make a Difference? Bringing Parliaments and Civil Society into Patent Governance*, pp. 129–157. See also Christopher May, *Why IPRs Are a Global Political Issue*, pp. 277–283, in ed. David Vaver, *Intellectual Property Rights – Critical Concepts in Law*, Vol. IV, (Routledge 2006), Vol. I.

# INDEX

Lightning Source UK Ltd.
Milton Keynes UK
UKHW020225151218
334055UK00006B/92/P

9 781107 421035